D0152597

BASIC ECOLOGY

Eugene P. Odum

Callaway Professor of Ecology
and Director of the Institute of Ecology,
University of Georgia

SAUNDERS COLLEGE PUBLISHING

Philadelphia New York Chicago
San Francisco Montreal Toronto
London Sydney Tokyo Mexico City
Rio de Janeiro Madrid

Address orders to:
383 Madison Avenue
New York, NY 10017

Address editorial correspondence to:
West Washington Square
Philadelphia, PA 19105

Text Typeface: Caledonia
Compositor: Bi-Comp, Inc.
Acquisitions Editor: Michael Brown
Project Editor: Lynne Gery
Copy Editor: Michael Fare
Managing Editor & Art Director: Richard L. Moore
Design Assistant: Virginia A. Bollard
Text Design: Terri Jackson
Cover Design: Larry Didona
New Text Artwork: Tom Mallon and Mary Helen Fink
Production Manager: Tim Frelick
Assistant Production Manager: Maureen Read

Cover credit: Landsat 3 image of San Francisco, California, area. Courtesy of National Aeronautics and Space Administration.

Library of Congress Cataloging in Publication Data

Odum, Eugene Pleasants, 1913–
Basic ecology.

Updated ed. of pt. 1 of: Fundamentals of ecology. 3rd ed. 1971.
Bibliography: p.
Includes index.

1. Ecology. I. Title.
QH541.0312 1983   574.5   82-60633

ISBN 0-03-058414-0

BASIC ECOLOGY   ISBN 0-03-058414-0

Library of Congress catalog card number 82-60633.

345 016 98765432

CBS COLLEGE PUBLISHING
Saunders College Publishing
Holt, Rinehart and Winston
The Dryden Press

Preface

When the third edition of *Fundamentals of Ecology* was completed in 1971, I assumed that it would be the last major revision. In general, textbooks in successive editions suffer from the "dinosaur syndrome"; that is, they grow ever larger, more encyclopedic, and less useful for the student, especially the beginner. Then, like those prehistoric monsters, they rapidly become extinct. To avoid this fate, the publisher and I decided that I would prepare a less detailed text with a new title, one suitable for a one-semester or one-quarter undergraduate course. Subsequently, we would update *Fundamentals of Ecology*.

Basic Ecology is Part 1 of *Fundamentals of Ecology*, updated and very extensively rewritten in light of new findings and the increased environmental awareness of the public during the 1970s. In illustrating basic principles a special effort has been made to select new examples that relate to human affairs. This approach should make ecology more interesting and exciting. Part 2 of *Fundamentals* has been condensed and included as an appendix entitled "Major Natural Ecosystem Types of the Biosphere."

As with my other books, a whole-to-the-parts progression is followed. Although the ecosystem level of organization is thus emphasized, population ecology is by no means slighted; the two chapters dealing with populations and communities give these areas extensive coverage. Though it is likely that the undergraduate student will not have time to consult many of the extensive references and suggested readings that are included in every section, these citations can provide material for special reports should the instructor choose to require such.

During the past ten years or so, ecology has become more and more an integrated discipline that links the natural and the social sciences. While ecology retains a strong and basic root in the biological sciences, it is no longer just a biological subject. Ecology is a "hard" science in that ecological research involves the concepts and tools of mathematics, chemistry, physics, and so on. But it is also a "soft" science in that human behavior has a lot to do with the structure and function of ecosystems. Ecology as an integrated natural-social science has a tremendous potential for application to human affairs, since real-world situations almost always involve a natural science component and a social, economic, and political component. The two cannot be dealt with separately if one expects to find lasting solutions to critical problems.

As with my previous books, this one is very much a product of students and colleagues who have been associated with Georgia's Institute of Ecology over the past 30 years or so. I won't try to list everyone by name since the list would fill several pages.

When it comes to energy economics and environment, there is no one in the whole world with the innovative expertise of my brother, Howard T. Odum. The material on energetics (especially Chapter 3) reflects the strong influence his ideas have had on my own thinking. I am also indebted to my son, William E. Odum, who teaches environmental science at the University of Virginia. In addition to his own research, which I have cited, he and his colleagues have been most helpful in my effort to select material that appeals to the current generation of students.

Without the understanding and encouragement of my wife, Martha Ann, I would never have been able to face the task of another revision of a complex subject that is evolving and changing in scope and emphasis almost daily. She also helped with the index, which we consider an important component since it also serves as a glossary.

My special thanks go to Mrs. Julia Fortson for her dedicated work on the manuscript. She never complained when yet another chapter had to be retyped.

I am most appreciative of the encouragement, dedication, and patience of the staff at Saunders, especially editors Michael Brown and Lynne Gery, editorial assistant Margaret Mary Kerrigan, and illustrator Tom Mallon.

What you find in this book is due in no small way to many excellent suggestions of outside reviewers from academia, especially Peter Rich, Nelson Marshall, Daniel Stern, Elliot J. Tramer, Frank Trama, Gregory Gillis, Roger C. Anderson, Alan P. Covich, and Steve Carpenter.

Eugene P. Odum

Contents

1
Introduction: The Scope of Ecology

1. Ecology—Its Relation to Other Sciences and Its Relevance to Civilization

The word "ecology" is derived from the Greek *oikos*, meaning "household," and *logos*, meaning "study." Thus, the study of the environmental house includes all the organisms in it and all of the functional processes that make the house habitable. Literally, then, ecology is the study of "life at home" with emphasis on "the totality or pattern of relations between organisms and their environment," to cite one of the definitions in Webster's Unabridged Dictionary.

The word "economics" is also derived from the Greek root *oikos*. Since *nomics* means "management," economics translates as "the management of the household," and accordingly, ecology and economics should be companion disciplines. Unfortunately, many people view ecologists and economists as adversaries with antithetical visions. Later, this text will consider the confrontation that results because each discipline takes too narrow a view of its subject and the special effort made to bridge the gap between them.

Ecology was of practical interest early in human history. To survive in primitive society, all individuals needed to know their environment, i.e., the forces of nature and the plants and animals around them. Civilization, in fact, began coincidentally with the use of fire and other tools to modify the environment. Because of technological achievements, we seem to depend less on the natural environment for our daily needs; we forget our continuing dependence on nature. Also, economic systems of whatever political ideology value things made by human beings that primarily benefit the individual, but place little value on the goods and services of nature that benefit us as a society. Until there is a crisis, we tend to take for granted natural goods and services; we assume they are unlimited or somehow replaceable by technological innovations, despite evidence to the contrary.

The great paradox is that industrialized nations have succeeded by temporarily uncoupling humankind from nature through exploitation of finite, naturally produced fossil fuels that are rapidly being depleted. Yet civilization still depends on the natural environment, not only for energy and materials but also for vital life-support processes such as air and water cycles. The basic laws of nature have not been repealed; only their complexion and quantitative relations have changed, as the world's human population and its prodigious consumption of energy have increased our power to alter the environment. Accordingly, our survival depends on knowledge and intelligent action to preserve and enhance environmental quality by means of harmonious rather than disruptive technology.

Like all phases of learning, the science of ecology has had a gradual, if spasmodic, development during recorded history. The writings of Hippocrates, Aristotle, and other philosophers of ancient Greece clearly contain references to ecological topics. However, the Greeks did not have a word for ecology. The word "ecology" is of recent origin, having been first proposed by the German biologist Ernst Haeckel in 1869. Before this, many great men of the biological renaissance of the eighteenth and nineteenth centuries had contributed to the subject, even though the word "ecology" was not in use. For example, in the early 1700s, Anton van Leeuwenhoek, best known as the premier microscopist, also pioneered the study of food chains and population regulation (Egerton, 1968), and the writing of the English botanist Richard Bradley revealed that he had a good understanding of biological productivity (Egerton, 1969). All three of these subjects are important areas of modern ecology.

As a recognized, distinct field of science, ecology dates from about 1900, but only in the past decade has the word become part of the general vocabulary. At first, the field was rather sharply divided into plant and animal ecology, but the biotic community concept of F. E. Clements and V. E. Shelford, the food chain and material cycling concepts of Raymond Lindeman and G. E. Hutchinson, and the whole lake studies of E. A. Birge and Chauncy Juday, among others, helped establish basic theory for a unified field of general ecology. The work of these pioneers will be cited often in subsequent chapters.

What can best be described as a worldwide environmental awareness movement burst upon the scene during two years, 1968 to 1970. Suddenly, it seemed, everyone became concerned about pollution, natural areas, population growth, and food and energy consumption, as indicated by the wide coverage of environmental concerns in the popular press. The increase in public attention had a profound effect on academic ecology. Before the 1970s, ecology was viewed largely as a subdivision of biology. Ecologists were staffed in biology departments, and courses were generally found only in the biological science curricula. Although ecology remains strongly rooted in biology, it has emerged from biology as an essentially new integrative discipline that links physical and biological processes and forms a bridge between the natural sciences and the social sciences (Odum, 1977). Many colleges now offer campus-wide courses and have separate majors, departments, or institutes of ecology. Some of the larger universities offer advanced interdisciplinary degrees in ecology. While the scope of ecology was expanding, the study of how individuals and species interact and use resources intensified. Many core concepts in what can be termed the "evolutionary approach" to ecology were originated by the late Robert MacArthur, as detailed in Chapters 6 and 7.

2. Levels of Organization Hierarchy

Perhaps the best way to delimit modern ecology is to consider the concept of **levels of organization**, visualized as a sort of "biological spectrum," as shown in Figure 1–1. Community, population, organism, organ, cell, and gene are widely used terms for major biotic levels shown in hierarchical arrangement from large to small. Actually, the "levels" spectrum, like a radiation spectrum or a logarithmic scale, theoretically can be extended infinitely in both directions. **Hierarchy** means "an arrangement into a graded series" (Webster's Collegiate Dictionary). Interaction with the physical environment (energy and matter) at each level produces characteristic functional

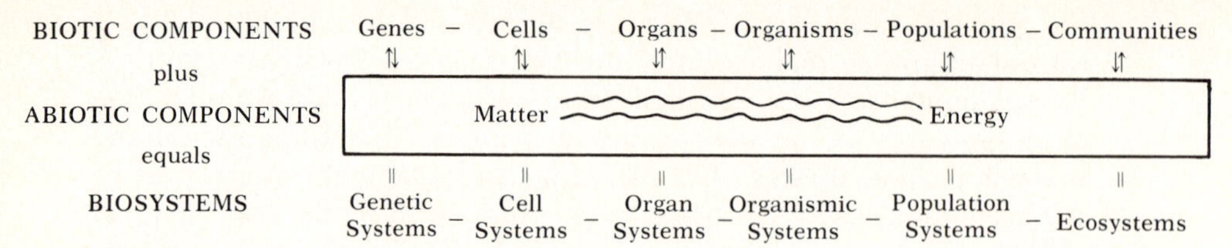

Figure 1–1 Levels-of-organization spectrum. Ecology focuses on the right-hand portion of the spectrum, that is, the levels of organization from organisms to ecosystems.

systems. A **system** consists of "regularly interacting and interdependent components forming a unified whole" (Webster's Collegiate Dictionary) or, from a different point of view, "a set of mutual relationships constituting an identifiable entity, real or postulational" (Laszlo and Margenau, 1972). Systems containing living components (biological systems or biosystems) may be conceived of or studied at any level, as illustrated in Figure 1–1, or at any intermediate position convenient or practical for analysis. For example, host-parasite systems or a two-species system of mutually linked organisms (such as the fungi-algae partnership that constitutes the lichen) are intermediate levels between population and community.

Ecology is largely concerned with the right-hand portion of this spectrum, that is, the system levels beyond that of the organism. In ecology, the term **population**, originally coined to denote a group of people, is broadened to include groups of individuals of any one kind of organism. Likewise, **community**, in the ecological sense (sometimes designated as "biotic community"), includes all the populations occupying a given area. The community and the nonliving environment function together as an ecological system or **ecosystem. Biocoenosis** and **biogeocoenosis** (literally, life and earth functioning together), terms frequently used in the European and Russian literature, are roughly equivalent to community and ecosystem, respectively. **Biome** is a convenient term in wide use for a large regional or subcontinental biosystem characterized by a major vegetation type or other identifying landscape aspect, as, for example, the temperate deciduous forest biome. The largest and most nearly self-sufficient biological system is often designated the **biosphere** or **ecosphere**, which includes all the earth's living organisms interacting with the physical environment as a whole to maintain a steady-state system intermediate in the flow of energy between the input of the sun and the thermal sink of space. By **steady state,** we mean a self-adjusting equilibrium or balanced condition relatively immune to at least small-scale disturbances.

Hierarchical theory provides a convenient framework for subdividing and dealing with complex situations or extensive gradients. As Novikoff (1945) has pointed out, there is both continuity and discontinuity in the evolution of the universe. Development may be viewed as continuous because it involves never-ending change, but it is also discontinuous because it passes through a series of different levels of organization. As we shall discuss in Chapter 3, the organized state of life is maintained by a continuous but stepwise flow of energy. Thus, dividing up a graded series, or hierarchy, into components is in many cases arbitrary, but sometimes subdivisions can be based on natural discontinuities. Because each level in the biosystem spectrum is "integrated" or interdependent with other levels, there can be no sharp lines or breaks in a functional sense, not even between organism and population. The individual organism, for example, cannot survive for long without its population any more than the organ would be able to survive for long as a self-perpetuating unit without its organism. Similarly, the community cannot exist without the cycling of materials and the flow of energy in the ecosystem. This argument is applicable to the mistaken notion discussed previously that human civilization can exist separately from the natural world.

For more on hierarchy theory, see articles in Pattee (1973).

3. The Emergent Property Principle

An important consequence of hierarchical organization is that as components, or subsets, are combined to produce larger functional wholes, new properties emerge that were not present at the level below. Accordingly, an **emergent property** of an ecological level or unit cannot be predicted from the study of the components of that level or unit. **Nonreducible properties**, that is, properties of the whole not reducible to the sum of the properties of the parts, is another way to express the concept. Though findings at any one level aid in the study of the next level, they never completely explain the phenomena occurring at the next level, which must itself be studied to complete the picture.

Two examples, one from the physical realm and one from the ecological realm, will suffice to illustrate emergent properties. When hydrogen and oxygen are combined in a certain molecular configuration, water is formed, a liquid with properties utterly different from those of its gaseous components. When certain algae and coelenterate animals evolve together to produce a coral, an efficient nutrient-cycling mechanism is created that enables the combined system to maintain a high rate of productivity in waters with a very low nutrient content. Thus, the fabulous productivity and diversity of coral reefs are emergent properties found only at the level of the reef community.

Feibleman (1954) has theorized that at least one new property emerges with each integration of subsets into a new set. Salt (1979) suggests that a distinction be made between **emergent properties**, as defined previously, and **collective properties**, which are summations of the behavior of components. Both are properties of the whole, but the collective properties do not involve new or unique characteristics resulting from the functioning of the whole unit. Birth rate is an example of a collective property, since it is merely a sum of individual births within a designated time period, expressed as a fraction or percent of the total number of individuals in the population. New properties emerge because the components interact, not because the basic nature of the components is changed. Parts are not "melted down," as it were, but integrated to produce unique new properties. Simon (1973) has shown mathematically that integrative hierarchies evolve more rapidly from their constituents than nonhierarchical systems with the same number of elements; they are also more resilient in response to disturbance. Theoretically, when hierarchies are decomposed to their various levels of subsystems, the latter can still interact and reorganize to achieve a higher level of complexity (Laszlo, 1972).

Simon (1973) illustrates these concepts with a parable about two watchmakers: one uses a modular approach (constructs stable subunits that are then assembled to make the watch) and one does not use such a hierarchical method. When both watchmakers are interrupted frequently by telephone calls, the second watchmaker finds it difficult to pick up where he left off when he returns to the bench, because his as-yet-unorganized parts have fallen into disarray. The first watchmaker completes his watch sooner because his work has not been "set back" so greatly by the interruptions (i.e., the disturbances).

Some attributes, obviously, become more complex and variable as one proceeds from left to right along the levels-of-organization hierarchy (Figure 1–1), but often other attributes become less complex and less variable as one goes from the small to the large unit. Because homeostatic mechanisms, that is, checks and balances, forces and counterforces, operate throughout, the amplitude of oscillations tends to be reduced, as smaller units function within larger units. Statistically, variance of the whole is less than the sum of the variance of the parts. For example, the rate of photosynthesis of a forest community is less variable than that of individual leaves or trees within the community, because when one part slows down, another may speed up to compensate. When one considers both the emergent properties and the increasing homeostasis that develop at each level, not all component parts must be known before the whole can be understood. This is an important point, because some contend that it is useless to try to work on complex populations and communities when the smaller units are not yet fully understood. Quite the contrary; one may begin study at any point in the spectrum, provided adjacent levels, as well as the level in question, are considered, since, as already noted, some attributes are predictable from parts (collective properties), but others are not (emergent properties). Ideally, as Patten (1978) points out, a system-level study is itself a threefold hierarchy: system, subsystem (next level below), and suprasystem (next level above). For more on emergent properties, see Henle (1942), Bergmann (1944), Lowry (1974), and Edson et al. (1981).

Each biosystem level has emergent properties and reduced variance as well as a summation of attributes of its subsystem components. The old folk wisdom about the forest being more than just a collection of trees is indeed a first working principle of ecology. While the philosophy of science has always been holistic in seeking to understand phenomena as a whole, in recent years the practice of science has become increasingly reductionist in seeking to understand phenomena by detailed study of smaller and smaller components. Laszlo and Margenau (1972) see in the history of science an alternation of reductionist and holistic thinking (reductionism-constructionism and atomism-holism are other pairs of words used to contrast philosophical approaches). The law of diminishing returns may very well be involved here, since excessive effort in any one direction eventually necessitates direction in the other (or another).

The reductionist approach that has dominated science and technology since Isaac Newton has contributed much good. For example, research at the cellular and molecular level is establishing a firm basis for the future cure and prevention of cancer at the level of the organism. However, cell-level science will contribute very little to the well-being or survival of human civilization if we understand the higher levels of organization so inadequately that we can find no solutions to population overgrowth, social disorder, pollution, and other forms of societal and environmental cancer. Both holism and reductionism must be accorded equal value, and simultaneously, not alternatively. Ecology, an emerging science, seeks synthesis, not separation. The revival of the holistic disciplines may be due at least partly to the public's dissatisfaction with the specialized scientist who cannot respond to the large-scale problems needing urgent attention. (Historian Lynn White's 1980 essay is recommended reading on this viewpoint.) Accordingly, we shall discuss ecological principles at the ecosystem level, with appropriate attention to organism, population and community subsets, and the biosphere supraset. This is the philosophical basis for the organization of chapters in this text.

Fortunately, in the past ten years, technological advances have allowed us to deal quantitatively with large, complex systems such as the ecosystem. Tracer methodology, mass chemistry (spectrometry, colorimetry, chromatography), remote sensing, automatic monitoring, mathematical modeling, and computer technology are providing the tools. Technology is, of course, a two-edged sword; it can be the means of understanding the wholeness of man and nature or of destroying it.

4. About Models

If ecology is to be discussed at the ecosystem level, for reasons already indicated, how can this complex and formidable system level be dealt with? We begin by describing simplified versions that encompass only the most important or basic properties and functions. Since, in science, simplified versions of the real world are called **models**, it is appropriate now to introduce this concept.

A model (by definition) is a formulation that mimics a real-world phenomenon and by which predictions can be made. In simplest form, models may be verbal or graphic (informal). Ultimately, however, models must be statistical and mathematical (formal) if quantitative predictions are to be reasonably good. For example, a mathematical formulation that mimics numerical changes in a population of insects and that predicts the numbers in the population at some time would be considered a biologically useful model. If the population in question is a pest species, the model could have an economically important application.

Computer-simulated models permit one to predict probable outcomes as parameters in the model are changed, as new parameters are added, or as old ones are removed. Thus, a mathematical formulation can often be "tuned" by computer operations to improve the "fit" to the real-world phenomenon. Above all, models summarize what is understood about the situation modeled and thereby delimit aspects needing new or better data or new principles. When a model does not work—when it poorly mimics the real world—computer operations often can provide clues to the refinements or changes needed. Once a model proves to be a useful mimic, opportunities for experimentation are unlimited, since one can introduce new factors or perturbations and see how they would affect the system. Even when a model inadequately mimics the real world, which is often the case in the early stages of development, it remains an exceedingly useful teaching and research tool if it reveals key components and interactions that merit special attention.

Contrary to the feeling of many who are skeptical about modeling complex nature, information about only a relatively small number of variables is often a sufficient basis for effective models because "key factors" and "emergent" and other "integrative properties," as discussed in Sections 2 and 3, often dominate or control a large percentage of the action. Watt (1963), for example, states that "We do not need a tremendous amount of information about a great many variables to build revealing mathematical models." Though the mathematical aspects of modeling are a subject for advanced texts, we should review the first steps in model building.

Modeling usually begins with the construction of a diagram, or "graphic model," which is often a box or compartment diagram, as illustrated in Figure 1–2. Shown are two properties, P_1 and P_2, that interact as I to produce or affect a third property, P_3, when the system is driven by an energy source, E. Five flow pathways, F, are shown, with F_1 representing the input and F_6 the output for the system as a whole. Thus, at a minimum, there are four ingredients or components for a working model of an ecological situation, namely, (1) an energy source or other outside **forcing function**; (2) properties, called **state variables** by systems analysts; (3) **flow pathways** showing where energy flows or material transfers connect properties with each other and with forces; and (4) interactions or **interaction functions** where forces and properties interact to modify, amplify, or control flows or create new "emergent" properties.

Figure 1–2 could serve as a model for the production of photochemical smog in the air over Los Angeles. In this case, P_1 could represent hydrocarbons and P_2 nitrogen oxides, two products of automobile exhaust emission. Under the driving force of sunlight energy E, these interact to produce photochemical smog P_3. In this case, the interaction function I is a synergistic or augmentative one in that P_3 is a more serious pollutant for humans than is P_1 or P_2 acting alone.

Alternatively, Figure 1–2 could represent a grassland ecosystem in which P_1 are the green plants, which convert sun energy E to food. P_2 might represent a herbivorous animal that eats plants, and P_3 an omnivorous animal that can eat either the herbivores or the plants. In

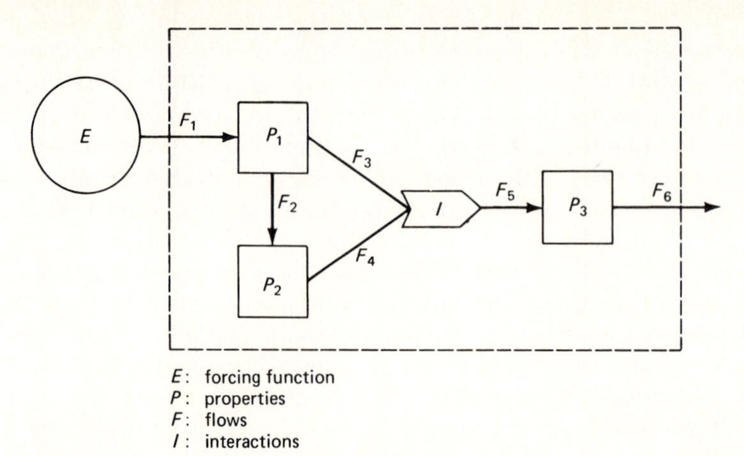

Figure 1–2 A compartment diagram showing four basic components of primary interest in modeling ecological systems.

this case, the interaction function I could represent several possibilities. It could be a "no-preference" switch if observations in the real world showed that the omnivore P_3 eats either P_1 or P_2, according to availability. Or, I could be specified to be a constant percentage value if it was found that the diet of P_2 was composed of, say, 80 percent plant and 20 percent animal matter, irrespective of the state of P_1 and P_2. Or, I could be a seasonal switch if P_3 feeds on plants during one part of the year and animals during another season. Or, it could be a threshold switch if P_3 greatly prefers animal food and switches to plants only when P_2 is reduced to a low level.

Figure 1–3 is a simplified diagram of a system that features a strong feedback or control loop in which "downstream" output, or

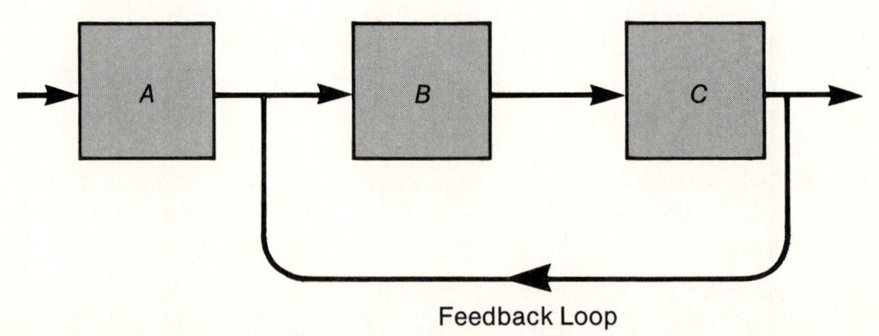

Figure 1–3 A compartment model with a feedback or control loop that transforms a linear system into a partially circular one.

some part of it, is fed back or recycled to affect or perhaps control "upstream" components. The feedback loop could represent predation by "downstream" organisms (C) that reduce and thereby tend to control the growth of "upstream" herbivores or plants (B and A) in the food chain. Or, the diagram could represent a desirable economic system in which resources (A) are converted into useful goods and services (B) with the production of wastes (C) that are recycled and used again in the conversion process ($A \rightarrow B$), thus reducing the waste output of the system. By and large, natural ecosystems have a circular or loop design rather than a linear structure. (Feedback and cybernetics, the science of controls, are discussed in Chapter 2.)

A general systems model with an internal feedback loop is shown in Figure 1–4. Two kinds of inputs, external (Z) and internal (ZX), act on the entity in question to maintain its state or to produce a new system state and new outputs in a period of time. In theory, the internal loop will tend to maintain an organized state despite disrupting external inputs.

Good model definition should include three dimensions: (1) the space to be considered (how the system is bounded), (2) subsystems (components) judged to be important in overall function, and (3) time interval to be considered. Once an ecosystem, ecological situation, or problem has been properly defined and bounded, a testable hypothesis or series of hypotheses is developed that can be rejected or ac-

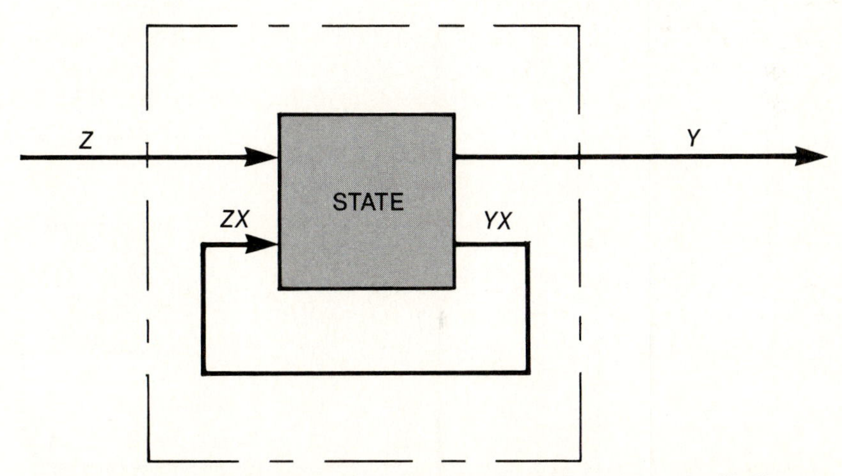

Figure 1–4 A general systems model with inputs, Z, and outputs, Y. The state of the system and its behavior over time depend on the interaction of external input Z with internal feedback loop input ZX. (After Mesarovic and Takahara, 1975.)

cepted, at least tentatively, pending further experimentation or analysis. For more on ecological modeling, see Hall and Day (1979) and Meadows (1982).

In the following chapters, the paragraphs headed by the word "**Statement**" are, in effect, "word" models of the ecological principle in question. In many cases, graphic models are also presented, and in some cases, simplified mathematical formulations are included. Most of all, this text attempts to provide the principles, simplifications, and abstractions that one must deduce from the real world before one can understand and deal with situations and problems or construct mathematical models of them.

2
The Ecosystem

1. Concept of the Ecosystem

Statement

Living organisms and their nonliving (abiotic) environment are inseparably interrelated and interact upon each other. Any unit (a biosystem) that includes all the organisms that function together (the biotic community) in a given area interacting with the physical environment so that a flow of energy leads to clearly defined biotic structures and cycling of materials between living and nonliving parts is an ecological system or **ecosystem.**

The ecosystem is the basic functional unit in ecology, since it includes both organisms and abiotic environment, each influencing the properties of the other and both necessary for maintenance of life as we have it on the earth. This level of organization must be of primary concern for us if society is to begin implementing holistic solutions for the problems now emerging at the level of the biome and the biosphere.

Since ecosystems are open systems, consideration of both the **input environment** and the **output environment** is an important part of the concept.

Explanation

The term "ecosystem" was first proposed in 1935 by the British ecologist A. G. Tansley, but, of course, the concept is by no means so recent. Allusions to the idea of the unity of organisms and environment (as well as the oneness of humans and nature) can be found as far back in written history as one might care to look. Not until the late 1800s did formal statements begin to appear, interestingly enough, in a parallel manner in the American, European, and Russian ecological literature. Thus, Karl Mobius in 1877 wrote (in German) about the community of organisms in an oyster reef as a "biocoenosis," and in 1887 S. A. Forbes, an American, wrote his classic essay on the lake as a "microcosm." The pioneering Russian, V. V. Dokuchaev (1846–1903) and his chief disciple, G. F. Morozov (who specialized in forest ecology),* emphasized the concept of the "biocoenosis," a term later expanded by Russian ecologists to "geobiocoenosis" (Sukachev, 1944). No matter what environment they studied (whether freshwater, marine, or terrestrial), biologists around the turn of the century began to consider the idea that nature does function as a system. It was not until a general systems theory was developed half a century later by Bertalanffy (1950, 1968) and others that ecologists, notably Hutchinson (1948a), Margalef (1958a), Watt (1966), Patten (1966, 1971), Van Dyne (1969), and H. T. Odum (1971), began to develop the definitive, quantitative field of **ecosystems ecology.** The extent to which ecosystems actually operate as general systems in the manner of well-understood physical systems and whether they are self-organizing in the manner of organisms are matters of continuing research and debate. The utility of the systems approach in solving real-world environmental problems is just now receiving serious attention.

Some other terms that have been used to express the holistic viewpoint, but that are not necessarily synonymous with ecosystems, include "holocoen" (Friederichs, 1930), "biosystem" (Thienemann,

* Dokuchaev's chief work, reprinted in Moscow in 1948, was *Uchenie o zonax prirody* (*Teaching About the Zones of Nature*). Morozov's chief book is *Uchenie o lese* (*Teaching About Forests*). We are indebted to Dr. Roman Jakobson, Professor of Slavic Languages at Harvard University, for information on these two works, which are little known in the United States.

1939), "bionenert body" (Vernadsky, 1945), and "holon" (Koestler, 1969).

The components and processes that make an ecosystem functional are shown in Figure 2–1. The interaction of the three basic components, namely, (1) the community, (2) the flow of energy, and (3) the cycling of materials, are diagrammed as a simplified compartment model with the general features discussed in the preceding chapter (see Figures 1–2 and 1–3). Energy flow is one-way; some of the incoming solar energy is transformed and upgraded in quality (that is, converted into organic matter, a more concentrated form of energy than sunlight) by the community, but most of it is degraded and passes through and out of the system as low-quality heat energy (heat sink). Energy can be stored, then "fed back," or exported, as shown in the diagram, but it cannot be reused. The physical laws governing the behavior of energy are considered in detail in Chapter 3. In contrast with energy, materials, including the nutrients necessary for life (carbon, nitrogen, phosphorus, and so on), and water can be used over and over again. The efficiency of recycling and the magnitude of imports and exports of nutrients vary widely with the

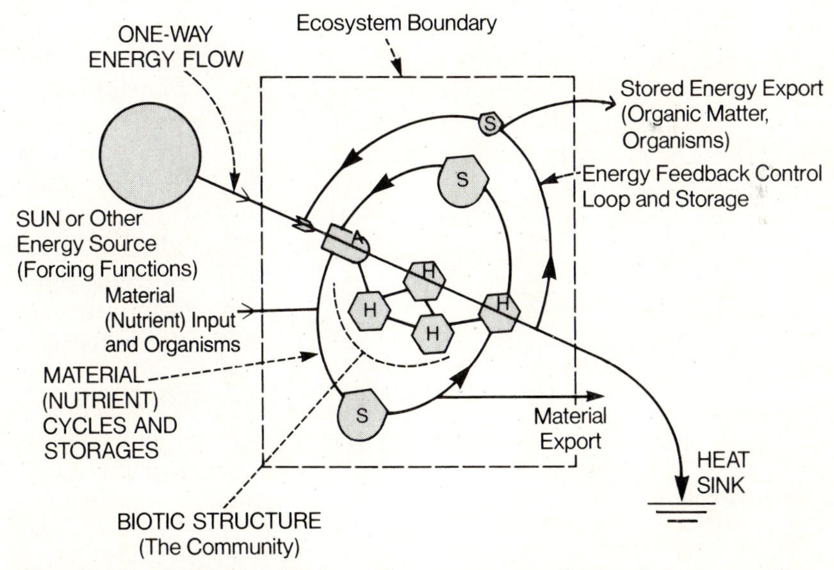

Figure 2–1 A functional diagram of an ecosystem. Energy flow, material cycles, the community, and feedback control loops are the four major components. The community is shown as a network or food web of autotrophs (A) and heterotrophs (H). Storages are indicated by S.

type of ecosystem, as noted in our earlier commentary on coral reefs (see pages 5 and 6).

Figure 2–1 introduces a new dimension to the compartment model diagram. Each "box" in the diagram is given a distinctive shape that indicates its general function according to an "energy language" devised by H. T. Odum (1967, 1971). As detailed in the figure legend, circles are energy sources, tank-shaped modules are storages, bullet-shaped modules are autotrophs (green plants capable of transforming sun energy to organic matter), and hexagons are heterotrophs (organisms requiring ready-made food). In the functional diagram of Figure 2–1, the community is depicted as a "food web" of autotrophs and heterotrophs linked together with appropriate energy flows, nutrient cycles, and storages. Food webs will be discussed in more detail in Chapter 3.

All ecosystems, even the ultimate biosphere, are open systems: there is a necessary inflow and outflow of energy. Of course, ecosystems below the level of the biosphere are also open in varying degrees to material flows and to the immigration and emigration of organisms. Accordingly, an important part of the ecosystem concept is recognizing that there is both an **input environment** and an **output environment** that are coupled and essential for the ecosystem to

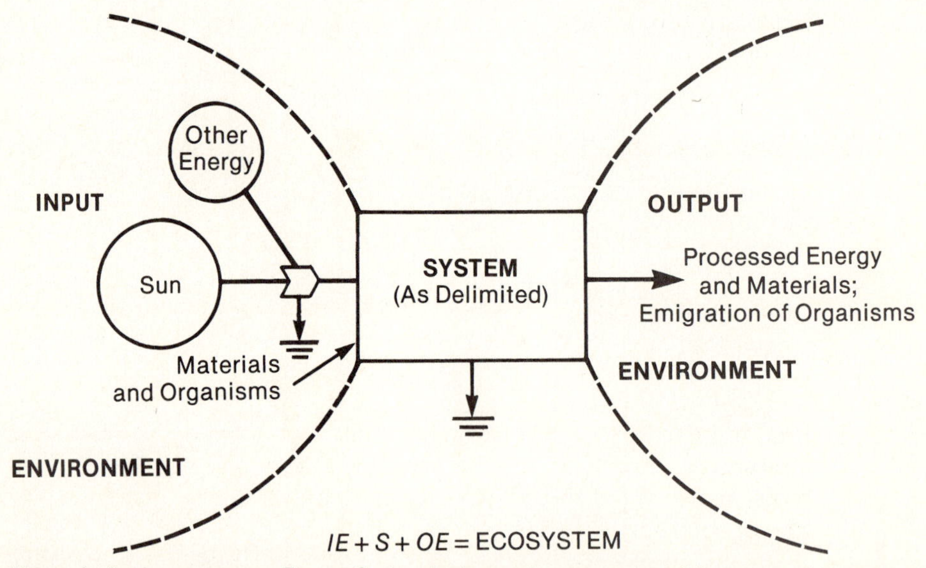

Figure 2–2 In contrast to Figure 2–1, which emphasizes internal functions, this ecosystem model emphasizes the external environment, which must be considered an integral part of the ecosystem concept (Concept based on Patten, 1978).

function and maintain itself. Figure 2–2 emphasizes this feature: a conceptually complete ecosystem includes an input and output environment along with the system as delimited, or ecosystem = $IE + S + OE$. This scheme solves the problem of where to draw lines around an entity that one wishes to consider, because it does not matter very much how the box portion of the ecosystem is delimited. Often, natural boundaries, such as a lake shore or forest edge, or political ones, such as city limits, make convenient boundaries, but limits can just as well be arbitrary so long as they can be accurately designated in a geometric sense. The box is not all there is to the ecosystem, because if the box were an impervious container its living contents (lake or city) would not survive such enclosure. A functional or real-world ecosystem must have an input life line and, in most cases, a means of exporting processed energy and materials.

The extent of the input and output environment varies extremely and depends on several variables, for example, (1) size of the system (the larger, the less dependent on externals), (2) metabolic intensity (the higher the rate, the greater the input and output), (3) autotrophic-heterotrophic balance (the greater the imbalance, the more externals required to balance), and (4) stage and development (young systems differ from mature systems, as detailed in Chapter 8). Thus, a large, forested mountain range has much smaller input-output environments than does a small stream or a city. These contrasts are brought out in the discussion of examples of ecosystems, Section 7 of this chapter.

2. The Structure of the Ecosystem

Statement

From the standpoint of **trophic structure** (from *trophe* = nourishment), an ecosystem is two-layered: it has (1) an upper **autotrophic** (= self-nourishing) **stratum** or "green belt" of chlorophyll-containing plants or plant parts in which fixation of light energy, use of simple inorganic substances, and the buildup of complex organic substances predominate, and (2) a lower **heterotrophic** (= other-nourishing) **stratum** or "brown belt" of soils and sediments, decaying matter, roots, and so on, in which utilization, rearrangement, and decomposition of complex materials predominate. From a biological viewpoint, it is convenient to recognize the following components as constituting the ecosystem: (1) **inorganic substances** (C, N, CO_2, H_2O, and others) involved in material cycles; (2) **organic compounds** (proteins, carbohydrates, lipids, humic substances, and so on) that link biotic and abiotic; (3) **air, water,** and **substrate environment** including the **climate regime** and other physical factors; (4) **producers**, autotrophic organisms, largely green plants, that can manufacture food from simple inorganic substances; (5) **macroconsumers** or **phagotrophs** (from *phago* = to eat), heterotrophic organisms, chiefly animals, that ingest other organisms or particulate organic matter; (6) **microconsumers, saprotrophs** (from *sapro* = to decompose), **decomposers,** or **osmotrophs** (from *osmo* = to pass through a membrane), heterotrophic organisms, chiefly bacteria and fungi, that obtain their energy either by breaking down dead tissues or by absorbing dissolved organic matter extruded by or extracted from plants or other organisms. The decomposing activities of saprotrophs release inorganic nutrients that are usable by the producers; they also provide food for the macroconsumers and often excrete hormone-like substances that inhibit or stimulate other biotic components of the ecosystem.

Another useful two-category subdivision of heterotrophs suggested by Wiegert and Owen (1971) is **biophages,** organisms that consume other living organisms, and **saprophages**, organisms that feed on dead organic matter.

Explanation

In Section 1, the function of ecosystems was considered; now their structure will be discussed. Figure 2–3 compares the profile, or cross section, of a pond ecosystem and a grassland ecosystem. The layers and components, as listed in the Statement, are labeled in these diagrams.

One of the universal features of all ecosystems, whether terrestrial, freshwater, marine, or human-engineered (for example, agricultural), is the interaction of the autotrophic and heterotrophic components, as outlined in the Statement. The organisms responsible for the processes are partially separated in space; the greatest autotrophic metabolism occurs in the upper "green belt" stratum where light energy is available. The most intense heterotrophic metabolism occurs in the lower "brown belt" where organic matter accumulates in soils and sediments. Also, the basic functions are partially separated in time, since there may be a considerable delay in heterotrophic utilization of the products of autotrophic organisms. For example, photosynthesis predominates in the canopy of a forest ecosystem. Only a part, often only a small part, of the photosynthate is immediately and directly used by the plant and by herbivores and parasites, which feed on foliage and other actively growing plant tissue. Much

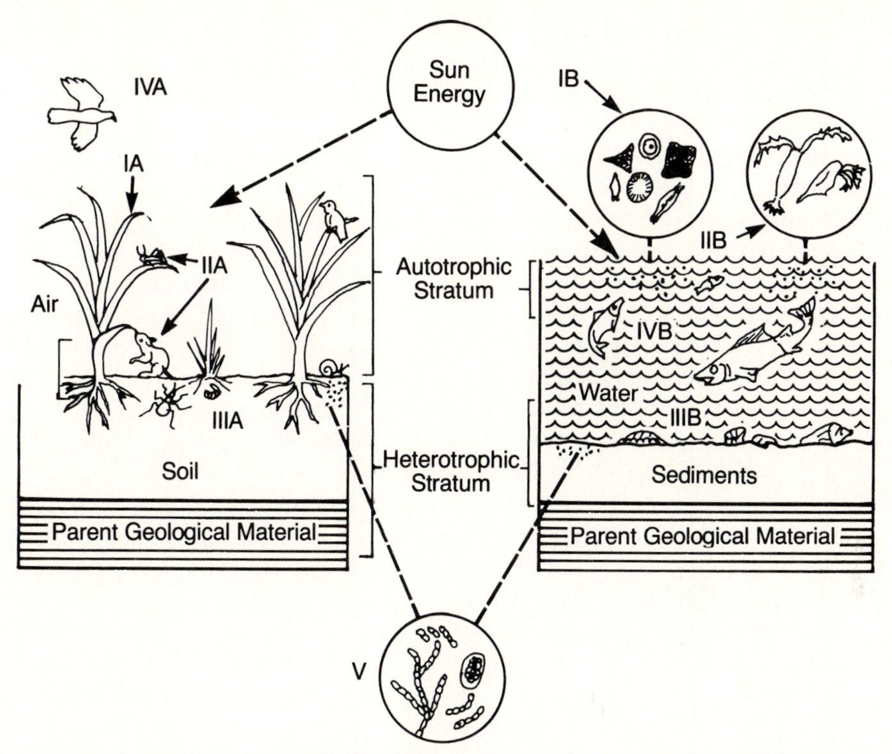

Figure 2–3 Comparison of the structure of a terrestrial (a grassland) and an aquatic ecosystem (lake or sea). Necessary units for function are sun (and other) energy input; water; nutrients (basic abiotic inorganic and organic compounds) in soils, sediments, and water; and autotrophic and heterotrophic organisms that comprise the biotic food webs. Terrestrial and aquatic systems are similar in operation, but the species are largely different. Also, green plants (phytoplankton) are small (often microscopic) in deep-water systems and large in terrestrial and some shallow-water ecosystems. I. Autotrophs: (A) grasses and forbs, (B) phytoplankton. II. Grazing herbivores (plant eaters): (A) insects and mammals in the grassland, (B) zooplankton in water column. III. Detritivores: (A) soil invertebrates on land, (B) bottom invertebrates in water. IV. Carnivores: (A) birds and others on land, (B) fish in water. V. Saprovores: bacteria and fungi of decay.

of the synthesized material (leaves, wood, and stored food in seeds and roots) escapes immediate consumption and eventually reaches the litter and soil (or the equivalent sediments in aquatic ecosystems), which together constitute a well-defined heterotrophic system. Weeks, months, or years (or many millennia, in the case of fossil fuels now being rapidly consumed by human societies) may pass before all the accumulated organic matter is utilized.

The term "organic detritus" (= product of disintegration, from the Latin *deterere*, "to wear away") is borrowed from geology, in which it is traditionally used to designate the products of rock disintegration. As used in this text, "detritus," unless otherwise indicated, refers to all the organic matter involved in the decomposition of dead organisms. Detritus seems the most suitable of many terms that have been suggested to designate this important link between the living and the inorganic world (Odum and de la Cruz, 1963). Rich and Wetzel (1978) suggest that the dissolved organic matter that leaks out of or is extracted by saprotrophs from both living and dead tissues be included under the heading of "detritus" since it has a similar function. Environmental chemists use a shorthand designation for the two physically different products of decomposition as follows: POM is particulate organic matter; DOM is dissolved organic matter. The role of POM and DOM in food chains is reviewed in Chapter 3.

Abiotic components that limit and control organisms are discussed in detail in Chapter 5, and the role of organisms in controlling the abiotic environment is considered later in this chapter. As a general principle, from the operational standpoint, the living and nonliving parts of ecosystems are so interwoven into the fabric of nature that it is difficult to separate them; hence, operational or functional classifications do not sharply distinguish between biotic and abiotic.

Most of the vital elements (carbon, nitrogen, phosphorus, and so on) and organic compounds (carbohydrates, proteins, lipids, and so on) are not only found inside and outside of living organisms but are also in a constant state of flux between living and nonliving states. Some substances, however, appear to be unique to one or the other state. The high energy storage material ATP (adenosine triphosphate), for example, is found only inside living cells (or at least its existence outside is very transitory), whereas **humic substances**, which are resistant end-products of decomposition (see page 39), are never found inside cells, yet are a major and characteristic component of all ecosystems. Other key biotic complexes, such as the genetic material DNA (deoxyribonucleic acid) and the chlorophylls, occur both inside and outside organisms but become nonfunctional when outside the cell.

The three living components (producers, phagotrophs, and saprotrophs) may be thought of as the three functional kingdoms of nature since they are based on the type of nutrition and the energy source used. These ecological categories should not be confused with taxonomic kingdoms, although there are certain parallels, as pointed out by Whittaker (1969) and as shown in Figure 2–4. In Whittaker's

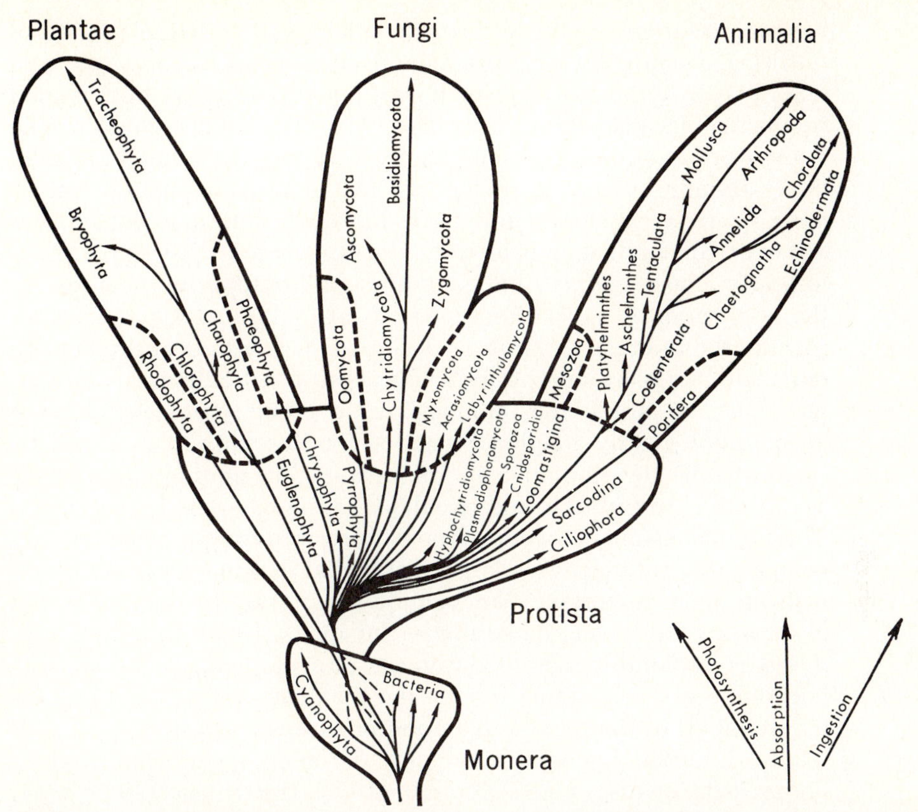

Figure 2–4 A five-kingdom system based on three levels of organization—the procaryotic (kingdom Monera), eucaryotic unicellular (kingdom Protista), and eucaryotic multicellular and multinucleate. On each level there is divergence in relation to three principal modes of nutrition—the photosynthetic, absorptive, and ingestive. Many biology and microbiology texts list four kingdoms by combining the "lower Protista" (i.e., Monera) with the "higher Protista" to form the "Protista." Evolutionary relations are much simplified, particularly in the Protista. Only major animal phyla are entered, and phyla of the bacteria are omitted. The Coelenterata comprise the Cnidaria and Ctenophora; the Tentaculata comprise the Bryozoa, the Brachiopoda, the Phoronida, and in some treatments the Entoprocta. (From Whittaker, 1969.)

arrangement of the phyla into an evolutionary "family tree," all three types of nutrition are found in the Monera and Protista, while the three higher branches, namely "plants," fungi, and "animals," specialize as "producers," "absorbers" (saprotrophs), and "ingestors" (phagotrophs), respectively. The ecological classification is one of function rather than of species as such. Some species occupy intermediate positions in the series, and others can shift their mode of nutrition according to environmental circumstances. Separation of heterotrophs into large and small consumers is arbitrary but justified in practice because of the very different study methods required. The heterotrophic microorganisms (bacteria, fungi, and others) are relatively immobile (usually imbedded in the medium being decomposed), are very small, and have high rates of metabolism and turnover. Specialization is more evident biochemically than morphologically; consequently, one cannot usually determine their role in the ecosystem by such direct methods as looking at them or counting their numbers. Organisms designated as macroconsumers obtain their energy by heterotrophic ingestion of particulate organic matter. These are largely the "animals" in the broad sense. They tend to be morphologically adapted for active food seeking or food gathering, with the development of complex sensory-neuromotor as well as digestive, respiratory, and circulatory systems in the higher forms. The microconsumers, or saprotrophs, were often designated as "decomposers," but recent work has shown that in some ecosystems animals are more important than bacteria or fungi in the decomposition of organic matter (see, for example, Johannes, 1968). Consequently, it seems preferable not to designate any particular organisms as "decomposers" but rather to consider "decomposition" as a process involving all of the biota and abiotic processes as well.

For additional general discussions of the ecosystem concept, see Forbes's classic essay (1887), Tansley (1935), Hutchinson (1948, 1964, 1967a), Evans (1956), Cole (1958), E. P. Odum (1969a, 1972), and H. T. Odum (1971, Chapter 1). Schultz (1967) and Van Dyne (1969) discuss the concept from the standpoint of resource management, Stoddard (1965) from the viewpoint of a geographer, Duncan (1964) from the standpoint of the sociologist, and Vayda and Rappaport (1968) from the viewpoint of the anthropologist. For historical reviews of the ecosystem concept, see Major (1969) and Golley (1982).

Every student of ecology and, indeed, every citizen should read Aldo Leopold's "The Land Ethic" (1949), an eloquent, often quoted and reprinted essay on the special relevance of the ecosystem concept. We all also should reread *Man and Nature* (written in 1864, reprinted in 1965) by the Vermont prophet George Perkins Marsh, who analyzed the causes of the decline of ancient civilizations and forecast a similar doom for modern ones unless an "ecosystematic" view of the world is taken. Especially recommended is Russell's excellent 1968 review of Marsh's work, as viewed from today's perspective, and Flader's 1979 review of the Leopold philosophy.

3. The Study of Ecosystems

Statement

Ecologists approach the study of large, complex ecosystems, such as lakes and forests, in two ways: (1) the **holological** (from *holos* = whole) in which inputs and outputs are measured, collective and emergent properties of the whole are assessed (as discussed in Chapter 1, Section 3), and then the component parts are investigated as needed; and (2) the **merological** (from *meros* = part) in which the major parts are studied first and then integrated into a whole system. Recently, ecologists have resorted more and more to two additional approaches, involving experimental and modeling techniques.

Explanation and Examples

In his 1964 essay "The Lacustrine Microcosm Reconsidered," the eminent American ecologist G. E. Hutchinson speaks of E. A. Birge's 1915 work on the heat budget of lakes as pioneering the holological approach. Birge concentrated his study on measuring the inflows and outflows of energy to and from the lake rather than focusing on what happens in the lake. Essentially, the ecosystem is treated as a **black box**, which may be defined as any unit whose function can be evaluated without specifying the internal contents. Hutchinson contrasts this method of study with the component or merological approach of Stephen Forbes in his classic 1887 paper (referred to in Section 1), in which approach "we discourse on parts of the system and try to build up the whole from them."

The contrasting holistic and reductionist approaches are complementary and not antagonistic aspects for a given descriptive level. Components cannot be distinguished if there is no "whole" or "system" to abstract from, and there cannot be a whole unless there are constituent parts (recall our definition of "system," page 4). In practice, one's approach depends on the objective of the study, and especially on the degree with which the parts are coupled. When constituents are strongly coupled, emergent properties will likely reveal themselves only at the level of the whole. Such important attributes could be missed if only the merological approach was taken. Above all, a given organism may behave quite differently in different systems, and this variability has to do with how the organism is coupled with other components. Many insects, for example, are destructive pests in an agricultural habitat but not in their natural habitat where parasites, competitors, predators, or chemical inhibitors keep them under control.

Sometimes the best way to gain insight into an ecosystem is to experiment with it, i.e., disturb it in some manner in the hope that the response will clarify hypotheses that one has deduced from observation. In recent years, "stress ecology" or "perturbation (from *perturbare* = to disturb) ecology" has become an important research field (see international symposium edited by Barrett and Rosenberg, 1982). Besides manipulating the real thing, one can also gain insight by manipulating models, as very briefly discussed in Chapter 1. As you read this text, watch for examples of all of these approaches.

4. The Biological Control of the Geochemical Environment: The Gaia Hypothesis

Statement

Individual organisms not only adapt to the physical environment, but by their concerted action in ecosystems they also adapt the geochemical environment to their biological needs. Thus, communities of organisms and their input and output environments develop together as ecosystems. The fact that the chemistry of the atmosphere and the earth's strongly buffered physical environment are utterly different from conditions on any other planet in our solar system has led to the **Gaia Hypothesis**, which holds that organisms, especially microorganisms, have evolved with the physical environment to provide an intricate control system that keeps earth's conditions favorable for life (Lovelock, 1979).

Explanation

Although everyone knows that the abiotic environment ("physical factors") controls the activities of organisms, not everyone recognizes that organisms influence and control the abiotic environment in many important ways. The physical and chemical nature of inert materials are constantly being changed by organisms that return new compounds and energy sources to the environment. The actions of marine organisms largely determine the content of the sea and of its bottom oozes. Plants growing on a sand dune build up a soil radically different from the original substrate. A South Pacific coral atoll is a striking example of how organisms modify the abiotic environment. From simple raw material of the sea, whole islands are built because of the activities of animals (corals) and plants. Organisms control the very composition of our atmosphere.

That extension of biological control to the global level is the basis for James Lovelock's **Gaia Hypothesis** (*Gaia,* the Greek word for "earth goddess"). Lovelock, a physical scientist, inventor, and engineer, has teamed with microbiologist Lynn Margulis to explain the Gaia Hypothesis in a series of articles and a recent book (Lovelock, 1979; Lovelock and Margulis, 1973; Margulis and Lovelock, 1974, 1975; Lovelock and Epton, 1975). They conclude that the earth's atmosphere, with its unique high oxygen–low carbon dioxide content and the moderate temperature and pH conditions of earth's surface, cannot be accounted for without the critical buffering activities of early life forms and the continued coordinated activity of plants and microbes that dampen fluctuations in physical factors that would occur in the absence of well-organized living systems. For example, ammonia produced by organisms maintains a pH in soils and sediments that is favorable to a wide variety of life. Without this organismic output, the pH of the earth's soil could become so low that it would preclude all but a very few kinds of organisms.

Table 2–1 contrasts the earth's atmosphere with a hypothetical atmosphere devoid of life and with the atmosphere of Mars, where, if there is life, it certainly is not in control. In other words, the earth's atmosphere did not just develop by chance interaction of physical forces into a life-sustaining condition, and then life evolved to adapt to this condition. Rather, organisms from the very beginning played the major role in the development and control of a geochemical environment favorable to themselves. Lovelock and Margulis envision the microorganism web of life operating in the "brown belt" as an intricate control system that functions like a chemostat, somewhat

Table 2–1 Comparison of Atmospheric and Temperature Conditions of Mars, Venus, Earth, and a Hypothetical Earth Without Life*

	Mars	Venus	Earth Without Life	Earth As Is
Atmosphere				
Carbon dioxide	95%	98%	98%	0.03%
Nitrogen	2.7%	1.9%	1.9%	79%
Oxygen	0.13%	Trace	Trace	21%
Surface temperature °C	−53	477	290 ± 50	13

* After Lovelock, 1979.

analogous to an environmental control device that keeps a large skyscraper livable. This control system ("Gaia") makes the earth one complex but unified cybernetic system (see Section 6 of this chapter). All of this is very much a "hypothesis." No actual control network has yet been demonstrated to the satisfaction of many skeptical scientists, although a strong biological influence on the atmosphere is accepted by most. Lovelock admits that the "search for Gaia" may be long and difficult, since hundreds of processes would have to be involved in an integrated control mechanism of such magnitude.

Humans, of course, more than any other species, attempt to modify the physical environment to meet their immediate needs, but in doing so are increasingly shortsighted. Biotic components necessary for our physiological existence are being destroyed, and global balances are beginning to be perturbed. Since we are heterotrophs and phagotrophs who thrive best near the end of complex food chains, we depend on the natural environment no matter how sophisticated our technology. Our great cities are only parasites in the biosphere when we consider what we have already designated as **life-support resources**, namely, air, water, fuel, and food. The bigger and the more technologically advanced the cities, the more they demand from the surrounding countryside and the greater the danger of damaging the natural environment "host."

Lovelock's Gaia Hypothesis suggests the importance of discovering and preserving the controls that enable the biosphere to adjust to at least certain amounts of non-point source pollution such as carbon dioxide, heat, sulfur, nitrogenous oxides, and so on. Accordingly, besides striving to reduce pollution by every means possible, human beings must also preserve the integrity and the large scale of the life-support buffer system.

Examples

One of the classic papers that every student of ecology should read is Alfred Redfield's summary essay, published in 1958, entitled "The Biological Control of Chemical Factors in the Environment." Redfield marshals the evidence to show that the oxygen content of the air and the nitrate in the sea are produced and largely controlled by organic activity and, furthermore, that the quantities of these vital components in the sea are determined by the biocycling of phosphorus. This system is as intricate and beautifully organized as a fine watch, but unlike a watch, the marine times regulator was not built by engineers and is, comparatively speaking, little understood. Lovelock's little 1979 book, referred to previously, essentially extends Redfield's hypothesis to the global level and should also be read. See also Jantsch's *The Self-Organizing Universe* (1980).

The Copper Basin at Copperhill, Tennessee, impressively demonstrates the result when the number of organisms is so reduced that a system's organization is destroyed, and the area is subject to the extremes of physical forces. In these regions, sulfuric acid fumes from copper smelters exterminated all of the rooted plants over a wide area. A type of smelting, known as "roasting," involves the igniting of great piles of ore, green wood, and coke. These piles then smoldered continuously, giving off the acidic fumes. Most of the soil eroded, leaving a spectacular desert that looks like a landscape on Mars, as shown in Figure 2–5. Also, large areas nearby were deforested to provide wood for the "roasting." Although improved smelting methods have reduced the emission of fumes, vegetation has failed to return in the most severely eroded areas, and peripheral areas have recovered very slowly. Artificial reforestation using heavy fertilization with minerals or sewage sludge has succeeded somewhat. Pine seedlings inoculated with symbiotic root fungi that assist the tree in extracting nutrients from impoverished soils (see Figure 7–10) are surviving on their own as the large input of fertilizer is used up. At least such experiments demonstrate that locally damaged ecosystems can be restored, but only with great effort and expense. Everyone should visit Copperhill as part of his or her general education, or visit a badly eroded or stripmined area and ask: How much will it cost "us taxpayers" to rehabilitate such land, and how much of the damage was needless and could have been prevented?

Important economic and political lessons can be learned from Copperhill. When a single industry uses up all the life-support capacity of a large area, and in this case destroys a part of it, no further economic development in that area is possible. No other industry or business not directly connected with the resident industry will come in, because no possible environmental support exists for anything else. People living in the area are mired in an unhealthy environment and the "one industry" syndrome of political domination and cultural stagnation. Also, little or none of the profits from exploitative industry of this type remains in the area; the money is exported to other areas where economic development is still possible. Historians James Cobb and Thomas Dyer (1979) describe the political struggle between Georgia and Tennessee over attempts to control the pollution at Copperhill that continued for nearly three quarters of a century. Georgia, which is "downwind" from the operation, early pushed for environmental and health protection, while Tennessee, which reaped most of the economic benefits, continually resisted any attempts to

A

B

Figure 2–5 *A*. The Copper Basin at Copperhill, Tennessee, suggests what land without life would be like. A luxuriant forest once covered this area until fumes from smelters killed all of the vegetation. Although fumes are no longer released by modern methods of ore preparation, the vegetation has not become reestablished. (Photo courtesy of U.S. Forest Service.) *B*. Farmland in Mississippi ruined by soil erosion. Such abuses leave abandoned houses and poor people. (Photo courtesy of U.S. Forest Service.)

change the status quo. Only after years of court battles, including, finally, U.S. Supreme Court action, did the copper company begin to modernize its operation to reduce the intense "acid rain." Today, of course, such an extreme "point source" pollution would be acted upon much more quickly; however, more dilute but far more extensive "acid rain" spreading from large industrial areas is causing a crisis in northern Europe, the eastern United States and Canada, and elsewhere. Accordingly, the political struggle between environmental protection and short-term profits continues, but on a regional and global level, not just a local level. Conflicts will undoubtedly become increasingly bitter until a majority of people become aware not only of the dangers of letting pollution get out of control but also of the corrective or preventive technology that can be applied before this happens.

Garrett Hardin (1968) has aptly called this "The Tragedy of the Commons." The commons are resources such as the air, the sea, or public land that everybody is free to use. Unless society acts to set up agreed-upon restrictions, overuse and subsequent deterioration are almost inevitable. Hardin and Boden have recently edited a book (1977) entitled *Managing the Commons,* in which various authors suggest ways to avoid "The Tragedy of the Commons."

Prospects and techniques for restoring damaged ecosystems are covered in recent symposia edited by Cairns et al. (1977), Holdgate and Woodman (1978), and Cairns (1980). Repairing the biosphere will likely become big business, and this could be good news for faltering world economics. In any event, quality and repair of the biosphere will have to be emphasized more, since most of us cannot escape to space colonies, which are not yet practical (see page 70).

5. Global Production and Decomposition

Statement

Every year approximately 10^{17} grams (about 100 billion tons) of organic matter is produced on the earth by photosynthetic organisms. An approximately equivalent amount is oxidized back to CO_2 and H_2O during the same time interval as a result of the respiratory activity of living organisms. *But, the balance is not exact.**

* Vallentyne, 1962.

Over most (but not all) of geological time (at least since the beginning of the Cambrian period 600 million to 1 billion years ago), a very small but significant fraction of the organic matter produced is incompletely decomposed in anaerobic (anoxic) sediments or completely buried and fossilized without being respired or decomposed. This excess organic production, over respiration, is regarded as a major reason for a decrease in CO_2 and a buildup of oxygen in the atmosphere to the high levels of recent geological times. Thus, evolution and continued survival of the higher forms of life became possible. About 300 million years ago, especially excess production formed the fossil fuels that made the industrial revolution possible. During the past 60 million years, shifts in biotic balances coupled with variations in volcanic activity, rock weathering, sedimentation, and solar input have resulted in an oscillating steady state in CO_2/O_2 atmospheric ratios. Oscillations in atmospheric CO_2 were associated with and presumably caused alternate warming and cooling of climates. During the past half century, human agroindustrial activities have significantly raised the CO_2 concentration in the atmosphere, which, because of the potential for climate alteration, poses a serious global problem.

Explanation

The Kinds of Photosynthesis and Producer Organisms Chemically, the photosynthetic process involves the storage of a part of the sunlight energy as potential or "bound" energy of food. The general equation of the oxidation-reduction reaction can be written as follows:

$$CO_2 + 2H_2A \xrightarrow[\text{energy}]{\text{light}} (CH_2O) + H_2O + 2A,$$

the oxidation being

$$2H_2A \rightarrow 4H + 2A$$

and the reduction being

$$4H + CO_2 \rightarrow (CH_2O) + H_2O.$$

For green plants in general (algae, higher plants) A is oxygen; water is oxidized with release of gaseous oxygen, and the carbon dioxide is reduced to carbohydrate (CH_2O) with release of water. In bacterial photosynthesis, on the other hand, H_2A (the "reductant") is not water but either an inorganic sulfur compound, such as hydrogen sulfide (H_2S) in the green and purple sulfur bacteria (Chlorobacteriaceae and Thiorhodaceae, respectively), or an organic compound, as in the purple and brown nonsulfur bacteria (Athiorhodaceae). Consequently, *oxygen is not released* in these types of bacterial photosynthesis.

The **photosynthetic bacteria** are largely aquatic (marine and freshwater) and in most situations play a minor role in the production of organic matter. However, they can function under unfavorable conditions for the general run of green plants, and they do cycle certain minerals in aquatic sediments. The green and purple sulfur bacteria, for example, are important in the sulfur cycle (see Figure 4–5). They are obligate anaerobes (able to function only in the absence of oxygen), and they occur in the boundary layer between oxidized and reduced zones in sediments or water where there is light of low intensity. Tidal mudflats are good places to observe these bacteria because they often form distinct pink or purple layers just under the upper green layers of mud algae (in other words, at the very upper edge of the anaerobic or reduced zone where light but not much oxygen is available). In a study of Japanese lakes, Takahashi and Ichimura (1968) found that photosynthetic sulfur bacteria accounted for only 3 to 5 percent of the total annual production in most lakes, but in stagnant lakes rich in H_2S these bacteria accounted for up to 25 percent of the total photosynthesis. In contrast, the nonsulfur photosynthetic bacteria are generally facultative anaerobes (able to function with or without oxygen). They can also function as heterotrophs in the absence of light, as can many algae. Bacterial photosynthesis, then, can be helpful in polluted and eutrophic waters, and hence is being increasingly studied, but it is no substitute for the "regular" oxygen-generating photosynthesis on which the world depends.

Recent discoveries that higher plants differ in biochemical pathways for carbon dioxide reduction (the reduction portion of the equation on page 30) have important ecological implications. In most plants, carbon dioxide fixation follows a **C_3 pentose phosphate** or **Calvin cycle**, which for many years was the accepted scheme for photosynthesis. Then, in the 1960s, several plant physiologists, notably Hatch and Slack in Australia, contributed to the discovery that certain plants reduce carbon dioxide in a different manner according to a **C_4 dicarboxylic acid cycle**. The latter plants have large chloroplasts in the bundle sheaths around the leaf veins, a distinctive morphological feature that had been noted a century earlier but had not been suspected as an indicator of a major physiological characteristic. More important, plants with the dicarboxylic acid cycle respond differently to light, temperature, and water. For the purposes of discussion of the ecological implications, the two photosynthetic types are designated **C_3 and C_4 plants**.

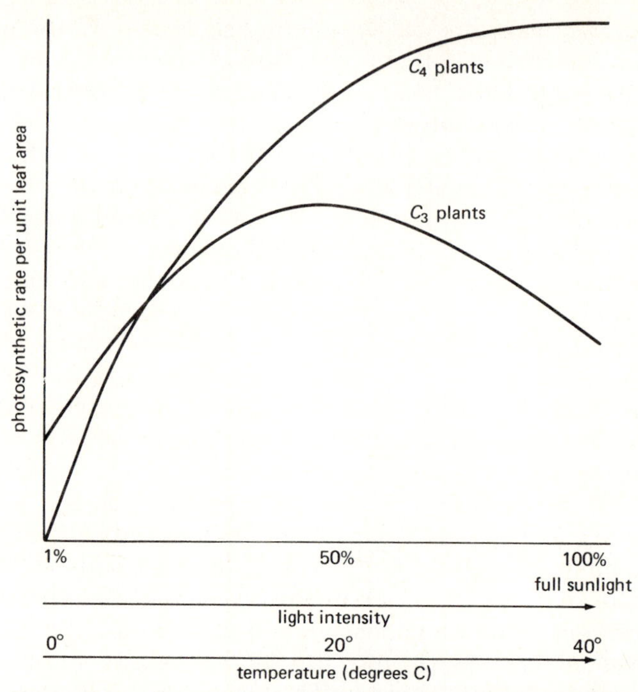

Figure 2–6 Comparative photosynthetic response of C_3 and C_4 plants to increasing light intensity and temperature. See text for explanation.

Figure 2–6 contrasts the response of C_3 and C_4 plants to light and temperature. The former tend to peak in photosynthetic rate (per unit of leaf surface) at moderate light intensities and temperatures and to be inhibited by high temperatures and the intensity of full sunlight. In contrast, C_4 plants are adapted to high light and high temperature, greatly exceeding the production of C_3 plants under these conditions. They also use water more efficiently, requiring generally less than 400 grams of water to produce 1 gram of dry matter, compared with the 400- to 1000-gram water requirement of C_3 plants. In addition, they are not inhibited by high oxygen concentration as are C_3 species. One reason C_4 plants are more efficient at the high end of the light-temperature scales is that there is little photorespiration, that is, the plant's photosynthate is not respired away as light intensity rises. Although this has not yet been checked out thoroughly, some C_4 plants seem to be more resistant to grazing by insects (see Caswell et al., 1973), perhaps because they tend to have a lower protein content. On the other hand, Haines and Hanson (1979) report that detritus made from C_4 plants was a richer food source for consumers in a salt marsh than is detritus originating from C_3 plants.

Species with a C_4-type photosynthesis are especially numerous in the grass family (Gramineae) but occur in many other families of both monocots and dicots. As would be expected, C_4 species dominate the vegetation in deserts and grasslands in warm temperature and tropical climates and are rare in forests and in the cloudy north where low light intensities and temperatures predominate. Table 2–2 illustrates how the proportion of C_4 species increases along a gradient from the cool, moist prairies of the midwestern United States to the hot, dry deserts of the southeast and also how C_4/C_3 proportions differ with the seasons in temperate deserts. It is not surprising that crabgrass, that legendary pest of suburban lawns, turns out to be a C_4 species, as are a number of other weeds that thrive in man-made warm open spaces. For more on comparison of C_3 and C_4 plants, see Black (1971), Bjorkman and Beery (1973), and Beery (1975).

Despite their lower photosynthetic efficiency at the leaf level, C_3 plants account for most of the world's photosynthetic production, presumably because they are more competitive in mixed communities where there are shading effects, where light, temperature, and so on are average rather than extreme (note from Figure 2–6 that C_3 plants outperform C_4 plants under low light and temperature conditions). This appears to be another good example of the whole-is-not-the-sum-of-the-parts principle. Survival of the fittest in the real world does not always go to the species that are merely physiologically superior under optimum conditions in monoculture, but rather to those that are superior in multiculture under varying conditions that are not always optimum. To put it another way, what is efficient in isolation is not necessarily efficient in the community, where interaction among species is vital in natural selection.

Table 2–2 **Percentage of C_4 Species in an East-West Transect of United States Grasslands and Deserts**

	% C_4 Species
Tall grass prairie	50
Mixed-grass grassland	67
Short-grass grassland	100
Desert-summer annuals	100
Desert-winter annuals	0

Plants that humans now depend on for food, such as wheat, rice, potatoes, and most vegetables, are for the most part C_3 types, since most crops suitable for intensive, mechanized agriculture were developed in north-temperature countries. Crops of tropical origin, such as corn, sorghum, and sugar cane, are C_4 plants. More C_4 varieties should obviously be developed for use in irrigated deserts and in the tropics.

Another recently discovered photosynthetic mode especially adapted to deserts is known as CAM, for crassulacean acid metabolism. Several desert succulent plants, including the cacti, keep their stomates closed during the hot daytime and open them in the cool of the night. Carbon dioxide, absorbed through the leaf openings, is stored in organic acids (hence the name) and not "fixed" until the next day. This delayed photosynthesis greatly reduces water loss during the day, thereby enhancing the succulent plant's ability to maintain water balance and water storage.

Microorganisms called **chemosynthetic bacteria** are considered to be **chemolithotrophs** because they obtain their energy for carbon dioxide assimilation into cellular components not by photosynthesis but by the chemical oxidation of simple inorganic compounds, for example, ammonia to nitrite, nitrite to nitrate, sulfide to sulfur, and ferrous to ferric iron. They can grow in the dark, but most require oxygen. The sulfur bacterium *Thiobacillus,* often abundant in sulfur springs, and the various nitrogen bacteria, which are important in the nitrogen cycle, are examples. For the most part, chemolithotrophs are involved in carbon recovery rather than in primary production, since their ultimate energy source is organic matter produced by photosynthesis.

Recently, however, unique deep-sea ecosystems based entirely on chemosynthetic bacteria not dependent on a photosynthate source have been discovered. These are located in totally dark areas where the sea floor is spreading, creating vents from which hot, mineral-rich, sulfurous water escapes. Here, various marine animals, including foot-long clams and strange 10-foot-long seaworms, obtain their energy from bacteria that metabolize sulfide, and possibly other inorganic compounds along with CO_2 and O_2 (for pictures, see Ballard and Grassle, 1979). Some of these animals feed directly on sulfur bacteria; others apparently cultivate the bacteria in their guts. Here, indeed, is an age-old geothermally powered ecosystem, since the heat of the earth's core produces the reduced sulfur compounds that are the energy source for this ecosystem. It is a curious exception to the general rule that light and green plants initiate the food-making process (Karl et al., 1980).

Because they can function in the dark recesses of sediments, soil, and ocean bottoms, the chemosynthetic bacteria not only recover mineral nutrients but, as the Russian hydrobiologist Sorokin (1966) has pointed out, they also rescue energy (carbon recovery, as noted previously) that would otherwise be unavailable to consumers.

Most species of the higher plants (spermatophytes) and many species of algae require only simple inorganic nutrients and are therefore completely autotrophic. Some species of algae, however, require a single complex organic "growth substance," which they themselves cannot synthesize. Still, other species require one, two, three, or many such growth substances and are therefore partly heterotrophic. The term **auxotrophic** (in the sense of auxiliary sources) is often used for intermediates between autotrophy and heterotrophy. (See reviews by Provasoli, 1958; Hutner and Provasoli, 1964; and Lewin, 1963.) In the land of the midnight sun in northern Sweden, Rodhe (1955) has presented evidence to indicate that during the summer, phytoplankton in lakes are producers; during the long winter night (which may last for several months), when they apparently are able to utilize the accumulated organic matter in the water, they are consumers.

On the global scale, among the evolutionally higher forms of life, the distinction between autotrophs and heterotrophs is clearcut, and gaseous oxygen is essential for survival. But many species and varieties among the lowly microorganisms—the bacteria, fungi, and more primitive algae and protozoa—are not so specialized. Rather, they are adapted to be intermediate or to shift between autotrophy and heterotrophy, with or without oxygen.

Types of Decomposition (Catabolism) and Decomposers In the world at large, the heterotrophic process of decomposition approximately balances the autotrophic metabolism. If decomposition is considered in the broad sense as "any energy-yielding biotic oxidation," then several types of decomposition roughly parallel the types of photosynthesis when oxygen requirements are considered:

1. Aerobic respiration—gaseous (molecular) oxygen is the electron acceptor (oxidant).
2. Anaerobic respiration—gaseous oxygen not involved. An inorganic compound other than oxygen or an organic compound is the electron acceptor (oxidant).
3. Fermentation—also anaerobic, but the organic compound oxidized is also the electron acceptor (oxidant).

Type 1, aerobic respiration, is the reverse of the "regular" photosynthesis and is the process by which organic matter (CH_2O) is decomposed back to CO_2 and H_2O with a release of energy. All of the higher plants and animals and most of the Monerans and Protistans (see Figure 2–4) obtain their energy for maintenance and for the formation of cell material in this manner. Complete respiration yields CO_2, H_2O, and cell material, but the process may be incomplete, leaving energy-containing organic compounds to be used later by other organisms, as is the case in Types 2 and 3.

As a way of life, respiration without O_2 is largely restricted to the saprophages such as bacteria, yeasts, molds, and protozoa, although it occurs as a dependent process within certain tissues of higher animals. The methane bacteria are good examples of obligate anaerobes that decompose organic compounds with the production of methane (CH_4) through reduction of either organic or carbonate carbon (thus employing both Types 2 and 3 metabolisms). The methane gas, often known as swamp gas, rises to the surface where it can be oxidized, or, if it catches fire, it may become a UFO (unidentified flying object)! The methane bacteria are also involved in the breakdown of forage within the rumen of cattle and other ruminates. As we deplete supplies of natural gas and other fossil fuels, these microbes may be domesticated to produce methane on a large scale from manure or other organic sources.

Desulfovibrio bacteria are ecologically important examples of Type 2 anaerobic respiration, because they reduce SO_4 in deep sediments and in anoxic waters such as the Black Sea to H_2S gas. The H_2S gas can rise to shallow sediments or surface waters where it can be acted on by other organisms (the photosynthetic bacteria, for example). Yeasts, of course, are well-known examples of fermentors (Type 3). They are not only commercially important but also abundant in soils, where they help decompose plant residues.

As already indicated, many kinds of bacteria are capable of both aerobic and anaerobic respiration (for example, facultative anaerobes), but the end-products of the two reactions will be different, and the amount of energy released will be much less under anaerobic conditions. Figure 2–7 shows the results of an interesting study in which the same species of bacterium, *Aerobacter*, was grown under anaerobic and aerobic conditions with glucose as the carbon source. When oxygen was present, almost all of the glucose was converted into bacterial biomass and CO_2, but in the absence of oxygen, decomposition was incomplete, a much smaller portion of the glucose ended up as cell carbon, and a series of organic com-

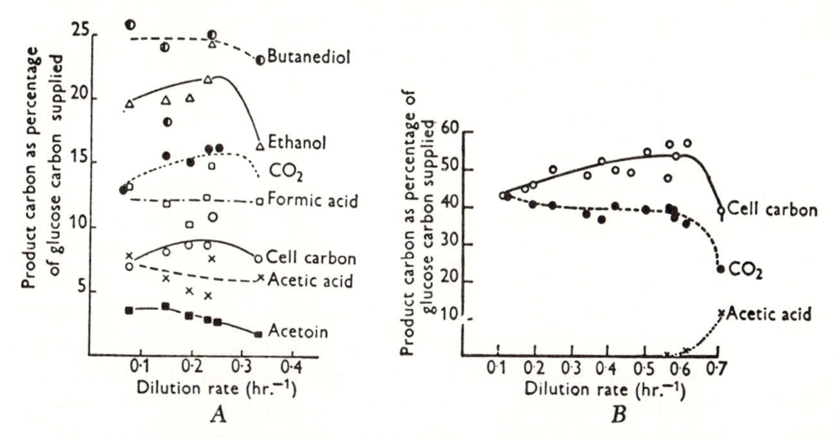

Figure 2–7 The decomposition of glucose by the bacterium *Aerobacter* under anaerobic (*A*) and aerobic (*B*) conditions. Note that under aerobic conditions decomposition is complete and 40 to 50 percent of the original carbon of glucose is converted into bacterial biomass. However, under anaerobic conditions only about 15 percent of glucose carbon is converted and a number of incompletely decomposed organic substances remain in the environment. The decline in cell carbon and CO_2 and the beginning of acetic acid production in *B* indicate that oxygen is beginning to decrease, with the result that the situation shown in *A* is beginning to develop. (After Pirt, 1957.)

pounds was released into the environment. Additional bacterial specialists would be required to oxidize these. When the rate of input of organic detritus into soils and sediments is high, bacteria, fungi, protozoa, and other organisms create anaerobic conditions by using up the oxygen faster than it can diffuse into the media. Decomposition does not stop then but continues, often at a slower rate, provided an adequate diversity of anaerobic microbial metabolic types is present.

Although the anaerobic saprophages (both obligate and facultative) are inconspicuous components of the community, they are nonetheless important in the ecosystem because they alone can respire in the dark, oxygenless layers of soils and aquatic sediments. They "rescue" energy and materials that diffuse out of the depths and become available to the aerobes.

The anaerobic world represents the primodial world (since it is believed that the earliest lifeforms were anaerobic procaryotes) (see Figure 2–4) on which the later aerobic world was superimposed. Rich (1978) describes the two-step evolution of life as follows:

Pre-Cambrian life evolved as the free energy from lengthening electron transport increased, i.e. the *quality* of energy available to organisms increased. In the second part, the realm of conventional evolution, the energetic value of a unit of organic matter was fixed (ultimate electron accepter = oxygen) and life evolved in response to the *quantity* of energy available to organisms.

In today's world, the reduced inorganic and organic compounds produced by anaerobic microbial processes serve as carbon and energy reservoirs for photosynthetically fixed energy. When later exposed to aerobic conditions, the compounds serve as substrates for aerobic chemolithotrophs and heterotrophs. Accordingly, the two life styles are intimately coupled and function together for mutual benefit. A sewage disposal system, which is a human-engineered decomposing subsystem, depends on the partnership between anaerobic and aerobic saprophages for maximum efficiency.

Decomposition: An Overview Decomposition results from both abiotic and biotic processes. For example, prairie and forest fires are not only major limiting or controlling factors, as will be discussed later, but they are also "decomposers" of detritus, releasing large quantities of CO_2 and other gases to the atmosphere and minerals to the soil. Fire is an important, even necessary process in so-called fire-type ecosystems, in which physical conditions are such that microbial decomposers do not keep up with organic production. The grinding action of freezing and thawing and water flow also break down organic materials. By and large, however, the heterotrophic microorganisms or saprophages ultimately act on the dead bodies of plants and animals. This kind of decomposition, of course, is the result of the process by which bacteria and fungi obtain food for themselves. Decomposition, therefore, occurs through energy transformations within and between organisms and is an absolutely vital function. If it did not occur, all the nutrients would soon be tied up in dead bodies, and no new life could be produced. Within the bacterial cells and the fungal mycelia are sets of enzymes necessary to carry out specific chemical reactions. These enzymes are secreted into dead matter; some of the decomposition products are absorbed into the organism as food, and other products remain in the environment or are excreted from the cells. No single species of saprotroph can completely decompose a dead body. However, populations of decomposers prevalent in the biosphere consist of many species that, by their sequential action, can completely decompose a body. Not all parts of the bodies of plants and animals are broken down at the same rate. Fats, sugars, and proteins are decomposed readily, but plant cellulose, lignin of wood, and the chitin, hair, and bones of animals are acted on very slowly. Figure 2–8A shows a comparison of the rate of decomposition of dead marsh grass and fiddler crabs placed in nylon-mesh "litter bags" in a Georgia salt marsh. Note that most of the animal remains and about 25 percent of the dry weight of the marsh grass were decomposed in about two months, but the remaining 75 percent of the grass, largely cellulose, was acted on more slowly. After ten months, 40 percent of the grass remained, but all of the crab remains had disappeared from the bag. As the detritus becomes finely divided and escapes from the bag, the intense activities of microorganisms often result in protein enrichment, as shown in Figure 2–8B, thus providing a more nutritious food for detritus-feeding animals (Odum and de la Cruz, 1967; Kaushik and Hynes, 1968). The graphic model of Figure 2–9 shows that the decomposition of forest litter (leaves and twigs) is very much influenced by lignin (resistant woody polysaccharide) content and climatic conditions.

The more resistant products of decomposition end up as **humus** (or **humic substances**), which, as already indicated, is a universal component of ecosystems. It is convenient to recognize three stages of decomposition: (1) the formation of particulate detritus by physical and biological action accompanied by release of dissolved organic matter, (2) the relatively rapid production of humus and release of additional soluble organics by saprotrophs, and (3) the slower mineralization of humus.

The slowness of decomposition of humic substances is a factor in the decomposition lag and oxygen accumulation that has been stressed. In general appearance, humus is a dark, often yellow-brown, amorphous or colloidal substance that is difficult to characterize chemically. No great difference in physical properties or chemical structure exists between humic substances in geographically scattered or biologically different terrestrial ecosystems, but recent studies suggest that marine humic materials have a different origin and hence a different structure.

In chemical terms, humic substances are condensations of aromatic compounds (phenols) combined with the decomposition products of proteins and polysaccharides. A model for the molecular structure of humus derived from lignocellulose is shown in Figure 2–10. The phenolic type of benzene ring and the side-chain bonding make these compounds recalcitrant to microbial decomposition. Fission of these structures apparently requires special deoxygenase enzymes (Gibson, 1968), which may not be present in the common soil and water

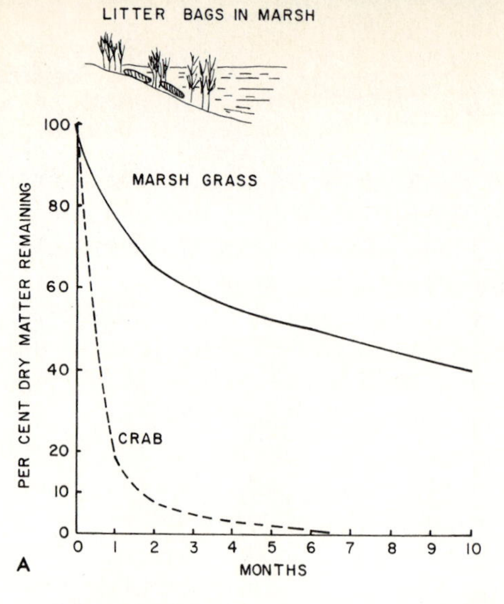

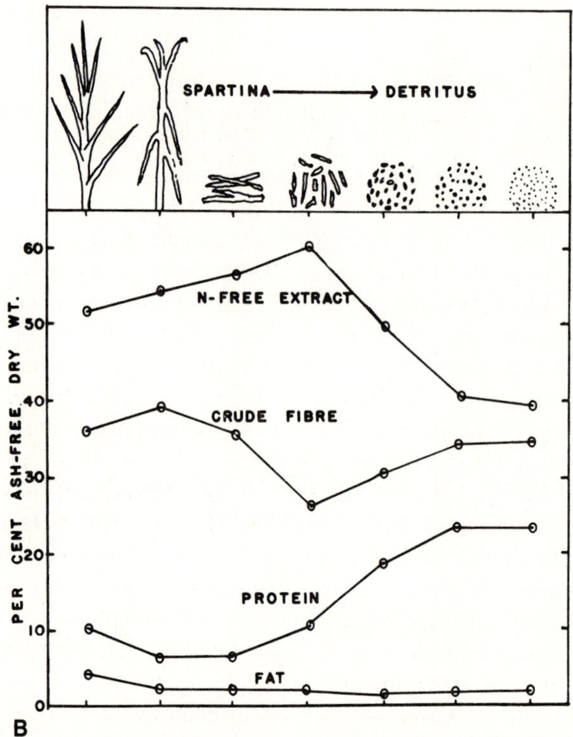

Figure 2–8 Two aspects of decomposition in a Georgia salt marsh. *A*. Decomposition in terms of percentage of dead marsh grass (*Spartina alterniflora*) and fiddler crabs (*Uca pugnax*) remaining in nylon-mesh litter bags placed in the marsh where they would be subjected to daily tidal inundation. *B*. Protein enrichment resulting from microbial activity on marsh grass detritus in the final stages of particulate breakdown. (After E. P. Odum and de la Cruz, 1967.) Deposit-feeding animals (snails, worms, clams, shrimp, fish) often feed selectively on the more nutritious small particles (see Newell, 1965, and W. E. Odum, 1968a).

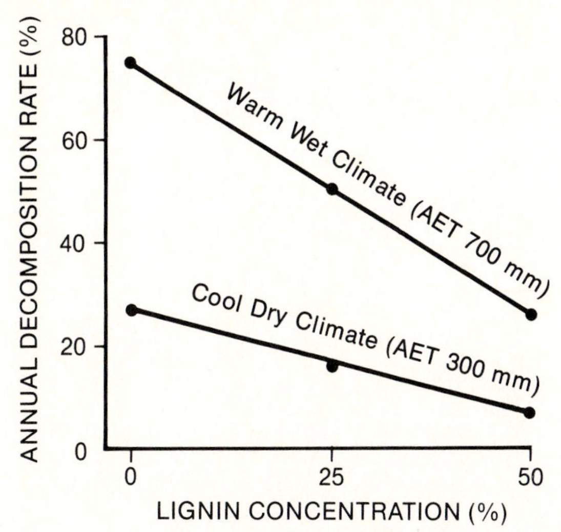

Figure 2–9 Decomposition of forest litters as a function of lignin content and climate. AET = actual evapotranspiration. (After Meentemeyer, 1978.)

saprotrophs. Ironically, many of the toxic materials that humans are now adding to the environment, such as herbicides, pesticides, and industrial effluents, are derivatives of benzene and are causing serious trouble because of their low degradability and their toxicity.

The pioneer microbial ecologist Winogradsky proposed in 1925 the concept that organisms decomposing fresh organic matter are an ecologically separate flora from those that decompose humus. He called these groups **zymogenous** and **autochthonous**, respectively (Winogradsky, 1949, p. 473). To this day, however, it is uncertain whether humus is broken down by special organisms with special enzymes, by abiotic chemical processes, or by both. The study of humus has progressed slowly because it does not lend itself to conventional analysis in the chemical laboratory. More *in situ* studies are needed, similar to those described by Tribe (1963), who observed formation of humus in material placed between two glass slides, which could be periodically removed from their positions in the soil for microscopic study and chemical analysis.

In becoming fossilized, organic matter goes through a two-step process: (1) humification, mainly aerobic and relatively rapid, and (2) carbonification, mostly anaerobic and very slow. The latter goes from peats to lignites to soft coals to hard coals, with carbon concentration increasing at each step (Hartenstein, 1981).

Detritus, humic substances, and other organic matter undergoing decomposition are important for soil fertility. In moderate quantity,

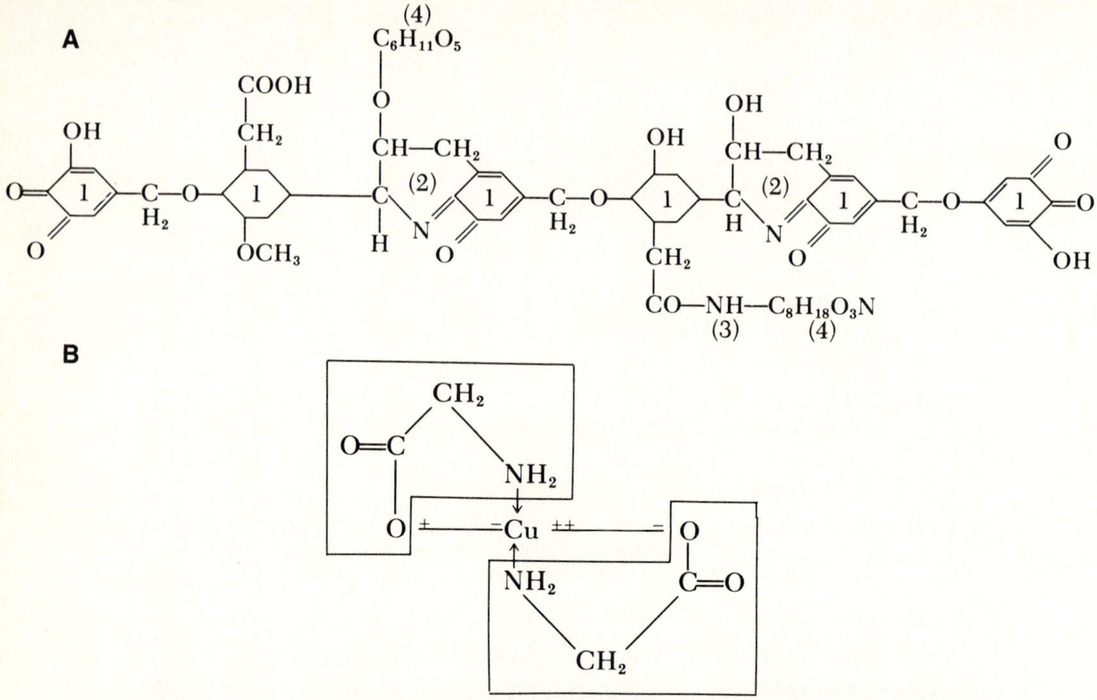

Figure 2–10 *A*. A possible model for a humic acid molecule illustrating the following key features: the aromatic or phenolic benzene rings (labeled 1), cyclic nitrogen (2), nitrogen side chains (3), and carbohydrate residues (4), all of which make humic substances difficult to decompose. *B*. A picture model of chelation. A copper ion (Cu) held in "crab claws" by a pair of covalent (→) and ionic (−+) linkages between two molecules of an amino acid, glycine.

these materials provide a favorable texture for plant growth. Furthermore, many organics form complexes with minerals that greatly affect the biological availability of the minerals. For example, **chelation** (from *chele* = claw, referring to grasping), a complex formation with metal ions, keeps the element in solution and nontoxic compared with the inorganic salts of the metal. Figure 2–10B shows how a copper ion can be held in "crab claws" by pairs of covalent (→) and ionic (−+) linkages between two molecules of glycine, an amino acid. Since industrial wastes are full of toxic metals, it is fortunate that chelators that are products of the natural decomposition of organic matter work to mitigate toxic effects on organisms. For example, the toxicity of copper to phytoplankton is correlated with the free ion (Cu^{++}) concentration, not total copper. Accordingly, a given amount of copper is less toxic in an inshore environment than in the open sea where there is less organic matter to complex the metal.

Studies are now showing that the phagotrophs, especially the small animals (protozoa, soil mites, collembolans, nematodes, ostracods, snails, and so on), are more important in decomposition than was previously suspected. When these microfauna are selectively removed, breakdown of dead plant material is greatly slowed, as shown by three experimental studies summarized in Figure 2–11. Although many detritus-feeding animals (detritivores) cannot actually digest the lignocellulose substrates, but obtain their food energy largely from the microflora in the material, they speed the decomposition of plant litter in a number of indirect ways: (1) by breaking down detritus into small pieces, thus increasing the surface area available for microbial action; (2) by adding proteins or growth substances (often in the animal's excretions) that stimulate microbial growth; and (3) by stimulating the growth and metabolic activity of these microbial populations in the eating of some of the bacteria and fungi. Furthermore, many detritivores are **coprophagic** (from *kopros* = dung); they regularly ingest fecal pellets after the pellets have been enriched by microbial activity in the environment (Newell, 1965; Frankenberg and Smith, 1967). For example, the "betsy beetle" (*Popilius*), which lives in decaying logs, uses its tunnels as a sort of external rumen where fecal pellets and chewed wood fragments become enriched by fungi and then reingested (Mason and Odum, 1969). Coprophagy in this case is an insect-fungus partnership that enables the beetle to utilize the food energy in wood and also hastens the decay of the log. In the sea, pelagic tunicates called "Salps," which feed by straining out microflora from the water, produce large fecal pellets that have been shown to provide a major food source for other marine animals, including fish.

Years ago, it was suggested that invertebrate animals were beneficial in sewage beds (Hawkes, 1963) because they presumably help break up the substrate material and speed up bacterial action. But there were few serious studies because engineers generally considered "worms" and so on a nuisance and assumed that bacterial action was all that mattered. Since 1975, several studies have indicated that phagotrophs can speed up sewage decomposition.

Although the mineralization of organic matter that provides plant nutrients has been stressed as the primary function of the decomposition, another function has been receiving more attention from ecologists. Apart from the importance of saprotrophs as food for other organisms, the organic substances released into the environment during decomposition may have profound effects on the growth of other

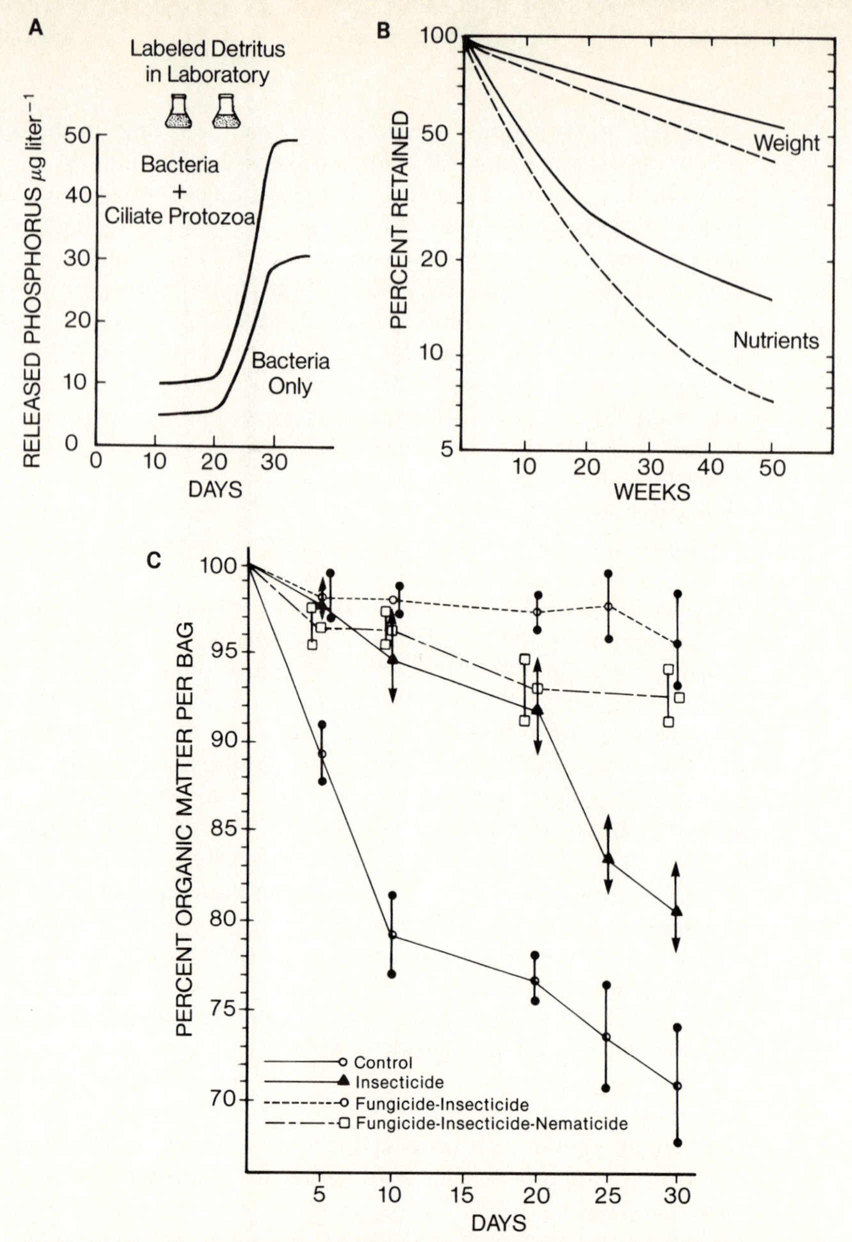

Figure 2–11 Three experimental demonstrations of the importance of small animals in decomposition of organic matter. *A*. Release of labeled phosphorus from marsh detritus is more rapid in a laboratory flask when protozoa as well as bacteria are present. (After Johannes, 1965.) *B*. Weight and nutrient loss from forest litter in litter bags is much slower (solid line) when microarthropods are killed with naphthaline, which does not affect bacteria or fungi. (After Crossley and Witkamp, 1964.) *C*. Loss of organic matter from buried litter bags in a grassland is greatly slowed when microathropods, nematodes, or fungi are selectively removed. (After Santos et al., 1981.)

organisms in the ecosystem. Julian Huxley in 1935 suggested the term "external diffusion hormones" for those chemical substances that exert a correlative action on the system via the external medium. Lucas (1947) has proposed the term "ectocrine" (or "exocrine," as preferred by some writers). The term "environmental hormone" also clearly indicates what is meant, but "secondary metabolites" is the label most often used for substances excreted by one species that affect others. These substances may be inhibitory, as in the antibiotic penicillin (produced by a fungus), or stimulatory, as in various vitamins and other growth substances, for example, thiamin, vitamin B_{12}, biotin, histidine, uracil, and others, many of which have not been identified chemically.

Although the saprotrophs seem most important in producing environmental hormones, the algae also release substances that have major effects on the structure and function of aquatic communities. Inhibitory leaf and root excretions of higher plants are also important in this regard. C. H. Muller and his associates speak of such excretions as "allelopathic substances" (from *allelon* = of each other; *pathy* = suffering), and they have shown that the metabolites interact complexly with fire in controlling desert and chaparral vegetation (Muller et al., 1968). In dry climates, excretions tend to accumulate and thus have more effect than under rainy conditions. Whittaker and Feeny (1971) and Rice (1974) have detailed the role of biochemical excretions in the development and structuring of communities.

In summary, degradation of organic matter is a long and complex process; it controls several important functions in the ecosystem. For example, it (1) recycles nutrients through mineralization of dead organic matter; (2) chelates and complexes mineral nutrients; (3) microbially recovers nutrients and energy; (4) produces food for a sequence of organisms in the detritus food chain; (5) produces secondary metabolites that may be inhibitory or stimulatory and are often regulatory; (6) modifies inert materials of the earth's surface to produce, for example, the unique earthly complex known as "soil;" and (7) maintains an atmosphere conducive to life of large biomass aerobes such as ourselves.

The Global Production-Decomposition Balance Despite nature's broad spectrum and great variety of functions, the simple autotroph-phagotroph-saprotroph classification is a good working arrangement for describing the ecological structure of a biotic community. "Production," "consumption," and "decomposition" are useful terms for describing overall functions. These and other ecological categories pertain to *functions and not necessarily to species as such*, since a particular species population may be involved in more than one basic function. As already noted, the evolutionally more advanced phyla of organisms tend to be restricted to a rather narrow range of functions. Individual species of bacteria, fungi, protozoa, and algae may be quite specialized metabolically, but collectively these lower phyla are extremely versatile and can perform just about any biochemical transformation. Although microorganisms are regarded as primitive, human beings and other "higher" organisms cannot live without what LaMont Cole has called the "friendly microbes" (Cole, 1966); they provide the "fine tuning" that maintains some degree of stability in the ecosystem since they can adjust quickly to changing conditions.

As emphasized in the statement, the delay in the complete heterotrophic utilization and decomposition of the products of autotrophic metabolism is one of the most important features of the biosphere, since fossil fuels have accumulated in the ground and oxygen in the atmosphere because of that delay. Of most immediate concern is that human activities are unwittingly, but very rapidly, speeding up decomposition (1) by burning the stored organic matter in fossil fuels, (2) by agricultural practices that increase the decomposition rate of humus, and (3) by worldwide deforestation and burning of wood (still the major energy source for the two thirds of the world's people who live in poor nations). All of this activity expels into the air the CO_2 stored in coal and oil and in trees and humus of deep forest soils. Although the amount of CO_2 diffused into the atmosphere by agroindustrial activities is still small compared with the total amount in circulation, the CO_2 concentration in the atmosphere has increased since 1900. Possible consequences for the modification of climate will be reviewed in Chapter 4.

6. The Cybernetic Nature and the Stability of Ecosystems

Statement

Besides energy flows and material cycles, as briefly described in Section 1 (and in more detail in Chapters 3 and 4), ecosystems are rich in information networks comprising physical and chemical communication flows that connect all parts and steer or regulate the system as a whole. Accordingly, ecosystems can be considered cybernetic (from *kybernetes* = pilot or governor) in nature, but control functions are internal and diffuse rather than external and specified as in human-engineered cybernetic devices. Redundancy—more than one species or component capable of performing a given function—also enhances stability. The degree to which stability is achieved varies widely, depending on the rigor of the external environment as well as on the efficiency of internal controls. There are two kinds of stability: **resistance stability** (ability to remain "steady" in the face of stress) and **resilience stability** (ability to recover quickly); the two may be inversely related.

Explanation and Examples

The very elementary principles of cybernetics are modeled in Figure 2–12, which compares a goal-seeking automatic control system with specified external control as in a mechanical device (*A*) with a nonteleologic system with diffuse subsystem regulation as in ecosystems (*B*). In any case, control depends on **feedback**, which occurs when part of the output feeds back as input. When this feedback input is positive (like compound interest, which is allowed to become part of the principal), the quantity grows. **Positive feedback** is deviation-accelerating and, of course, necessary for growth and survival of organisms. However, to achieve control—for example, to prevent the overheating of a room or the overgrowth of a population—there must also be **negative feedback**, or deviation-counteracting input. The energy involved in a negative feedback signal is extremely small compared with energy flow through the system, whether it be a household-controlled heating system, an organism, or an ecosystem. *Low-energy components that have very much amplified high-energy feedback effects are major characteristic features of cybernetic systems.*

The science of cybernetics, as founded by Norbert Wiener (1948), embraces both inanimate and animate controls. Mechanical feedback mechanisms are often called **servomechanisms** by engineers, while biologists use the phrase **homeostatic mechanisms** to refer to organismic systems. Homeostasis (from *homeo* = same; *stasis* = standing) at the organism level is a well-known concept in physiology, as outlined in Walter B. Cannon's classic book, *The Wisdom of the Body* (1932). In servomechanisms and organisms, a distinct mechanical or anatomical "controller" has a specified "set point" (Figure 2–12A). In the familiar household heating system, the thermostat controls the furnace; in a warm-blooded animal, a specific brain center controls body temperature. In contrast, the interplay of material cycles and energy flows, along with subsystem feedbacks in large ecosystems, generates a self-correcting homeostasis with no outside

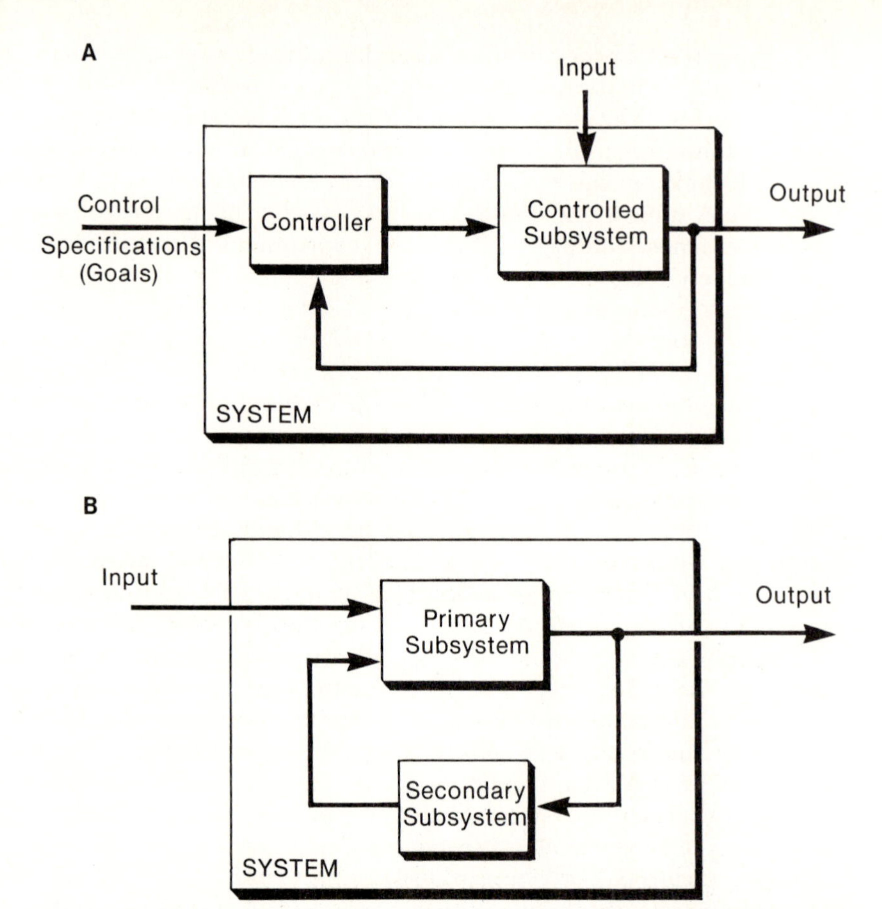

Figure 2–12 Feedback control systems. *A*. Model appropriate for human-made automatic control systems and homeostatic, goal-seeking organismal systems. *B*. Model appropriate for nonteleological systems, including ecosystems, where control mechanisms are internal and diffuse, involving interaction between primary and secondary subsystems. (After Patten and Odum, 1981.)

control or set point required (Figure 2–12B). As already described in part, control mechanisms operating at the ecosystem level include microbial subsystems that regulate the storage and release of nutrients, behavioral mechanisms, and predator-prey subsystems that control population density, to mention just a few examples. Because ecosystems are controlled somewhat differently than mechanical or organismic systems, whether or to what extent ecosystems are cybernetic has been controversial and much discussed in the literature. (For contrasting views, see Engelberg and Boyarsky, 1979, and Patten and Odum, 1981.)

One difficulty in perceiving cybernetic behavior at the ecosystem level is that components at the ecosystem level are coupled in networks by various physical and chemical messengers that are analogous to but far less visible than nervous or hormonal systems of organisms. Simon (1973) has pointed out that "bond energies," which link components, become more diffuse and weaker with an increase in space and time scales. At the ecosystem scale, these weak but very numerous bonds of energy and chemical information have been called the "invisible wires of nature" (H. T. Odum, 1971), and the phenomenon of organisms responding dramatically to low concentrations of substances is more than just a weak analogy to hormonal control. Low-energy causes producing high-energy effects are ubiquitous in ecosystem networks; two examples will suffice to illustrate. Tiny insects known as parasitic hymenoptera in a grassland ecosystem account for only a very small portion (often less than a tenth of 1 percent) of the total community metabolism, yet they can have a very large controlling effect on total primary energy flow (production) by the impact of their parasitism on herbivorous insects. In a cold spring ecosystem model, described by Patten and Auble (1979), a small biomass of carnivores improbably regulates bacteria through a feedback loop in which only 1.4 percent of the energy input to the system is fed back to the detrital substrate of the bacteria. In diagrams of ecological systems (see Figures 1–3, 1–4, 2–1, and 3–12) this phenomenon is commonly shown as a reverse loop in which "downstream" energy of low quantity is fed back to an "upstream" component with a greatly augmented effect in controlling the whole system. This type of amplified control, by virtue of position in a network, is exceedingly widespread and indicates the intricate global feedback structure of ecosystems. Through evolutionary time, such interactions have stabilized ecosystems by preventing boom-and-bust herbivory, catastrophic predator-prey oscillations, and so on. Although, as already noted, feedback control at the level of the biosphere is a controversial subject, it follows naturally from what is known at the ecosystem level.

In addition to feedback control, redundancy in functional components also contributes to stability. For example, if several species of autotrophs are present, each with a different temperature operating range, the rate of photosynthesis of the community as a whole can remain stable despite changes in temperature. Beyers (1962) has demonstrated this kind of homeostasis in a microcosm experiment. Figure 2–13 compares negative feedback control with redundancy

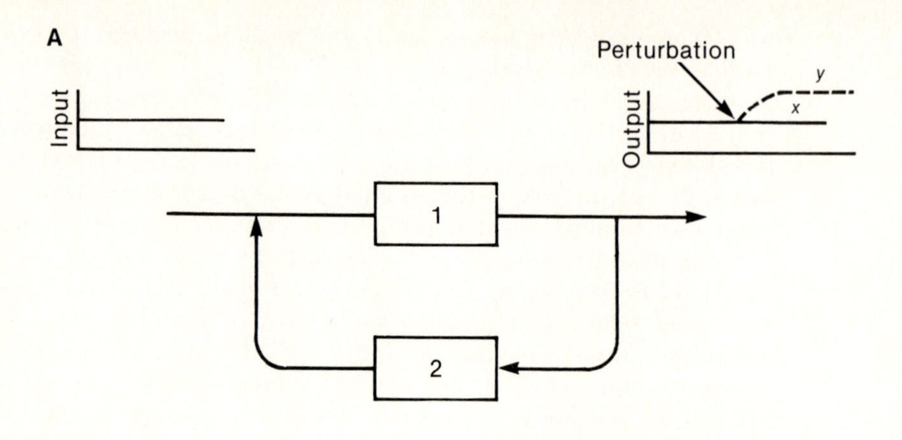

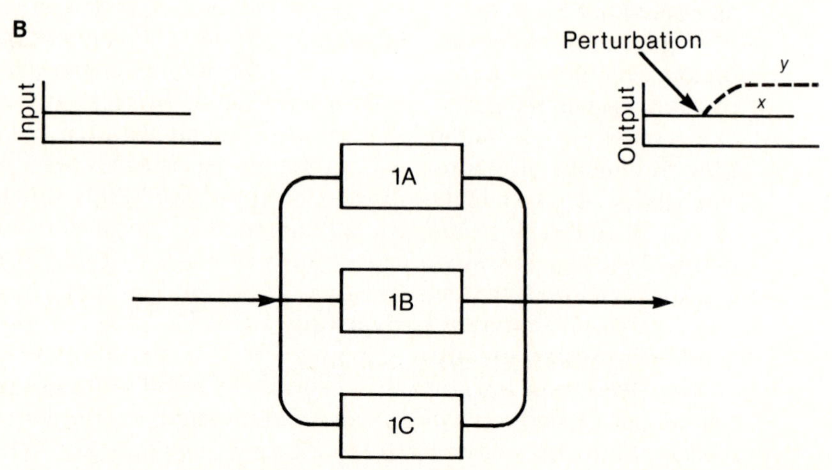

Figure 2–13 Two contrasting mechanisms for maintaining stability in an ecosystem: A, by negative feedbaack, and B, by redundancy in functional components (congeneric homotaxis). In both cases output, x, remains unchanged following a perturbation, in contrast to a deflected output, y, as would occur in the absence of a controlled response. (After Hill and Wiegert, 1980.)

control, or what Hill and Durham (1978) and Hill and Wiegert (1980) call "congeneric homotaxis." In Figure 2–13B, three components (A, B, C) having a similar function are arranged in parallel to show how they can compensate for one another by providing alternate pathways for energy and material flows. In this manner a controlled response to a disturbance can be achieved without feedback.

Homeostatic mechanisms have limits beyond which unrestricted positive feedback leads to death unless adjustment can be made. As stress increases, the system, although controlled, may not be able to return to exactly the same level as before. In fact, C. S. Holling (1973) has developed a widely accepted theory that populations and, by inference, ecosystems have more than one equilibrium state and often return to a different one after a disturbance. Remember how CO_2 introduced into the atmosphere by human activities is largely, but not quite, absorbed by the carbonate system of the sea and other carbon storages, but as the input increases, new equilibrium levels in the atmosphere are higher. In this case, even a slight shift may have far-reaching effects. On many occasions, really good homeostatic control comes only after a period of evolutionary adjustment. New ecosystems (such as a new type of agriculture) or new host-parasite assemblages tend to oscillate more violently and are more likely to develop overgrowths, compared with mature systems in which the components have had a chance to adjust jointly to one another.

The stability actually achieved by a specific ecosystem depends not only on its evolutionary history and on efficiency of internal controls but also on the nature of the input environment and perhaps also on complexity. Generally, ecosystems tend to become more complex in benign physical environments than when subjected to stochastic (random, unpredictable) input disturbances such as storms. Functional complexity seems to enhance stability (more potential feedback loops) more than does structural complexity (for example, see Van Voris et al., 1980), but cause-and-effect relationships between complexity and stability are little understood. Some years ago, the late Robert MacArthur suggested that a diversity of species of organisms should enhance the stability of the biotic community, but species diversity per se has not proved to be strongly correlated with stability. However, as already discussed, the theory of redundancy or congeneric homotaxis suggests that a moderate diversity of species, each capable of performing key functions, should contribute to controlled responses.

Part of the difficulty in dealing with the concepts of homeostasis and stability is semantic. A dictionary definition of the term "stability" is, for example, "The property of a body that causes it, when disturbed from a condition of equilibrium, to develop forces or moments that restore the original condition" (Webster's Collegiate Dictionary). This seems straightforward enough, but in practice "stability" assumes different meanings in different professions (such as engineering, ecology, or economics), especially when one is trying to measure and quantify it. Accordingly, confusion abounds in the lit-

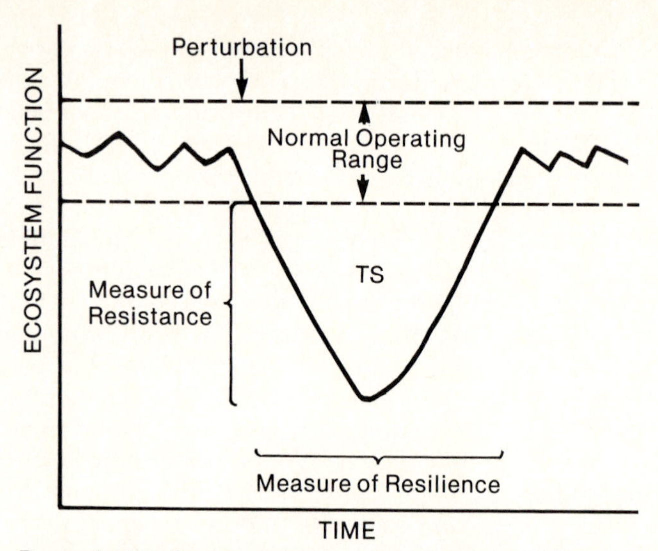

Figure 2–14 Resistance and resilience stability. When a perturbation (disturbance or stress) causes a major ecosystem function to deviate from the normal operating range, the degree of deviation is a measure of relative resistance, while time required for recovery is a measure of relative resilience. The area under the curve is a relative measure of total stability (TS). (After Leffler, 1978.)

erature, and a full discussion of stability theory is beyond the scope of this text. However, for an ecological perspective, two "kinds" of stability can be contrasted, as shown in Figure 2–14.

Resistance stability indicates the ability of an ecosystem to resist perturbations (disturbances) and maintain its structure and function intact. **Resilience stability** indicates ability to recover when the system is disrupted by a perturbation. Growing evidence shows that these two kinds of stability may be mutually exclusive, or to put it in other words, it is difficult to develop both at the same time. Thus, a California redwood forest is quite resistant to fire (thick bark and other adaptations), but if it does burn, it will recover very slowly or perhaps never. In contrast, California chaparral vegetation (see page 282) is very easily burned (little resistance stability) but recovers quickly in a few years (excellent resilience stability). In general, ecosystems in benign physical environments can be expected to exhibit more resistance and less resilience stability, and the opposite is true in uncertain physical environments.

In summary, an ecosystem is not equivalent to an organism, since it has emergent properties of its own. In other words, an ecosystem is a supraorganismic level organization but not a superorganism;

nor is it like an industrial complex (an atomic power plant, for example). It does have one thing in common with these systems: built-in cybernetic behavior.

Because of the evolution of the central nervous system, *Homo sapiens* has gradually become the most powerful organism, at least as far as the ability to modify the operation of ecosystems is concerned. The human brain requires only an extremely small amount of energy to crank out all sorts of powerful ideas. Most of our thinking so far has involved positive feedback that promotes the expansion of power, technology, and exploitation of resources. Ultimately, the quality of human life and environment will likely be degraded unless adequate negative feedback controls can be established.

Social critic Lewis Mumford (1967) in a famous essay pleads for "quality in control of quantity," which eloquently states the cybernetic principle of low-energy causes producing high-energy effects. So important is our role becoming as "a mighty geological agent" that Vernadsky (1945) suggested that we think of the "noosphere" (from Greek *noos*, mind), or the world dominated by the human mind, as gradually replacing the biosphere, the naturally evolving world, which has existed for billions of years. Though the human mind is a low quantity–very high quality energy "device" with great control potential, the time probably has not yet come for the noosphere, since one must assume that we now not only are wise enough to understand the results of all our actions but also can operate and maintain the biospheric life-support system or replace it with a completely artificial environment. When you have finished reading this book, I am sure you will agree that we cannot yet manage our life-support system, especially since proven natural processes work so well (and are so inexpensive).

7. Examples of Ecosystems

One of the best ways to begin studying ecology is to study a small pond and a meadow or old-field where the basic features of ecosystems can be conveniently examined and the nature of aquatic and terrestrial ecosystems can be contrasted. Any area exposed to light, even a lawn, a window flower box, or a laboratory-cultured microcosm, can be observed for the beginning study of ecosystems, provided that the physical dimensions and biotic diversity are not so great as to make observations of the whole difficult. Consider seven examples: a pond, a meadow, a watershed, a laboratory microecosystem, a spacecraft, a city, and an agroecosystem.

The Pond and the Meadow

The inseparability of living organisms and the nonliving environment is at once apparent with the first sample collected. Plants, animals, and microorganisms not only live in the pond and meadow, but they also modify the chemical nature of the water, soil, and air that composes the physical environment. Thus, a bottle of pond water or a scoopful of bottom mud or meadow soil is a mixture of living organisms, both plant and animal, and inorganic and organic compounds. Some of the larger animals and plants can be separated from the sample for study or counting, but it would be difficult to separate completely the myriad of small living things from the nonliving matrix without changing the character of the water. True, one could autoclave the sample of water, bottom mud, or soil so that only nonliving material remained, but this residue would then no longer be pond water or soil. It would have entirely different appearances and characteristics.

The basic components of an aquatic and a terrestrial ecosystem are discussed next.

Abiotic Substances Abiotic substances include basic inorganic and organic compounds, such as water, carbon dioxide, oxygen, calcium, nitrogen, sulfur, and phosphorus salts, amino and humic acids, and others. A small portion of the vital nutrients is in solution and immediately available to organisms, but a much larger portion is held in reserve (the "storages" shown in the functional diagram of Figure 2–1) in particulate matter as well as in the organisms themselves. As Hayes (1951) has expressed it, a pond or lake "is not, as one might think, a body of water containing nutrients, but an equilibrated system of water and solids, and under ordinary conditions nearly all of the nutrients are in a solid state." Much the same can be said for the soil-water-biomass complex of a terrestrial system. In a New Hampshire forest, for example, about 90 percent of nitrogen is stored in soil organic matter, 9.5 percent is in biomass (wood, roots, leaves) and only about 0.5 percent is in a soluble, quickly available form in soil water (Bormann et al., 1977).

The rate of release of nutrients from the solids, the solar input and the cycle of temperature, day length, and other climatic conditions are the most important processes, which daily regulate the rate of function of the entire ecosystem.

To assess fully the chemistry of the environment, extensive laboratory analysis of samples is necessary, but quite a bit of insight can be obtained by simple measurements with water analysis kits and soil test kits that are widely available. Some are designed for the professional, others for the amateur scientist and home gardener. Pond water or soil water is treated with appropriate chemicals to produce a color that is specific for the factor being measured. The intensity or shade of color is then compared with a color scale or measured in a portable photoelectric colorimeter or spectrophotometer to estimate concentration. The relative acidity or alkalinity, as indicated by pH and parts-per-million total alkalinity, often determines what kinds of organisms are present. Even a "deluxe" swimming pool water test kit can be used for these two measurements. Acidic soils and waters (pH less than 7) are usually characteristic of regions underlain with igneous and metamorphic rocks; "hard" or alkaline waters and soils occur in regions with limestone and related substrates.

Producer Organisms In a pond the producers may be of two main types: (1) rooted or large floating plants (**macrophytes**) generally growing in shallow water only and (2) minute floating plants, usually algae, called **phytoplankton** (from *phyto* = plant; *plankton* = floating) (Figure 2–3, IB), distributed throughout the pond as deep as light penetrates. In abundance, the phytoplankton gives the water a greenish color; otherwise, these producers are not visible to the casual observer, and their presence is not suspected by the layperson. Yet, in large, deep ponds and lakes (as well as in the oceans), phytoplankton is much more important than rooted vegetation in the production of basic food for the ecosystem. In the grassland, and terrestrial communities, in general, the reverse is the case: rooted plants dominate (Figure 2–3, IA), but small photosynthetic organisms such as algae, mosses, and lichens also occur on soil, rocks, and stems of plants. Where these substrates are moist and exposed to light, these microproducers may substantially contribute to organic production.

Macroconsumer Organisms The primary macroconsumers or **herbivores** (Figure 2–3, IIA and IIB) feed directly on living plants or plant parts. In the pond are two types: **zooplankton** (animal plankton) and **benthos** (= bottom forms), paralleling the two types of producers. Herbivores in the grassland also come in two sizes, the small plant-feeding insects and other invertebrates and the large grazing rodents and hoofed mammals. The secondary consumers or **carnivores**, such as predacious insects and game fish in the pond, and predatory insects, spiders, birds, and mammals in the grassland (Figure 2–3, IVA and IVB) feed on the primary consumers or on other secondary consumers (thus making them tertiary consumers). Another important type of consumer is the **detritivore** (IIIA and IIIB), which subsists on the "rain" of organic detritus from autotrophic layers above and along with herbivores provide food for carnivores. Many, if not most, detritivorous animals obtain much of their food energy by digesting the microorganisms that colonize detritus particles.

Saprotrophic Organisms (Figure 2–3, V) The bacteria, flagellates, and fungi are distributed throughout the ecosystem, but they are especially abundant in the mud-water interface of the pond and the litter-soil junction in the grassland accumulates. Though a few of the bacteria and fungi are pathogenic in that they will attack living organisms and cause disease, the majority attack only after the organism dies. Important groups of microorganisms also form mutually beneficial associations with plants, even to the extent of becoming an integral part of roots and other plant structures (see Chapter 7). When temperature conditions are favorable, the first stages of decomposition occur rapidly. Dead organisms do not retain their identification for very long but are soon broken up into pieces by the combined action of detritus-feeding animals and microorganisms. Some of their nutrients are released for reuse. As already noted, the resistant fraction of detritus, such as cellulose, lignin (wood), and humus, endures and provides a spongy texture for soil and sediments that makes a good habitat for plant roots and many tiny creatures. Some of the latter convert atmospheric nitrogen to forms usable by plants (nitrogen fixation; see Chapter 4) or do other chores to benefit the whole ecosystem.

The partial stratification into an upper production zone and a lower decomposition-nutrient regeneration zone can be illustrated by measuring oxygen diurnal metabolism in the water column of a pond. A "light-and-dark bottle" technique may be used for this purpose and also for charting energy flow in the system. As shown in Figure 2–15, samples of water from different depths are placed in paired bottles; one (the dark bottle) is covered with black tape or aluminum foil to exclude all light. Other water samples are "fixed" with reagents so that the original oxygen concentration at each depth can be determined.* Then the string of paired dark and light bottles is suspended

* The Winkler method is the standard procedure for oxygen measurement in water. It involves fixation with $MnSO_4$, H_2SO_4, and alkaline iodide, which releases elemental iodine in proportion to oxygen. The iodine is titrated with sodium thiosulphate (the "hypo" used to fix photographs) at a concentration calibrated to estimate milligrams of oxygen per liter, which, conveniently, is also grams per m^3 and parts per million (ppm). Electronic methods employing oxygen electrodes have now been perfected and are especially useful when continued monitoring of oxygen changes is desirable. For details on methods, see the latest edition of *Standard Methods* published by American Public Health Association, New York, NY.

Figure 2–15 Measuring the metabolism of a pond with light-and-dark bottles. *A*. Filling a pair of light and dark (black) bottles with water collected at a specific depth with a water sampler (the cylindrical instrument with rubber stoppers at each end). *B*. Lowering a string of dark and light bottles to the depth at which the water was collected. The white plastic jug will serve as a float. See text for further explanation of the method. The energetics of the pond shown in these pictures is discussed and modeled in Chapter 3. (Photos by K. Kay for Institute of Ecology, University of Georgia.)

in the pond so that the samples are at the same depth from which they were drawn. After 12 hours, the string of bottles is removed, and the oxygen concentration in each sample is determined and compared with the original concentration. The decline of oxygen in the dark bottle indicates the amount of respiration by producers and consumers (i.e., the total community) in the water, whereas change of oxygen in the light bottle reflects the net result of oxygen consumed by respiration and oxygen produced by photosynthesis. Adding respiration and net production together, or subtracting final oxygen concentration in the dark bottle from that in the light bottle (provided that both bottles had the same oxygen concentration to begin with), gives an estimate of the total or gross photosynthesis (food production) for the time period, since the oxygen released is proportional to dry matter produced.

With a light-and-dark bottle experiment in a shallow, fertile pond on a warm, sunny day, one might expect an excess of photosynthesis over respiration in the top 2 or 3 meters, as indicated by a rise in oxygen concentration in the light bottles. Below 3 meters, the light intensity in a fertile pond is usually too low for photosynthesis, so only respiration occurs in the bottom waters. The point in a light gradient at which plants can just balance food production and use (zero change in light bottle) is called the **compensation level** and marks a convenient functional boundary between the autotrophic stratum (**euphotic zone**) and the heterotrophic stratum.

A daily production of 5 to 10 gm O_2/m^2 of water column and excess production over respiration would indicate a healthy condition for the ecosystem, since in the water column excess food is being produced that becomes available to bottom organisms as well as to all the organisms when light and temperature are not so favorable. If the hypothetical pond is polluted with organic matter, O_2 consumption (respiration) would greatly exceed O_2 production, resulting in oxygen depletion and (should the imbalance continue) eventual anaerobic (= without oxygen) conditions, which would eliminate fish and most other animals. In assaying the "health" of a body of water, we need not only to measure the oxygen concentration as a condition for existence but also to determine rates of change and the balance between production and use in the diurnal and annual cycle. Monitoring oxygen concentrations, then, conveniently allows one to feel the pulse of the aquatic ecosystem. Measuring the biochemical oxygen demand (BOD) of water samples incubated in the laboratory is also a standard method of pollution assay.

Enclosing pond water in bottles or other containers such as plastic spheres or cylinders has obvious limitations, and the bottle method used here as an illustration is not adequate for assaying the metabolism of the whole pond, since oxygen exchanges of bottom sediments and the larger plants and animals are not measured. The metabolism of land ecosystems can also be measured by enclosing them in containers (see Figure 2–17) and measuring the CO_2 (rather than O_2) exchanges, but the size of terrestrial biota and the need to "air condition" the container (to prevent heating by sun) make this approach generally unsatisfactory. Other methods for measuring the metabolism of ecosystems are discussed in Chapter 3.

Although aquatic and terrestrial ecosystems have the same basic structure and similar function, biotic composition and size of trophic components differ, as summarized in Table 2–3. The most striking contrast, as already noted, is in the size of the green plants. The autotrophs of land tend to be fewer but very much bigger, both as individuals and as biomass per unit area (Table 2–3). The contrast is especially impressive when one compares the open ocean, where

Table 2–3 Comparison of Density (Numbers/m²) and Biomass (as Grams Dry Weight/m²) of Organisms in Aquatic and Terrestrial Ecosystems of Comparable and Moderate Productivity

	Open-water Pond			Meadow or Old-field		
Ecologic Component	Assemblage	No./m²	Gm Dry Wt./m²	Assemblage	No./m²	Gm Dry Wt./m²
Producers	Phytoplanktonic algae	10^8–10^{10}	5.0	Herbaceous angiosperms (grasses and forbs)	10^2–10^3	500.0
Consumers in the autotrophic layer	Zooplanktonic crustaceans and rotifers	10^5–10^7	0.5	Insects and spiders	10^2–10^3	1.0
Consumers in the heterotrophic layer	Benthic insects, mollusks, and crustaceans*	10^5–10^6	4.0	Soil arthropods, annelids, and nematodes†	10^5–10^6	4.0
Large roving consumers (permeants)	Fish	0.1–0.5	15.0	Birds and mammals	0.01–0.03	0.3‡ 15.0§
Microorganism consumers (saprophages)	Bacteria and fungi	10^{13}–10^{14}	1–10‖	Bacteria and fungi	10^{14}–10^{15}	10–100.0‖

* Including animals down to size of ostracods.

† Including animals down to size of small nematodes and soil mites.

‡ Including only small birds (passerines) and small mammals (rodents, shrews, etc.).

§ Including two to three cows (or other large herbivorous mammals) per hectare.

‖ Biomass based on the approximation of 10^{13} bacteria = 1 gram dry weight.

phytoplankton are even smaller than in the pond, and the forest with its huge trees. Shallow water communities (edges of ponds, lakes, oceans, and marshes), grasslands, and deserts are intermediate between these extremes. In fact, the whole biosphere can be visualized as a vast gradient of ecosystems with the deep oceans at one extreme and the large forests at the other. The absence of large plants in the sea (except near shore) is the reason why food from the sea is most available to humans in animal rather than vegetable form.

Terrestrial autotrophs must invest a large part of their productive energy in supporting tissue, because the density (and hence lower supporting capacity) of air is much lower than that of water. This supporting tissue has a high content of cellulose and lignin (wood) and requires little energy for maintenance because it is resistant to consumers. Accordingly, plants on land contribute more to the structural matrix of the ecosystem than do plants in water, and the rate of metabolism per unit volume or weight of land plants is correspondingly much lower.

Turnover may be broadly defined as the ratio of throughput to content. Turnover can be conveniently expressed either as a rate fraction or as a "turnover time," which is the reciprocal of the rate fraction. Consider the productive energy flow as the throughput and the standing crop biomass (gm dry weight/m^2 in Table 2–2) as the content. If we assume that the pond and the meadow have a comparable gross photosynthetic rate of 5 gm/m^2/day, the turnover rate for the pond would be 5/5 or 1, and the turnover time would be one day. In contrast, the turnover rate for the meadow would be 5/500 or 0.01, and the turnover time would be 100 days. Thus, the tiny plants in the pond may replace themselves in a day when the pond metabolism is at a peak, whereas land plants are much longer-lived and turn over much more slowly (perhaps 100 years for a large forest). In Chapter 4, the concept of turnover will be especially useful in calculating the exchange of nutrients between organisms and environment.

The large structural mass of land plants results in large amounts of resistant fibrous detritus (leaf litter, wood) reaching the heterotrophic layer. In contrast, the "rain of detritus" in the phytoplankton system consists of small particles that are more easily decomposed and consumed by small animals. Therefore, larger populations of saprophagic microorganisms should be found in soil than in sediments under open water (Table 2–3). However, as already emphasized, numbers and biomass of very small organisms do not necessarily reflect their activity; a gram of bacteria can vary greatly in metabolic rate and turnover depending on the conditions. In contrast to the producers and microconsumers, numbers and weight of the macroconsumers tend to be more comparable in aquatic and terrestrial systems, if available energy in the system is the same. If large grazing animals on land are included, the numbers and biomass of large roving consumers, or "permeants," tend to be about equal in both systems (Table 2–3).

Table 2–3 is only a tentative model. Strange as it may seem, no one has yet made a complete census of any one pond or meadow (or any other outdoor ecosystem for that matter). Only approximations based on fragmentary information gleaned from many sites are now possible. Even in the simplest of natural ecosystems, the number and variety of organisms and the complexity of the coupling of assemblages are bewildering. As would be expected, more is known about the large organisms (trees, birds, fish, and so on) than about the small ones, which are not only more difficult to see but require technically difficult functional methods of assay. Likewise, temperature, rainfall, and other "macrofactors" have been measured, but very little is known about micronutrients, vitamins, detritus, antibiotics, and other difficult-to-assay "microfactors," which nonetheless are important to maintaining ecological balances. Developing better inventory techniques challenges the next generation of ecologists, because curiosity is no longer the only motivation for analyzing nature. Our existence is being threatened by our abysmal ignorance about running a balanced ecosystem.

In both land and aquatic ecosystems, a large part of solar energy is dissipated in the evaporation of water, and only a small part, generally less than 5 percent, is fixed by photosynthesis. However, the role of this evaporation in moving nutrients and in maintaining temperatures is different. These contrasts are considered in Chapters 3 and 4. For every gram of CO_2 fixed in a grassland or forest ecosystem, as much as 100 grams of water must be moved from the soil, through the plant tissues, and transpired (= evaporated from plant surfaces). No such massive use of water is associated with production of phytoplankton or other submerged plants.

For interesting discussions comparing and contrasting land and water ecosystems, see Smith (1969), Wiegert and Owen (1971), and Margalef (1979).

The Watershed or Catchment Basin*

Although the biological components of the pond and the meadow seem self-contained, they are both actually very open systems that

* In this text, the term "watershed" is considered synonymous with "catchment basin," the term generally used in Europe, where "watershed" often refers only to the slopes or sides of a basin (not including the streams or lakes in the valley).

are parts of larger watershed systems. Their function and relative stability over the years are very much determined by the rate of inflows or outflows of water, materials, and organisms from other parts of the watershed. A net inflow of materials often occurs when bodies of water are small or when outflow is restricted. If organic material from sewage or industrial wastes, for example, cannot be assimilated, the rapid accumulation of such materials may destroy the system. The phrase **cultural eutrophication** (= cultural enrichment) is becoming widely used to denote organic pollution resulting from human activities. Not only do soil erosion and loss of nutrients from a disturbed forest or poorly managed cultivated field impoverish these ecosystems, but such outflows will likely have "downstream" eutrophic or other impacts. Therefore, the whole drainage basin, not just the body of water or patch of vegetation, must be considered as the minimum ecosystem unit when it comes to human interests. The ecosystem unit for practical management must then include for every square meter or acre of water at least 20 times an area of terrestrial watershed.* In other words, fields, forests, bodies of water, and towns, linked together by a stream or river system (or sometimes an underground drainage network), interact as a practical ecosystem-level unit for both study and management. For a picture of a watershed manipulated and monitored for experimental study, see Figure 2–16.

The watershed concept helps put many of our problems and conflicts in perspective. For example, the cause of and the solutions for water pollution are not to be found by looking only into the water; usually, bad management of the watershed destroys our water resources. The entire drainage or catchment basin must be considered as the management unit. The Everglades National Park in south Florida is a good example of this need to consider the whole drainage basin. Although large in area, the park does not now include the source of the fresh water that must drain southward into the park if it is to retain its unique ecology. The Everglades National Park, therefore, is vulnerable to reclamation and agricultural and urban developments north and east of the park boundary, which could divert or pollute its "life blood." The future of the park depends on the public's awareness of water requirements and the necessity of not diverting all the water to the urbanized "Gold Coast" of Florida. The park and urban coast can coexist only if they share the water within their

* The ratio of water surface to watershed area varies widely and depends on rainfall, geological structure of underlying rocks, and topography.

Figure 2–16 Experimental watersheds at the Coweeta Hydrologic Laboratory in the mountains of western North Carolina. All the trees have been cleared from the watershed in the center of the picture in order to compare water input (rainfall) and output (stream runoff) with that of the undisturbed forested watersheds on either side. Inset shows the V-notch weir and recording equipment used to measure the amount of water flowing out of each watershed area. (Photo courtesy of U.S. Forest Service.)

common basin. For more on the coupling of land and water ecosystems, see Likens and Bormann, 1974, and Hasler, 1975.

Microcosms

Little self-contained worlds, or microcosms, in bottles or other containers can simulate in miniature the nature of ecosystems. Such setups can be considered microecosystems. Some examples of microcosms are illustrated in Figures 2–17, 2–18, and 2–19. Completely

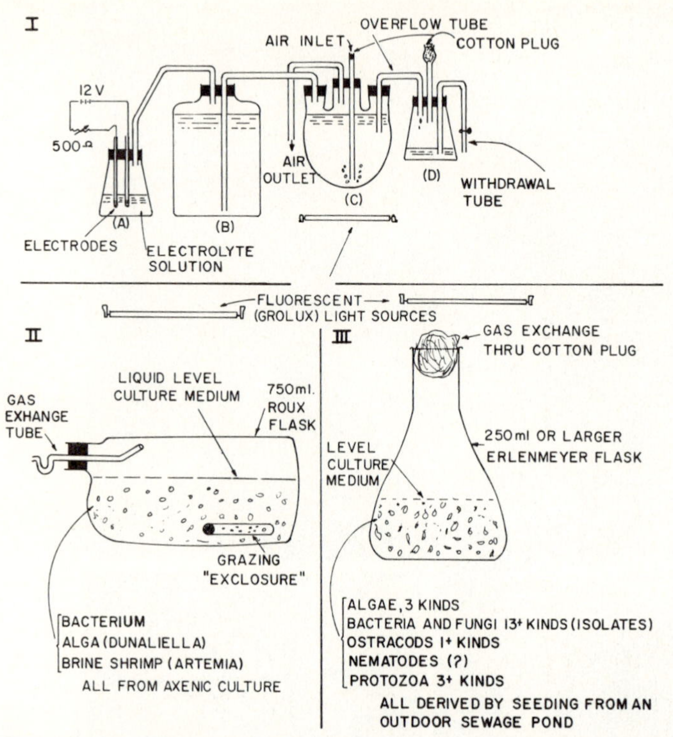

Figure 2–17 Three types of laboratory microecosystems. *I*. A simple, inexpensive **chemostat** in which a flow of culture media (B) through the culture chamber (C) and into an overflow vessel (D) is regulated by adjusting the electric current fed into an electrolysis pump (A). In the **turbidostat** a steady-state regulation is accomplished by a sensor, placed within the cultured community, that responds to the density (turbidity) of the organisms (internal regulation as contrasted with the external "constant input" regulation of the chemostat). (After Carpenter, 1969.) *II*. A gnotobiotic or "defined" microcosm containing three species from axenic (i.e., "pure") culture. The tube provides an area in which algae can multiply free from grazing by the shrimp (hopefully, preventing "overgrazing"). (After Nixon, 1969.) *III*. A microcosm "derived" from an outdoor system by multiple seeding (see Beyers, 1963). System I is "open," and Systems II and III are "closed" to material flows but open to the input of light energy and gas exchange with the atmosphere. Equilibrium in the closed systems, if achieved, results from nutrient cycle regulation by the community rather than by mechanical control devices (as in the chemostat or turbidostat).

closed systems that require only light energy (miniature biospheres, as it were) are very difficult to achieve on a small scale (see next section). Experimental microcosms generally range from partially closed systems open to gaseous exchange with the atmosphere but closed to exchange of nutrients and organisms to (at the opposite

A

B

Figure 2–18 Large, outdoor experimental ecosystems, or "mesocosms." *A*. A series of tall cylinders on the shore of Narragansett Bay, Rhode Island, that simulate conditions and communities in the shallow marine bay, thus providing opportunities for observing the effects of experimental alterations such as the introduction of pollutants. *B*. Floating plastic containers (which extend well below the photo zone) large enough to support most components of the water column (including small fish) in a bay in British Columbia.

extreme) very open systems involving assemblages of organisms maintained in various kinds of chemostats and turbidostats with regulated inflow and outflow of both nutrients and organisms (Figure 2–17A). Well-designed microcosms may exhibit most, if not all, basic

ORNL DWG 76-18158

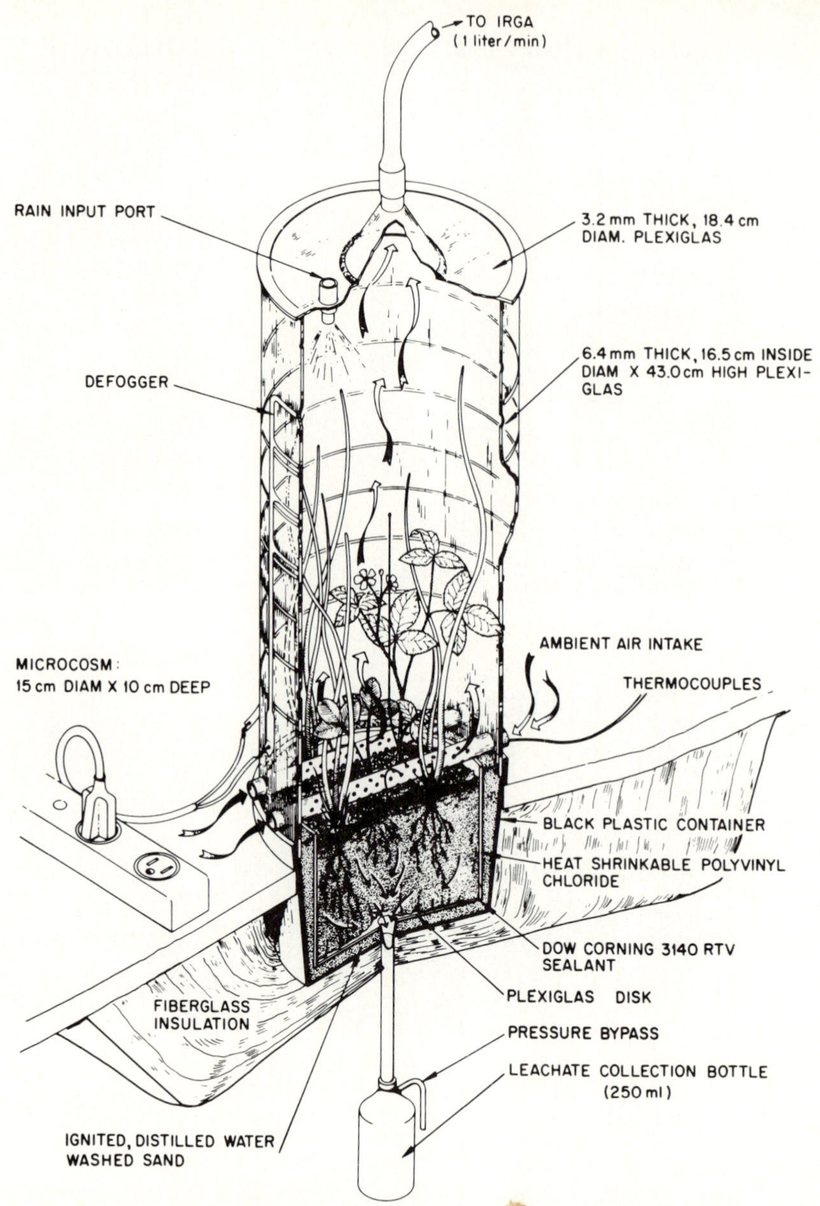

Figure 2–19 A terrestrial laboratory microcosm in which a core of vegetation and soil from an old-field community is enclosed in a transparent container. The metabolism of the community is monitored by continuous measurement of CO_2 flux by means of an infra-red gas analyzer (IRGA). Since stress often disrupts monitoring of mineral cycling, the leachate from the bottom of the soil column can be used as a measure of the impact of the stress. (From Van Voris et al., 1980.)

functions and trophic structures of an outdoor ecosystem, but of necessity, the variety and size of components are greatly reduced. Advantages for study and experimentation include discrete boundaries and ease of replication and manipulation. In a real sense, microcosms are living, working models (simplifications) of nature, but they should not be thought of as duplicates of any real-world ecosystem.

Two basic types of biological microcosms may be distinguished: (1) microecosystems derived directly from nature by multiple seeding of culture media with environmental samples, and (2) systems built up by adding species from "pure" or axenic cultures (free from other living organisms) until the desired combinations are obtained. The former systems represent nature "stripped down" or "simplified" to those organisms that can survive and function together for a long time within the limits of the container, the culture medium, and the light-temperature environment imposed by the experimenter. Such systems, therefore, are usually intended to simulate some specific outdoor situation. For example, the microcosm shown in Figure 2–17 III is derived from a sewage pond; the example in Figure 2–19 is derived from an old-field community. One problem with derived microecosystems is that their exact composition, especially the bacteria (Gorden et al., 1969), is difficult to determine. The ecological use of derived or "multiple-seeded" systems was pioneered by H. T. Odum and his students (Odum and Hoskins, 1957; Beyers, 1963).

In the second approach, defined systems are built up by adding previously isolated and carefully studied components. The resulting cultures are often called **gnotobiotic** (see Dougherty, 1959, for a discussion of terminology) because the exact composition, down to the presence or absence of bacteria, is known. Gnotobiotic cultures have been used mostly to study the nutrition, biochemistry, and other aspects of single species or strains or to study the interactions of two species (see Chapter 7). Recently, ecologists have experimented with more complex polyaxenic cultures to devise self-contained ecosystems (Nixon, 1969; Taub, 1969, 1974).

These contrasting approaches to the laboratory microecosystem parallel the two longstanding ways (i.e., holological versus merological) ecologists have attempted to study lakes and other large systems of the real world.

A common misconception exists about the "balanced" fish aquarium. An approximate balance of gases and food can be achieved in an aquarium if the ratio of fish to water and plants remains small. In 1851, Robert Warington "established that wondrous and admirable balance between the animal and vegetable kingdoms" in a 12-

gallon aquarium, using a few goldfish, snails, and lots of eelgrass (*Vallisneria*), as well as a diversity of associated microorganisms. He not only clearly recognized the reciprocal role of fish and plants but also correctly noted the importance of the snail detritivore "in decomposing vegetation and confevoid mucus," thus "converting that which would otherwise act as a poisonous agent into a rich and fruitful pablum for vegetable growth." Most amateur attempts to balance aquaria fail because far too many fish are stocked for the available resources (diagnosis: an elementary case of gross overpopulation). Table 2–2 shows that for complete self-sufficiency, a medium-sized fish requires many cubic meters of water and attendant food organisms. Since "fish-watching" is the usual motivation for keeping aquaria in the home, office, or school, supplemental food, aeration, and periodic cleaning are necessary if large numbers of fish are to be crowded into small spaces. The home fish culturist, in other words, is advised to forget about ecological balance and leave the self-contained microcosm to the student of ecology. Fish and human beings require more room than you might think!

Large outdoor tanks for aquatic systems and various kinds of enclosures of terrestrial habitat represent increasingly used experimental setups that are intermediate between laboratory culture systems and the real world of nature. These outdoor experimental setups could be considered **mesocosms** (middle-sized worlds) to contrast them with microcosms and real-world macrocosms. Figure 2–18A shows a series of large cylinders that serve as marine mesocosms on the shore of Narragansett Bay in Rhode Island. Water with the same chemical composition and salinity as the bay is gently agitated by a plunger to create the kind of turbulence occurring in marine bays subjected to tides and other currents. The bottom of the cylinder is filled with sediment from the bay; water and organisms are exchanged with the bay with a turnover time of one month for water in each cylinder. According to Pilsom and Nixon (1980), these mesocosms track very well the seasonal changes in organism behavior and community metabolism (production and respiration) that occur naturally in the bay.

For a review of the early ecological work with microcosms and a discussion of the balanced aquarium controversy, see Beyers (1964). For more recent studies, see the symposium volumes edited by Witt and Gillett (1978) and Giesy (1980).

Both indoor and outdoor model ecosystems provide useful tools for estimating tentatively or preliminarily the effect of pollutants or other experimentally imposed disturbances related to human activity. For example, oil introduced into the marine mesocosms shown in

Figure 2–18A (to simulate a moderate oil spill) proved to be more toxic to consumers than to producers. Consequently, the phytoplankton increased in the absence of grazing pressure by the reduced population of zooplankton: the treated cylinder was noticeably greener than untreated controls. This alteration in trophic structure was, as might be expected, followed by increased microbial decomposition, lasting until organic matter in the oil and excess algal growth were broken down. The floating mesocosms shown in Figure 2–18B were set up specifically to test for the effects of various pollutants.

A general pattern seems to be emerging from these experiments. The marine systems responded to low-level concentration of toxic substances the way natural communities respond to nutrient shortages. Thus, chronic pollution produces an effect similar to that of "starving" the system; the system's productivity and general performance are reduced.

Microecosystem research is also proving useful in testing of various ecological hypotheses generated from observing nature. For example, the terrestrial microcosm pictured in Figure 2–19 was designed to test the hypothesis that functional diversity enhances a system's ability to resist, or recover, or both, from acute stress. The results of such an experiment indicated that this was the case (Van Voris et al., 1980).

In the next several chapters, the ways in which microecosystem research has helped establish and clarify basic ecological principles will be described.

Spacecraft as an Ecosystem

Perhaps the best way to visualize the ecosystem is to think about space travel. When we leave the biosphere, we must take with us a sharply delimited, enclosed environment that can supply all vital needs by using sunlight as the energy input from space. For journeys of a few days or weeks, such as to the moon and back, we do not need a completely self-sustaining ecosystem, since sufficient oxygen and food can be stored, and since CO_2 and other waste products can be fixed or detoxified for short periods. For long journeys, such as trips to the planets or to establish space colonies, we must design a more closed or regenerative spacecraft that includes all vital abiotic substances and the means to recycle them. The vital processes of production, consumption, and decomposition must also be performed in a balanced manner by biotic components or their mechanical substitutes. In a very real sense, the self-contained spacecraft is a human microecosystem.

The life-support modules for all manned spacecraft so far launched have been storage types; water and atmospheric gasses have been partly regenerated by physiochemical means in some cases. The possibility of coupling humans and microorganisms, such as algae and hydrogen bacteria, has been considered but found unworkable. Large organisms, especially for food production, considerable diversity, and, above all, large volumes of air and water will be required for a truly regenerative ecosystem that could survive for long periods in space without resupply from earth (recall our previous comment about the large amount of room needed by a fish or a man). Accordingly, something akin to conventional agriculture and other large plant communities will have to be included.

The critical problem is how the buffering capacity of the atmosphere and the oceans, which stabilizes the biosphere as a whole, will be provided. For every square meter of land surface on earth, more than 1000 cubic meters of atmosphere and almost 10,000 cubic meters of ocean, plus large volumes of permanent vegetation, are available as sinks, regulators, and recyclers, as shown in Figure 2–20. Obviously, for space-living, some of this buffering work will have to be accomplished mechanically, using solar energy (and perhaps atomic energy). Two recent reviews prepared for the National Aeronautics and Space Administration (NASA) conclude that "It is a moot point whether an artificial ecosystem, totally closed to the entry or exit of mass, fully recycling, and completely regulated by its biological components can be constructed" (MacElroy and Averner, 1978). "A safe, reliable closed ecological life support system could not be built today with existing technology (even for earth-based use)" (Sperlock and Modell, 1978). Miniaturizing the biospheric life-support system for use in a space colony will be not only difficult but also expensive, if one considers all the fossil fuel needed to shuttle the components into space.

Though we do not know how to build a human mesocosm, or whether we could afford it even if we knew how, enthusiasts for space colonization, such as physicist Gerald O'Neill in his book *The High Frontier* (1977), predict that within the next century millions of people will be living in space colonies supported by a carefully selected biota, free from pests and other unwanted or unproductive organisms that earthbound people contend with. Successful colonization of the "high frontier" (according to its proponents) would permit the continued growth of human population and affluence long after

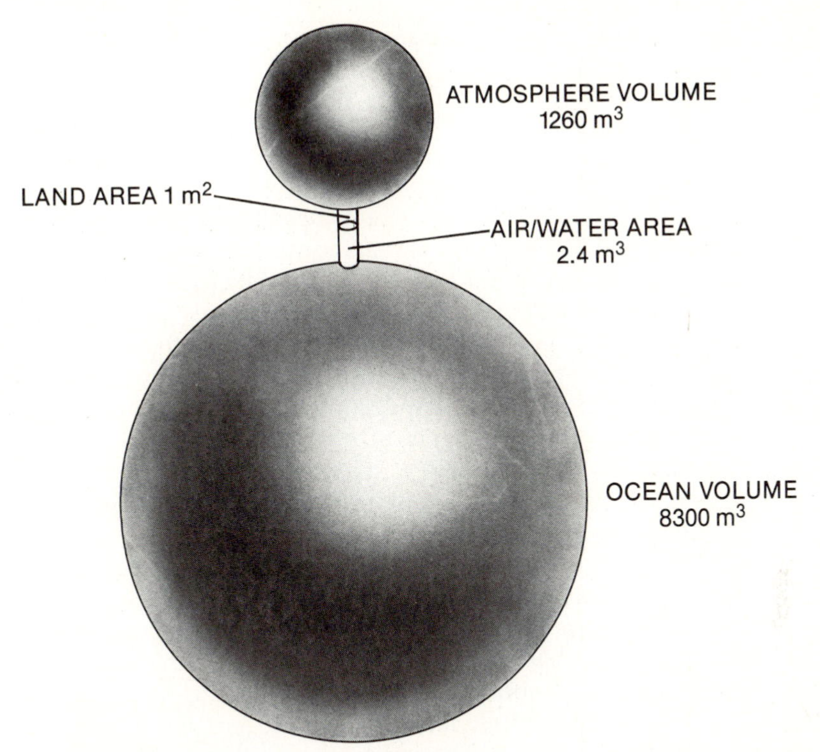

Figure 2–20 Relative volumes of atmosphere and ocean that act as buffers to 1 m² of land. Not shown is the large volume of terrestrial vegetation that also contributes to the capacity of the biosphere to buffer disturbances. (After MacElroy and Averner, 1978.)

such growth is no longer possible within the confines of the earth. Solar energy and mineral wealth of moons and asteroids could be exploited to support such growth. As a start, O'Neill projects assembling a rotating (to create gravity) tube-like station 4 miles in diameter and 20 miles long containing 500 square miles of soil surface. About half the area would be devoted to intensive agriculture. Two million people would occupy such a "small" space colony. With only 0.16 acre per person and the limited air and water buffer, it seems doubtful that such a density of people could be maintained without an umbilical cord to earth. Also, social, economic, political, and pollution problems within such a satellite would be formidable indeed. The extent to which sociopolitical forces shape and limit human life and growth on earth will be considered later in the text.

The City—A Heterotrophic Ecosystem

As already noted (page 17), a city,* especially an industrialized one, is an incomplete or heterotrophic ecosystem dependent on large areas outside it for energy, food, fiber, water, and other materials. As shown in Figure 2–21, the city differs from a natural heterotrophic ecosystem, such as an oyster reef, because it has (1) a much more intense metabolism per unit area, requiring a larger inflow of concentrated energy (currently supplied mostly by fossil fuels); (2) a large input requirement of materials, such as metals for commercial and industrial use, above and beyond that needed to sustain life itself; and (3) a larger and more poisonous output of waste products, many of which are synthetic chemicals more toxic than their natural progenitors. Thus, the input and the output environments (recall Figure 2–2, page 16) are relatively much more important to the urban system than is the case for an autotrophic system such as a forest.

Actually, most metropolitan districts even in dry areas have a substantial greenbelt or autotrophic component of trees, shrubs, areas of grass, and, in many cases, lakes and ponds, but the organic production of this green component does not appreciably support the people and machines that so densely populate the urban-industrial area. Without the huge inflows of food, fuel, electricity, and water, the machines (automobiles, factories, and so on) would stop working. People would soon starve or have to emigrate. The urban forests, grasslands, and parks, of course, have tremendous esthetic and recreational value, and they function to attenuate temperature extremes, reduce noise and other pollution, provide habitats for song birds and other small animals, and so on. But, the labor and fuel expended in watering, fertilizing, mowing, pruning, removing wood and leaves, and other work required to maintain the city's private and public greenbelts add to the energy (and money) cost of living in the city. Table 2–4 compares the residential "woodland" of Madison, Wisconsin, with an adjacent natural, undisturbed forest. About 30 percent of the residential district is covered with concrete, buildings, or other "impervious" surfaces, but by occupied area, the residential forest has a much greater variety of plant species and is more productive owing to human horticulture and, especially, to fertilizer and water

* In this section, the term "city" is used synonymously with the geographers' term "standard metropolitan district (SMD)," which includes industrial areas and residential suburbs that often extend far beyond official city limits. Cities adjoining each other or overlapping (as, for example, the "twin cities" of Minneapolis and St. Paul, Minnesota) are generally included in a single SMD.

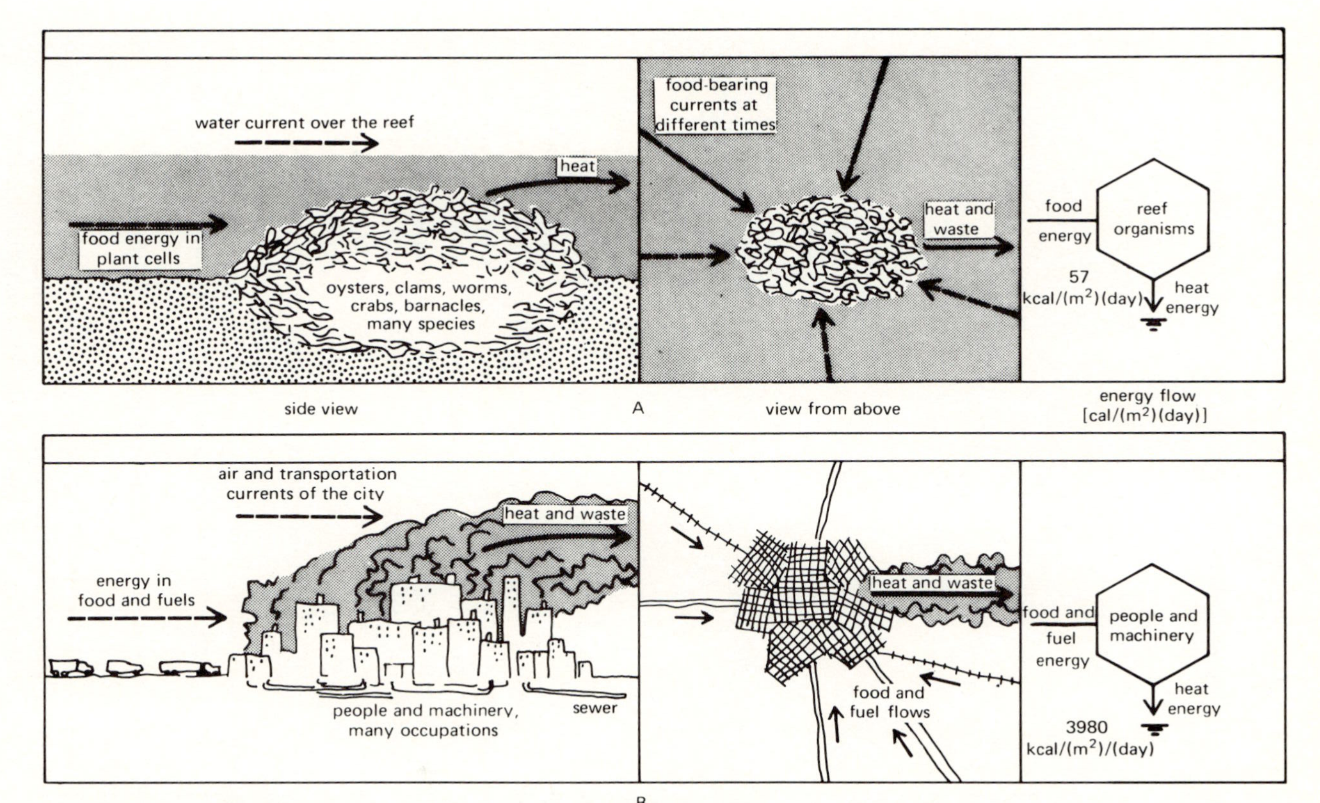

Figure 2–21 Heterotrophic ecosystems. *A*. One of nature's "cities"—an oyster reef that is dependent on the inflow of food energy from a large area of surrounding environment. *B*. Industrialized city maintained by a huge inflow of fuel and food with a correspondingly large outflow of waste and heat. Energy requirement, on a square-meter basis, is about 70 times that of the reef, or about 4000 kcal/day, which comes to about 1 1/2 million kcal per year. (After H. T. Odum, 1971. Courtesy of the author and John Wiley & Sons.)

Table 2–4 Comparison of Natural and Adjacent Residential Woodland in Madison, Wisconsin*

	Natural Forest	Urban Forest
Number of tree species	10	75
Number of shrub species	20	74
Tree biomass (average above-ground dry weight)	27 kg/m²	10 kg/m²
Annual net production total vegetation	812 gm/m²	719 gm/m²
Net production, occupied-area basis (excluding 30% area without vegetation in residential area)	812 gm/m²	1027 gm/m²
Fertilizer added	0	120 lb/acre
Annual export of organic matter	0	497 gm/m²
Irrigation water	0	Large (exact amount unknown)

* *Source:* Unpublished manuscript, "Terrestrial Primary Production of Adjacent Urban and Natural Ecosystems," by G. J. Lawson, Grant Cottam, and Orie L. Loucks, 1973.

subsidies. Little or no organic matter is exported from the natural forest, but more than half the annual growth of the residential vegetation is exported to waste dumps or land fills in the form of wood, leaves, and grass clippings. Society would be better served if this organic matter could be incorporated into agricultural soils and home gardens. Fortunately, there is a trend in this direction.

In a study of a suburban lawn, Falk (1976) found that its inputs and outputs compared rather well with a cornfield or a natural, moist prairie, although birds ate 30 times more insects and seeds than in a comparable prairie. Annual energy subsidy, such as labor, gasoline, fertilizer, and so on, was estimated to be 528 kcal/m², about the same as required to produce a good corn crop. Suburbanites, then, do not live by bread alone, since they are willing to expend as much energy, by area, on a lawn as they do to put food on the table. And, of course, the sale of lawn-care equipment and supplies is big business in suburban America.

Rapid urbanization and growth of cities during the past half-century has changed the face of the earth probably more than anything else resulting from human activity throughout history. The extent of urbanization and urban influence in the United States is shown by the two maps in Figure 2–22. In the upper map, all shaded

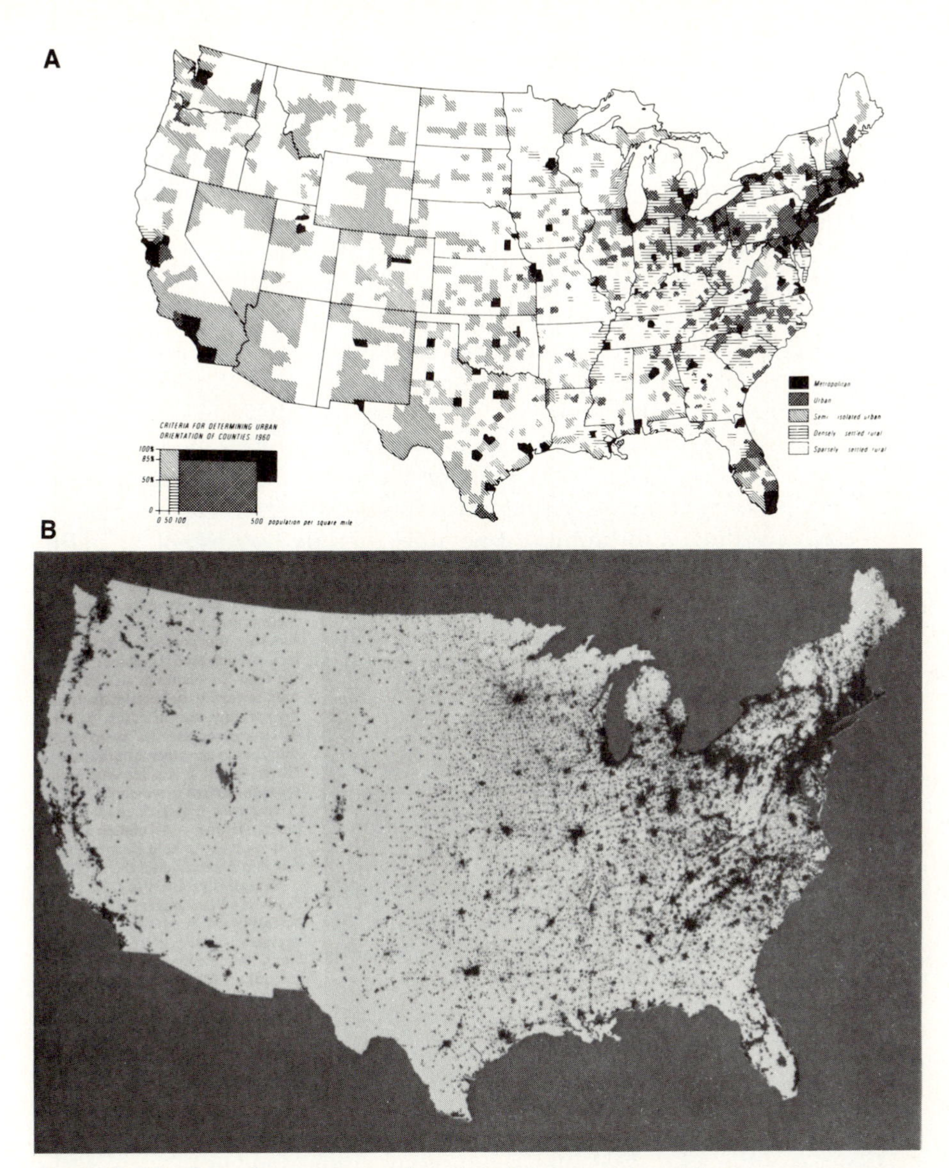

Figure 2–22 The urbanization of America. A large and rapidly increasing area has a density of 50 persons per square mile or higher (all shaded areas in A) and, more important, a high density of energy use as shown in B, where dark areas represent light visible at night by satellite.

areas have a density of 50 people or more per square mile. The lower map is a reverse color composite photograph taken at night by satellite: dark areas are cities, suburbs, and densely settled rural areas lit up with electric lights. This map is essentially an energy-density map insofar as the use of electricity is concerned. Metropolitan-level density now extends in continuous strips from Boston to Washington, D.C., from Pittsburgh to Cleveland and Detroit, along the western and southern shores of Lake Michigan, the east coast of Florida, and portions of the California coast.

Even in the economically poor countries, cities are growing much faster than the general population. Cities do not cover such a large area of the terrestrial landscape—only 1 to 5 percent worldwide. Cities do alter the nature of waterways, forests, grasslands, and croplands, not to mention the atmosphere and the oceans, because of their impact on extended input and output environment. A city may affect a distant forest not only directly because of air pollution or demand for wood products but also indirectly by altering forest management. For example, a large demand for paper products produces very strong economic pressure to convert a natural multispecies, multiaged forest into an even-aged, one-species plantation especially adapted to pulp production.

In energy consumption, cities are "hot-spots," as shown by the dramatic nighttime satellite photo of Figure 2–22B. As will be documented in greater detail in Chapter 3, an acre or hectare of a metropolitan district consumes 1000 or more times as much energy as similar areas of rural environment. The resultant heat, dust, and other air pollutants make the climate of cities noticeably different from that of surrounding countryside. In general, cities are warmer (Figure 2–23) and cloudier, with less sunshine and more drizzle and fog than adjacent rural areas. Urban construction (especially suburban, "second-home," and industrial plant developments) has become the leading cause of soil erosion in the United States, threatening to reverse the hard-earned soil conservation gains that have been achieved in rural areas. Ecologically sound conservation of soil on farms resulted from coordinated efforts by federal agencies and land-grant universities after the "dust bowl" and soil erosion disasters of the 1930s. The urban ecosystem demands comparable effort on its behalf.

Figure 2–23 Washington, D.C., as an urban heat island, as shown by warmer conditions in spring and fall. *A*, Average dates of earliest freezing temperatures in the fall. *B*, Average dates of latest freezing temperatures in the spring. (Data prepared by C. A. Woollum, U.S. Weather Bureau.) ▶

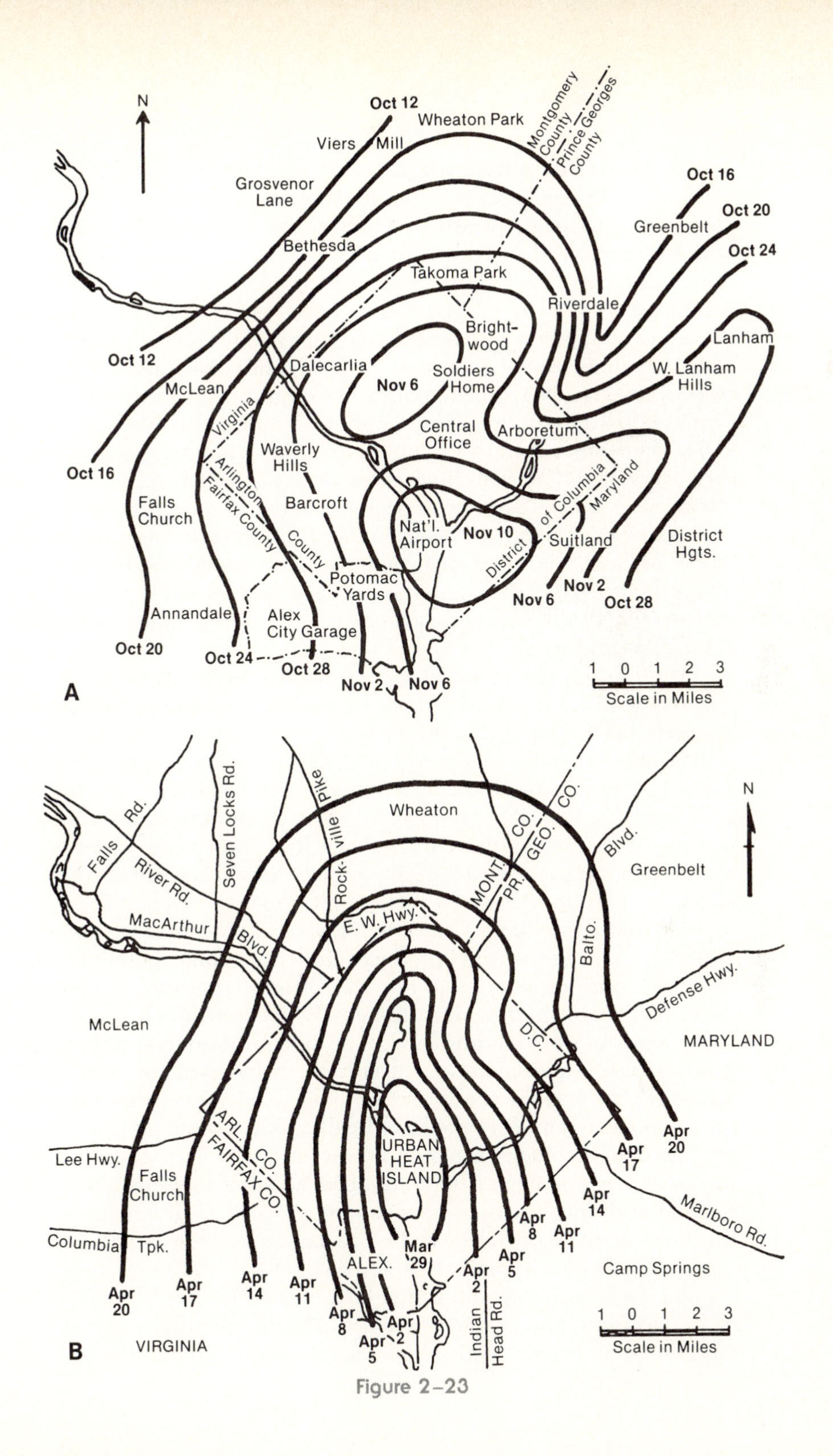

N
Oct 12
Wheaton Park
Viers Mill
Montgomery County
Prince Georges County
Grosvenor Lane
Oct 16
Oct 20
Oct 24
Greenbelt
Bethesda
Takoma Park
Riverdale
Bright-wood
Lanham
W. Lanham Hills
Oct 12
Dalecarlia
Nov 6
Soldiers Home
McLean
Virginia
Central Office
Arboretum
Waverly Hills
Oct 16
Arlington
Fairfax County
Barcroft
County
District of Columbia
Maryland
Falls Church
Nat'l. Airport
Nov 10
Suitland
District Hgts.
Potomac Yards
Nov 6
Nov 2
Oct 28
Annandale
Alex City Garage
Oct 20
Oct 24
Oct 28
Nov 2
Nov 6
1 0 1 2 3
Scale in Miles
A
Falls Rd.
Seven Locks Rd.
Rock-ville Pike
Wheaton
MONT. CO.
PR. GEO. CO.
N
River Rd.
Blvd.
Greenbelt
MacArthur Blvd.
E. W. Hwy.
Balto.
Defense Hwy.
McLean
D.C.
MARYLAND
ARL. CO.
FAIRFAX CO.
URBAN HEAT ISLAND
Lee Hwy.
Falls Church
Apr 20
Apr 17
Apr 14
Marlboro Rd.
Columbia Tpk.
Apr 8
Apr 11
Mar 29
Apr 5
ALEX.
Apr 2
Camp Springs
Apr 20
Apr 17
Apr 14
Apr 11
Apr 8
Apr 2
Apr 5
Indian Head Rd.
1 0 1 2 3
Scale in Miles
B
VIRGINIA

Figure 2–23

Even sparsely settled regions (the white areas in Figure 2–22A) can be greatly affected by distant cities, because from such areas come food, minerals, water, and other urban necessities. It can be very far downwind and downstream that water and air pollutants are finally dissipated.

The size of a United States city's input environment can be estimated by noting that food from 2 acres of agricultural land, paper and wood products from 1 acre of forest land, and 2000 gallons of water per day are the approximate per capita requirement for an American urban-industrial district. Thus, a city of 1 million people occupying 100 square miles requires many more square miles to feed, clothe, and water it—some 2 million acres or 3125 square miles for food alone. Two billion gallons of water per day requires a very large watershed, even in areas of heavy rainfall. In the arid western United States, only vast mountainous watersheds can provide water for large cities such as Denver, Tucson, and Los Angeles, and practically all surface water in this area is used or committed (legal water rights not yet exercised).

Though the city's output environment is generally smaller, the geographical extent of acid rain originating from industrial regions (see page 29) suggests that this component of the urban ecosystem is expanding. Cities in less developed countries have a less intense "urban metabolism," a correspondingly lower energy consumption, and consequently smaller input and output environments. But the lack of treatment facilities for sewage and industrial wastes often results in a more severe "close-in" impact than that exerted by cities in technologically advanced regions.

The modern city is a parasite of the rural environment, since, as currently managed, the city produces little or no food or other organic materials, purifies no air, and recycles little or no water or inorganic materials. From another view, the city is, theoretically at least, symbiotic with the surrounding landscape since it produces and exports goods and services, money, and culture that enrich the rural environment in return for goods and services received. The city or metropolitan district as delimited politically, or as shown on maps, does not have an "ecology" separate from that of the surrounding countryside. Urban ecology is too often conceived of in social science as dealing with interrelations between people and environment within the built-up part of the city. This is far too narrow a concept. To perceive the city as it really is and to deal with its problems, one must expand one's thinking and action far beyond city limits. Only if the extensive input and output environments are included can a city be considered an ecosystem in the complete sense. One of the unfortunate barriers to such a common-sense approach is the well-ingrained, often vicious political dichotomy between urban and rural districts that promotes constant political "warfare" and short-term self-interest to the detriment of the longer-term well-being of the states and nation as a whole. Until urban and rural leaders can put common interests above special interests, the management of the city as a functional ecosystem will exist in theory only.

For more on cities and their plight, see Saarinen (1943), Davis (1965), Lowry (1967), Henderson (1974), Ward (1976), and Newland (1980).

Agroecosystems

Agroecosystems (short for agricultural ecosystems) have an energy dependence and an output impact on distant regions, as do cities. Unlike cities, of course, agroecosystems have an autotrophic component or greenbelt as an integral part. Agroecosystems differ from natural or seminatural solar-powered ecosystems, such as lakes, forests, and so on, in three basic ways: (1) the auxiliary energy that augments or subsidizes the solar energy input is under the control of man, and consists of human and animal labor, fertilizers, pesticides, irrigation water, fuel-powered machinery, and so on; (2) the diversity of organisms is greatly reduced (again by human management) to maximize yield of a specific food or other product; and (3) the dominant plants and animals are under artificial selection rather than natural selection. In other words, agroecosystems are designed and managed to channel as much conversion of solar and other energy as possible into edible products by a twofold process: (1) by employing auxiliary power to do maintenance work that in natural systems would be accomplished by solar energy, thus allowing more solar energy to be converted directly into food, and (2) by genetic selection of food plants and domestic animals adapted to optimize yield in the specialized energy-subsidized environment. As in all intensive and specialized land use, there are costs as well as benefits, including soil erosion, pollution from pesticide and fertilizer runoff, high cost of fuel subsidies, and increased vulnerability to weather changes and pests.

Approximately 10 percent of the world's ice-free land area is cropland, converted mostly from natural grasslands and forests but also from deserts and wetlands. Another 20 percent of land area is pasture, designed for animal rather than plant production. Thus, about 30 percent of the terrestrial world is devoted to agriculture in the broadest sense. Recent comprehensive analyses of the world food situation emphasize that all the best land, that is, land most easily farmed by existing technology, is now in use. Extending agriculture to additional and less suitable terrain will be very costly and may require new types of agroecosystems (see President's Science Advisory Committee report, 1967, and National Academy of Science report, 1977).

At the risk of oversimplifying, one can divide agroecosystems into two broad types:

1. Pre-industrial agriculture—self-sufficient and labor intensive (human and animal labor provide the energy subsidy); provides food for the farmer (and family) and for sale or barter in local markets.
2. Intensive mechanized, fuel-subsidized agriculture (machines and chemicals provide the energy subsidy)—produces food exceeding local needs for export and trade, thus making food a commodity and a major market force in the economy rather than just life-support goods and services.

About 60 percent of world croplands are in the pre-industrial category, a large proportion of them in the less developed countries of Asia, Africa, and South America that have large human populations. A great variety of types have been adapted to local soil, water, and climate conditions, but for the purposes of general discussion, three types predominate: (1) pastoral systems, (2) shifting or swidden (slash-and-burn) agriculture, and (3) flood-irrigated and other permanent nonmechanized systems.

Pastoralism involves herding cattle or other domestic animals in arid and semiarid regions (especially the savanna and grassland regions of Africa) with people subsisting on livestock products such as milk, meat, and hides. Shifting agriculture, once practiced throughout the world, is still widely practiced in forested areas of the tropics. After patches of forest have been cut and debris burned (or sometimes left lying on the ground as mulch) crops are cultivated for a few years until nutrients are used up and leached out of soil. Then the site is abandoned to be rejuvenated by natural regrowth of the forest. Permanent, nonmechanized agriculture has persisted for centuries in southeast Asia and elsewhere, feeding millions of people surprisingly well. The most productive of these agroecosystems are subsidized by flood irrigation, either naturally by seasonal floods along rivers and on fertile deltas or by artificially controlled flooding as in the ancient canal-irrigated paddy rice culture.

Contrary to the label "primitive" or "subsistence" often applied to them, pre-industrial systems may be quite sophisticated, and they often function in harmony with natural ecosystems. These systems can be very efficient in terms of food produced per unit of energy expended. Rappaport (1967, 1971), for example, found that about 24 million kcal of food is produced on a hectare with 1.4 million kcal of human labor expended in cultivating elaborate mixed-crop food gardens in New Guinea, a 16 to 1 ratio of output to input calories. In contrast, many mechanized agroecosystems consume as much energy (or more) in subsidies as they produce in food. If longevity is considered a measure of success, then paddy rice culture, which has continued for thousands of years in the Philippines and elsewhere, must be considered worthy of preservation, especially since modern rice culture, which depends heavily on cheap fuel and fertilizer, may have difficulty surviving beyond the age of petroleum.

Pre-industrial systems, however, even well-adapted, permanent, and energy-efficient ones, do not produce enough surplus food to feed huge cities unless the farming "input" area is very large and the means of transporting, preserving, and storing food from distant areas is efficient. Also, many "man hours" are required, as any successful home gardener who grows his or her own food (with a small excess for neighbors or to sell or barter) can attest. Thus, nonindustrialized agriculture can efficiently conserve energy but is less productive in terms of food produced per farmer and generally lower in yield per area in comparison with intensive, mechanized farming. As is so often the case, gaining one advantage means giving up another advantage; all benefits have their costs.

As already indicated, mechanized agriculture has capitalized on the availability of relatively inexpensive fuel, fertilizers and agricultural chemicals (both of which require large amounts of fuel to make), and, of course, advanced technology, not only on the farm but in genetics, food processing, and marketing. In an incredibly short time (relative to the long sweep of agricultural history) farming in the United States and other industrialized countries has changed from small farms with a large percent of people making their living in rural areas to only 4 percent of the population farming ever larger tracts and producing more food on less land. The yield from fuel-subsidized agroecosystems on about 40 percent of world croplands has at least temporarily provided a respite from the desperate race between human population growth and food production. The race threatens to become grim again as the cost of the subsidies rises and as more and more countries cannot feed themselves and are forced to import from the very few countries, such as the United States, that have a surplus to export.

While the number of farmers in the developed world has declined dramatically, the number of farm animals has not, and the intensity of production of animal products has increased, paralleling that of crops. Thus, grain-fed beef replaces grass-fed, and chickens are bred and managed as egg- or meat-producing machines encased in wire cages under artificial light and plied with growth-promoting food mixtures and drugs. In the United States, most of the corn crop (and also other grains and soybeans) is fed to domestic animals, which feed, or perhaps overfeed, the affluent rather than the poor of this world.

When thinking about population pressure on the environment and resources, one should not forget that not only are there many more domestic animals than people worldwide, but animals also consume about five times more calories than do people. This does not include pets, which also consume a lot of food. Fortunately, some of these calories, in grass, for example, are not directly edible by humans. New Zealand has the world's highest ratio of domestic animals to people, about 37 to 1 population equivalents. Thus, although demographers often cite New Zealand as an underpopulated region, the environment is strongly affected by humans through the grazing of sheep.

The development of intensive, mechanized agriculture in the midwestern United States from 1830 to 1980 and the ecological implications of the changes in land use are summarized in Table 2–5. This table summarizes what is sometimes called the twentieth century revolution in American agriculture. Since both intensity of energy subsidy and yield may have peaked, meaning that more of the same may run into the law of diminishing returns (or negative feedback), one can expect some changes in agricultural strategies in the future, as suggested in the bottom lines of Table 2–5.

For more on agroecosystems, see Spedding (1975), Harper (1974), Cox and Atkins (1979), and Phillips, et al (1980).

8. The Classification of Ecosystems

Statement

Ecosystems can be classified by either functional or structural characteristics. A classification based on the quantity and quality of the energy input "forcing function" is an example of a useful functional scheme. Vegetation and/or major, stable physical features provide the basis for the widely used **biome** classification (a term introduced in Chapter 1, page 4).

Table 2–5 History of the Development of Intensive Agriculture in Midwestern United States*

1833–1934	90% of prairie, 75% of wetlands, and all forest land on good soils converted to cropland and pastures, leaving natural vegetation restricted largely to steep land and shallow, unfertile soils.
1934–1961	Intensification of farming associated with fuel and chemical subsidies, mechanization, an increase in crop specialization, and monoculture. Total cropland acreage decreased and forest cover increased 10% as more food was harvested on fewer acres.
1961–1980	Increase in energy subsidy, size of farm, and farming intensity on best soils, with emphasis on continuous culture of grain and soybean cash crops (with a decrease in crop rotation and fallowing). Much of grain grown for export trade. Yields per unit area of many crops peaked during this period. Increasing losses of farmland to urbanization and soil erosion. Also decline in water quality due to excessive fertilizer and pesticide runoff.
1980, trends for future	Increase in energy efficiency, use of crop residues for mulch and silage, multiple cropping, limited-till (less plowing), integrated control of pests, and other practices that conserve soil, water, and expensive fuels and reduce air and water pollution. Special carbohydrate crops developed for fuel alcohol production. State and regional land use plans enacted to stem loss of good soils to erosion and urbanization.

* Data for 1833 to 1980 after Auclair, 1976; "trends for future" is the author's optimistic assessment.

Explanation

Although classification of ecosystems is not to be considered a discipline in itself as is the classification of organisms (i.e., taxonomy), the human mind seems to require some kind of orderly categorization when it comes to dealing with a large variety of entities, like books in a library. Ecologists have not agreed upon any one classification for ecosystem types or even what would be a proper basis for it, and this is as it should be. Many approaches serve useful purposes.

Energy provides an excellent basis for a functional classification since it is a major common denominator for all ecosystems, natural and human-managed alike. Conspicuous, ever-present structural macrofeatures are the basis for the widely used biome classification. In terrestrial environments, vegetation usually provides such a macrofeature that "integrates," as it were, the organisms with climate, water, and soil conditions. In aquatic environments where plants are often inconspicuous, a dominant physical feature, such as "standing water," "running water," "marine continental shelf," and so on, provides a basis for recognizing major types of ecosystems.

Examples

An energy-based classification is outlined in Table 3–17 and will be discussed after the basic laws of energy behavior are outlined in the next chapter. A biome classification based on structural macrofeatures is outlined in Table 2–6. Terrestrial biomes are based on natural or original conditions of vegetation; aquatic ecosystem types are based on geological and physical structure. The 16 major types of ecosystems, as listed in Table 2–6, represent the matrix in which humans

Table 2–6 **Major Natural Ecosystem Types and Biomes of the Biosphere**

Terrestrial biomes
- Tundra: arctic and alpine
- Boreal coniferous forests
- Temperate deciduous forests
- Temperate grassland
- Tropical grassland and savanna
- Chaparral: winter rain–summer drought regions
- Desert: herbaceous and shrub
- Semievergreen tropical forest: pronounced wet and dry seasons
- Evergreen tropical rain forest

Freshwater ecosystem-types
- Lentic (standing water): lakes, ponds, and so on
- Lotic (running water): rivers, streams, and so on
- Wetlands: marshes and swamp forests

Marine ecosystem-types
- Open ocean (pelagic)
- Continental shelf waters (inshore water)
- Upwelling regions (fertile areas with productive fisheries)
- Estuaries (coastal bays, sounds, river mouths, salt marshes, and so on)

imbedded their civilizations. From another view, the table is a list of the major life-support biotic communities.

A map and a more detailed description of the major types of ecosystems, together with photographs, are included in the Appendix. These maps, pictures, and descriptions will be referred to in connection with discussion of general principles in the chapters that follow.

3
Energy in Ecological Systems

1. Review of Fundamental Concepts Related to Energy: The Entropy Law

Statement

Energy is defined as the ability to do work. The behavior of energy is described by the following laws. The **first law of thermodynamics**, or the **energy conservation law**, states that energy may be transformed from one type into another but is neither created nor destroyed. Light, for example, is a form of energy, for it can be transformed into work, heat, or potential energy of food, depending on the situation, but none of it is destroyed. The **second law of thermodynamics**, or the **entropy law**, may be stated in several ways, including the following: No process involving an energy transformation will spontaneously occur unless there is a degradation of the energy from a concentrated form into a dispersed form. For example, heat in a hot object will spontaneously tend to become dispersed into the cooler surroundings. The second law of thermodynamics may also be stated as follows: Because some energy is always dispersed into unavailable heat energy, no spontaneous transformation of energy (light, for example) into potential energy (protoplasm, for example) is 100 percent efficient. **Entropy** (from *en* = in; *trope* = transformation) is a measure of unavailable energy resulting from transformations; the term is also used as a general index of the disorder associated with energy degradation.

Organisms, ecosystems, and the entire biosphere possess the essential thermodynamic characteristic: they can create and maintain a high state of internal order, or a condition of low entropy (a low amount of disorder or unavailable energy in a system). Low entropy is achieved by continually and efficiently dissipating energy of high utility (light or food, for example) to energy of low utility (heat, for example). In the ecosystem, "order" in a complex biomass structure is maintained by the total community respiration, which continually "pumps out disorder." Ecosystems and organisms are, accordingly, open, nonequilibrium, thermodynamic systems that exchange energy and matter with the environment continuously to decrease internal but increase external entropy (thus conforming to the thermodynamic laws).

Explanation

The fundamental concepts of physics outlined in the preceding paragraph are the most important of natural "laws" that apply to just about everything. So far as is known, no exceptions and no technological innovations can "break" these laws of physics. Any system of man or nature that does not conform is indeed doomed. The two laws of thermodynamics are illustrated by energy flow through an oak leaf as shown in Figure 3–1.

Various forms of life are all accompanied by energy changes, even though no energy is created or destroyed (first law of thermodynamics). The energy that enters the earth's surface as light is balanced by the energy that leaves the earth's surface as invisible heat radiation. The essence of life is the progression of such changes as growth, self-duplication, and synthesis of complex relationships of matter. Without energy transfers, which accompany all such changes, there could be no life and no ecological systems. Civilization is just one of the remarkable natural proliferations that depend on the continuous inflow of the concentrated energy. Should civilization become a closed system in its inability to obtain and store enough high-utility energy, it will soon become disorderly, as dictated by the second law.

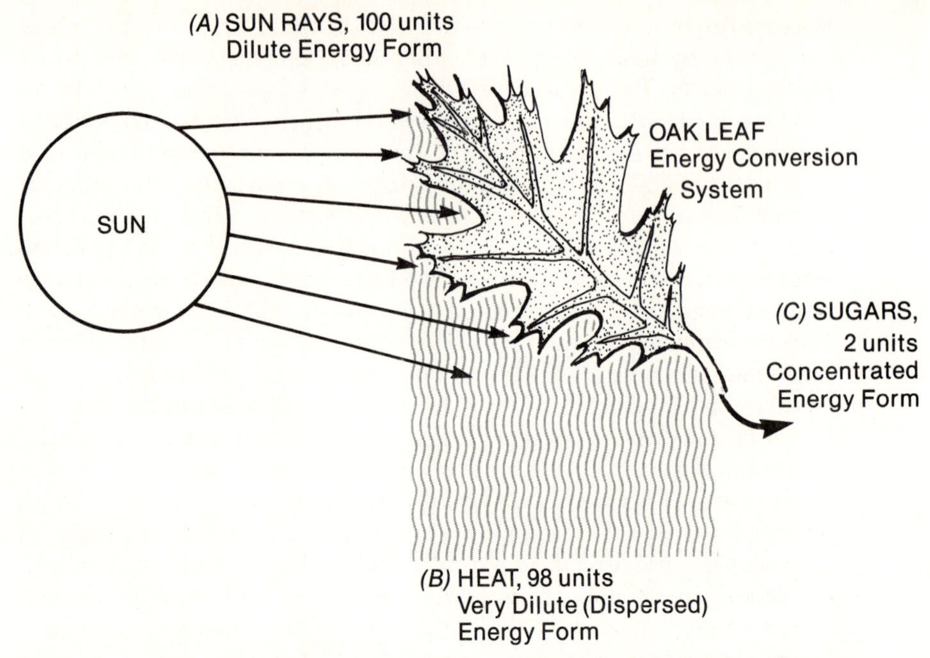

Figure 3–1 Illustration of the two laws of thermodynamics—conversion of sun energy to food (sugars) energy by photosynthesis. $A = B + C$ (first law); C is always less than A because of dissipation during conversion (second law).

Ecologists probe how light is related to ecological systems and how energy is transformed within the system. Thus, the relationships between producer plants and consumer animals, between predator and prey, not to mention the numbers and kinds of organisms in a given environment, are all limited and controlled by the flow of energy from concentrated to dispersed forms. Ecologists are especially concerned with how fuel, atomic energy, and other forms of concentrated energy are transformed in industrial societies. Thus, the same basic laws that govern nonliving systems, such as electric motors or automobiles, also govern all types of ecosystems. The difference is that living systems use part of their internally available energy for self-repair and for "pumping out" the disorder; machines have to be repaired and replaced by use of external energy. In our enthusiasm for machines, we forget that a considerable amount of energy resources must be reserved at all times for reducing entropy created by their operation.

When light is absorbed by some object, which becomes warmer as a result, the light energy has been transformed into another kind of energy: heat energy. Heat energy comprises the vibrations and mo-tions of the molecules that make up an object. The absorption of the sun's rays by land and water causes hot and cold areas, ultimately leading to the flow of air, which may drive windmills and perform work such as the pumping of water against the force of gravity (as in a well). Thus, in this case, light energy passes to heat energy of the land to **kinetic** energy of moving air, which accomplishes the work of raising water. The energy is not destroyed by lifting of the water; instead, it becomes **potential energy**, because the latent energy inherent in having the water elevated can be transformed into some other type of energy by allowing the water to fall back down the well. Energy required to generate a flow of kinetic energy has been called **embodied energy.** In the preceding example, the embodied energy of wind equals the solar energy required to generate the wind.

As indicated in previous chapters, food resulting from the photosynthesis of green plants contains potential energy, which changes to other types when food is utilized by organisms. Since the amount of one type of energy is always equivalent to a particular quantity of another type into which it is transformed, we can calculate one from the other. Energy that is "consumed" is not actually used up. Rather, it is converted from a state of high utility to a state of low utility. Gasoline in an automobile's tank is indeed used up as gasoline; however, the energy in it is not destroyed but converted into forms no longer usable by the automobile.

The second law of thermodynamics deals with the transfer of energy toward an ever less available and more dispersed state. As far as the solar system is concerned, the ultimate dispersed state is one in which all energy ends up in the form of evenly distributed heat energy. The degradation process has often been spoken of as "the running down of the solar system." Whether the tendency for energy to be leveled applies to the universe as a whole is not yet known.

At present the earth is far from a stable state of energy, because vast potential energy and temperature differences are maintained by the continual influx of light energy from the sun. However, the process of going *toward* the stable state is responsible for the succession of energy changes that constitute natural phenomena on the earth. The situation is like that of a person on a treadmill: that person never reaches the end of the treadmill, but the effort to do so results in well-defined processes. Thus, when the sun's energy strikes the earth, it tends to be degraded into heat energy. Only a very small portion of the light energy absorbed by green plants is transformed into potential or food energy; most of it goes into heat, which then passes out of the plant, the ecosystem, and the biosphere (Figure 3–1). The rest of the biological world obtains its potential chemical energy from organic substances produced by plant photosynthesis or microorganism chemosynthesis. An animal, for example, consumes chemical potential energy of food and converts a large part into heat to enable a small part of the energy to be reestablished as the chemical potential energy of new protoplasm. At each step in the transfer of energy from one organism to another, a large part of the energy is degraded into heat. However, entropy is not all negative. As the quantity of available energy declines, the quality of the remainder may be greatly enhanced, as discussed in Section 5 of this chapter.

Over the years, many theorists (for example, Brillouin, 1949) have been bothered by the fact that functional order maintained within living systems seems to defy the second law. Ilya Prigogine (1962), who won a Nobel Prize for his work in nonequilibrium thermodynamics, resolved this apparent contradiction by showing that self-organization and creation of new structures can and does occur in systems that are far from equilibrium and have well-developed "dissipative structures" that pump out the disorder (see Nicolis and Prigogine, 1977). The respiration of the highly ordered biomass is the "dissipative structure" in an ecosystem.

H. T. Odum (1967), building on the concepts of A. J. Lotka (1925) and E. Schrödinger (1945), places the thermodynamic principles in the ecological context in the following manner. Antithermal maintenance is the first priority in any complex system of the real world. The continual work of pumping out "disorder" is necessary if internal "order" is to be maintained in the presence of thermal vibrations in any system with a temperature above absolute zero. In the ecosystem the ratio of total community respiration to total community biomass (R/B) can be considered as the maintenance-to-structure ratio, or as a thermodynamic order function. This "Schrödinger ratio" is an ecological turnover, a concept introduced in Chapter 2 (see page 60). If R and B are expressed in calories (energy units) and divided by absolute temperature, the R/B ratio becomes the ratio of entropy increase of maintenance (and related work) to entropy of ordered structure. The larger the biomass, the greater the maintenance cost; but if the size of the biomass units (individual organisms, for example) is large (such as large trees in a forest) the antithermal maintenance per unit of biomass structure is decreased. One of the theoretical questions being debated is whether nature maximizes the ratio of structure to maintenance metabolism or whether energy flow itself is maximized.

Although entropy in the technical sense relates to energy, the word is also used in a broader sense to refer to degradation of materials. Freshly made steel represents a low entropic (high utility) state of iron; a rusting automobile body represents a high entropic (low utility) state. Accordingly, a highly entropic civilization is characterized by degrading energy, rusting machinery, rotting water pipes, and eroding soil. (Does this sound like your city?) Constant repair is one of the costs of high-energy civilizations.

In Table 3–1, the basic units of energy are defined, and useful conversion factors and reference points are listed. There are two

Table 3–1 Units of Energy, and Some Useful Ecological Approximations

A—Units of Potential Energy

calorie or gram-calorie (cal or gcal)—heat energy required to raise 1 cubic centimeter of water 1 degree Centigrade (at 15°C)

kilocalorie or kilogram-calorie (kcal)—heat energy to raise 1 liter of water 1 degree Centigrade (at 15°C) = 1000 calories

British Thermal Unit (BTU)—heat energy to raise 1 pound of water 1 degree Fahrenheit

joule (J)—work energy to raise one kilogram to height of 10 centimeters (or one pound to approximately 9 inches) = 0.1 kilogram-meters

foot-pound (ft-lb)—work energy to raise one pound, one foot = 0.138 kilogram-meters.

B—Units of Power (Energy-Time Units)

watt (W) (the standard international unit of power) = 1 joule per second = 0.239 cal per second

kilowatt-hour (KW hr) (the standard unit of electric power) = 1000 watts per hour = 3.6×10^9 watts

horsepower (hp) = 550 foot-pounds per second

C—Energy Unit Conversion Chart

From \ To	BTU	kcal	ft-lb	Joules	hp-hr	KW hr
BTU	—	1/4	778	1055	.00039	.00029
kcal	4	—	3090	4200	.00156	.0012
ft-lb	.0013	.00032	—	1.4	5.0×10^{-7}	3.8×10^{-7}
Joules	.00095	.00024	0.74	—	3.7×10^{-7}	2.8×10^{-7}
hp-hr	2500	640	2,000,000	2,700,000	—	0.75
KW hr	3400	860	2,700,000	3,600,000	1.3	—

Table continued on following page

Table 3–1 (*continued*)

D—Reference Values (Averages or Approximations)

Purified foods, kcal/gm dry wt.: carbohydrate 4; protein 5; lipid 9.2

	Dry Wt. (kcal/gm)	Ash-Free Dry Wt. (kcal/gm)
Biomass*		
Terrestrial plants (total)	4.5	4.6
seeds only	5.2	5.3
Algae	4.9	5.1
Invertebrates (excl. insects)	3.0	5.5
Insects	5.4	5.7
Vertebrates	5.6	6.3

Phytoplankton production: 1 gm carbon = 2.0 + gm dry matter = 10 kcal

Daily food requirements (at nonstressing temperatures)
Human: 40 kcal/kg live body wt. = 0.04 kcal/gm (about 3000 kcal/day for 70-kg adult)
Small bird or mammal: 1.0 kcal/gm live body wt.
Insect: 0.5 kcal/gm live body wt.

Gaseous exchange—caloric coefficients in respiration and photosynthesis

% Carbohydrate in Dry Matter Respired or Synthesized	Oxygen (kcal/liter)	Carbon Dioxide (kcal/liter)
100	5.0	5.0
66	4.9	5.5
33	4.8	6.0
0 (fat only)	4.7	6.7

E—Energy Content of Fossil Fuels (Round Figures)
1 gram coal = 7.0 kcal = 28 BTU
1 pound coal = 3200 kcal = 12.8×10^3 BTU
1 gram gasoline = 11.5 kcal = 46 BTU
1 gallon gasoline = 32,000 kcal = 1.28×10^5
1 cubic foot natural gas = 250 kcal = 1000 BTU = 1 therm
1 barrel crude oil (42 gallons) = 1.5×10^6 kcal = 5.8×10^6 BTU

* Since most living organisms are two-thirds or more water and minerals, 2 kcal/gm live (wet) weight is a very rough approximation for biomass in general.

classes of basic units: potential energy units without respect to time (A) and power or rate units with time built into the definition (B). Interconversions take on the time factor of the power unit; thus, 1 KW hr = 860 kcal/hr. And, of course, Class A units become power units if a time period is added (i.e., BTU per hour, day, or year).

The behavior of energy in ecosystems can be conveniently termed the **energy flow** because, as we have seen, energy transformations are "one-way," in contrast to the cyclic behavior of materials. Later in this chapter, the total energy flow that constitutes the energy environment of the biosphere will be considered, and then that portion of the total energy flow that passes through the living components of the ecosystem will be analyzed. Finally, energy quality and an energy-based classification of ecosystems will be studied to demonstrate better that energy is the common denominator for all kinds of systems, whether natural or designed by humans.

For more on how thermodynamic theory relates to biology and ecology, see Prigogine, Nicolis, and Babloyantz (1972); Lephowski (1979); Wesley (1974); and H. T. Odum (1971, Chapter 2).

2. Energy Environment

Statement

Organisms at or near the surface of the earth are constantly irradiated by solar radiation and long-wave thermal radiation flux from nearby surfaces. Both contribute to the climatic environment (temperature, evaporation of water, movement of air and water), but only a small fraction of the solar radiation can be converted by photosynthesis into energy for the biotic components of the ecosystem. Extraterrestrial sunlight reaches the biosphere at a rate of 2 gcal/cm²/min* but is attenuated exponentially as it passes through the atmosphere; at most, 67 percent (1.34 gcal/cm²/min) may reach the earth's surface at noon on a clear summer day (Gates, 1965). Solar radiation is further attenuated and the spectral distribution of its energy greatly altered as it passes through cloud cover, water, and vegetation. The daily input of sunlight to the autotrophic layer of an ecosystem varies mostly between 100 and 800, averaging about 300 to 400 gcal/cm² (= 3000 to 4000 kcal/m²) for an area in the temperate zone such as the United States (Reifsnyder and Lull, 1965). The 24-hour flux of heat energy within an ecosystem (or received by exposed organisms) can be several times as much as or considerably less than incoming solar radiation. The variation in total radiation flux within different strata of the ecosystem, as well as from one season or site to another on the earth's surface, is enormous, and the distribution of individual organisms responds accordingly.

* The "solar constant."

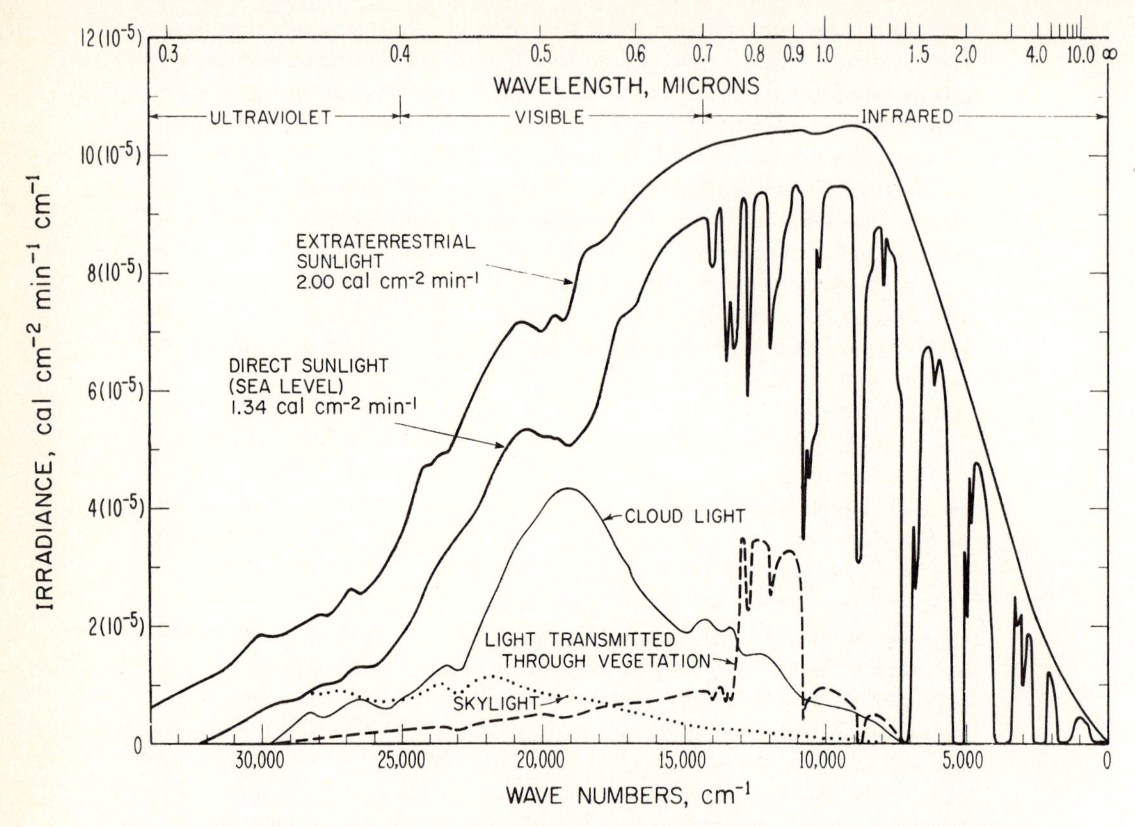

Figure 3–2 Spectral distribution of extraterrestrial solar radiation, of solar radiation at sea level for a clear day, of sunlight from a complete overcast, and of sunlight penetrating a stand of vegetation. Each curve represents the energy incident on a horizontal surface. (From Gates, 1965.)

Explanation

In Figure 3–2, the spectral distribution of extraterrestrial solar radiation, coming in at a constant rate of 2 gcal/cm²/min (±3.5 percent), is compared with (1) solar radiation actually reaching sea level on a clear day, (2) sunlight penetrating a complete overcast (cloudlight), and (3) light transmitted through vegetation. Each curve represents energy incident on a horizontal surface. In hilly or mountainous country, south-facing slopes receive more solar radiation and north-facing slopes receive much less than do horizontal surfaces; this results in striking differences in local climates (microclimates) and vegetation.

Radiation penetrating the atmosphere is attenuated exponentially by atmospheric gases and dust but to varying degrees depending on the frequency or wavelength. Shortwave ultraviolet radiation below 0.3 μ is abruptly terminated by the ozone layer in the outer atmosphere (about 18 miles or 25 km altitude), which is fortunate because such radiation is lethal to exposed protoplasm. Adsorption in the atmosphere broadly reduces visible light and irregularly reduces infrared radiation. Radiant energy reaching the surface of the earth on a clear day is about 10 percent ultraviolet, 45 percent visible, and 45 percent infrared (Reifsnyder and Lull, 1965). Visible radiation is least attenuated as it passes through dense cloud cover and water, which means that photosynthesis (which is restricted to the visible range) can continue on cloudy days and at some depth in clear water. Vegetation absorbs the blue and red visible wavelengths and the far infrared strongly, the green less strongly, the near infrared very weakly. Because the green and near infrared are reflected by vegetation, these spectral bands are used in aerial and satellite remote sensing and photography to reveal patterns of natural vegetation, condition of crops, presence of diseased plants, and so on.

The cool, deep shade of the forest is due to absorption of the visible and far infrared radiation by the foliage overhead. Chlorophyll especially absorbs the blue and red light (0.4- to 0.5-μ and 0.6- to 0.7-μ bands, respectively); water in the leaves and the water vapor around them absorb the far infrared heat energy. Green plants thus efficiently absorb the blue and red light most useful in photosynthesis. By rejecting, as it were, the near infrared band, in which the bulk of the sun's heat energy is located, leaves of terrestrial plants avoid lethal temperatures. In addition, leaves are cooled by evaporation, and aquatic plants are, of course, "water-cooled." Light as a limiting and controlling factor for organisms is discussed in Chapter 5.

Thermal radiation, the other component of the energy environment, comes from any surface or object at a temperature above absolute zero. This includes not only soil, water, and vegetation, but also clouds, which contribute a substantial amount of heat energy radiated downward into ecosystems. For example, temperatures on a cloudy winter night often remain higher than on a clear night. The "greenhouse effect" of reradiation and heat retention already mentioned briefly will be considered in Chapter 4 in connection with the CO_2 theory of climate. The longwave radiation fluxes, of course, are incident at all times and come from all directions, whereas the solar component is directional and is present only during the daytime. Thus, during a 24-hour period in the summer, an animal in the open or a leaf of a plant could be subjected to a total 24-hour upward and downward thermal radiation flux several times that of the direct downward solar input (1660 gcal/cm^2 compared with 670 gcal/cm^2 in an example presented by Gates, 1963). Furthermore, thermal radiation is absorbed by biomass to a greater degree than is solar radiation. The daily variation is of great ecological significance. In places like deserts or alpine tundras, the daytime flux is much greater than the nighttime flux, whereas in deep water or in the interior of a tropical forest (and, of course, in caves), the total radiation environment may be practically constant throughout the 24-hour period. Water and biomass, then, tend to reduce fluctuations in the energy environment and thus make conditions less stressful for life, another example of mitigation at the ecosystem level (recall Chapter 2, Section 4).

Although the total radiation flux determines the conditions of existence for organisms, the integrated direct solar radiation to the autotrophic stratum—the sun energy received by green plants over days, months, and the year—is of greatest interest for productivity and cycling of nutrients within the ecosystem. This primary energy input drives all biological systems. The average daily solar radiation received each month in five regions of the United States is shown in Table 3–2. In addition to latitude and season, cloud cover is a major factor, as shown in the comparison between the humid Southeast and the arid Southwest United States. A range between 100 and 800 gcal/cm^2/day would probably cover most of the world most of the time, except in the polar regions or arid tropics. There, conditions are so extreme anyway that little biological output is possible. Therefore, for most of the biosphere, the radiant energy input is of the order of 3000 to 4000 kcal/m^2/day and 1.1 to 1.5 million kcal/m^2/year. The fate of this annual inflow of solar radiation is summarized in Table 3–3 as percentages dissipated in major processes.

Of special interest is the so-called **net radiation** at the surface of the earth, "the difference between all downward streams of radiation minus all upward streams of radiation" (Gates, 1962). Between latitudes 40° north and south the annual net radiation is 1 million kcal/m^2/year over the oceans and 0.6 million kcal/m^2/year over the continents (Budyko, 1955). This tremendous bundle of energy is dissipated in the evaporation of water and the generation of thermal winds (two major solar-powered processes with high embodied energy) and eventually passes into space as heat, so the earth as a whole may remain in an approximate energy balance. The differing impact of evaporative energy on terrestrial and aquatic ecosystems has already been mentioned (see page 61). Also, as already pointed out, any factor that delays the outward passage of this energy will cause the temperatures of the biosphere to rise.

Table 3–2 Solar Radiation Received Regionally Over the United States on a Unit Horizontal Surface*

	Average Langleys (gcal/cm^2) per Day				
	Northeast	Southeast	Midwest	Northwest	Southwest
January	125	200	200	150	275
February	225	275	275	225	375
March	300	350	375	350	500
April	350	475	450	475	600
May	450	550	525	550	675
June	525	550	575	600	700
July	525	550	600	650	700
August	450	500	525	550	600
September	350	425	425	450	550
October	250	325	325	275	400
November	125	250	225	175	300
December	125	200	175	125	250
Mean-gcal/cm^2/day	317	388	390	381	494
Mean-kcal/m^2/day (round figures)	3200	3900	3900	3800	4900
Estimated kcal/m^2/year (round figures)	1.17×10^6	1.42×10^6	1.42×10^6	1.39×10^6	1.79×10^6

* After Reifsnyder and Lull, 1965.

Table 3–3 Energy Dissipation of Solar Radiation as Percentages of Annual Input into the Biosphere*

	Percent
Reflected	30
Direct conversion to heat	46
Evaporation, precipitation	23
Wind, waves, and currents	0.2
Photosynthesis	0.8

Tidal energy—about 0.0017 percent of solar
Terrestrial heat—about 0.5 percent of solar

* After Hulbert, 1971.

The solar component is usually measured by **pyrheliometers** or **solarimeters**, which employ a thermopile, a junction of two metals that generates a current proportional to the input of light energy. Instruments that measure the total flux of energy at all wavelengths are termed **radiometers.** The **net radiometer** has two surfaces, upward and downward, and records the difference in energy fluxes. Airplanes and satellites equipped with thermal scanners can quantitatively sense heat rising from earth surfaces. Pictures generated from such imagery show cities as "heat islands," the location of water bodies, contrasting microclimates (as in north- and south-facing ravines), and many other useful aspects of the energy environment. Cloud covers interfere with this kind of remote sensing much less than in the case of visual imagery.

Finally, the units of irradiance energy, namely gcal/cm^2 (also called the langley) and kcal/m^2, should not be confused with units of illumination, namely the foot candle (1 foot candle = 1 lumen/ft^2) and the lux (1 lux = 1 lumen/m^2 = approximately 0.1 foot candle), which apply only to the visible spectrum. Although one cannot convert irradiance rate to illuminescence accurately because of the variation in the brightness of different spectral regions, 1 gcal/cm^2/min of sunlight multiplied by 6700 will give an approximate illuminance on a horizontal surface expressed as foot candles (Reifsnyder and Lull, 1965).

The fate of solar energy coming into the biosphere is summarized in Table 3–3. Although only about 1 percent is converted into food and other biomass, the 70 percent or so that goes into heat, evaporation, precipitation, wind, and so on is not wasted, because these energies create a livable temperature and drive weather systems and water cycles, all necessary for life on earth, as detailed in Chapters 4 and 5. Though energy from tides and the earth's heat may provide useful local sources of energy for people, little is available globally. A lot of heat is deep in the earth (so-called geothermal power), but to tap it would require very energy-expensive deep drilling in most parts of the world.

Of interest to the ecologist are David Gates' several excellent summaries of our energy environment (Gates, 1962, 1963, 1965, 1965a, 1971, and 1980).

3. Concept of Productivity

Statement

The **primary productivity** of an ecological system, community, or any part thereof is defined as the rate at which radiant energy is converted by photosynthetic and chemosynthetic activity of producer organisms (chiefly green plants) to organic substances. It is important to distinguish the four successive steps in the production process as follows:

1. **Gross primary productivity** is the total rate of photosynthesis, including the organic matter used up in respiration during the measurement period. This is also known as "total photosynthesis" or "total assimilation."

2. **Net primary productivity** is the rate of storage of organic matter in plant tissues exceeding the respiratory use by the plants during the period of measurement. This is also called "apparent photosynthesis" or "net assimilation." In practice, the amount of respiration is usually added to measurements of "apparent" photosynthesis as a correction to estimate gross production.

3. **Net community productivity** is the rate of storage of organic matter not used by heterotrophs (that is, net primary production minus heterotrophic consumption) during the period under consideration, usually the growing season or a year.

4. Finally, the rates of energy storage at consumer levels are referred to as **secondary productivities.** Since consumers use only food materials already produced, with appropriate respiratory losses, and convert to different tissues by one overall process, secondary productivity should not be divided into "gross" and "net" amounts. The total energy flow at heterotrophic levels, which is analogous to gross production of autotrophs, should be designated "assimilation" and not "production."

In all these definitions, the term "productivity" and the phrase "rate of production" may be used interchangeably. Even when the term "production" designates an amount of accumulated organic matter, a time element is always assumed or understood, for example, a year in agricultural crop production. To avoid confusion, one should always state the time interval. In accord with the second law of thermodynamics, as stated in Section 1, the energy flow decreases at each step, as listed, by the heat loss occurring with each transfer of energy from one form to another.

High rates of production, in both natural and cultured ecosystems, occur when physical factors are favorable and especially when energy subsidies from outside the system reduce the cost of maintenance (note "other energy" input, Figure 2–2). Such energy subsidies may be the work of wind and rain in a rain forest, tidal energy in an estuary, or the fossil fuel, animal, or human work energy used in cultivating a crop. In evaluating the productivity of an ecosystem, one must consider the nature and magnitude not only of the **energy drains** resulting from climatic, harvest, pollution, and other stresses that divert energy away from the production process but also of the

energy subsidies that enhance it by reducing the respiratory heat loss (i.e., the "disorder pump-out") necessary to maintain the biological structure.

Explanation

The key word in the preceding definitions is *rate*. The time element must be considered, that is, the amount of energy fixed in a given time. Biological productivity thus differs from "yield" in the chemical or industrial sense. In industry, the reaction ends with the production of a given amount of material; in biological communities, the process is continuous in time, so a time unit must be designated, for example, the amount of food manufactured per day or per year. In more general terms, productivity of an ecosystem refers to its fertility or "richness." Though a rich or productive community may have more organisms than a less productive community, this is not so if organisms in the productive community are removed or "turn over" rapidly. For example, a fertile pasture being grazed by livestock is likely to have a much smaller standing crop of grass than a less productive pasture not being grazed at the time of measurement. *Biomass or standing crop present at any given time should not be confused with productivity.* Students of ecology often confuse these two quantities. Usually, one cannot determine the primary productivity of a system or the production of a population component simply by counting and weighing ("censusing") the organisms present at any one moment, although net primary productivity can be estimated from standing crop data when organisms are large and living materials accumulate over a period of time without being consumed (as in cultivated crops, for example).

The different kinds of production, the important distinction between gross and net primary production, and their relationship to the solar energy input are explained in Tables 3–4 and 3–5. Only about half of total radiant energy (mostly in the visible portion) is absorbed, and at most about 5 percent (10 percent of the absorbed) can be converted as gross photosynthesis under the most favorable conditions. Then, plant respiration appreciably reduces—at least 20 percent, usually about 50 percent—the food available for heterotrophs.

Table 3–4 is a generalized model for long-term energy transfers, that is, energy transfers over the annual cycle or longer. During the peak of the growing season, especially during the long summer days of the north, as much as 10 percent of the total daily solar input may be converted into gross production, and up to 75 to 80 percent of this may remain as net primary production during a 24-hour period. Even

Table 3–4 Relationships Between Solar Energy Input and Primary Productivity

A. Percentage Transfers

Steps	1 Total Solar Radiant Energy	2 Absorbed by Autotrophic Stratum	3 Gross Primary Production	4 Net Primary Production (Available to Heterotrophs)
Maximum	100	50	5	4
Average favorable condition	100	50	1	0.5
Average for biosphere	100	<50	0.2	0.1

B. Percentage Efficiencies

Step	Maximum	Average Favorable Condition	Average for Whole Biosphere
1–2	50	50	<50
1–3	5	1	0.2
2–3	10	2	0.4
3–4	80	50	50
1–4	4	0.5	0.1

C. On a kcal/m^2/year Basis (Round Figures)

Radiant Energy		Gross Primary Production	Net Primary Production
1,000,000	Maximum	50000	40000
	Average fertile regions*	10000	5000
	Open oceans and semi-arid regions†	1000	500
	Mean for biosphere‡	2000	1000

* Moisture, nutrients, and temperature not strongly limiting; auxiliary energy input (see text for explanation).

† Moisture, nutrients, or temperature strongly limiting.

‡ Based on estimate of 10^{18} kcal gross productivity and 5×10^8/km^2 area of the whole biosphere (see Table 3–7).

Table 3–5 **Channeling of the Energy of Gross Production in a Soybean (Glycine max) Crop Ecosystem: Hypothetical Annual Budget***

Energy Flow	Percent Gross Production Utilized	Percent Gross Production Remaining
1. Plant respiration	25	
Theoretical net primary production		75
2. Symbiotic microorganism (nitrogen-fixing bacteria and mycorrhizal fungi†)	5	
Net primary production allowing for needs of beneficial symbionts		70
3. Root nematodes, phytophagous insects, and pathogens	5‡	
Net community production allowing for minimum primary consumption by "pests"		65
4. Beans harvested by man (export)	32	
Stems, leaves, and roots remaining in the field		33
5. Organic matter decomposed in soil and litter	33	
Annual increment		0

* Adapted from Gorden, 1969.

† Mutualistic fungi that aid mineral uptake by roots (see page 397).

‡ Low percent possible only with energy subsidy by man (fossil fuel, human and/or animal labor involved in cultivation, pesticide application).

under the most favorable conditions, however, these high daily rates cannot yet be maintained over the annual cycle or cannot achieve such high yields over large areas of farmland, as is evident when they are compared with the annual yields actually obtained nationwide or worldwide (see Table 3–8).

Concept of Energy Subsidy As already noted in Chapter 2, Section 7, high productivity and high net/gross ratios in crops are maintained by virtue of large energy inputs involved in cultivation, irrigation, fertilization, genetic selection, and insect control. Fuel used to power farm machinery is just as much an energy input as sunlight; it can be measured as calories or horsepower diverted to heat in performance of the work of crop maintenance. In the United States, the fuel energy input into agriculture increased tenfold between 1900 and the 1970s, from about 1 to 10 calories per calorie of food harvested (see Steinhart

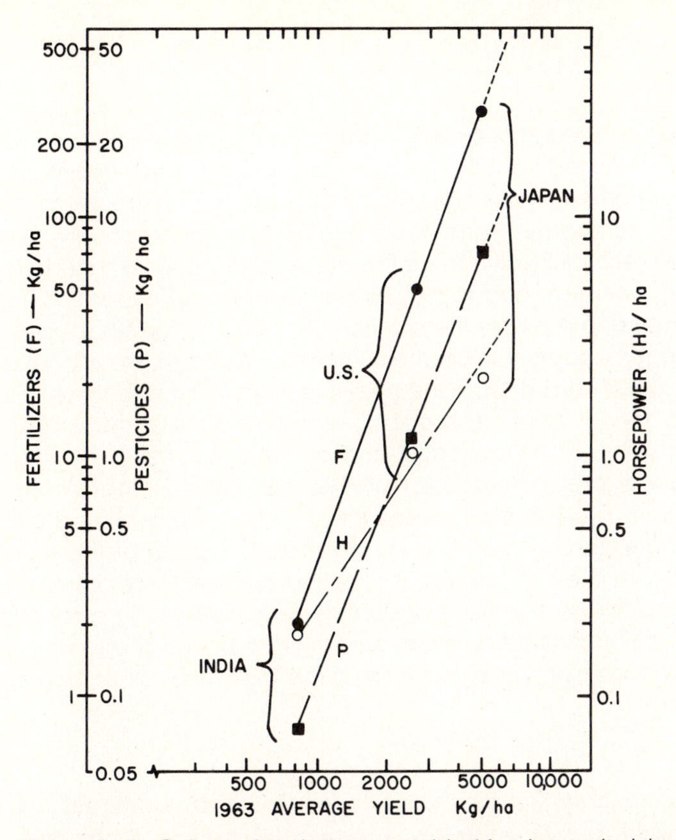

Figure 3–3 Relationships between yield of food crops (in kilograms dry weight per hectare) and requirements for fertilizer (F), pesticides (P), and horsepower (H) used in cultivation and harvest of crops. Regressions based on data from three countries (United States, India, Japan). Note that doubling the yield of food requires a tenfold increase in use of fertilizers, pesticides, and animal or machine power. (Graph prepared from data in *The World Food Problem*, A Report of the President's Science Advisory Committee, Panel on World Food Supply. The White House, 1967; volume III, pages 141, 143, and 180.)

and Steinhart, 1974). The general relationship between inputs of fertilizers, pesticides, and work energy is shown in Figure 3–3; doubling of crop yield requires approximately a tenfold increase in all these inputs. Genetic selection for increased food-to-fiber ratio is the other way crop yields have been increased. The ratio of grain to straw dry weight for wheat and rice, for example, has been increased from 50 percent to almost 80 percent during this century.

H. T. Odum was one of the first ecologists to state the vital coupling between energy input, selection, and agricultural productivity.

In 1967 he wrote the following:

> Man's success in adapting some natural systems to his use has essentially resulted from the process of applying auxiliary work circuits into plant and animal systems from such energy rich sources as fossil and atomic energy. Agriculture, forestry, animal husbandry, algal culture, etc. all involve huge flows of auxiliary energy that do much of the work that had to be self-served in former systems. Of course, when one provides the auxiliary support, the former species are no longer adapted since their internal programs would have them continue to duplicate the previous work effort and there would be no savings. Instead, species which do not have the machinery to self-serve have the edge and are selected either by man or by natural processes of survival. Domestication in the extreme produces "organic matter machines," such as egg makers and milk-producing cows which can hardly stand up. All of the self-serving work of these organisms is supplied by new routes controlled and directed by man from auxiliary energy sources. In a real way the energy for potatoes, beef, and plant produce of intensive agriculture is coming in large part from the fossil fuels rather than from the sun.

Any energy source that reduces the cost of internal self-maintenance of the ecosystem, and thereby increases the amount of other energy that can be converted to production is called an **auxiliary energy flow** or an **energy subsidy.**

High temperatures (and high water stress) generally require the plant to expend more of its gross production energy in respiration. Thus, it costs more to maintain the plant structure in hot climates, although, as noted in Chapter 2, Section 5, C_4 plants have evolved a photosynthesis that partly circumvents this restraint imposed by hot and dry climates. The general relationship between gross and net production as a function of latitude is shown in Figure 3–4. Though this graph refers to natural vegetation, the tendency for p_n/p_g ratios to be lower in the tropics than in temperate zones also applies to C_3 crops such as rice (see Best, 1962).

Natural communities that benefit from natural energy subsidies are those with the highest gross productivity. The role of tides in coastal estuaries was mentioned earlier; a marsh benefiting from an optimal tidal or other water flow subsidy (which replaces part of the respiratory and other maintenance energy that would otherwise be

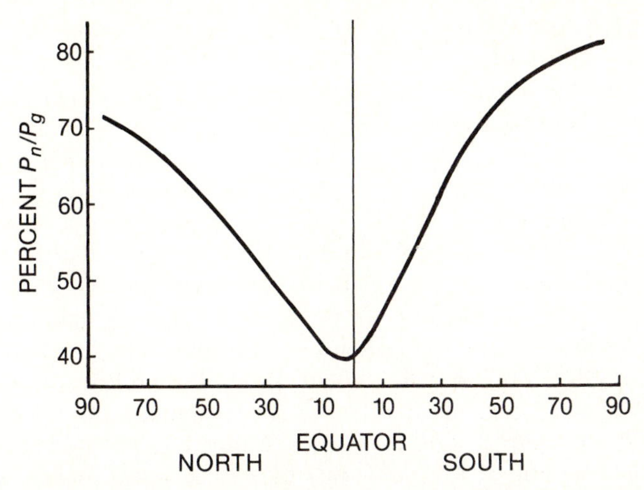

Figure 3–4 The percent of gross primary production that ends up as net primary production in natural vegetation varies with latitude. The trend is from less than 50 percent in equatorial regions to 60 to 70 percent at high latitudes. (Graphic model based on data of Box, 1978.)

diverted to mineral cycling and transportation of food and wastes) has about the same gross productivity as an intensively farmed Iowa corn field. The complex interaction of wind, rain, and evaporation in a tropical rain forest is another example of a natural energy subsidy that enables the leaves to make optimum use of the high solar input of the tropical day.

As a general principle, gross productivity of cultured ecosystems does not exceed that found in nature. We do, of course, increase productivity by supplying water and nutrients in areas where these are limiting (such as in deserts and grasslands). Most of all, however, we increase net primary and net community production through energy subsidies that reduce both autotrophic and heterotrophic consumption and thereby increase the harvest for ourselves. The "green revolution" involves development by genetic selection of crop varieties with high food/fiber ratios adapted to respond to massive energy, irrigation, and nutrient subsidies. Without these inputs, the "miracle rice" and other new varieties do not do as well as traditional varieties requiring no such subsidies. Those who think that we can upgrade the agricultural production of undeveloped countries simply by sending seeds and a few agricultural advisors are naive. The necessary energy-rich subsidies become more difficult to provide as the fossil fuels decline in availability and increase in cost. The green revolution has so far benefited the rich countries more than the poor ones. Bouvonder (1979) analyzed the impact of the green revolution in India and concluded, "The green revolution has made poor farmers poorer and the government has to initiate some action to rectify its adverse socioeconomic consequences."

There is one other important point to be made about the general concept of energy subsidy. A factor that under one set of environmental conditions or level of intensity acts as a subsidy may under another environmental condition or at a higher level of input act as an energy drain that reduces productivity. Too much of a good thing can be as serious a stress as too little. For example, evapotranspiration may be an energy stress in dry climates but an energy subsidy in the humid climates (see H. T. Odum and Pigeon, 1970). Flowing water systems, such as the Florida spring listed in Table 3–6, tend to be more fertile than standing water systems, but not if the flow is too abrasive or irregular. The gentle ebb and flow of tides in a salt marsh, a mangrove estuary, or a coral reef contributes tremendously to the high productivity of these communities, but tides crashing against a northern rocky shore subjected to ice in winter and heat in summer can be a tremendous drain. Swamp and riverine forests subjected to regular flooding during the winter and early spring dormant period have a much higher production rate than those flooded continually or for long periods into the growing season (see E. P. Odum, 1978).

In agriculture, plowing the soil helps in the north but not in the south, where the resulting rapid leaching of nutrients and loss of organic matter can severely stress subsequent crops. The trend toward "no-till" agriculture as a way of reducing these drains was noted in Chapter 2, Section 7. Finally, certain types of pollution, such as treated sewage, can act as a subsidy or as a stress, depending on the rate and periodicity of input. Treated sewage released into an ecosystem at a steady but moderate rate can increase productivity, but massive irregular dumping can almost completely destroy the system as a biological entity.

The concept of a subsidy-stress gradient is illustrated in Figure 3–5. If the input perturbation (from *perturbare* = to disturb) is poisonous, the response must be negative. If, however, the input involves usable energy or materials, productivity or other measures of performance may be enhanced, which is what is meant by a subsidized ecosystem. As the level of subsidy input is increased, the ability of the system to assimilate can reach saturation; performance will then decline, as shown in the model. Response of crop yield to increasing nitrogen fertilization is a good illustration of the subsidy-stress pattern as shown in Figure 3–5B.

In Table 3–6, selected ecosystems are listed in sequence from crop-type, rapid-growth systems to mature steady-state systems. This

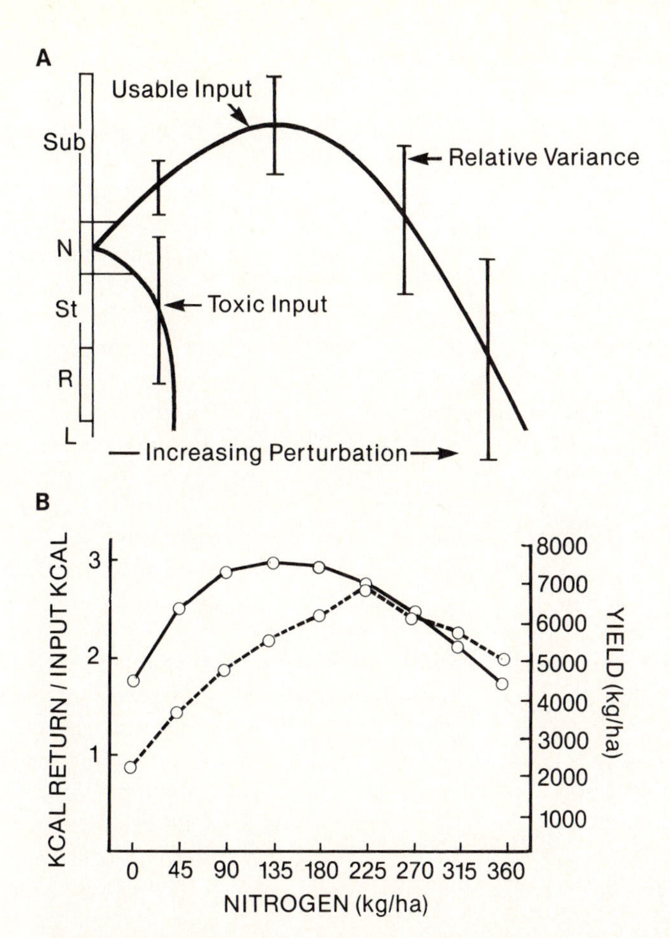

Figure 3–5 Subsidy-stress curves. *A*. Generalized curve showing how increasing inputs of energy or materials can bring about a deviation from normal operating range (N). If the input is usable, basic functions, such as productivity, may be enhanced (subsidy effect—Sub) at moderate levels of input and then depressed with increasing inputs (stress effect—St). If the input is toxic, functions will be depressed, with the likelihood that the community will be replaced by a more tolerant one or eliminated altogether. R = replacement; L = lethal. *B*. Subsidy-stress effects of increasing nitrogen fertilization on a corn crop (phosphorus fertilizer remaining constant). Solid line: efficiency curve, kcal return (harvest) per unit of input. Broken line: yield curve, kg/hectare. Note that the efficiency curve peaks at a lower rate of fertilization than the yield curve. (After Pimentel et al., 1973; from data of Munson and Doll, 1959.)

arrangement illustrates several important points about the relationships between gross primary production (GPP), net primary production (NPP), and net community production (NCP). Rapid growth or "bloom-type" systems (i.e., rapid production for short periods) such

Table 3–6 **Annual Production and Respiration as kcal/m²/year in Growth-Type and Steady-State Ecosystems**

	Alfalfa Field (U.S.A.)*	Young Pine Plantation (England)†	Medium-Aged Oak Pine Forest (New York)‡	Large Flowing Spring (Silver Springs, Florida)§	Mature Rain Forest (Puerto Rico)‖	Coastal Sound (Long Island, N.Y.)¶
Gross primary production (GPP)	24400	12200	11500	20800	45000	5700
Autotrophic respiration (R_A)	9200	4700	6400	12000	32000	3200
Net primary production (NPP)	15200	7500	5000	8800	13000	2500
Heterotrophic respiration (R_H)	800	4600	3000	6800	13000	2500
Net community production (NCP)	14400	2900	2000	2000	Very little or none	Very little or none
Ratio NPP/GPP (percent)	62.3	61.5	43.5	42.3	28.9	43.8
Ratio NCP/GPP (percent)	59.0	23.8	17.4	9.6	0	0

* After Thomas and Hill, 1949. Heterotrophic respiration estimated as 5 percent loss from insects and disease organisms.

† After Ovington, 1961. Mean annual production, 0–50 years, one-species plantation. GPP estimated from measurement of respiratory losses in young pines by Tranquillini, 1959. Part of NCP harvested (exported) as wood by humans.

‡ After Woodwell and Whittaker, 1968. 45-year-old natural regeneration following fire; no timber harvest by humans.

§ After H. T. Odum, 1957.

‖ After H. T. Odum and Pigeon, 1970.

¶ After Riley, 1956.

Note: Conversion factors from dry matter and carbon to kcal as in Table 3–1. All figures rounded off to the nearest 100 kcal.

as the alfalfa field tend to have a high net primary production and, if protected from consumers, a high net community production. Reduced heterotrophic respiration, of course, can result either from evolved self-protection mechanisms (natural systemic insecticides or cellulose production, for example) or from outside assistance such as energy subsidies. In steady-state communities, the gross primary production tends to be dissipated by the combined autotrophic respiration (R_A) and heterotrophic respiration (R_H), so little or no net community production is at the end of the annual cycle. Furthermore, communities with large biomasses or "standing crops," such as the rain forest, require so much autotrophic respiration for maintenance that there tends to be a low NPP/GPP ratio (Table 3–6). As a matter of fact, it is difficult, if not impossible, to distinguish by measurement between autotrophic and heterotrophic respiration in ecosystems such as forests. Thus, oxygen consumption or CO_2 production by a large tree trunk or tree root system is due as much to the respiration of associated microorganisms (many of which are beneficial to the tree) as to the living plant tissues. Consequently, the estimates of autotrophic respiration, and thereby the estimate of net primary production obtained by subtracting R_A from GPP, for the terrestrial communities listed in Table 3–6 are rough approximations of more theoretical than practical value. This point needs to be emphasized because a number of summary papers on primary production (Whittaker and Likens, 1973; Lieth and Whittaker, 1975; Reichle et al., 1975) compare all kinds of communities ranging from low biomass aquatic communities and crops to high biomass forests by net primary production when, in fact, net community production (i.e., dry matter accumulation in the community) was being compared.

Remember the difference between behavior of energy and behavior of materials: materials circulate; energy does not. Nitrogen, phosphorus, carbon, water, and other materials that constitute living things circulate through the system in a variable and complex manner. On the other hand, energy is used once by a given organism, is converted into heat, and is lost to the ecosystem. Thus, there is a nitrogen cycle (nitrogen may circulate many times between living and nonliving entities) but there is no energy cycle. Energy can be stored, conserved (i.e., transformed more efficiently), and "fed back" from one part of the system to another (see Chapter 2, Section 6), but it cannot be recycled in the manner of water and minerals. There can be no living closed thermodynamic systems. Every living component, whether organism or ecosystem, must have a continuous inflow of energy from its input environment. Many people, including political leaders and decision-makers, often seem blissfully unaware of this fact of life.

The Distribution of Primary Production The vertical distribution of primary production and its relation to biomass are illustrated in Figure 3–6. In these diagrams, the forest (Figure 3–6A), in which turnover time (ratio of biomass to production) is measured in years, is compared with the sea (Figure 3–6B), in which turnover time is measured in days. Even if we considered only the green leaves, which compose 1 to 5 percent of total forest biomass, as comparable to the phytoplankton, replacement time would still be longer in the forest. In the more fertile inshore waters, primary production is concentrated in the

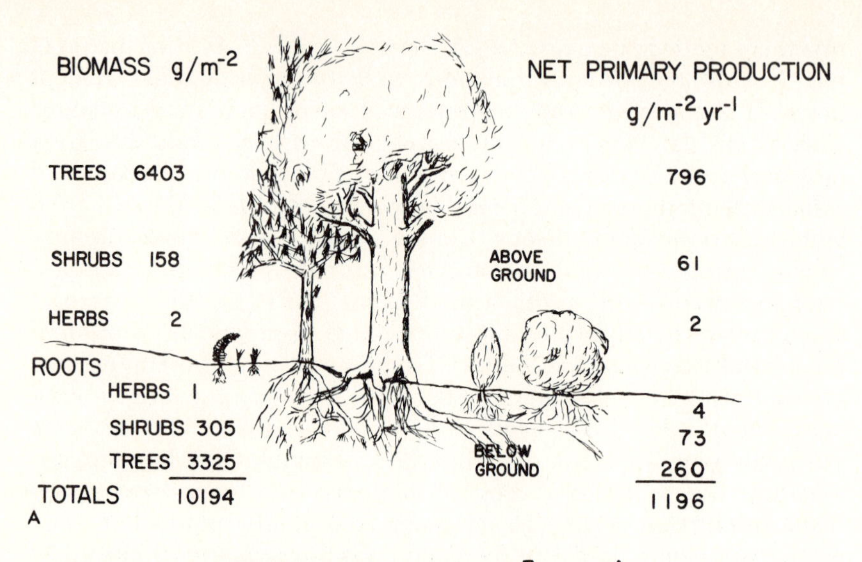

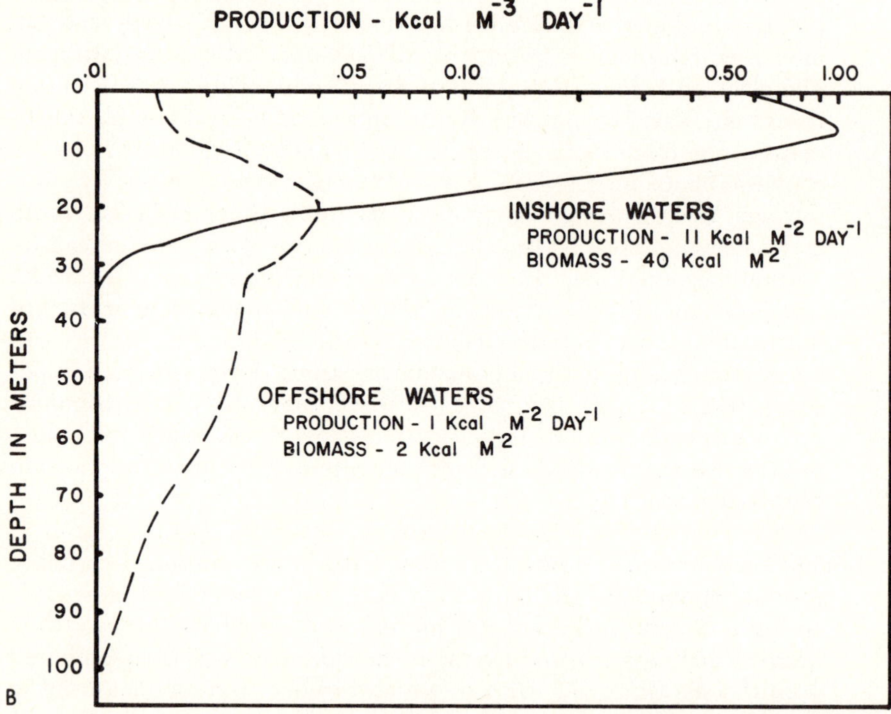

Figure 3–6 Comparison of the vertical distribution of primary production and biomass in the forest (*A*) and in the sea (*B*). These data also contrast the rapid turnover in the sea (*B*/*P* ratio is two to four days in this illustration) with slower turnover of the forest (*B*/*P* ratio is nine years). (*A* based on data of Whittaker and Woodwell, 1969, for young oak-pine forest; *B* based on data of Currie, 1958, for the northeast Atlantic.)

upper 30 meters or so; in the clearer but poorer waters of the open sea, the primary production zone may extend down to 100 meters or more. This is why coastal waters appear dark greenish and the ocean waters blue. In all water, the peak of photosynthesis tends to occur just under the surface because the circulating phytoplankton are "shade-adapted" and are inhibited by full sunlight. In the forest, where the photosynthetic units (the leaves) are permanently fixed in space, treetop leaves are sun-adapted, and understory leaves are shade-adapted (see Figure 3–8).

The attempt to estimate the rate of organic production, or primary productivity, of the world's solar-powered natural systems has an interesting history. In 1862 Justus Liebig, the pioneer agricultural chemist and plant nutritionist, based an estimate of the dry-matter production of the global land area on a single sample, a green meadow. Interestingly, his estimate of approximately 10^{11} metric tons a year is very close to Lieth's and Whittaker's estimate of 118×10^9 tons a year for continental areas (see *Primary Productivity of the Biosphere*, Lieth and Whittaker, 1975). Their figure was based on measurements of many types of vegetation, using models, computer mapping, and other modern techniques. On the other hand, Gordon Riley, in 1944, overestimated ocean productivity by basing his estimate on measurements in fertile inshore waters. Not until the 1960s, after the introduction of the carbon-14 measurement technique, was the very low productivity of most of the open ocean recognized. Since the oceans cover about 2.5 times the area of the land, it was natural to assume, as did Riley, that marine ecosystems fixed more total solar energy than did terrestrial systems. Actually, land seems to outproduce the sea, perhaps by as much as 2 to 1.

Conservative estimates of gross primary productivity of major ecosystem types, estimates of the areas occupied by each type, and the total gross productivity for land and water are listed in Table 3–7. For the estimated mean values for large areas, *productivity varies by about two orders of magnitude (100-fold), from 200 to 20,000 kcal/m²/year, and the total gross production of the world is on the order of 10^{18} kcal/year*. The general pattern of distribution of world productivity is diagrammed in Figure 3–7.

A very large part of the earth is in the low-production category because either water (in deserts, grasslands) or nutrients (in the open ocean) are strongly limiting. Naturally fertile areas that receive natural energy subsidies are found chiefly in river deltas, estuaries, coastal upwelling areas and areas of rich glacial till, and wind-transported or volcanic soils in regions of adequate rainfall. There is a general correlation between evapotranspiration (and, to a lesser ex-

Table 3-7 Estimated Gross Primary Production (Annual Basis) of the Biosphere and Its Distribution Among Major Ecosystems

Ecosystem	Area (10^6 km²)	Gross Primary Productivity (kcal/m²/year)	Total Gross Production (10^{16} kcal/year)
Marine*			
Open ocean	326.0	1000	32.6
Coastal zones	34.0	2000	6.8
Upwelling zones	0.4	6000	0.2
Estuaries and reefs	2.0	20000	4.0
Subtotal	362.4	–	43.6
Terrestrial†			
Deserts and tundras	40.0	200	0.8
Grasslands and pastures	42.0	2500	10.5
Dry forests	9.4	2500	2.4
Boreal coniferous forests	10.0	3000	3.0
Cultivated lands with little or no energy subsidy	10.0	3000	3.0
Moist temperate forests	4.9	8000	3.9
Fuel-subsidized (mechanized) agriculture	4.0	12000	4.8
Wet tropical and subtropical (broadleaved evergreen) forests	14.7	20000	29.0
Subtotal	135.0	–	57.4
Total and average (Colum 2) for biosphere (not including ice caps) (round figures)	500.0	2000	100.0

* Marine productivity estimated by multiplying Ryther's (1969) net carbon production figures by 10 to get kcal, then doubling these figures to estimate gross production and adding an estimate for estuaries (not included in his calculations).

† Terrestrial productivity based on Lieth and Whittaker's net production figures doubled for low biomass systems and tripled for high biomass systems (which have high respiration) as estimates of gross productivity. Tropical forests have been upgraded in light of recent studies, and the industrialized (fuel-subsidized) agriculture of Europe, North America, and Japan has been separated from the preindustrial agriculture characteristic of a large percentage of the world's cultivated lands.

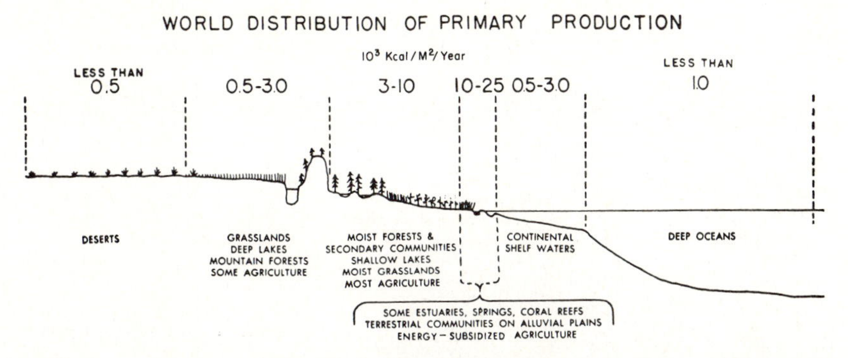

Figure 3–7 The world distribution of primary production in terms of annual gross production (in thousands of kilocalories per square meter) of major ecosystem types. Only a relatively small part of the biosphere is naturally fertile. (After E. P. Odum, 1963.)

tent, rainfall per se) and productivity on land and an inverse correlation between productivity and depth in lakes and the ocean.

Favorable sites in each broad ecosystem type can produce twice (or more) the mean values shown (see Table 3–6). For all practical purposes, a level of 50,000 kcal/m²/year can be considered the upper limit for gross photosynthesis. Until it can be shown conclusively that photosynthetic conversion of light energy can be substantially altered without endangering the balance of other, more important life-cycle resources, we should plan to live within this limit. Most agriculture shows up low annually, because annual crops are productive for less than half the year. Double cropping, that is, raising crops that produce throughout the year, can approach the gross productivity of the best of natural communities. Recall that net primary production will average about half of gross productivity and that the "yield to man" of crops will be one third or less of the gross productivity.

Chlorophyll and Primary Production Gessner, in 1949, observed that chlorophyll, which actually develops "per square meter," tends to be similar in diverse communities. This finding indicates that the content of the green pigment in whole communities is more uniform than in individual plants or plant parts. The whole is not only different from the parts but cannot be explained by them alone. Intact communities containing various plants, young and old, sunlit and shaded, are integrated and adjust, as fully as local limiting factors allow, to the incoming solar energy, which, of course, impinges on the ecosystem on a "square meter" basis.

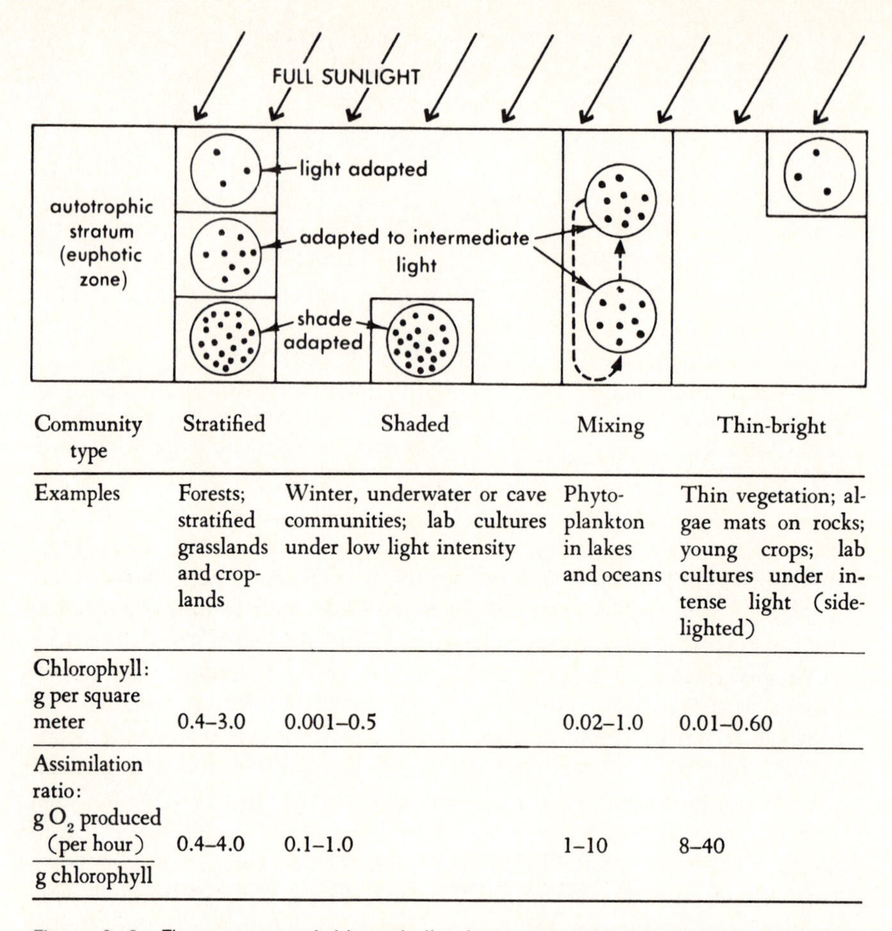

Community type	Stratified	Shaded	Mixing	Thin-bright
Examples	Forests; stratified grasslands and crop-lands	Winter, underwater or cave communities; lab cultures under low light intensity	Phyto-plankton in lakes and oceans	Thin vegetation; al-gae mats on rocks; young crops; lab cultures under in-tense light (side-lighted)
Chlorophyll: g per square meter	0.4–3.0	0.001–0.5	0.02–1.0	0.01–0.60
Assimilation ratio: g O_2 produced (per hour) / g chlorophyll	0.4–4.0	0.1–1.0	1–10	8–40

Figure 3–8 The amounts of chlorophyll to be expected in a square meter of four types of communities. The relation of area-based chlorophyll and photosynthetic rate is also indicated by the ratio between chlorophyll and oxygen production. (From E. P. Odum, 1963; after H. T. Odum, McConnell, and Abbott, 1958.)

Figure 3–8 is a model prepared by H. T. Odum, McConnell, and Abbott (1958); it shows the amount of chlorophyll to be expected per square meter in four types of ecosystems that cover the range found in nature. The dots in the diagrams indicate the relative concentration of chlorophyll per cell (or per biomass). The relation of total chlorophyll to the photosynthetic rate is indicated by the **assimilation ratio**, or rate of production per gram of chlorophyll, shown as grams O_2/hour/gram chlorophyll in the bottom row of numbers below the diagrams.

Shade-adapted plants or plant parts tend to have a higher concentration of chlorophyll than light-adapted plants or plant parts; this property enables them to trap and convert as many scarce light photons as possible. Consequently, use of light is highly efficient in shaded systems, but the photosynthetic yield and the assimilation ratio are low. Algae cultures grown in weak light in the laboratory often become shade-adapted. The high efficiency of such shaded systems has sometimes been mistakenly projected to full sunlight condition by those who would feed humankind from mass cultures of algae. When light input is increased to obtain a good yield, the efficiency goes down, as it does in any other kind of plant.

Total chlorophyll is highest in stratified communities, such as forests, and is generally higher on land than in water. In a given light-adapted system, the chlorophyll in the autotrophic zone self-adjusts to nutrients and other limits. Consequently, if the assimilation ratio and the available light are known, gross production can be estimated by extracting pigments and then measuring the chlorophyll concentration with a spectrophotometer. For example, Ryther and Yentsch (1957) found that marine phytoplankton at light saturation has a reasonably constant assimilation ratio of 3.7 gm of carbon assimilated per hour per gram of chlorophyll. Calculated production rates based on this ratio and on chlorophyll-light measurements were very similar to those obtained by other methods. Since that time, the chlorophyll method has been widely used to estimate primary production of the sea and other large bodies of water.

Human Use of Primary Production Primary production of food for human beings is summarized in Tables 3–8 and 3–9. Yields and estimated net primary production (P_n) of the major food crops in the developed and undeveloped countries are compared with world averages in Table 3–8. Because of variation in water content, crops are best compared by caloric content of harvested weights (second column of Table 3–8). A developed country is defined as one with a per capita gross national product (GNP) of more than $1000. About 30 percent of all people live in such countries, which also tend to have a low rate of population growth (1 percent per year or less). In contrast, 65 percent of the world's people live in undeveloped countries that have a per capita GNP of less than $300, often less than $100, and also have a high growth rate (more than 2 percent per year). Undeveloped countries have low per-hectare food production because they are too poor to afford the energy subsidies necessary for high yields. These two masses of humanity are sharply divided—that is, the distribution of per capita income and food production (per unit area) is strongly

Table 3–8 Annual Yields (as of 1977) of Edible Food and Estimated Net Primary Production (P_n) of Major Food Crops of Varying Protein Content at Three Levels of Energy Subsidy*

	Harvest Weight† (kg/ha)	Calorie Content‡ (kcal/m²)	Estimated P_n§ (kcal/m²)
Wheat (about 12% protein)			
1. Netherlands	5200	1715	5150
2. India	1400	460	1380
3. World average	1700	560	1680
Corn (about 10% protein)			
1. U.S.A.	5700	2000	6000
2. India	1100	385	1150
3. World average	2950	1030	3090
Rice (about 10% protein)			
1. Japan	6200	2230	6690
2. Brazil	1450	520	1560
3. World average	2600	950	2850
White potatoes (about 2% protein)			
1. U.S.A.	29250	2650	5300
2. India	11500	1050	2100
3. World average	14000	1250	2500
Sweet potatoes (about 2% protein)			
1. Japan	19500	1750	3500
2. Indonesia	7800	700	1400
3. World average	9650	875	1750
Soybeans (about 30% protein)			
1. Canada	2550	1025	3075
2. Indonesia	1300	525	1575
3. World average	1550	620	1860
Crude sugar (less than 1% protein)			
1. Hawaii (cane)	18362	6740	67400
1a. Netherlands (beets)	6950	2570	25700
2. Cuba (cane)	4920	1820	18200
3. World average (beets and cane)	4140	1435	14350

* (1) Fuel-subsidized industrial agriculture (U.S., Canada, Europe, Japan); (2) little or no fuel subsidy (undeveloped nations); (3) world average.

† Data from *Production Yearbook*, Vol. 31, 1977. Food and Agricultural Organization (FAO), United Nations.

‡ Conversion to kcal/gm harvested weight as follows: wheat, 3.3; corn, 3.5; rice, 3.6; soybeans, 4.0; potatoes, 0.9; crude sugar, 3.7. (See USDA Handbook No. 8.)

§ Established on basis of 3 × edible portions for grains and soybeans, 2 × for potatoes, and 10 × for sugar values as shown in Column 2.

bimodal—since only about 5 percent of people live in what might be called "transitional countries."

The world average crop production is much closer to the lower than the upper range (as is evident from Table 3–8), and in the undeveloped countries, yields are rising no faster than the population, since each year more countries have to import food. As of 1980, only North America and Australia had large food surpluses for export. In the past decade the yield of all the crops in all nations listed in Table 3–8 and world averages have increased an average of about 25 percent; however, the gap between rich and poor countries has not narrowed. For cereal grains, the gap was 1630 kg/ha in 1969 and 1740 kg/ha in 1977 (U.N. *Production Yearbook* 31). Cropland has increased about 15 percent worldwide during the same period, but in Europe, the United States, and Japan, the harvested area has decreased. In a desperate effort to stay above subsistence, undeveloped countries have increased food supply as much by increasing land under cultivation as by increasing yields. If such a trend continues, more and more marginal land will be cultivated at increasing cost and risk of environmental degradation. Furthermore, protein, rather than total calories, tends to limit the diet in the undeveloped world. Under equivalent conditions, the yield of a high-protein crop, such as soybean, must always be less (in total calories) than that of a carbohydrate crop, such as sugar (compare the means of these crops in Table 3–8, and also note that potatoes yield more but have a lower protein content than grains).

Table 3–9A is a more generalized model of food production at three levels that exist today and also at a theoretical level that might be obtained with algae culture or hydroponics (soilless culture in greenhouses) supported by massive subsidies of energy and money. As of 1980, the estimated 4.3 billion people in the world each required about 1 million kcal per year, or a total of 4.3×10^{15} kcal of food energy needed to support the human "biomass." The 6.7×10^{15} kcal of food estimated to be harvested worldwide (Table 3–8B) is inadequate because of poor distribution, waste, and low quality. The origin of human food is shown in Table 3–9B. Only about 1 percent of food comes from the sea, and most of that is of animal origin for reasons already explained (small size and rapid turnover of plants preclude accumulation of harvestable biomass). The ratio of plant to animal food from land agriculture is about 4 to 1. Food harvest comes to about 1 percent of the net or 0.5 percent of the gross primary production of the biosphere. But there is much more to be considered than just the food intake of humans. For example, there is the huge population of domestic animals (cows, pigs, horses, poultry, sheep),

Table 3–9 Yield (Production) of Human Food

A. Edible Portion of Net Primary Production on a Unit Area Basis

Level of Agriculture	kg Dry Matter/ha/year	kcal/m²/year
Food-gathering culture	0.4–20	0.2–10
Agriculture without energy (fuel) subsidy	50–2000	25–1000
Energy-subsidized grain agriculture	2000–20,000	1000–10,000
Theoretical maximum energy-subsidized culture of algae or other continuous cultures	20,000–80,000	10,000–40,000

B. Portion (by Harvest Weight) of Human Food of Plant and Animal Origin Obtained from Land and Sea*

	Total 10^6 Metric Tons	Percent Total	Percent Plant	Percent Animal
Ocean	73	2	0.1	99.1
Land	3300	98	78	22
Total	3373 = 6746×10^{12} kcal†			

* Based on United Nations 1978 *Statistical Yearbook;* round-figure average for three years, 1975 to 1977.

† 10^6 metric tons wet weight converted to 10^{12} kcal by multiplying by 2, since 10^6 mt = 10^{12} gm, and 1 gm wet weight = 2 kcal (see Table 3–1).

most of which are direct consumers of primary production, not only from agricultural lands but also from "wildlands" (grasslands, forests). The standing crop of livestock in the world at large is about five times that of human beings in terms of equivalent food requirements. Borgstrom (1965, 1979) uses the expression "livestock population equivalent" to denote domestic animal biomass equal to a person in food needs. The ratio of livestock to people varies from 43 to 1 in New Zealand to 0.6 to 1 in Japan (where fish largely replace terrestrial meat in the diet), with a world average of 4.7 (Borgstrom, 1979). One might say that the general ecology of the landscape, not to mention the culture and the economy, is determined by the sheep in New Zealand and by the fish in Japan. In large areas of the United States, the landscape ecology and culture are dominated by beef cattle and the macho cowboy!

There are still other uses of primary production, namely for fiber (e.g., cotton) and fuel. For more than half the world's population, wood is the chief fuel used for cooking, heating, and light industry. In the poorest countries, wood is burned much faster than it can be grown, so forests are turned into shrublands and then into deserts.

Echholm (1975) has aptly labeled the shortage of firewood as the "other energy crisis" (oil, of course, is the crisis most talked about). He notes that in the African countries of Tanzania and Gambia per capita fuel-wood consumption is about 1.5 tons per year, and 99 percent of the population uses wood as fuel.

In North America and other regions with large standing stocks of vegetation, more people have become interested in using biomass from both forest and agricultural lands as fuel to supplement or replace dwindling supplies of petroleum. Among the options available are (1) planting fast-growing trees (pines, sycamores, poplars, among others) harvested on short rotation (clearcut and replanted in ten years or less), the so-called "fuel forests"; (2) using limbs and other parts of trees not suitable for lumber or paper that are now left in the woods to decompose; (3) reducing pulp demand by recycling paper and using pulp wood instead for home heating and electricity generation; (4) using agricultural plant and animal wastes (manure) to produce methane gas or alcohol; and (5) growing crops such as sugar cane and corn specifically for alcohol production to be used to fuel internal combustion engines.

Though all these options can certainly ease fuel shortages for the short term, all except the third have drawbacks in that the quality of soils could be adversely affected, or competition between food and fuel for arable land could worsen the world's already critical food situation. Agronomist Hans Jenny (1980) points out that the one-third or so organic residues remaining after crop or tree harvest (note Table 3–5) are extremely important in maintaining the fertility and water-holding capacity of soils. He cites European studies showing that removal of forest litter and crop residues results in a loss of soil texture and reduced yields; increased use of mineral fertilizers can partly restore the yields but not the soil texture. Jenny thus argues against the indiscriminate conversion of biomass and organic wastes to fuel, because in the long run the "humus capital" is more valuable than fuel, especially since there are other sources of fuel but not of humus.

Converting high-quality food such as corn to alcohol fuel does not make good ecological sense. Several recent studies (Weisz and Marshall, 1979; Chambers et al., 1979; Hopkinson and Day, 1980) have shown that as much or more high-quality energy is required to produce the alcohol, resulting in little or no net energy gain (see Section 6 of this chapter). Brown (1980) estimates that it would take 8 acres to grow enough grain to run the average American automobile for one year, whereas such an area could feed ten to 20 people. For the most part, gasohol (mixture of gasoline and alcohol) is being marketed in the grain belt of the United States because there is a surplus of grain that cannot be eaten (by humans or animals) or sold on the world market (the hungry cannot buy). Such a situation is not likely to continue.

From the holistic, long-term view, the use of primary production as fuel could replace only a small portion of current petroleum use, since biomass production worldwide amounts to only about 1 percent of total solar radiation.

Biomass fuel from special crops grown for that purpose, or perhaps from materials that are truly being wasted or cannot be used as food (as noted, litter in the forest and mulch on the field are not "waste," but serve a vital function in maintaining ecosystem order against the thermodynamic drain) can perhaps supplement other sources in regions with productive forests, surplus land, and low human populations. Excessive enthusiasm for this form of solar energy in the industrial nations will only place them in the same plight as today's undeveloped nations.

The human impact on the biosphere may be seen in another way. Human density is now one person to about 3.25 hectares (8 acres) of land (i.e., 4.3×10^9 people on 14.0×10^9 hectares of land). When the domestic animals are included, the density is one population equivalent to about 0.65 hectare (i.e., 21.5×10^9 population equivalents on 14.0×10^9 hectares of land). This is less than 2 acres for every person and person-sized domestic animal consumer. If the population doubles in the next century, and if we wish to continue to eat and use animals, there will be only about 1 acre (0.3 hectare) to supply all the needs (water, oxygen, minerals, fibers, biomass fuels, living space, as well as food) for each 50-kilogram consumer, and this does not include pets and wildlife that contribute so much to the quality of human life.

Within a decade, expert panels commissioned by the President's Science Advisory Committee (PSAC) and the National Academy of Science (NAS) have twice assessed the "World Food and Nutrition Problem." The first assessment (PSAC) was published in 1967 (three volumes) and the second (NAS) in 1977 (six volumes). These reports (see also Barr, 1981) are "cautiously optimistic" that global famine can be avoided if the United States and the world "dig in now," which means doubling food production by the end of this century, improving use and distribution of supplies, and sharply reducing the human birth rate. Well-known and highly respected writers Lester Brown (1975, 1978a), Paul Ehrlich (1977), and Georg Borgstrom (1979) are more pessimistic that these "tall orders" can be met anytime soon. All agree that partisan politics and self-serving economics are the chief obstacles. Most agroecologists believe that too much emphasis has been placed on the monoculture of annuals. It makes good ecological and also common sense to consider increasing the diversity of crops, multiple cropping, limited till procedures (less disturbance of soil structure), and use of perennial species. For more on relationships between energy and food production, see J. N. Black (1971) and Pimentel et al. (1973, 1975, 1976).

4. Food Chains, Food Webs, and Trophic Levels

Statement

The transfer of food energy from the source in autotrophs (plants) through a series of organisms that consume and are consumed is called the **food chain.** At each transfer, a proportion (often as much as 80 or 90 percent) of the potential energy is lost as heat. Therefore, the shorter the food chain, or the nearer the organism to the beginning of the chain, the greater the energy available to that population. Food chains are of two basic types: the **grazing food chain**, which, starting from a green plant base, goes to grazing herbivores (i.e., organisms eating living plant cells or tissue) and on to carnivores (i.e., animal eaters); and the **detritus food chain**, which goes from nonliving organic matter into microorganisms and then to detritus-feeding organisms (detritivores) and their predators. Food chains are not isolated sequences; they are interconnected. The interlocking pattern is often spoken of as the **food web.** In complex natural communities, organisms whose nourishment is obtained from the sun by the same number of steps are said to belong to the same trophic level. Thus, green plants (the producer level) occupy the first trophic level, plant-eaters the second level (the primary consumer level), primary carnivores the third level, and secondary carnivores the fourth level (the tertiary consumer level). *This trophic classification is one of function and not of species as such.* A given species population may occupy one or more than one trophic level according to the source of energy actually assimilated. The energy flow through a trophic level equals the total assimilation (A) at that level, which, in turn, equals the production (P) of biomass and organic matter plus respiration (R).

Explanation

Food chains are more or less vaguely familiar to everyone, since we eat the big fish that ate the little fish that ate the zooplankton that ate the phytoplankton that fixed the sun's energy; or we may eat the cow that ate the grass that fixed the light energy; or we may use a much shorter food chain by eating the grain that fixed the sun's energy. In the latter case, human beings function as primary consumers at the second trophic level. In the grass-cow-human food chain, we function at the third trophic level (secondary consumer). In general, humans tend to be both primary and secondary consumers since our diet most often comprises mixtures of plant and animal food. Accordingly, energy flow is divided between two or more trophic levels in proportion to percent of plant and animal food eaten.

The lay person usually fails to recognize that potential energy is lost at each food transfer. Only a small portion of the available solar energy is fixed by the plant in the first place. Consequently, the number of consumers, such as people, that can be supported by a given output of primary production very much depends on the length of the food chain; each link in our traditional agricultural food chain decreases the available energy by about one order of magnitude (order of 10). Therefore, fewer people can be supported when large amounts of meat are part of the diet. Meat will disappear, or be very much reduced, if there are a great many mouths to feed with a given primary production base.

The principles of food chains and the workings of the two laws of thermodynamics are clarified by the flow diagrams shown in Figures 3–9, 3–10, and 3–11. In these diagrams the boxes represent successive trophic levels, and the "pipes" or lines connecting them depict the energy flow in and out of each level. Energy inflows balance outflows, as required by the first law of thermodynamics, and each energy transfer is accompanied by dispersion of energy into unavailable heat (i.e., respiration), as required by the second law.

A very simplified model of energy flow for the three trophic levels is presented (Figure 3–9). This diagram introduces standard notations for the different flows and illustrates how the energy flow is greatly reduced at each successive level regardless of whether the total flow (I and A) or the components P and R are considered. Also shown are the double metabolism of producers (i.e., gross and net production) and the approximately 50 percent absorption–1 percent conversion of light at the first trophic level (compare Table 3–4). Secondary productivity (P_2 and P_3 in the diagram) is about 10 percent at successive consumer trophic levels, although efficiency tends to be

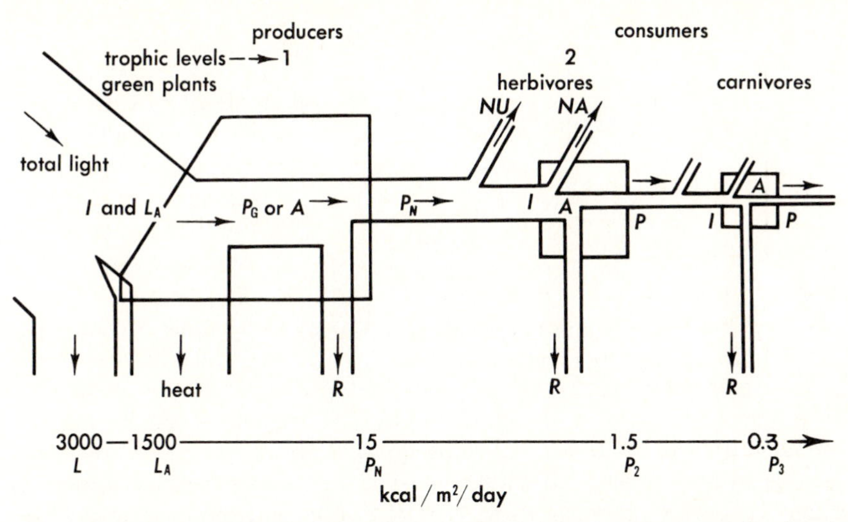

Figure 3–9 A simplified energy flow diagram depicting three trophic levels (boxes numbered 1, 2, 3) in a linear food chain. Standard notations for successive energy flows are as follows: I = total energy input; L_A = light absorbed by plant cover; P_G = gross primary production; A = total assimilation; P_N = net primary production; P = secondary (consumer) production; NU = energy not used (stored or exported); NA = energy not assimilated by consumers (egested); R = respiration. Bottom line in the diagram shows the order of magnitude of energy losses expected at major transfer points, starting with a solar input of 3000 kcal per square meter per day. (After E. P. Odum, 1963.)

higher, say 20 percent, at the carnivore levels, as shown in Figure 3–9. Where the nutritional quality of the energy source is high (for example, photosynthate withdrawn or exuded directly from plant tissues), transfer efficiencies can be much higher. However, since both plants and animals produce a lot of hard-to-digest organic matter (cellulose, lignin, and chitin) together with chemical inhibitors (that discourage would-be consumers), the average transfers between whole trophic levels average 20 percent or less.

In Figure 3–10, the grazing and detritus food chains are shown as separate flows in a Y-shaped, or two-channel, energy flow diagram. This model is more realistic than the single-channel model because (1) it conforms to the basic stratified structure of ecosystems; (2) direct consumption of living plants and utilization of dead organic matter is usually separated in both time and space; and (3) the macroconsumers (phagotrophic animals) and the microconsumers (saprotrophic bacteria and fungi) differ greatly in size-metabolism relations and in techniques required for study.

The portion of net production energy that flows down the two pathways varies in different kinds of ecosystems and often varies

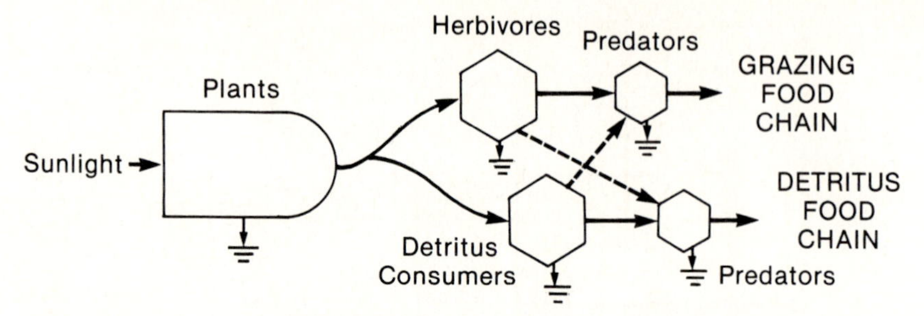

Figure 3–10 The Y-shaped energy flow model showing linkage between the grazing and detritus food chains.

seasonally or annually in the same ecosystem. In some shallow water (as in Riley's classic 1956 study of Long Island Sound) and in a heavily grazed pasture or grassland, 50 percent or more of the net production may pass down the grazing path. In contrast, marshes, oceans, forests, and indeed most natural ecosystems operate as detrital systems in that 90 percent or more of autotrophic production is not consumed by heterotrophs until the leaves, stems, and other plant parts die and become "processed," as it were, into particulate and dissolved organic matter in water, sediments, and soils. As emphasized in Chapter 2, delayed consumption in this manner increases the structural complexity as well as the storage and buffering capacities of ecosystems.

In all ecosystems, the grazing and detritus food chains are interconnected, so shifts in flows can occur quickly in response to forcing function inputs from outside the system. Not all food eaten by grazers is actually assimilated; some—undigested material in feces, for example—is diverted to the detritus route. The impact of the grazer on the community depends on the rate of removal of living plant material, not just on the amount of energy in the food that is assimilated. Marine zooplankton commonly "graze" more phytoplankton than they can assimilate; the excess is egested to the detritus food chain. As already discussed in the section on primary productivity, direct removal of more than 30 to 50 percent of the annual plant growth by terrestrial grazing animals or by mowing makes the ecosystem less able to resist future stress.

The many mechanisms in nature that control or reduce grazing are as impressive as humanity's past ability to control domestic grazing animals is unimpressive. Overgrazing has contributed to the decline of past civilizations. The choice of words is important here. Overgrazing, by definition, is detrimental, but what constitutes over-

Table 3–10 **Grazing Intensity Model***

	Intensity of Grazing		
	Light (cow/21 acres)	Moderate (cow/15 acres)	Heavy (cow/9 acres)
Mean cow standing crop at end of growing season (lb/acre)	321	405	648
Mean weight of individual cows at end of growing season (lb)	1003	942	912
Mean weight at weaning of calves produced (lb)	382	363	354
Forage utilization; percent above-ground vegetation grazed during growing season	37%	46%	63%
Condition of range at end of the nine-year study†	Improved	Unchanged	Deteriorated

* Data based on a nine-year study at Cottonwood, South Dakota, reported by Johnson et al. (1951).

† Based on relative abundance of "decreaser" plants, that is, palatable species that tend to decrease with increasing grazing pressure.

grazing in different kinds of ecosystems is only now being defined in terms of energetics as well as in terms of long-range economics.

Table 3–10 summarizes the results of a long-term study of the impact of three different cattle-grazing rates on natural range land in the Great Plains. During the first several years of a nine-year study, the yield of meat was highest on the heavily stocked range, where steers removed more than 50 percent of annual net production, but after the nine years, the range was so badly deteriorated that all stock had to be removed (yield became zero). In contrast, not only were yield and good range condition maintained in the low and moderate stocking levels, but "quality" of stock was higher, as shown by the larger size of individual calves at the end of the season. The question of optimum versus maximum consumption is discussed further in the subsequent section on carrying capacity.

Undergrazing can also be detrimental. In the complete absence of direct consumption of living plants, detritus could accumulate faster than microorganisms could decompose it, thereby delaying mineral recycling and perhaps making the system vulnerable to fires.

Energy flows originating from nonliving organic materials involve several distinct food chain pathways, as shown in Figure

3–11A. What was labeled the detritus pathway in Figure 3–10 is subdivided into three flows in the graphic model of Figure 3–11. One flow, often the dominant one, originates with particulate organic matter (POM); the other two paths start with dissolved organic matter (DOM). Fungi called mycorrhiza (see Figure 7–12), aphids, and parasites and pathogens actively extract photosynthate directly from the plant's vascular system or tissues, whereas the great majority of saprotrophic microorganisms utilize the DOM, which extrudes or "leaks out" from cells, roots, and so on. Recent studies have shown that these two DOM pathways may account for a large percent of total energy flow in the ocean and in forests with well-developed mycorrhizal systems.

Two distinct subsystem food chains are largely restricted to terrestrial or shallow water ecosystems, as shown in Figure 3–11A: a **granivorous food chain** originating from seeds, high-quality energy sources that are major food items for animals as well as humans, and the **nectar food chain** originating from the nectary of flowering plants that depend on insects and other animals for pollination. The intricate and mutualistic relationships that have evolved between plants and pollinators and plants and granivores are discussed in Chapter 6.

Finally, Figure 3–11B shows yet another way to picture food chain routes, this one especially applicable to aquatic environments. The anaerobic pathway discussed in some detail in Chapter 2 is shown as a separate route along with the direct grazing, the DOM, and the POM flows. All four pathways of energy flow are important, but just how much primary energy goes down each of the four routes in different kinds of ecosystems is unknown. Recent work by Jorgenson and Fenchel (1974), Rich and Wetzel (1978), and Howarth and Teal (1980) indicates that the reduced sulfur food chain is a major energy flow pathway in salt marshes and other shallow water ecosystems. More important, ecologists understand little of how natural and anthropogenic forces influence the energy partitioning among these or other routes. These are major research challenges for the immediate future.

Resource Quality The quality of the resource is as important a consideration as the quantity of energy flow involved in the different food chains (E. P. Odum and Biever, 1983). For example, the resource quality of the photosynthate extracted by mycorrhizal fungi is much greater than that of dead leaves in terms of ease of assimilation. Accordingly, transfers along the mycorrhizal pathway are rapid, and the assimilation efficiency is high. It is also important to note that all food chains have feedback potential in that consumers often transport nutrients

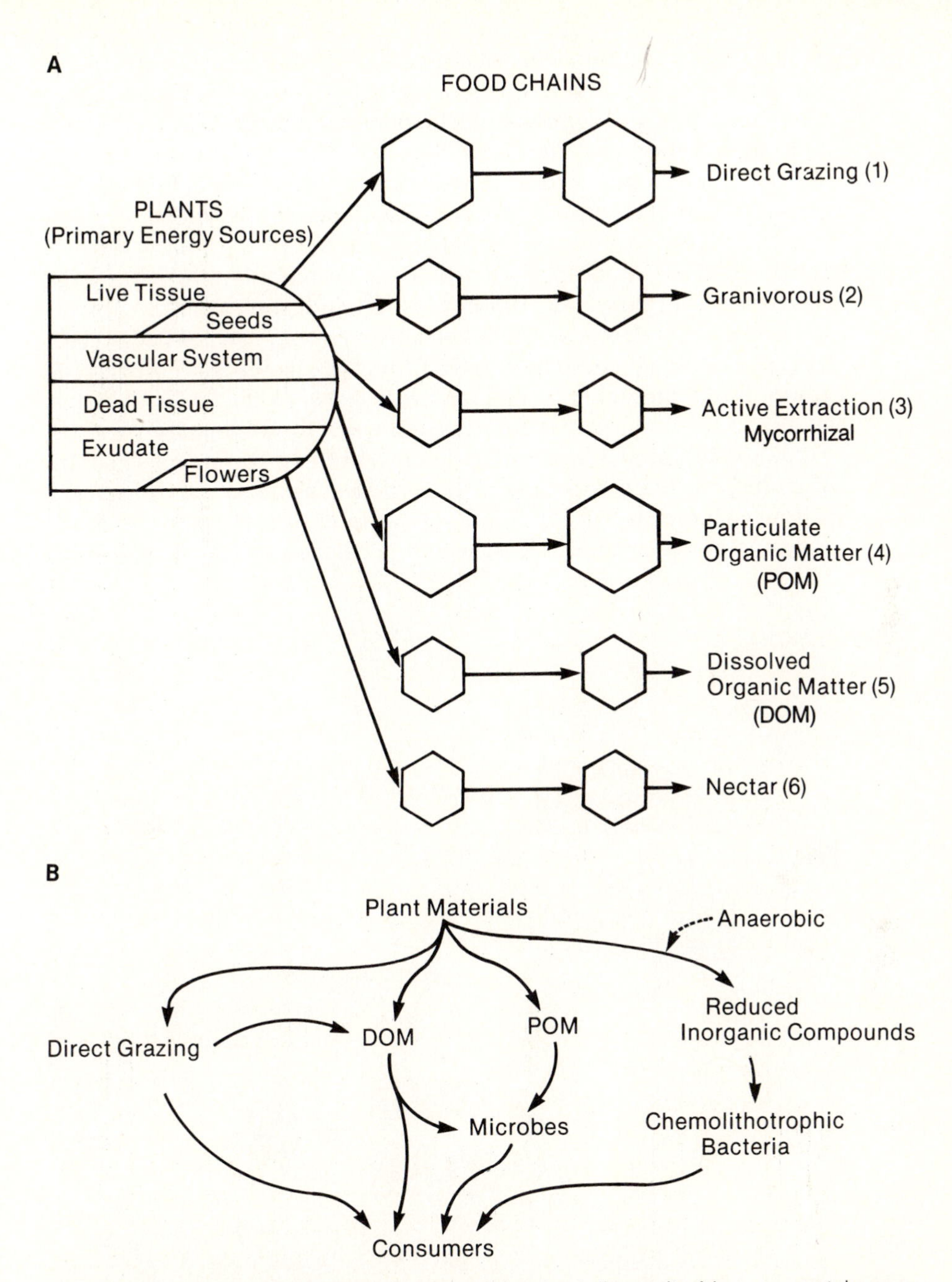

Figure 3–11 Multi-channel food chain models especially applicable to terrestrial ecosystems (*A*) and to aquatic ecosystems (*B*). DOM = dissolved organic matter; POM = particulate organic matter.

and disseminules or produce hormones that affect the plant source, often in a beneficial manner. Returning to the mycorrhizal example, these fungi transport mineral nutrients to plant roots in return for the high-quality food obtained from the plant. The producer as well as the consumer benefits from the allocation. More about this kind of mutualism later in this section and in Chapter 7, Section 4.

A Universal Energy Flow Model What is the basic component of an energy flow model? Figure 3–12 presents what might be called a "universal" model, one that is applicable to any living component whether it be plant, animal, microorganism, individual, population, or trophic group. Linked together, such graphic models can depict food chains, as already shown, or the bioenergetics of an entire ecosystem. In Figure 3–12, the shaded box labeled *B* represents the living structure of biomass of the component. Although biomass is usually measured as some kind of weight (living [wet] weight, dry weight, or ash-free weight), biomass should be expressed in calories so that relationships

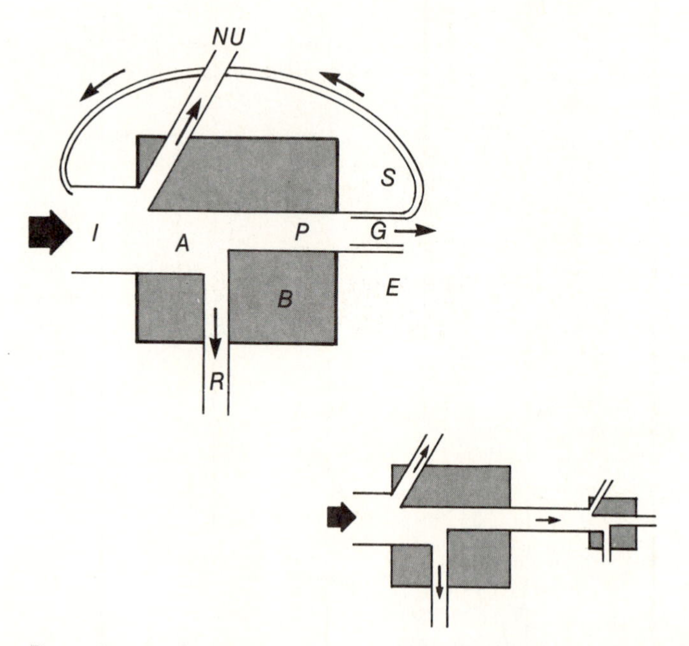

Figure 3–12 Components for a "universal" model of ecological energy flow. I = input or ingested energy; NU = not used; A = assimilated energy; P = production; R = respiration; B = biomass; G = growth; S = stored energy; E = excreted energy. See text for explanation. (After E. P. Odum, 1968.)

between the rates of energy flow and the instantaneous or average standing-state biomass can be established. The total energy input or intake is indicated by *I* in Figure 3–12. For strict autotrophs, this is light; for strict heterotrophs, it is organic food. As discussed in Chapter 2, some species of algae and bacteria can use both energy sources, and many may require both in certain proportions. A similar situation holds for certain invertebrate animals and lichens, which contain mutualistic algae. In such cases, the input flow in the energy flow diagram can be subdivided accordingly to show the different energy sources, or the biomass can be subdivided into separate boxes if one wishes to keep everything in the same box at the same energy level (i.e., the same trophic level).

Such flexibility in usage can be confusing. Again, the concept of trophic level is not primarily intended for categorizing species. Energy flows stepwise through the community according to the second law of thermodynamics, but a given population of a species may be (and very often is) involved in more than one step or trophic level. Therefore, the universal model of energy flow illustrated in Figure 3–12 can be used in two ways. The model can represent a species population, in which case the appropriate energy inputs and links with other species would be shown as a conventional species-oriented food-web diagram (see Figure 3–13). Or, the model can represent a discrete energy level, in which case the biomass and energy channels represent all or parts of many populations supported by the same energy source. Foxes, for example, usually obtain part of their food by eating plants (fruit, for example) and part by eating herbivorous animals (such as rabbits or field mice). A single box diagram could be used to represent the whole population of foxes if intrapopulation energetics were to be stressed. On the other hand, two or more boxes (such as shown in the lower right of Figure 3–12) would be employed should the metabolism of the fox population be divided into two trophic levels according to the proportion of plant and animal food consumed. In this way the fox population can be placed into the overall pattern of energy flow in the community. When an entire community is modeled, one cannot mix these two usages unless all species happen to be restricted to single trophic levels, as is the case in an African grass-zebra-lion food chain.

So much for the problem of the source of the energy input. Not all of the input into the biomass is transformed; some of it may simply pass through the biological structure, such as when food is egested from the digestive tract without being metabolized or when light passes through vegetation without being fixed. This energy component is indicated by *NU* (not utilized). That portion used or assimilated is indicated by A in the diagram. The ratio between A and I, i.e., the efficiency of assimilation, varies widely. It may be very low, as in light fixation by plants or food assimilation in detritus-feeding animals, or very high, as when animals or bacteria consume high-energy food such as sugars and amino acids.

In autotrophs, the assimilated energy (A) is, of course, gross production or gross photosynthesis. The analogous component (the A component) in heterotrophs, as already pointed out (see page 99), represents food already produced somewhere else. Therefore, the term "gross production" should be restricted to primary or autotrophic production. In higher animals, the term "metabolized energy" is often used for the A component (Kleiber and Dougherty, 1934; Kendeigh, 1949).

A key feature of the model is the separation of assimilated energy into the P and R components. That part of the fixed energy (A) burned and lost as heat is designated respiration (R); that portion transformed to new or different organic matter is designated production (P). This is net production in plants or secondary production in animals. The P component is energy available to the next trophic level, compared with the NU component still available at the same trophic level.

The ratio between P and R and between B and R varies widely and is ecologically significant, as explained in Section 1 of this chapter as well as in Chapter 2 (pages 59 and 61). In general, the proportion of energy going into respiration (i.e., maintenance energy) is large in populations of large organisms, such as people and trees, and in communities with a large standing crop biomass. R increases when a system is stressed. Conversely, the P component is relatively large in active populations of small organisms, such as bacteria or algae, in youthful, rapidly growing communities, and in systems benefiting from energy subsidies. The relevance of P/R ratios to food production for humans was mentioned in Section 3 of this chapter and will be noted again in Chapter 8.

Production may take a number of forms, as already modeled in Figure 3–11A. Three subdivisions are shown in Figure 3–12: G refers to growth or additions to the biomass; E to assimilated organic matter excreted, secreted or extracted; and S to storage, as in the accumulation of fat, which may be reassimilated at some later time. The reverse S flow shown in Figure 3–12 may also be considered a work loop since it depicts that portion of production needed to ensure a future input of the new energy, for example, reserve energy used by a predator in the search for prey or energy in the feces or excretion of a grazer that fertilizes the plants on which the grazer feeds.

Examples

Four examples should suffice to illustrate major features of food chains, food webs, and trophic levels. First, in the far north, in the region known as tundra, only relatively few kinds of organisms have become successfully adapted to low temperatures. Food chains and food webs are thus relatively simple. The pioneer British ecologist Charles Elton realized this early and during the 1920s and 1930s studied the ecology of arctic lands. He was one of the first to clarify the principles and concepts relating to food chains (Elton, 1927). Plants on the tundra, the reindeer lichens (or "moss") (*Cladonia*), grasses, sedges, and dwarf willows provide food for the caribou of the North American tundra and for its ecological counterpart, the reindeer of the Old World tundra. These animals, in turn, are preyed upon by wolves and humans. Tundra plants are also eaten by lemmings—shaggy-haired voles with short tails and a bearlike appearance—and the ptarmigan or arctic grouse. Throughout the long winter, as well as during the brief summer, the arctic white fox and the snowy owl and other birds prey largely on the lemming. Any radical change in the numbers of lemming affects the other trophic levels, because the alternative choices of food are few. That is why the numbers of some groups of arctic organisms fluctuate greatly from superabundance to near extinction. The same has often happened to human civilizations that depended on a single or on relatively few local food items, as, for example, in the Irish potato famine.

Food chains, of course, function on small as well as large scales. Food webs delineated for small organisms of a stream community and for the North Sea, one of the most studied and heavily fished

Unlike the wild caribou, domesticated reindeer do not migrate. In Lapland, reindeer are herded from one place to another to prevent overgrazing, but herding is not a part of the culture of Alaskan Indians and Eskimos (since caribou did their own herding). As a result, reindeer introduced into Alaska tend to remain in one place or are fenced in; they severely overgraze many areas and reduce the carrying capacity for caribou as well. This is a good example of failing to include a major component, namely the pasture rotation procedure, when organisms are introduced into a new area. Introduced animals often become severe pests when their natural or anthropogenic control mechanisms are not also introduced.

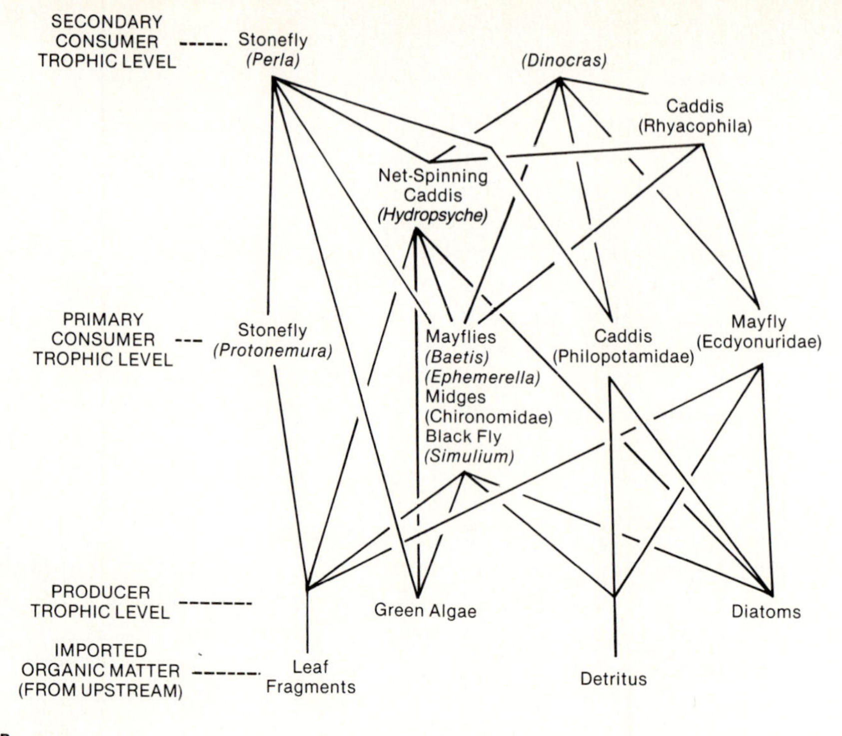

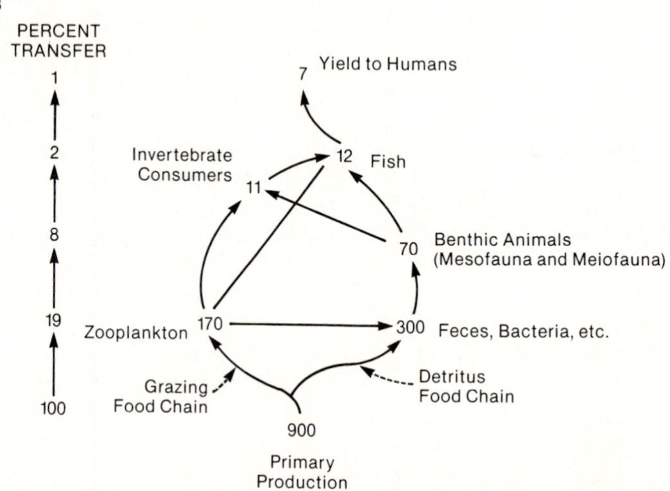

Figure 3–13 *A.* A portion of a food web in a small stream community in South Wales. The diagram illustrates (1) the interlocking of food chains to form the food web, (2) three trophic levels, (3) the fact that organisms, such as *Hydropsyche,* may occupy an intermediate position between major trophic levels, and (4) an "open" system in which part of the basic food is "imported" from outside the stream. (Redrawn from Jones, 1949.) *B.* North Sea food web in terms of kilocalories per square meter transferred up the grazing and detritus food chains. (Based on Steele, 1970.)

areas of the world, are shown in Figure 3–13. Figure 3–13A illustrates the interlocking of food chains to form a food web, three trophic levels, and two sources of primary energy, that produced within the system (by the algae) and that imported from outside (leaf detritus). The diagram also shows how some organisms, such as net-spinning caddis, derive their food energy from more than one trophic level. The North Sea diagram (Figure 3–13B) shows how the estimated annual primary productivity of 900 kcal/m^2 is divided between grazing (zooplankton) and detritus (benthos) pathways that support fish production, 1 percent of which is consumed by people.

A farm pond managed for sport fishing, thousands of which have been built over the country, is an excellent example of food chains under fairly simplified conditions. Since a fish pond is supposed to provide the maximum number of fish of a particular species and a particular size, management procedures are designed to channel as much of the available energy as possible into the final product by restricting the producers to one group, the floating algae or phytoplankton. (Other green plants such as rooted aquatics and filamentous algae are discouraged.) Figure 3–14 is a compartment model of a sport-fishing pond in which transfers at each link in the food chain are quantitated in terms of kilocalories per square meter per year. In this model, only the successive inputs of ingested energy with respect to time, $i(t)$, are shown; losses during respiration and assimilation are not shown. The phytoplankton is fed upon in turn by the zooplankton crustacea in the water column, and the plankton detritus is taken by certain benthic invertebrates, notably bloodworms (chironomids), which are the preferred food of sunfishes; these in turn are fed upon by bass. The balance between the last two groups in the food chain (sunfish and bass) is very important to harvesting of fish by humans. A pond with sunfish as the only fish could actually produce a greater total weight or biomass of fish than one with bass and sunfish, but most of the sunfish would remain small because of the high reproduction rate and the competition for available food. Fishing by hook and line would soon be poor. Since the sportsman wants large fish, the final predator is necessary for a good sport-fishing pond. Emanuel and Mulholland (1976) have developed an environmental control model based on Welch's study that can be used to calculate the optimum fertilization rate for sport fishing.

Fish ponds are good places to demonstrate how secondary productivity is related to (1) the length of the food chain, (2) the primary productivity, and (3) the nature and extent of energy imports from outside the pond system. As shown in Table 3–11, the large lakes and the sea yield fewer fish per acre, or per square meter, than do small,

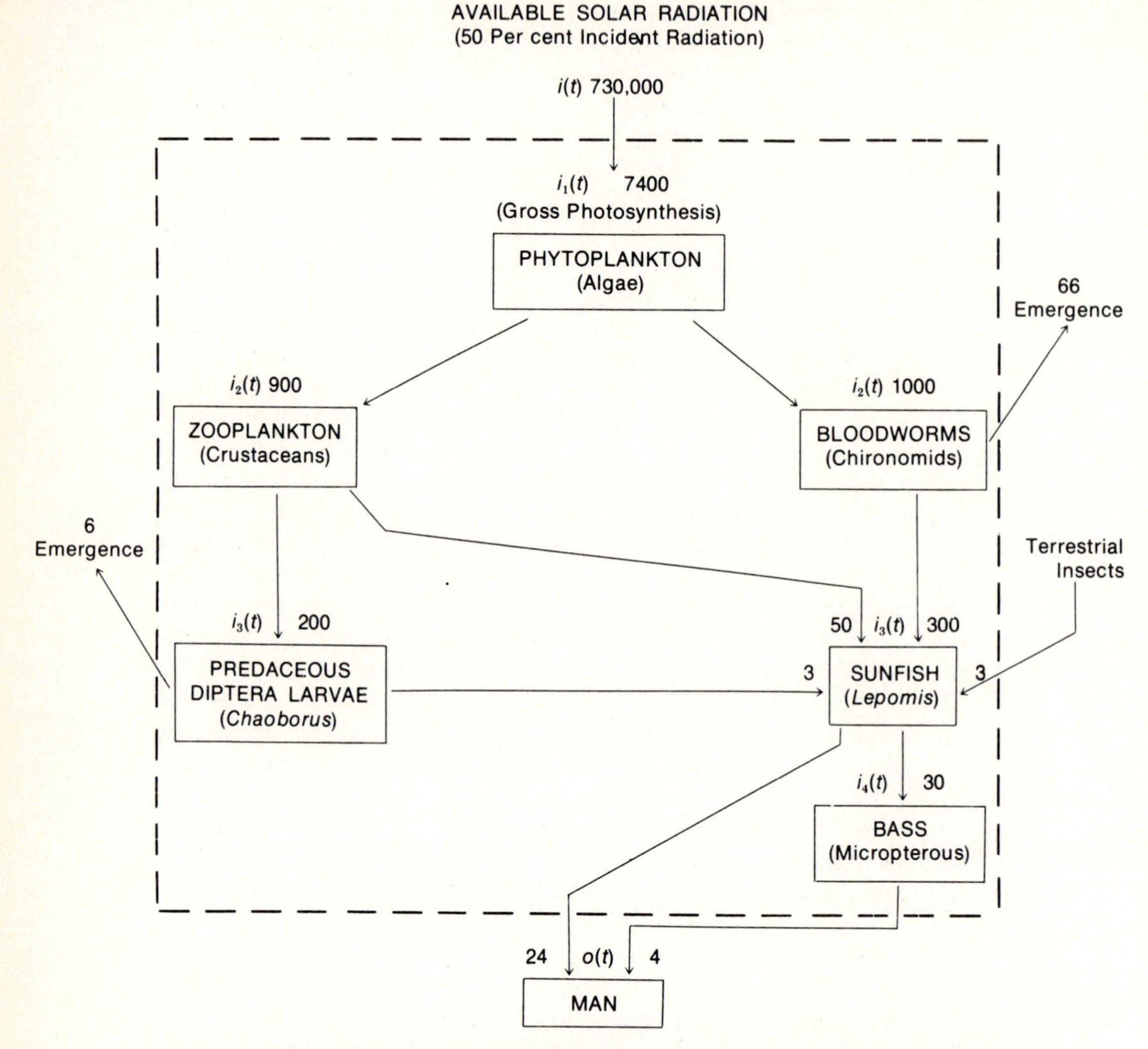

Figure 3–14 Compartment model of the principal food chains in a Georgia pond managed for sport fishing. Estimated energy inputs (i) with respect to time (t) are kilocalories per square meter per year. $i_1(t)$, $i_2(t)$, $i_3(t)$, and $i_4(t)$ represent ingested food energy at successive trophic levels; losses during assimilation and respiration are not shown. $o(t)$ is output from the pond in terms of caloric value of fish caught by humans. The model suggests an interesting possibility that fish production might be increased if the "side food chain" through *Chaoborus* were eliminated, but the possibility that this side chain enhances the stability of the system must also be considered. (Data from Welch, 1967, with his estimate of assimilated energy at the i_2 level changed to estimated ingested energy on the basis of a 60 percent assimilation efficiency for zooplankton and 40 percent for bloodworms.)

Table 3–11 Secondary Productivity as Measured in Fish Production

Ecosystem and Trophic Level	Human Harvest lb/acre/year	Human Harvest kcal/m²/year
I. Unfertilized Natural Waters		
Mixed Carnivores (Natural Populations)		
World marine fishery (average)*	1.5	0.3
North Sea†	27.0	5.0
Great Lakes‡	1–7	0.2–1.6
African lakes§	2–225	0.4–50
U.S. small lakes‖	2–160	0.4–36
Stocked Carnivores		
U.S. fish ponds (sports fishery)§	40–150	9–34
Stocked Herbivores		
German fish ponds (carp)¶	100–350	22–80
II. Peru Current Upwelling Area (Anchovies)		
Heavy Natural Fertilization#	1500	335
III. Artificially Fertilized Waters		
Stocked Carnivores		
U.S. fish ponds (sports fishery)**	200–500	45–112
Stocked Herbivores		
Philippine marine ponds (milkfish)§	500–1000	112–202
German fish ponds (carp)¶	1000–1500	202–336
IV. Fertilized Waters—Outside Food Added		
Carnivores		
One-acre pond, U.S.¶	2000	450
Herbivores		
Hong Kong§	2000–4000	450–900
South China§	1000–13500	202–3024
Malaya§	3500	785

* 60×10^6 metric tons harvested (FAO 1967 Prod. Handb.) from 360×10^6 km² total ocean.

† FAO Statistics.

‡ Rawson, 1952.

§ Hickling, 1948.

‖ Rounsefell, 1946.

¶ Viosca, 1936.

World's most productive natural fishery, 10^7 metric tons from 6×10^{10} m² (Ryther, 1969).

** Swingle and Smith, 1947.

fertilized, and intensively managed ponds, not only because primary productivity is less and food chains longer but also because only a part of the consumer population, the marketable species, is harvested in large bodies of water. Likewise, yields are several times greater when herbivores, such as carp, are stocked as when carnivores, such as bass, are harvested; the latter, of course, require a longer food chain. The high yields listed in Section IV of Table 3–11 are obtained by adding food from outside the ecosystem, that is, plant or animal products that represent energy fixed somewhere else. Actually, such yields should not be expressed by the area unless one adjusts the area to include the land from which the supplemental food was obtained. Many people, misinterpreting the high yields obtained in the Orient, think that these yields can be compared directly with fish pond yields in the United States, where outside food is not usually provided. As might be expected, fish culture depends on the human population density. Where people are crowded and hungry, ponds are managed for their yields of herbivores or detritus consumers; yields of 1000 to 1500 pounds per acre are easily obtainable without supplemental feeding. Where people are neither crowded nor hungry, sport fish are desired. Since these fish are usually carnivores produced at the end of a long food chain, yields are much less—100 to 500 pounds per acre. Finally, the 300 kcal/m^2/year fish yield from the most fertile natural waters or ponds managed for short food chains approaches the 10 percent conversion of net primary production to primary consumer production (compare Tables 3–4C and 3–6 with Table 3–10), as suggested by the generalized model of Figure 3–9.

Size of Organisms in Food Chains

Besides the operation of the second law of thermodynamics, size of food is one of the main reasons for the existence of food chains (see Elton, 1927) because there are usually rather definite upper and lower limits to the size of food that can efficiently support a given animal type. Size is also involved in the difference between a predator chain and a parasite chain. In the parasite chain, organisms at successive levels are smaller and smaller instead of being generally larger and larger. Thus, roots of vegetable crops are parasitized by nematodes, which may be attacked by bacteria or other smaller organisms. Mammals and birds are commonly parasitized by fleas, which in turn have protozoan parasites of the genus *Leptomonas*. However, from the energy standpoint, predator and parasite chains do not differ fundamentally, since both parasites and predators are "consumers." For this reason, in the energy flow diagrams, a parasite of a green plant would have the same position as a herbivore, whereas animal parasites would fall in the various carnivore categories. Theoretically, parasite chains should be shorter on the average than predator chains, since the metabolism per gram increases sharply with the diminishing size of the organism, resulting in a rapid decline in the biomass that can be supported.

The Detritus Food Chain

A good example of a detritus food chain based on mangrove leaves was described by W. E. Odum and Heald (1972, 1975). In southern Florida, leaves of the red mangrove (*Rhizophore mangle*) fall into the brackish waters at an annual rate of 9 metric tons per hectare (about 2.5 gm or 11 kcal/m^2/day) in areas occupied by stands of mangrove trees. Since only 5 percent of the leaf material is removed by grazing insects before leaf abscission, most of the annual net production becomes widely dispersed by tidal and seasonal currents over many square miles of bays and estuaries. As shown in Figure 3–15A, a key group of small animals, often called *meiofauna* (= diminutive animals), comprising only a few species but many individuals, ingest large quantities of the vascular plant detritus along with the associated microorganisms and smaller quantities of algae. The meiofauna in estuaries generally comprise small crabs, shrimp, nematodes, polychaete worms, small bivalves and snails, and, in less salty waters, insect larvae. The particles ingested by these algal-detritus consumers range from sizable leaf fragments to tiny clay particles on which organic matter has been sorbed. These particles pass through the guts of many individual organisms and species in succession (the process of coprophagy; see page 43), resulting in repeated removal and regrowth of microbial populations (or repeated extraction and reabsorption of organic matter) until the substrate has been exhausted. Protein enrichment (see page 40) presents a difficult technical problem that has not been solved (but see later section on isotopic tracers). Such information is not necessary to model the system, however, since one can consider the whole group as a convenient "black box," as shown in Figure 3–15B. So far as is now known, the model of Figure 3–15B can serve as well for a forest or grassland as it can for an estuary. The flow patterns would be expected to be the same; only the species would be different. Detrital systems enhance nutrient regeneration and recycling because plant, microbial, and animal components are tightly coupled so that nutrients are rapidly reabsorbed as soon as they are released.

Ecological Efficiencies

Ratios between energy flow at different points along the food chain are of considerable ecological interest. Such ratios, when expressed

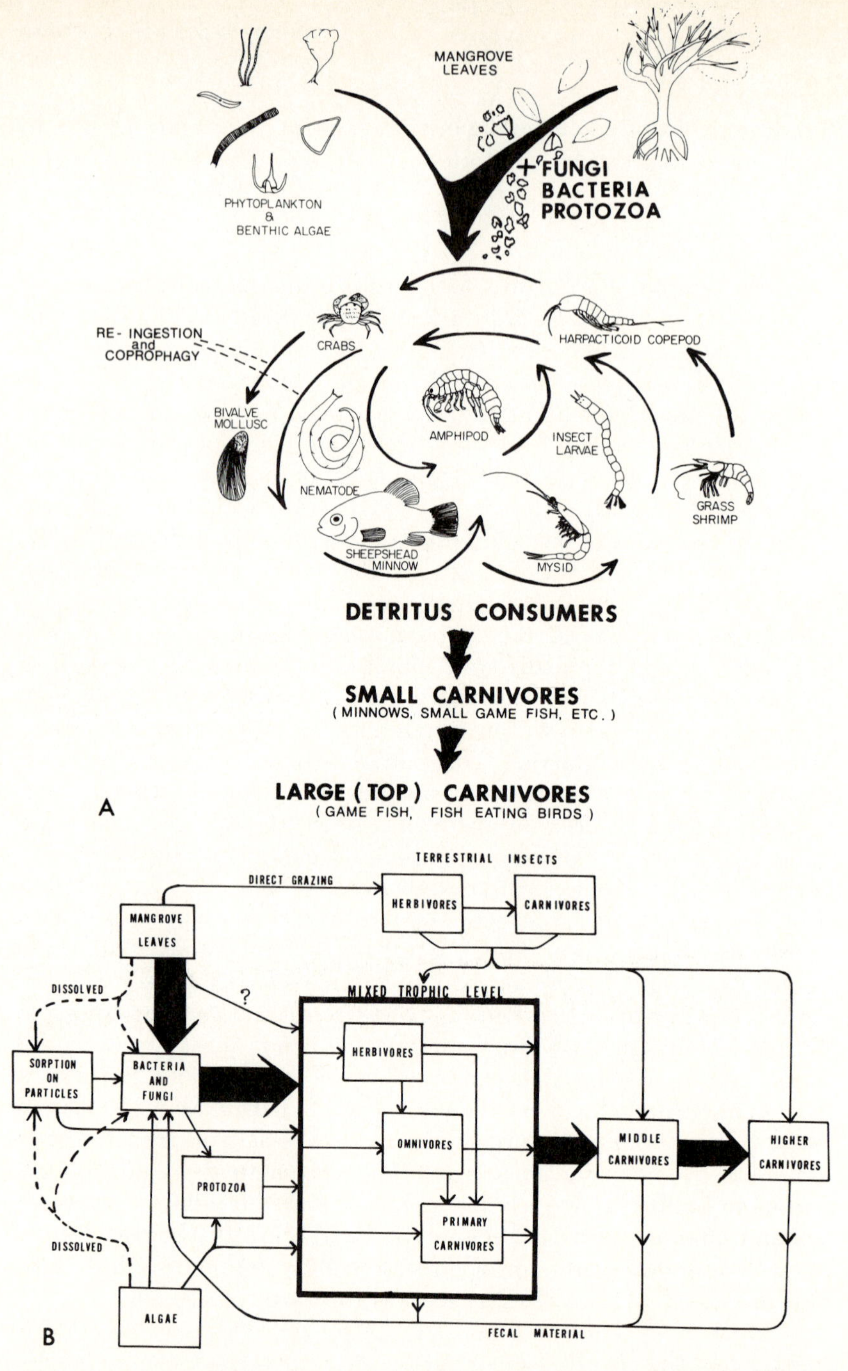

Figure 3–15 A detritus food chain based on mangrove leaves that fall into shallow estuarine waters of south Florida. Leaf fragments acted on by saprotrophs and colonized by algae are eaten and reeaten (coprophagy) by a key group of small detritus consumers, which in turn provide the main food for game fish, herons, storks, and ibis. A "picture model" of the food chain is shown in *A* and a "compartment model" in *B*. (Redrawn from W. E. Odum and Heald, 1975.)

Table 3–12 **Various Types of Ecological Efficiencies**

Ratio	Designation and Explanation
A. Ratios Between Trophic Levels	
$\frac{I_t}{I_{t-1}}$	Trophic level energy intake (or Lindeman's) efficiency. For the primary level this is $\frac{P_G}{L}$ or $\frac{P_G}{L_A}$
$\frac{A_t}{A_{t-1}}$	Trophic level assimilation efficiency — For the primary level P and A may be in terms of either L or L_A as above; $A_t/A_{t-1} = I_t/I_{t-1}$ for the primary level, but not for secondary levels.
$\frac{P_t}{P_{t-1}}$	Trophic level production efficiency — For the primary level P and A may be in terms of either L or L_A as above; $A_t/A_{t-1} = I_t/I_{t-1}$ for the primary level, but not for secondary levels.
$\frac{I_t}{P_{t-1}}$ or $\frac{A_t}{P_{t-1}}$	Utilization efficiencies
B. Ratios Within Trophic Levels	
$\frac{P_t}{A_t}$	Tissue growth or production efficiency
$\frac{P_t}{I_t}$	Ecological growth efficiency
$\frac{A_t}{I_t}$	Assimilation efficiency

Symbols are as follows (see Figure 3–9): L—light (total); L_A—absorbed light; P_G—total photosynthesis (gross production); P—production of biomass; I—energy intake; R—respiration; A—assimilation; NA—ingested but not assimilated; NU—not used by trophic level shown; t—trophic level; $t-1$—preceding trophic level.

as percentages, are often called ecological efficiencies. In Table 3–12, some of these ratios are listed and defined in terms of the energy flow diagram. For the most part, the ratios have meaning in reference to component populations as well as to whole trophic levels. Since the several types of efficiencies are often confused, defining exactly what relationship is meant is important; the energy flow diagram (Figures 3–6 and 3–9) helps clarify this definition.

Most important, efficiency ratios are meaningful in comparisons only when they are dimensionless, that is, when the numerator and denominator of each ratio are expressed in the same unit. Otherwise, statements about efficiency can be very misleading. For example, poultry farmers may speak of a 40 percent efficiency in the conversion of chicken feed to chickens (the P_t/I_t ratio in Table 3–12), but this is actually a ratio of "wet" chicken (worth about 2 kcal/gm) to dry feed (worth 4+ kcal/gm). The true growth efficiency in kcal/kcal in this case is more like 20 percent. Wherever possible, ecological efficiencies should be expressed in the "energy currency" (i.e., calories to calories).

The general nature of transfer efficiencies between trophic levels has already been discussed, that is, the 1 to 5 percent P_G/L, the 2 to 10 percent P_G/L_A, and the 10 to 20 percent production efficiencies between secondary trophic levels. Recent studies of oceanic food chains have indicated that trophic efficiencies higher than 20 percent are achieved when "concentrators" are present, such as pelagic tunicates (salps) that filter small organisms and particles and then produce fecal pellets, which are ingested and reingested by larger organisms (Pomeroy, 1979). Since the proportion of assimilated energy that must go to respiration is at least ten times higher in warm-blooded animals (endotherms), which maintain a high body temperature at all times, than in cold-blooded animals (exotherms), production efficiency, P/A, must be lower in warm-blooded species. Accordingly, the efficiency of trophic-level transfer should be higher in an invertebrate than in a mammalian food chain. As an example, transfer from moose to wolf on Isle Royale is about 1 percent, compared with a 10 percent transfer in a *Daphnia-Hydra* food chain (Lawton, 1981). Generally, an inverse relationship tends to exist between efficiency of tissue growth and efficiency of assimilation in animals (see E. P. Odum and Smalley, 1957; Welch, 1968). Herbivores tend to have higher P/A but lower A/I efficiencies than do carnivores (Kozlowski, 1968; May, 1979).

To many persons, the very low primary efficiencies characteristic of intact natural systems are puzzling in view of the relatively high efficiencies obtained in electric motors and other mechanical systems. This has naturally led many to consider ways of increasing nature's efficiency. Actually, the primary efficiencies of long-time, large-scale ecosystems are not directly comparable to short-time mechanical systems. For one thing, much fuel goes for repair and maintenance of living systems, and depreciation and repair are not included in calculating fuel efficiencies of engines. In other words, much energy (human or otherwise) other than fuel consumed in operation is required to build the machine and keep it running, repaired, and replaced. Engines and biological systems cannot fairly be compared unless all these energy costs are considered, because biological systems are self-repairing and self-perpetuating. Second, under certain conditions, more rapid growth per unit time probably has greater survival value than maximum efficiency in the use of energy. Thus, by a simple analogy, it might be better to reach a destination quickly at 50 mph than to achieve maximum efficiency in fuel consumption by driving slowly. Engineers should understand that any increase in the efficiency of a biological system will be obtained at the expense of maintenance: a gain from increasing the efficiency will be lost in increased cost, not to mention the danger of increased disorder that may result from stressing the system. As already noted, such a point of diminishing returns may now have been reached in the industrialized agroecosystem (see also Section 8). The quality of the energy source is also an important factor in determining efficiency and is discussed in detail in Section 5.

Role of Consumers in Food Web Dynamics

Animals and other consumers are not just passive "eaters" along the food chain. In satisfying their own energy requirements, they often exert a positive feedback on "upstream" trophic levels. Through natural selection, predators and parasites become adapted not only to avoid destroying their food sources but also in many cases to ensure or even increase continued well-being of their prey. In theory, then, there are not only negative feedback controls as discussed in Chapter 2, Section 6, but also positive feedback effects. However, only during the past decade have specific cases been documented to show that consumers can positively affect primary production. Several examples will suffice to illustrate.

McNaughton (1976) has demonstrated that the grazing of the great herds of migratory East African antelopes, coupled with dry-season fires, increases the rate of return of nutrients to the soil. Regrowth is enhanced, and production of grass is increased during the subsequent rainy season. These interactions thus facilitate energy flow through all of the system. This beautiful symbiosis between grass and grazers that has evolved in what Sinclair and Norton-Griffiths (1979) call the Serengeti Ecosystem unfortunately works only on a large scale, since the most numerous grazers migrate long distances. This efficient plant-animal partnership will be difficult to preserve in the face of rapid growth of the human population in Africa. Restricting the animals to game parks will disrupt the symbiosis and likely produce overgrazing (recall our previous discussion of introduced reindeer in Alaska).

In an experimental greenhouse study, Dyer and Bokhari (1976) compared grass plants whose leaves were eaten by grasshoppers with plants from which the same number of leaves were removed by clipping. Regrowth was more rapid in plants grazed by the grasshoppers.

Apparently, a substance in the insect's saliva stimulates root growth and thus increases the plant's ability to regenerate new leaves (a similar effect has been reported for cattle grazing; see Reardon et al., 1974).

As an aquatic example, fiddler crabs of the genus *Uca* that feed on surface algae and detritus in coastal marshes "cultivate" their food plants in several ways. Their burrowing increases water circulation around the roots of marsh grass and brings oxygen and nutrients deep into the anaerobic zone. By constantly reworking the organically rich muds on which they feed, the crabs enhance conditions for the growth of benthic algae. Finally, egested sediment particles and fecal pellets provide substrates for growth of nitrogen-fixing and other bacteria that enrich the system (see Montague, 1980, for a diagram showing these and other positive feedback effects of crabs on primary production).

For reviews and more examples, see Vickery (1972), Chew (1974), Mattson and Addy (1975), O'Neill (1976), Owens and Wiegert (1976), and Kitchell et al. (1979).

The length of food chains is of some interest. Reducing the energy available to successive links obviously limits the length of food chains. However, availability of energy may not be the only factor, since long food chains often occur in infertile systems, such as oligotrophic lakes, and short ones are often found in very productive or eutrophic situations. Rapid production of nutritious plant material may invite heavy grazing, resulting in concentration of energy flow in the first two or three trophic levels. Eutrophication of lakes also shifts the planktonic food web from a phytoplankton–large zooplankton–game fish sequence to a microbial-detrital microzooplankton system not so conducive to support of recreational fisheries.

Pimm and Lawton (1977) suggest that living at the higher trophic levels increases the risk of extinction (uncertain food supply), a contention challenged by Saunders (1978) who reports that the longest food chains are found at the sea-land interface.

In 1960, Hairston, Smith, and Slobodkin proposed a "balance of nature" hypothesis that caused discussion and controversy among ecologists. They argued that since plants by and large accumulate a lot of biomass (the world *is* green), something is inhibiting grazing. That something, they theorized, is predators. Accordingly, primary consumers are limited by secondary consumers, and primary producers are thus resource- rather than grazer-limited. Subsequently, Smith (1969) and Fretwell (1977) suggested that odd-length and even-length chains differ in how primary production is controlled by the food chain exploiting it. Thus, if only plants and herbivores (primary consumers) are present, the plants will be limited and controlled by the grazing pressure, whereas the herbivores will be food-limited. If a predator level is present (resulting in an odd-length chain), then herbivores become predator-limited, and plants no longer being heavily grazed become resource-limited (i.e., limited by nutrients and water). If a fourth level (i.e., secondary predator or parasite) is added, plants again become grazer-limited. If this theory holds, a plot of standing crop biomass by trophic level should have peaks and valleys; the valleys (i.e., consumer-controlled levels) are 1,3,5–levels in even-numbered chains and 2,4,6–levels in odd-numbered chains. Such a situation has been reported for fish ponds.

Mechanisms other than consumers help control the use of primary production, for example, allelopathic chemicals produced by plants that inhibit heterotrophic consumption themselves. All of these theoretical control mechanisms are operating in the real world, but no one theory can explain everything.

Concentration of Toxic Substances Along Food Chains

The distribution of energy, of course, is not the only quantity influenced by food chain phenomena (as will be made evident in subsequent chapters). Some substances become concentrated instead of dispersed with each link in the chain. The **food chain concentration**, or as the popular press has it, **biological magnification**, is dramatically illustrated by the behavior of certain persistent radionuclides and pesticides.

The tendency for certain radionuclide byproducts of atomic fission and activation to become increasingly concentrated with each step in the food chain was first discovered at the Atomic Energy Commission's Hanford plant in eastern Washington in the 1950s. Extremely small (trace) amounts of radioactive iodine, phosphorus, cesium, and strontium released into the Columbia River were found to have become concentrated in the tissues of fish and birds. A concentration factor (amount in tissue/amount in water) of 2 million times was reported for radioactive phosphorus in the eggs of geese nesting on the islands in the river. Thus, what is considered "harmless" releases into the water can become highly toxic to the "downstream" components of the food chain.

An example of similar buildup of DDT is shown in Table 3–13. To control mosquitoes on Long Island, municipalities sprayed DDT for many years on the marshes. Insect control specialists tried to use spray concentrations that were not directly lethal to fish and other wildlife, but they failed to reckon with ecological processes and with

Table 3–13 An Example of Food Chain Concentration of a Persistent Pesticide, DDT*

	DDT Residues (ppm†)
Water	0.00005
Plankton	0.04
Silverside minnow	0.23
Sheephead minnow	0.94
Pickerel (predatory fish)	1.33
Needlefish (predatory fish)	2.07
Heron (feeds on small animals)	3.57
Tern (feeds on small animals)	3.91
Herring gull (scavenger)	6.00
Fish hawk (osprey) egg	13.8
Merganser (fish-eating duck)	22.8
Cormorant (feeds on larger fish)	26.4

* Data from Woodwell, Wurster, and Isaacson, 1967.

† Parts per million (ppm) of total residues, DDT + DDD + DDE (all of which are toxic), on a wet-weight, whole-organism basis.

the long-term toxicity of DDT residues. Instead of being washed out to sea, as some had predicted, the poisonous residues adsorbed on detritus, became concentrated in the tissues of detritus feeders and small fishes, and again concentrated in the top predators such as fish-eating birds. The concentration factor (ratio of parts per million in organism to parts per million in water) is about one-half million times for fish-eaters in the case shown in Table 3–13. In hindsight, a study of the detritus food chain model would indicate that anything sorbing readily on detritus and soil particles and dissolved in guts would become concentrated by the ingestion-reingestion process at the beginning of the detritus food chain. Such a buildup of DDT on detritus has been documented by W. E. Odum, Woodwell, and Wurster (1969). The magnification is compounded in fish and birds by the extensive deposition of body fat in which DDT residue accumulates. The widespread use of DDT ultimately wiped out whole populations of predatory birds such as the fish hawk (osprey), peregrine falcons, and pelicans and of detritus feeders such as fiddler crabs. Birds are especially vulnerable to DDT poisoning because DDT (and other chlorinated hydrocarbon insecticides as well) interferes with the formation of egg shells by causing a breakdown in steroid hormones (see Peakall, 1967; Hickey and Anderson, 1968). These fragile eggs then break before the young can hatch. Thus, very small amounts that are not lethal to the individual can be lethal to the population. Scientific documentation of this sort of frightening buildup (frightening because humans are also "top carnivores") and unanticipated physiological effects finally marshalled public opinion against the use of DDT and similar pesticides. DDT was banned in the United States in 1972. Dieldrin, another persistent chlorinated hydrocarbon, was banned in 1975. Both have also been outlawed in Europe, but unfortunately they are still being manufactured for export to countries where use is legal. Some of the bird populations (falcons, pelicans) decimated by chlorinated hydrocarbon pesticides have begun to recover as the use of these poisons has been reduced.

The principle of biological magnification (see review by Woodwell, 1967) should be considered in any waste management strategy. However, many nonbiological factors may either reduce or augment the concentration factor. Thus, humans ingest less DDT than fish hawks in part because food processing and cooking remove some of the materials. On the other hand, a fish is doubly endangered because it may become contaminated by direct absorption from the environment through the gills as well as by its foods.

Isotopic Tracers as Aids to the Study of Food Chains

Observation and examination of stomach contents have been the traditional means of determining what matter heterotrophs consume, but these methods are often not feasible, especially for small or reclusive animals and the saprotrophs (bacteria, fungi, and so on). In some cases, isotopic tracers can be used to track food webs in natural ecosystems where many species are interacting. Radioactive tracers have proved useful, for example, to determine what insects are feeding on what plants or what predators are feeding on what prey (see Odum and Kuenzler, 1963, and Wiegert et al., 1967, for examples of this approach). Ratios of stable carbon isotopes are proving especially useful in charting energy flow in food chains that are otherwise difficult to study. C_3 plants, C_4 plants, and algae have different $^{13}C/^{12}C$ ratios, which are carried over to whatever organism (animal or microbe) consumes the particular plant or plant detritus. In a study of estuarine food chains, Haines and Montague (1979) found that oysters fed largely on phytoplankton algae, whereas fiddler crabs used both benthic algae and detritus derived from marsh grass (a C_4 plant) for food. Peterson et al. (1980) suggest caution in the use of carbon isotopic ratios where bacteria are involved in converting plant detritus into animal food, since they find that bacterial action can alter the ratio.

Recently, it has been discovered that the ratio between deuterium (D_2) and ordinary hydrogen (H_2) in tissues and feces of animals corresponds to the ratio in food. In an intertidal habitat, species of seaweed proved to have different D/H ratios, so that one could determine which species was being grazed by snails (see Estep and Debrowski, 1980).

5. Energy Quality

Statement

Energy has quality as well as quantity. Not all calories (or whatever quantitative unit is employed) are equal, because the same quantities of different forms of energy vary widely in work potential. Highly concentrated forms such as oil have a higher work potential and, therefore, a higher quality than do more dilute forms such as sunlight; sunlight, in turn, has a higher quality than still more dispersed low-temperature heat. Energy quality is measured by the energy used in the transformation or, more specifically, by the amount of one type of energy required to develop another in a chain of energy transformations such as a food chain or a chain of energy conversions leading to the generation of electricity. Accordingly, as the quantity declines in a chain, the quality of what actually is converted (after appropriate thermodynamic dissipation) into the new form increases proportionally at each step. In other words, as quantity is degraded, quality is upgraded. A convenient quality factor can be defined as the number of calories of sunlight necessary to be dissipated to produce a calorie of a higher-quality form (food or wood, for example). Secondarily, the chemical structure of the energy source determines its quality as a food source for consumers. In comparing energy sources for direct use by humankind, one ought to consider quality as well as the amount available and, wherever possible, match the quality of the source with the quality of the use.

Explanation

The principle of energy quality is illustrated by the two flow diagrams in Figure 3–16. In a natural food chain, energy declines with each step from about 10^6 kcal/m^2 sun input to 100 or less at the predator (secondary consumer) level, as described in the previous section, *but* energy quality in terms of number of solar kilocalories dissipated increases from 1 to 10,000 (Figure 3–16A). Ten thousand

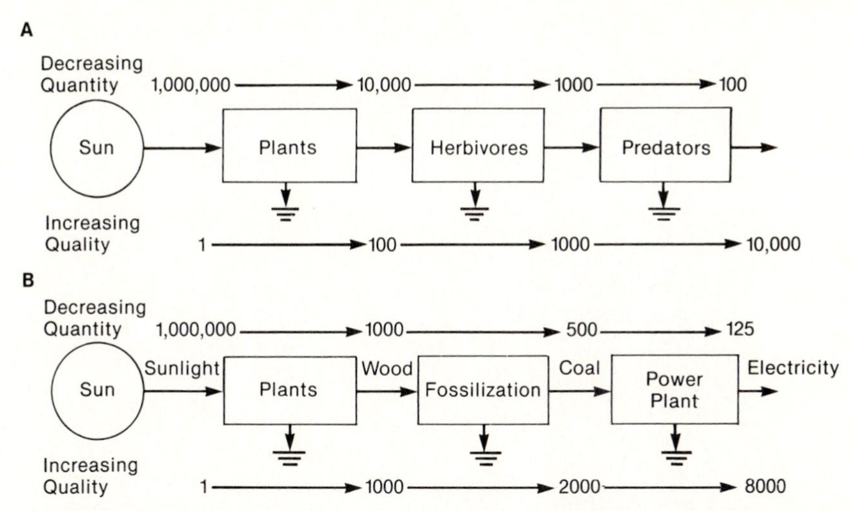

Figure 3–16 Increasing quality with decreasing quantity in two chains of energy transfers beginning with the sun. *A*. Food chain. *B*. Electric energy chain. (After H. T. Odum, 1979.)

kcal of sunlight are required to produce 1 kcal of predator. Or, 100 kcal of herbivore is required for every 1 kcal of predator. Accordingly, a small biomass of predators has an energy quality 100 times that of a similar biomass of herbivores. This higher quality is reflected in the control influence exerted by "downstream" units over "upstream" units, as discussed in previous sections. Essentially, energy quality is measured by distance from the sun, thermodynamically speaking. The actual energy flow at any level multiplied by the quality factor is known as the **embodied energy** of that component. Whether the potential energy in a component is available to a consumer depends on the **resource quality**—value as food, for example—as discussed in Section 4.

Figure 3–16B shows an energy chain leading to the generation of electricity. Energy quantity declines along the chain, but energy quality, the capacity to perform work, increases with each conversion. These relationships are summarized in Table 3–14 (first column). As shown, fossil fuels have a quality or work potential 2000 times that of sunlight. Thus, for solar power to do the work now being done by coal or oil, it must be upgraded or concentrated 2000 times. Society cannot shift from fossil fuels to solar energy for running automobiles and other machines that require high-quality energy *unless* the low-quality solar energy can be upgraded. This change requires expensive technology not yet developed. But, solar energy can be used

Table 3–14 **Energy Quality Factors***

Type of Energy	Solar Equivalent Calories	Fossil Fuel Equivalent Calories
Sunlight	1.0	0.0005
Plant production gross	100	0.05
Plant net production as wood	1000	0.5
Fossil fuel (delivered for use)	2000	1.0
Energy in elevated water	6000	3
Electricity	8000	4

* After H. T. Odum and E. C. Odum (1981).

directly without upgrading for low-quality jobs such as heating homes and other buildings. Matching the quality of source and use should be a major consideration in future national and global strategies if civilization is to conserve and efficiently use available sources. According to such a matching scenario, fossil fuels would be reserved for the high-quality demands of running machinery. Their high quality would not be wasted in a furnace to heat a house when the sun could do that job. Oil and coal would last much longer, giving time for development of other high-quality sources.

Since fossil fuels are now the primary basis for humanity's energy chains, quality factors should be expressed in terms of fossil fuel equivalent kilocalories as well as solar equivalents, as shown in the second column of Table 3–13. Energy and civilization are discussed further in Section 10.

6. Metabolism and Size of Individuals

Statement

The standing crop biomass (expressed as the total dry weight or total caloric content of organisms present at any one time) that can be supported by a steady flow of energy in a food chain depends considerably on the size of the individual organisms. The smaller the organism, the greater its metabolism per gram (or per calorie) of biomass, and the smaller the biomass that can be supported at a particular trophic level in the ecosystem. Conversely, the larger the organism, the larger the standing crop biomass. Thus, the amount of bacteria present at any one time would be very much smaller than the "crop" of fish or mammals, even though the energy use might be the same for both groups.

Explanation and Examples

The metabolism per gram of biomass of the small plants and animals such as algae, bacteria, and protozoa is immensely greater than the metabolic rate of large organisms such as trees and vertebrates. This applies to both photosynthesis and respiration. In many cases, the metabolically important parts of the community are not the few great, conspicuous organisms but the many organisms that are often invisible to the naked eye. Thus, the tiny algae (phytoplankton), weighing only a few pounds per acre at any one moment in a lake, can have as great a metabolism as a much larger volume of trees in a forest or hay in a meadow. Likewise, a few pounds of small crustacea (zooplankton) "grazing" on the algae can have a total respiration equal to that of many pounds of cows in a pasture.

The rate of metabolism of organisms, or group of organisms, is often estimated by measuring the rate at which oxygen is consumed (or produced, in the case of photosynthesis). The metabolic rate of an animal tends to increase as the two-thirds power of its volume (or weight) increases. The metabolic rate per gram biomass also decreases inversely as the length (Zeuthen, 1953; Bertalanffy, 1957; Kleiber, 1961). A similar relationship appears to exist in plants, although structural differences in plants and animals make direct comparisons of volume and length difficult. Relationships between body weight and respiration per individual and per unit weight are shown in Figure 3–17. The latter curve (Figure 3–17B) is important because it shows how weight-specific metabolic rate increases as the size of the individual decreases. Various theories about this trend have focused on diffusion processes: larger organisms have less surface area per gram through which diffusion processes might occur. However, the real explanation for the relationship between size and metabolism has not been agreed upon. Comparisons, of course, should be made at similar temperatures, because metabolic rates are usually greater at higher temperatures than at lower temperatures (except with temperature adaptation; see pages 233 and 234).

When organisms of the same general size are compared, the relationship shown in Figure 3–17 may not always hold. This is to be expected, since many factors secondary to size affect the rate of metabolism. For example, warm-blooded vertebrates have a greater respiration rate than do cold-blooded vertebrates of the same size.

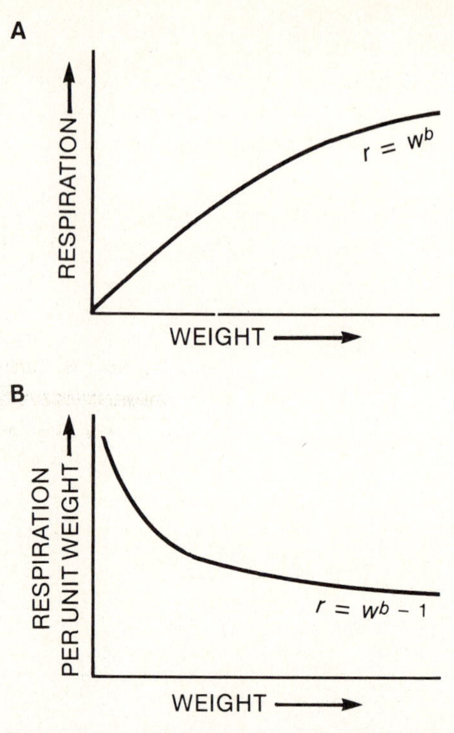

Figure 3–17 Relationship between respiration and body weight per individual (*A*) and respiration per unit weight and body weight (*B*). Exponent is generally within a range between 0.7 and 0.8. (Adapted from Agren and Axelsson, 1980.)

However, the difference is actually relatively small compared with the difference between a vertebrate and a bacterium. Thus, given the same amount of available food energy, the standing crop of cold-blooded herbivorous fish in a pond may be of the same order of magnitude as that of warm-blooded herbivorous mammals on land. However, as mentioned in Chapter 2, oxygen is less available in water than in air and is therefore more likely to be limiting in water. In general, aquatic animals seem to have a lower weight-specific respiratory rate than do terrestrial animals of the same size. Such an adaptation may well affect the trophic structure (see Misra et al., 1968).

In studying size-metabolism in plants, one often finds it difficult to decide what constitutes an "individual." Thus, a large tree can be regarded as one individual, but actually the leaves may act as functional individuals as far as size–surface area relationships are concerned (recall the concept of leaf area index). In a study of various species of seaweeds (large multicellular algae), we found that species with thin or narrow "branches" (and consequently a high surface-to-volume ratio) had a higher rate per gram biomass of food manufacture, respiration, and uptake of radioactive phosphorus from the water than did species with thick branches (E. P. Odum, Kuenzler, and Blunt, 1958). Thus, in this case, the "branches" or even the individual cells were functional individuals and not the whole plant, which might include numerous "branches" attached to the substrate by a single holdfast.

The inverse relationship between size and metabolism may also be observed in the ontogeny of a single species. Eggs, for example, usually show a higher metabolic rate per gram than the larger adults. In data reported by Hunter and Vernberg (1955), the metabolism per gram of trematode parasites was found to be ten times less than that of the small larval cercariae.

Remember, the weight-specific metabolic rate, not the total metabolism of the individual, decreases with increasing size. Thus, an adult human being requires more total food than a small child, but less food per pound of body weight.

7. Trophic Structure and Ecological Pyramids

Statement

The interaction of the food chain phenomenon (energy loss at each transfer) and the size-metabolism relationship results in communities having a definite **trophic structure**, which often characterizes a particular type of ecosystem (lake, forest, coral reef, pasture, and so on). Trophic structure may be measured and described either in terms of the standing crop per unit area or in terms of the energy fixed per unit area per unit time at successive trophic levels. Trophic structure and trophic function may be shown graphically by **ecological pyramids** in which the first or producer level forms the base, and successive levels form the tiers that make up the apex. Ecological pyramids may be of three general types: (1) the **pyramid of numbers,** in which the number of individual organisms is depicted; (2) the **pyramid of biomass**, based on the total dry weight, caloric value, or other measure of the total amount of living material; and (3) the **pyramid of energy**, in which the rate of energy flow and/or productivity at successive trophic levels is shown. The numbers and biomass pyramids can be inverted (or partly so); that is, the base may be smaller than one or more of the upper tiers if individual producer organisms are on aver-

A. PYRAMID OF NUMBERS. Individuals (exclusive of microorganisms and soil animals) per 0.1 hectare

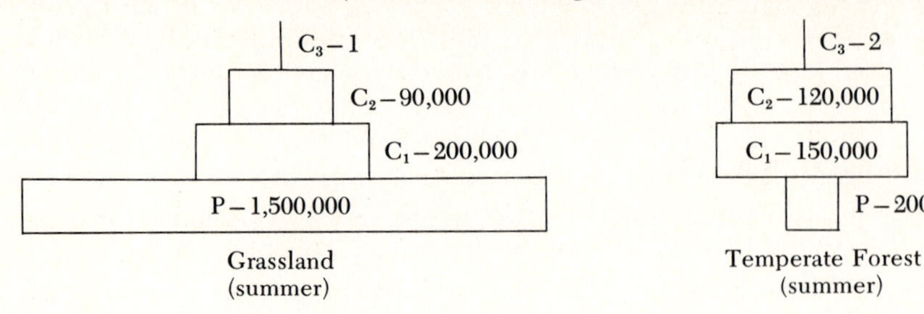

B. PYRAMID OF BIOMASS. Grams dry weight per square meter

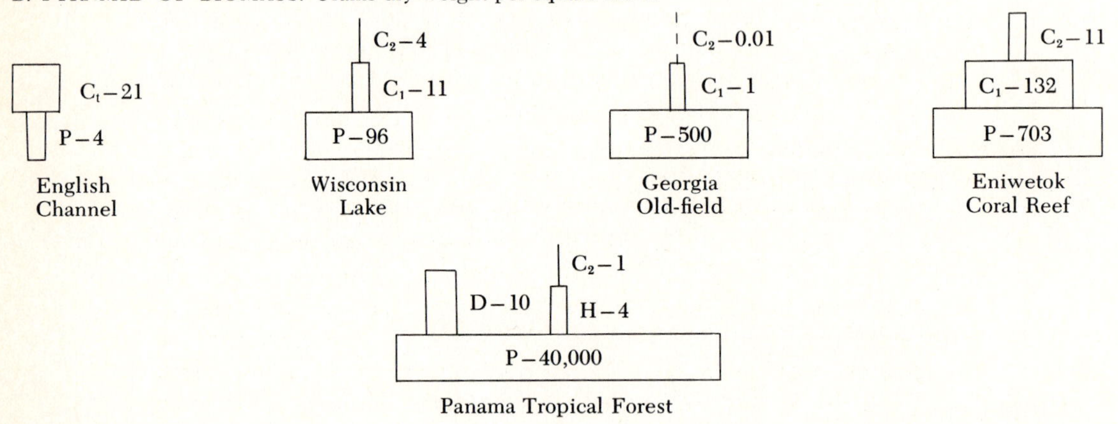

C. COMPARISON OF STANDING CROP AND ENERGY-FLOW PYRAMIDS FOR SILVER SPRINGS, FLORIDA

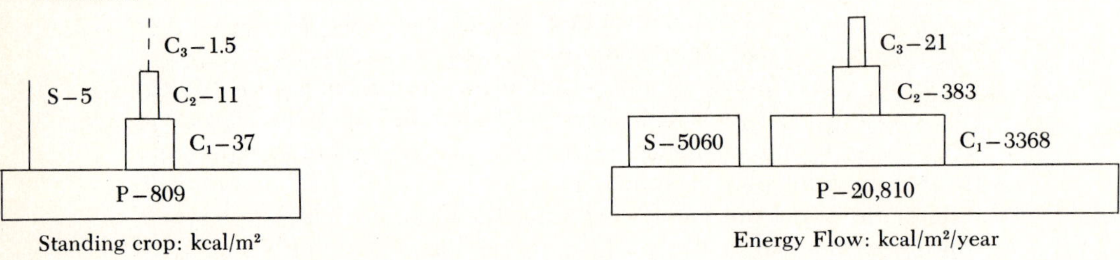

D. SEASONAL CHANGE IN BIOMASS PYRAMID IN THE WATER COLUMN (NET PLANKTON ONLY) OF AN ITALIAN LAKE. Milligrams dry weight per cubic meter

age larger than individual consumers. On the other hand, the energy pyramid must always have a true upright pyramid shape, provided all sources of food energy in the system are considered.

Explanation and Examples

The three kinds of ecological pyramids are illustrated in Figure 3–18. The pyramid of numbers is not very fundamental or instructive as an illustrative device, since the relative effects of the food chain and size factors are not indicated. The form of the numbers pyramid will vary widely with different communities, depending on whether producing individuals are small (phytoplankton or grass) or large (oak trees). Likewise, numbers vary so widely that it is difficult to show the whole community on the same numerical scale.

In general, the biomass pyramid gives a better picture of standing crop relationships for ecological groups as a whole. When the total weight of individuals at successive trophic levels is plotted, a gradually sloping pyramid may be expected as long as the size of the organisms does not differ greatly. However, if organisms of lower levels are much smaller on average than those of higher levels, the biomass pyramid may be inverted. For example, when producers are very small and consumers are large, the total weight of the latter may be greater at any one moment. In such cases, even though more energy is being passed through the producer trophic level than through consumer levels (which must always be the case), the rapid metabolism and turnover of the small producer organisms accomplish a larger output with a smaller standing crop biomass. Examples of inverted biomass pyramids are found most frequently in lakes and the sea. The plants (phytoplankton) usually outweigh their grazers (zooplankton) during periods of high primary productivity, as during the spring bloom, but at other times, as in winter, the reverse may be true. In most cases in lakes and the sea, secondary and tertiary con-

◀ Figure 3–18 Ecological pyramids of numbers, biomass, and energy in diverse ecosystems ranging from open-water types to large forests. P = producers; C_1 = primary consumers; C_2 = secondary consumers; C_3 = tertiary consumers (top carnivores); S = saprotrophs (bacteria and fungi); D = "decomposers" (bacteria, fungi, and detritivores). Pyramids are somewhat generalized, but each is based on specific studies as follows: *A*. Grassland plant data from Evans and Cain, 1952; animal data from Wolcott, 1937; temperate forest is based on Wytham woods, near Oxford, England, as summarized by Elton, 1966, and Varley, 1970. *B*. English Channel, Harvey, 1950; Wisconsin lake (Weber Lake), Juday, 1942; Georgia old-field, E. P. Odum, 1957; coral reef, Odum and Odum, 1955; Panama forest, Golley and Child (unpublished). *C*. Silver Springs, H. T. Odum, 1957. *D*. Italian lake (Lago Maggiore), Ravera, 1969.

sumers such as fish and shellfish are large and outweigh the phytoplankton producers.

Of the three types of ecological pyramids, the energy pyramid gives by far the best overall picture of the functional nature of communities. The number and weight of organisms that can be supported at any level in any situation depends not on the amount of fixed energy present at any one time in the level just below but rather on the *rate* at which food is being produced. In contrast with the numbers and biomass pyramids, which are pictures of the standing states (i.e., organisms present at any one moment), the energy pyramid depicts the rates of passage of food mass through the food chain. Its shape is not affected by variations in the size and metabolic rate of individuals, and, if all sources of energy are considered, it must always be "right side up" because of the second law of thermodynamics.

The concept of energy flow permits one not only to compare ecosystems with one another but also to evaluate the relative importance of populations within the biotic community portion of the ecosystem. Table 3–15 lists estimates of density, biomass, and energy flow rates for six populations differing widely in size of individual and in habitat. In this series, numbers vary 17 orders of magnitude (10^{17}) and biomass varies about 5 (10^5), whereas energy flow varies only about fivefold. Similarity of energy flow indicates that all six populations are functioning at approximately the same trophic level (primary consumers), even though neither numbers nor biomass indicates this. The ecological rule for this would be as follows: *Numbers*

Table 3–15 Density, Biomass, and Energy Flow of Five Primary Consumer Populations Differing in the Size of Individuals Composing the Population*

	Approximate Density (m^2)	Biomass (g/m^2)	Energy Flow ($kcal/m^2/day$)
Soil bacteria	10^{12}	0.001	1.0
Marine copepods (*Acartia*)	10^5	2.0	2.5
Intertidal snails (*Littorina*)	200	10.0	1.0
Salt marsh grasshoppers (*Orchelimum*)	10	1.0	0.4
Meadow mice (*Microtus*)	10^{-2}	0.6	0.7
Deer (*Odocoileus*)	10^{-5}	1.1	0.5

* After E. P. Odum, 1968.

Table 3–16 **Comparison of Total Metabolism and Population Density of Soil Microorganisms Under Conditions of Low and High Organic Matter***

	No Manure Added to the Soil	Manure Added to the Soil
Energy Dissipated:		
kilocalories × 10^6/acre/year	1	15
Average Population Density:		
(number/gram of soil)		
Bacteria, ×10^8	1.6	2.9
Fungi mycelia, ×10^6	0.85	1.01
Protozoa, ×10^3	17	72

* Data from Russell and Russell, 1950.

overemphasize the importance of small organisms, and biomass overemphasizes the importance of large organisms. Hence, neither can be used as a reliable criterion for comparing the functional role of populations that differ widely in size-metabolism relationships, although of the two, biomass is generally more reliable than numbers. However, energy flow (i.e., $P + R$) provides a more suitable index for comparing any and all components of an ecosystem.

The data in Table 3–16 illustrate further how the activities of decomposers and other small organisms may bear very little relation to the total numbers or biomass present at any one moment (see page 22). Note that a 15-fold increase in dissipated energy, which resulted from the addition of organic matter, was accompanied by less than a twofold increase in the number of bacteria and fungi. In other words, these small organisms merely turn over faster when they become more active and do not increase their standing crop biomass proportionally as do large organisms. The protozoa, being somewhat larger than the bacteria, have a somewhat greater increase in numbers.

As graphic devices, ecological pyramids can also be used to illustrate quantitative relationships in specific parts of ecosystems in which one might have a special interest, for example, predator-prey or host-parasite groups. As already indicated, a parasite pyramid of numbers would generally be reversed in contrast to biomass and energy pyramids. Unfortunately, entire populations of parasites and hyperparasites (parasites living on or in other parasites) have rarely been measured. One thing seems certain: one cannot take literally the well-known jingle by Jonathan Swift, or the whimsical diagram of Hegner (reproduced from *Big Fleas Have Little Fleas, or Who's Who Among the Protozoa,* by Robert Hegner, Baltimore, Williams & Wilkins, 1938):

Big fleas have little fleas
Upon their backs to bite 'em
And little fleas have lesser fleas
And so, ad infinitum.

The number of levels or steps in the parasite chain or pyramid is not "ad infinitum" but is definitely limited, both by size relations and by the second law of thermodynamics.

That trophic structure is a fundamental property that tends to be reconstituted when a particular community is acutely perturbed is suggested by Heatwale and Levins (1972). These authors examined the data of Simberloff and Wilson (1969), who studied recolonization of small mangrove islands in the Florida Keys from which all arthropods living in the foliage had been removed by a single dose of nonpersistent insecticide. They concluded that "Trophic structure reaches equilibrium independent of, and sooner than, species equilibrium." In other words, the ratio between herbivores and predators, for example, is reconstituted before all the species previously present have been able to recolonize the islands. Simberloff (1976) has challenged this conclusion that was based on his own data, so the hypothesis remains in need of further testing.

When an ecosystem is continuously stressed, the trophic structure is likely to be altered as the biotic components adapt to the chronic perturbation, as, for example, has occurred in the Great Lakes as a result of continuous pollution (see Schelske, 1977, for a review).

8. Complexity Theory, The Energetics of Scale, Law of Diminishing Returns, and Concept of Carrying Capacity

Statement

As the size and complexity of a system increase, the energy cost of maintenance tends to increase proportionally at a greater rate. Doubling in size usually requires more than a doubling in the amount of energy that must be diverted to reduce the increased entropy associated with maintaining the increased structural and functional complexity. There are **increasing returns to scale** or **economies of scale** associated with increase in size and complexity, such as increased quality and stability in the face of disturbances, but there are also **diminishing returns to scale** or **diseconomies of scale** involved in the increased cost of pumping out the disorder. These diminishing returns are inherent in large and complex systems and can be reduced by improved design that increases efficiency of energy transformation. However, they cannot be entirely mitigated. The law of diminishing returns applies to all kinds of systems. As an ecosystem becomes larger and more complex, the proportion of gross production that must be respired by the community to sustain it increases and the proportion that can go into further growth in size declines. When these inputs and outputs balance, the size cannot further increase. The amount of biomass that can be supported under these conditions is termed the **maximum carrying capacity.** Increasing evidence shows that the **optimum carrying capacity** sustainable over long periods in the face of environmental uncertainties is lower, by perhaps as much as 50 percent, than the theoretical maximum carrying capacity.

Explanation

Experience in dealing with physical networks such as telephone switchboards indicates that as the number of subscribers or calls, C, goes up, the number of needed switches, N, goes up, approaching a square of the number, viz.:

$$C = \frac{N(N-1)}{2}$$

In 1950, C. E. Shannon of the Bell Telephone Laboratory proved that a diseconomy of scale is an intrinsic feature of networks, and no method of construction, however ingenious, can avoid it. The best that has been achieved in switching networks is reduction of the diseconomy to something like N to the 1.5 power. See Pippenger (1978) for a discussion of complexity theory as applied to mechanical systems.

Whether this sort of diseconomy of scale is also an intrinsic feature of ecosystems is unknown, but at least some of the increased cost of complexity is balanced by benefits of the kind that economists call economies of scale. The metabolism per unit of weight decreases as the size of the organism or the biomass in a forest increases, so more structure can be maintained per unit of energy flow. Adding functional circuits and feedback loops can increase efficiency of energy use and recycling of materials and can increase resistance or resilience to disturbance. As emphasized in Chapter 1, the possibility that emergent properties involving symbiosis between organisms may develop increases overall efficiency. No matter what the adjustment may be, the total entropy that must be dissipated increases rapidly with increase in size, so that more and more of the total energy flow (gross production plus imports) must be diverted to respiratory maintenance, and less and less is available for new growth. When maintenance energy costs balance available energy, no further increase in size occurs, and the theoretical maximum carrying capacity has been reached.

Carrying capacity concepts can be clarified by a diagram such as Figure 3–19. Growth in size and complexity of populations, as well as ecosystems as a whole, often follows an S-shaped or sigmoid curve.

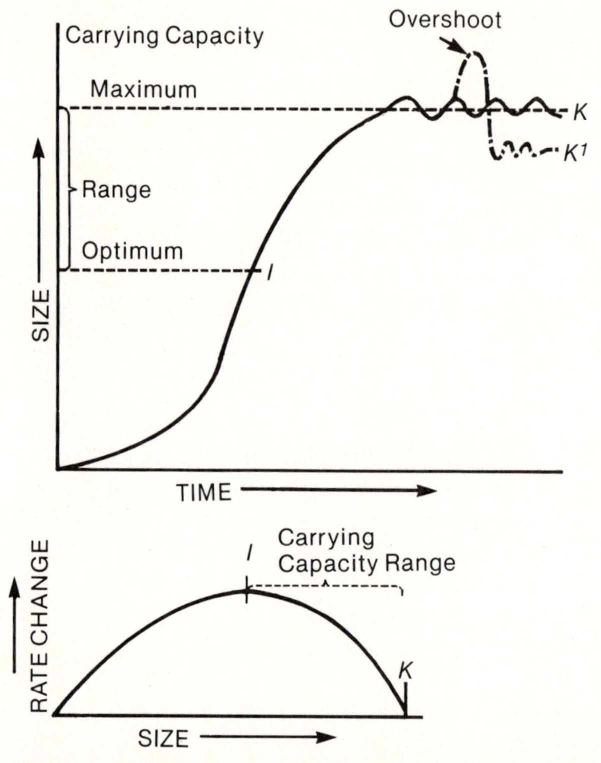

Figure 3–19 Carrying capacity in relation to sigmoid population growth. K represents the maximum density that can be supported with a given space and resource base. If density overshoots this level, K may be lowered (K^1) at least temporarily. I, the inflection point, represents the level of highest growth rate and is the theoretical optimum in terms of maximum sustainable yield for a game or fish population. The range between I and K represents a secure or desirable density, as explained in the text. (Modified from McCullough, 1979.)

Simple mathematical models of sigmoid growth are considered in Chapter 6. Two points on the growth curve need to be noted now: K, the upper asymptote, represents the maximum carrying capacity as defined in the Statement, and I, the inflection point where growth rate is highest, is shown by the lower diagram in Figure 3–19. The I level is often spoken of as the **maximum sustained yield** or **optimum density** by game and fish managers since, theoretically, the harvested biomass would be most rapidly replaced at this point.

The problem with maintaining the maximum or K level in the fluctuating environment of the real world is that overshoots are likely to occur either because the momentum of growth causes population size to exceed K or because a periodic reduction in resource availability (as during a drought, for example) reduces K at least temporarily. When an overshoot occurs and entropy exceeds the capacity of the system to dissipate it, a reduction in size or a "crash" must occur. If the productive capacity of the environment was damaged in the process, K itself may be lowered, at least temporarily, to a new level (K^1 in Figure 3–19). As discussed in Section 3, the global problem of feeding people is approaching that point where food needs equal maximum production capacity, given current technological, political, economic, and distributional constraints. Any widespread perturbation such as war, drought, or disease that reduces crop yields for even one year means severe malnutrition or starvation for millions living on the brink.

The margin of safety at the maximum carrying capacity level is very small. From the standpoint of long-term safety and stability, some level in the range between K and I (carrying capacity range as shown in Figure 3–19) would seem to represent the desirable carrying capacity.

Examples

An example of both increasing and diminishing returns relating to size of a city is diagrammed in Figure 3–20. As the city increases in size, wages tend to increase but air quality decreases. Other diminishing returns to scale as city size increases are listed by Sale (1978) as follows:

1. Higher transportation costs.
2. Massive unemployment during economic recessions.
3. Higher rate of employee sickness (chronic effects of air and other pollution).
4. Higher maintenance and service costs that rise faster than population.

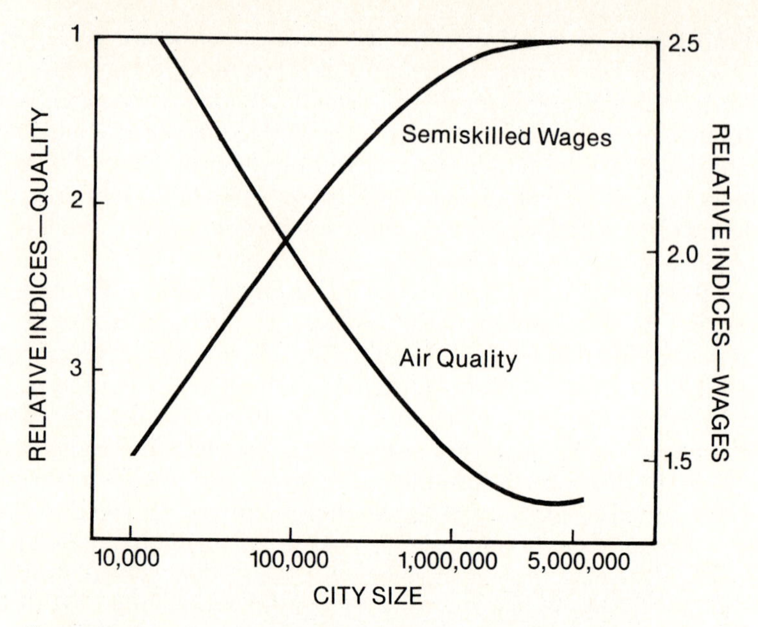

Figure 3–20 An example of economic increasing returns of scale accompanied by decreasing returns of scale in environmental quality as city size increases.

5. Higher costs of heating and cooling (the "heat island" effects of buildings and concrete).
6. Decline in quality of schools.
7. Increased crime.

A reasonable balance between the costs and benefits would seem to occur in a city of moderate size, around a population of 100,000. Of course, many complex factors would have to be considered in determining a theoretical optimum size. As noted previously, the urban-industrial ecosystem (page 72) depends greatly on the size and capacity of the available input and output environments necessary to maintain the city proper. Accordingly, cities probably have different optimum sizes depending on their location and the nature of their input and output environments.

Perhaps, ideally, a region would have one very large city with the cultural advantages that only a very large city can support, such as museums, symphonies, and major league sports, and also have many small cities and towns that provide the quality of life most people say they prefer. Citizens would have to accept that the large central city could not pay for itself and would need subsidies from state and federal governments, which could be considered payment for the economic and cultural benefits available to the whole state or region. If all urban centers were well buffered and spaced by greenbelts and agricultural lands that would provide the life-support input environments and reduce the diseconomies of supporting highly concentrated energy flows, and if cities were managed efficiently, the diseconomies could be reduced to a tolerable level.

A continuous or "strip" city, such as the one between Boston and Washington, D.C., on the east coast of the United States, severely affects the life-support environment (the output environment of one city becomes the input environment of the next city in line) and is increasingly costly to maintain, since all man-made structures age and have to be repaired and replaced. Wastes are disposed of by dumping them in the next city's back yard, clearly an example of the law of diminishing returns. In newly developing regions, perhaps, this syndrome of back-to-back cities can be avoided by urban planners.

Per capita taxes provide a good example of the network law of costs rising as a power function of size. As shown in Table 3–17, per capita state and local taxes are strongly correlated with population density, and especially with percent urbanization within the state. Thus, it costs a person about three times more in taxes to live in New York State than to live in Mississippi. Citizen "tax revolts" to the contrary, one cannot avoid higher taxes if one chooses to live in a large city and does not wish to see it become disorderly. It is the price one has to pay for high density and the economic and cultural values of cities.

Table 3–17 Rank Order Correlation Between Per Capita State and Local Taxes and Urbanization, Atlantic and Gulf Coast States of the United States

Range, 1976 taxes:	Highest—New York State	$1140
	Lowest—Mississippi	$ 455
Correlation coefficients per capita taxes and		
Population density (state as a whole)		$r = 0.78$; $z = 3.11$ (highly significant)
Size of largest metropolitan district		$r = 0.72$; $z = 2.67$ (highly significant)
Percent urban*		$r = 0.83$; $z = 3.31$ (highly significant)

* Percent of area of state in Standard Metropolitan Districts.

An excellent review of city size and quality of life was prepared by the Stanford Research Institute for the U.S. Senate in 1978 (published as a Congressional Hearing). This study concluded that the urban habitat maximizes economic functions to such a degree that social and environmental aspects of human existence are not being simultaneously maximized, and possibly cannot be (on the theory that contrary trends cannot be optimized at the same time and place). The part of the study using interviews and opinion polls suggested that if people really had a free choice, most would leave large cities, but economic factors do not allow such a free choice except for the wealthy. Other recommended readings on city size are ones by Henderson (1974) and by Hoch (1976).

Among the best recent studies about carrying capacity in the animal world discusses a deer herd in Michigan (McCullough, 1979). Six deer introduced into a 2-square-mile (5-square-kilometer or 500-hectare) fenced enclosure in 1928 increased to about 220 in the mid-1930s. When it became evident that the herd was damaging its environment by overbrowsing the vegetation, the population was reduced to about 115 by selective hunting and then maintained at this level up to the present. McCullough suggests that the ±200 level (2 hectare per deer) represents the *K*-carrying capacity level, and that deer populations tend to "track" this maximum level. Left to themselves, deer will increase right up to the limit of food or other vital resource. The ±100 number (about 4 hectares per deer), accordingly, represents the *I*-carrying capacity of optimum density (see Figure 3–17) that avoids overshoots, starvation stress, disease, possible damage to the habitat, and so forth. In this particular species, predation seems to be a function that favors quality over quantity. Other kinds of populations have evolved self-regulatory mechanisms that tend to maintain a below-maximum level (sometimes called the "secure carrying capacity"), as we shall see in Chapter 7.

A study of energy flow in ant colonies might reveal something about the energetics of scale and carrying capacity that is applicable to *Homo sapiens*. For example, the leaf-cutter ants, *Attica colombia*, which live in wet tropical forests, harvest fresh leaf sections from the vegetation and carry them into underground nests as a substrate for culturing fungi that provides their food. The fungal gardens are cared for and fertilized (partly by ant excretions) much as a human farmer cultivates his food crop. Lugo et al. (1973) have estimated the energy expenditures for all the major activities within the colony and conclude that carrying capacity (i.e., maximum size of colony) is reached when input of fuel calories (in the form of harvested leaves) balances the energy cost of the work involved in cutting and transporting leaves, maintaining trails, and cultivating the crops. At any one time in large colonies, Lugo et al. observed, 25 percent of ants were carrying leaves and 75 percent were maintaining trails and fungal gardens. When energy input is balanced by these maintenance costs, the colony stops growing. Reward feedback to other organisms, such as ash deposited by ants on the forest floor that increases leaf growth, increases efficiency and raises carrying capacity.

Estimating the carrying capacity for an agrarian civilization supported by subsistence agriculture (see page 80 for explanation of this term) is not too difficult, since input energy comes mostly from local resources and not from distant regions. For example, Mitchell (1979) has reported that density in the rural countryside in India is a linear function of rainfall, which determines crop yield in the absence of irrigation or other subsidy. He reports that 10 cm of rainfall supports 2 persons per hectare of harvested land, 100 cm supports 3 persons, 200 cm supports 4.5 persons, and 300 cm supports 6 persons. Another interesting study of agrarian carrying capacity is that of Pollard and Gorenstein (1980), who document the relation between maize production and human density in an early Mexican (Tarascan) civilization.

Estimating the carrying capacity for urban-industrialized societies is much more difficult, since such societies are supported by massive subsidies imported from afar and often drawn from storages accumulated before the advent of mankind, such as fossil fuels, (nonrenewable) underground water, virgin timber, and deep organic soils. All of these resources diminish with intensive use. One thing is certain: people, like deer, seem to "track" maximum or K-carrying capacity levels; our population tends to increase right up to or even beyond one limit after another (food and fossil fuels being the limits of concern at the moment). Reward feedback or other means of maintaining the optimum rather than the maximum levels are only weakly developed so far, perhaps for two reasons: (1) many people in developed countries believe that science and technology will continue to find substitutes for declining resources and continue to raise K, and (2) people in undeveloped countries often have a strong economic and social need for many children. Accordingly, the dangerous game of flirting with overshoots continues. These are good ecological reasons to control human growth, but the complex social, economic, and religious issues involved make such control very difficult.

Some resource scientists believe that global carrying capacity has already been exceeded (for example, Borgstrom whose 1969 book is entitled *Too Many*). The best projections for the human biomass are shown by the three sigmoid curves of Figure 3–21. Depending on

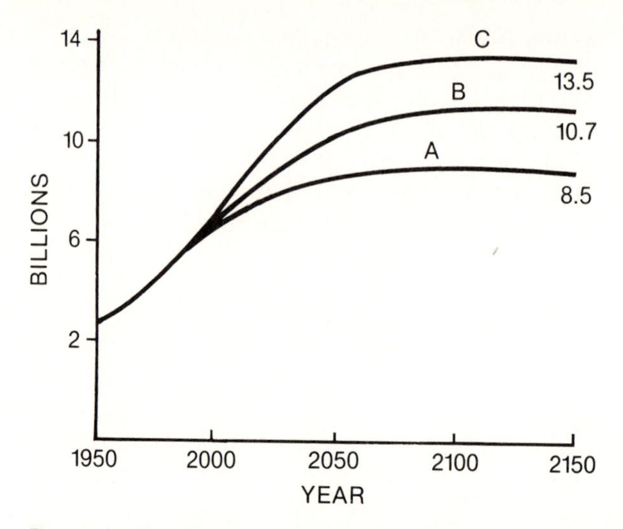

Figure 3–21 Three projections for leveling-off of global human population growth, if world attained replacement fertility (i.e., births balance deaths) in (*A*) years 2000–2006; (*B*) years 2020–2025; (*C*) years 2040–2045. (After Mauldin, 1980.)

when the trend of declining birth rates balances deaths, the global population will level off somewhere between 8.5 and 13.5 billion persons sometime in the next century. If this does not happen, a collapse to a lower density is a very real possibility. Anything over 10 billion persons would certainly reduce the *quality* of human existence, given current global resources and carrying capacity.

9. An Energy-Based Classification of Ecosystems

Statement

The source and quality of available energy determine to a greater or lesser degree the kinds and numbers of organisms, the pattern of functional and developmental processes, and the life style of human beings. Since energy is a common denominator and the ultimate forcing function in all ecosystems, whether designed by humans or by nature, it provides a logical basis for a "first order" classification. On this basis, it is convenient to distinguish four basic classes of ecosystems:

1. Natural unsubsidized solar-powered ecosystems.
2. Natural solar-powered ecosystems subsidized by other natural energies.

3. Human-subsidized solar-powered ecosystems.
4. Fuel-powered urban-industrial systems (source of energy from fossil fuels or other organic or nuclear fuels).

This classification is based on the input environment (see Figure 2–2), and it contrasts with and complements the biome classification (Chapter 2, Section 8) that is based on the internal structure of the ecosystem.

Explanation

The four major types of ecosystems classified according to source, level, and quantity of energy are described in Table 3–18 and accompanying diagrams. Ecosystems rely on two dissimilar sources of energy: the sun and chemical (or nuclear) fuels. Accordingly, one can conveniently distinguish between **solar-powered** and **fuel-powered** systems, while recognizing that both sources can be used in any given situation.

In comparing the major classes of ecosystems, we can speak of the energy flow as the **power density** (recall that a power unit is energy use or dissipation per unit of time—page 91), which indicates magnitude of potential or actual work performed in a unit-area of the ecosystem and also the amount of disorder or entropy that must be dissipated if the system is to remain viable. Note that in Table 3–18 two columns are under the heading of "power level"; one column shows range of energy quantity in kilocalories, the other shows solar-equivalent kilocalories corrected for quality, as described in Section 5 of this chapter.

The systems of nature that depend largely or entirely on the direct rays of the sun can be designated **unsubsidized solar-powered ecosystems** (Category 1 in Table 3–18). They are unsubsidized in the sense that there are few, if any, available auxiliary sources of energy to enhance or supplement solar radiation. The open oceans, great tracts of upland forests and grasslands, and large, deep lakes are examples of relatively unsubsidized solar-powered ecosystems. Frequently, they are subjected to other limitations as well, for example, a shortage of nutrients or water. Consequently, ecosystems in this broad category vary widely but are generally low powered and have a low productivity or capacity to do work. Organisms that populate such systems have evolved remarkable adaptations for living on scarce energy and other resources and using them efficiently.

Although the power density of natural ecosystems in this first category is not very impressive, and such ecosystems by themselves

Table 3–18 **Ecosystems Classified According to Source and Level of Energy**

	Annual Energy Flow (Power Level) ($kcal/m^2$)
1. Unsubsidized natural solar-powered ecosystems Examples: open oceans, upland forests. These systems constitute the basic life-support module for spaceship earth.	1000–10,000 (2000)*
2. Naturally subsidized solar-powered ecosystems Examples: tidal estuary, some rain forests. These are the naturally productive systems of nature that not only have high life-support capacity but also produce excess organic matter that may be exported to other systems or stored.	10,000–40,000 (20,000)*
3. Human-subsidized solar-powered ecosystems Examples: agriculture, aquaculture. These are food- and fiber-producing systems supported by auxiliary fuel or other energy supplied by humans.	10,000–40,000 (20,000)*
4. Fuel-powered urban-industrial systems Examples: cities, suburbs, industrial parks. These are our wealth-generating (also pollution-generating) systems in which fuel replaces the sun as the chief energy source. These are dependent (i.e., parasitic) on Classes 1–3 for life support and for food and fuel.	100,000–3,000,000 (2,000,000)*

* Numbers in parentheses are estimated round-figure averages, actually little more than guesses since the earth's ecosystems have yet to be inventoried in sufficient depth to calculate averages.

could not support a high density of people, they are nonetheless extremely important because of their huge extent (the oceans alone cover almost 70 percent of the globe). For humans, the aggregate of solar-powered natural ecosystems can be thought of, and certainly should be highly valued, as the basic life-support module that stabilizes and homeostatically controls spaceship earth. Here, large volumes of air are purified daily, water is recycled, climates are controlled, weather is moderated, and much other useful work is accomplished. A portion of the food and fiber needs of humans are also produced as a byproduct without economic cost or management effort by man. This evaluation, of course, does not include the priceless aesthetic values inherent in a sweeping view of the ocean, or the grandeur of an unmanaged forest, or the cultural desirability of green, open space.

When auxiliary sources of energy can be used to augment solar radiation, the power density can be raised considerably, perhaps by an order of magnitude. Recall from Section 2 that an **energy subsidy** is an auxiliary energy source that reduces the unit cost of self-maintenance of the ecosystem and thereby increases the amount of solar energy that can be converted to organic production. In other words, solar energy is augmented by nonsolar energy, freeing it for organic production. Such subsidies can be either natural or synthetic (or, of course, both). For the purpose of simplified classification, **naturally subsidized** and **human-subsidized solar-powered ecosystems** have been listed as Categories 2 and 3, respectively, in Table 3–18.

A coastal estuary is a good example of a natural ecosystem subsidized by the energy of tides, waves, and currents. Since the ebb and flow of water partly recycles mineral nutrients and transports food and wastes, the organisms in an estuary can concentrate their efforts, so to speak, on more efficient conversion of the sun's energy to organic matter. In a very real sense, organisms in the estuary are adapted to utilize tidal power. Consequently, estuaries tend to be more fertile than, say, an adjacent land area or pond that receives the same solar input but does not have the benefit of the tidal and other water-flow energy subsidy. Subsidies that enhance productivity can take many other forms, for example, wind and rain in a tropical rain forest, the flowing water of a stream, or imported organic matter and nutrients received by a small lake from its watershed.

Human beings, of course, learned early how to modify and subsidize nature for their direct benefit, and they have become increasingly skillful not only in raising productivity but more especially in channeling that productivity into food and fiber materials that are easily harvested, processed, and used. Agriculture (land culture) and aquaculture (water culture) are the prime examples of Category 3 of Table 3–18, the **human-subsidized solar-powered ecosystems.** High yields of food are maintained by large inputs of fuel (and, in more primitive agriculture, by human and animal labor) involved in cultivation, irrigation, fertilization, genetic selection, and pest control. Thus, tractor fuel, as well as animal or human labor, is just as much an energy input in agroecosystems as sunlight, and it can be measured as calories or horsepower expended not only in the field but also in processing and transporting food to the supermarket. As H. T. Odum (1971) has so aptly expressed it, the bread, rice, corn, and potatoes that feed the masses of people are "partly made of oil." This is why fuel or some comparable auxiliary energy is vital to food production.

In Table 3–18 the productivity, or power level, of natural and human-subsidized solar-powered ecosystems are listed as the same.

This evaluation is based on the observation that the most productive natural ecosystems and the most productive agriculture are at about the same level; 50,000 kcal $m^{-2} yr^{-1}$ seems to be about the upper limit for any plant-photosynthetic system in terms of continuous, long-term function. The real difference between these two classes of systems is the distribution of energy flow. People channel as much energy as possible into food they can use immediately; nature tends to distribute the products of photosynthesis among many species and products and to store energy as a "hedge" against bad times in what will be later discussed as "a strategy of diversification for survival."

The **fuel-powered ecosystem** (Category 4, Table 3–18), otherwise known as the urban-industrial system, is humanity's crowning achievement. Highly concentrated potential energy of fuel replaces, rather than merely supplements, the sun's energy. As cities are now managed, solar energy is not only unused within the city itself but becomes a costly nuisance by heating up the concrete and contributing to the generation of smog. Food, a product of solar-powered systems, is considered external since it is largely imported from outside the city. As fuel becomes more expensive, cities will likely become more interested in using solar energy. Perhaps a new class of ecosystem, the sun-subsidized, fuel-powered city, will begin. It may also be wise to develop a new technology designed to concentrate solar energy to a level where it might partially replace fuel, rather than merely supplement it.

Two properties of the fuel-powered system need to be emphasized. Most important is the enormous energy requirement of a densely populated urban-industrial area; it is at least two or three orders of magnitude greater than the energy flow that supports life in natural or seminatural solar-powered ecosystems. This is why many people can live together in a small space. The kilocalories of energy that flow annually through a square meter of an industrialized city are measured in the millions rather than thousands (Table 3–19). Thus, an acre of highly developed fuel-powered urban environment consumes a billion kilocalories (about 10^9) or more each year. A more dramatic way to view this energy demand is to consider per capita consumption. In 1970, 17.4×10^{15} kcal (69×10^{15} BTU) of fuel energy (including that required to generate electricity) were consumed in the United States, which, divided by 200 million people, comes to about 87 million kcal per person per year. Recall that only 1 million kcal per person is required for food energy. Thus, household, industrial, commercial, transportation, and other cultural activities in the United States use 86 times as much energy as that required for physiological needs (that is, food energy to power the body). In undevel-

Table 3–19 Energy Consumption Density Directly Related to Use of Fuels by Humans

Cities	($kcal\ m^{-2}\ year^{-1}$)*
Manhattan (New York City Center)	4.8×10^6
Tokyo	3.0×10^6
Moscow	1.0×10^6
West Berlin	1.6×10^5
Los Angeles	1.6×10^5
Large industrialized regions	
German industrial region	7.7×10^4
Los Angeles Basin	5.7×10^4
Japan (whole country)	2.3×10^4
United Kingdom	9.2×10^3
14 Eastern States, U.S.A.	8.4×10^3
United States (whole country)	1.8×10^3
World average	100

* Compare these figures with the solar energy that reaches the earth's surface, which is somewhere between 1 and 2×10^6 kcal m^{-2} $year^{-1}$, depending on latitude.

oped countries, of course, the situation is quite different. Consumption of fuel energy per capita in India and Pakistan is 1/50 and 1/100 times less, respectively, than in the United States. In such countries human and animal labor are still more important than machines, and a much larger proportion of the country's total energy flow involves food, fiber, and wood production.

10. Energy, Money, and Civilization

Statement

The history of civilization is very closely coupled with available energy sources. Hunters and gatherers lived as part of natural food chains in solar-powered ecosystems achieving highest density in naturally subsidized systems at coastal and riverine sites. When agriculture and aquaculture developed, carrying capacity was greatly increased as humans became more skillful in cultivating plants and domesticating animals and in subsidizing edible primary production. For many centuries, wood and other biomass provided the chief energy source; the great pyramids, cathedrals, cities, and farms were built with biomass-fueled muscle power, animal and human, though much of the latter was slave labor. This long period can be called the age of muscle power. Then came the current age of fossil fuels, which provide such a bountiful supply that the global population has doubled every half century or so. "Servant machines" powered by gasoline and electricity have gradually replaced animal and human labor (at least in the developed countries). Until recently, it seemed likely that as fossil fuels were exhausted, the third age of humankind would be the age of atomic energy. But pumping out the disorder associated with this source has so far proved troublesome, so the future is unpredictable. In considering potential sources, one must remember that, without exception, energy must be expended to develop and maintain a flow of usable energy from a source. Accordingly, the best sources are those promising the largest net energy yield, that is, the largest amount of usable work energy after the necessary energy costs are paid. Matching of quality of source and use is a second important consideration, as already noted in Section 5.

Money became an important force very early in civilization. Money represents a reverse flow to energy flow in that it flows out of cities and farms in exchange for energy and resources that flow in. Unlike energy, however, money circulates. In theory, at least, money can be converted to quality-corrected energy units (calories, for example) to establish a monetary value for goods and services of nature. A shortcoming of current economic systems, of whatever political ideology, is that they deal mostly with human-made goods and services, leaving the equally important natural life-support goods and services unpriced and undervalued (i.e., outside of, or external to, the monetary system). Ecologists and economists generally agree that closing the gap between market and nonmarket values (or correcting market failure when it comes to the goods and services of nature, to put it in other words) is urgent, since each of these two sets of values depends on the other.

Explanation

Civilization has progressed through the four ecosystem types outlined in Table 3–18. In the closing decades of the twentieth century, the part of the world that consumes petroleum and other fossil fuels on a large scale operates as a fuel-powered system, while another part, the so-called Third World, remains essentially dependent on biomass (food and wood) as the major energy source, supplemented by light fuel oils, and thus operates as a subsidized solar-powered system. As already noted, the difference in per capita income between high-energy and low-energy countries that creates worrisome social, economic, and political conflicts has been increasing rather than decreasing, despite worldwide efforts to close the gap.

At the first International Conference for Peaceful Uses of Atomic Energy, held in Geneva in 1955, the chairman of the conference, the late Homi J. Bhabha of India, described three ages of humankind: the age of muscle power, the age of fossil fuel, and the atomic age. Bhabha spoke eloquently of his belief that because of the universal availability of the atom, the coming of the atomic age would close the gap between rich and poor nations. The dream of equal and abundant energy for all from the atom is yet to materialize, because tapping the enormous potential of atomic energy has proved to have a far greater "disorder potential" than anticipated in 1955. Carroll Wilson, the first general manager of the U.S. Atomic Energy Commission, hit the nail on the head when, in an article (1979) entitled "What Went Wrong?", he wrote, "No one appeared to understand if the whole system does not hang together coherently, none of it might be acceptable." Until the entire cycle from raw material to waste disposal becomes "coherent" and new and better ways to tap energy from nuclear sources are devised, the coming of the atomic age is at least postponed. In the meantime, the world needs to consider seriously a return to solar power, coupled with more efficient (less wasteful) use of the remaining fossil fuels to prolong their availability for many years.

Since it takes energy to produce the quality and quantity of energy needed to run modern civilization, choices for the immediate future should be based on sources that promise the best yield with lowest entropy, as can be assessed by estimating net energy available after the energy costs of conversion have been satisfied. The concept of net energy is diagrammed in Figure 3–22. Net energy available from source S equals the energy flow, A, minus the amount of energy, B, that must be fed back to maintain the flow. B is sometimes called the "energy penalty" by engineers. Energy flows that generate more high quality than they use can be said to produce **net energy**. To be

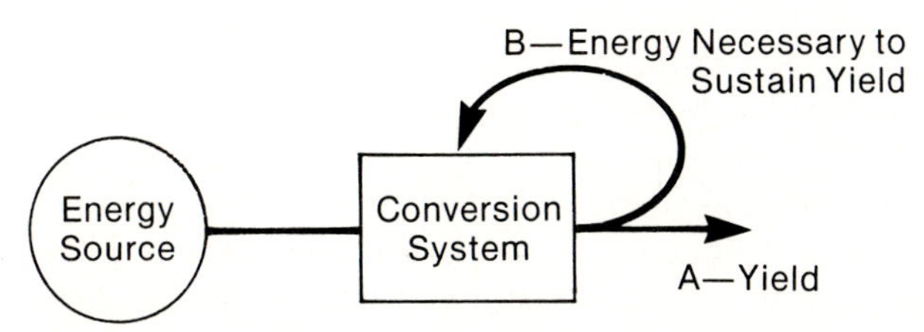

Figure 3–22 Concept of net energy. A (yield) must be greater than B (feedback energy necessary to maintain yield) in order for a source and its conversion system to yield net energy.

viable over the long haul, net energy should be at least twice the penalty, B, or to put it in other words, the **energy yield ratio** should be 2 or better. For example, if 10 units of fuel are required to extract 12 units of oil by deep drilling under the ocean floor, this source will not provide a lasting solution to energy shortages.

In other words, the question is not how much oil is deep in the earth or how much energy is released when uranium is fissioned, but rather how much high-quality energy will actually be available from such sources *after all the energy penalties associated with necessary entropy dissipation have been paid*, including protection of human health and preservation of the integrity of global life-support systems. Thus, as is the case with biotic production, the *net* and not the *gross* amount is of primary concern to the consumer. See H. T. and E. C. Odum, 1981, Chapter 6, for detailed analysis of these concepts.

Money can certainly be considered one of our most important inventions; it is now the basis for decision-making at all levels of society. Money and energy flows are closely coupled in that money is a counterflow to energy flow (Figure 3–23). As emphasized in the Statement, money circulates but energy does not. Nevertheless, money can be converted to approximate quality-corrected energy units and vice versa, since the cost of goods and services is closely related to how much energy has to be expended to produce them. Unfortunately, as shown in Figure 3–23A, money enters the picture only when a natural resource is converted into manufactured goods or human services, thus leaving unpriced the work of nature that sustains the whole resource. In the example shown, only the harvest and seafood-processing part of the production chain is valued in terms of money; all the energy and work performed by the estuary to sustain the crop and to provide other useful services such as recycling of air and water is entirely "external" to monetary systems. It has been estimated that if all work of the estuary useful to humans is cost-accounted in common-denominator energy currency and this currency is then converted to money, an acre of fertile estuary turns out to have many times the value of the estimate based only on the final harvest (Gosselink, Odum, and Pope, 1974). Although many economists are not happy with this approach (see Shabman and Batie, 1979), almost all agree that there is a serious "market failure" when it comes to allocation of natural resources. Kenneth Boulding (1962, 1964), Georgescu-Roegen (1971), and Rifkin (1980) have argued strongly for a more holistic economics (Boulding's "spaceship economics") and more attention to the role of the entropy law in economic transactions. Thus, although ways to close the gap between "market" (i.e., priced) and "nonmarket" (i.e., unpriced) values have

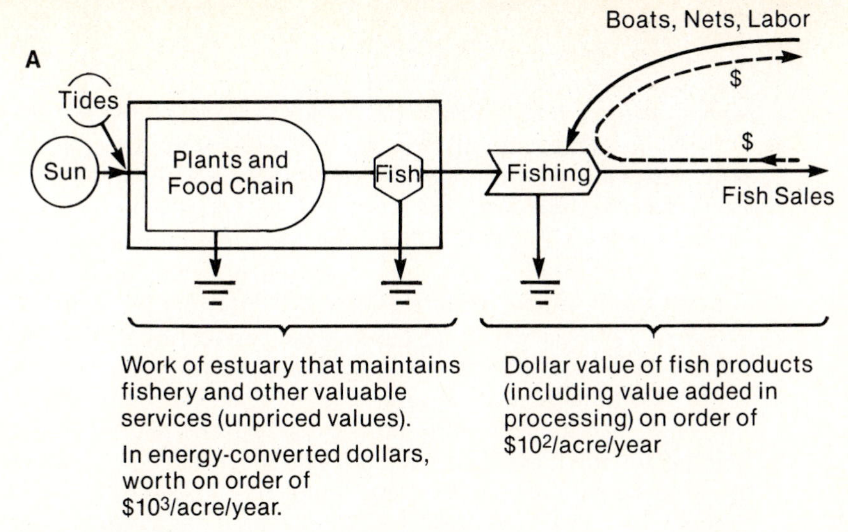

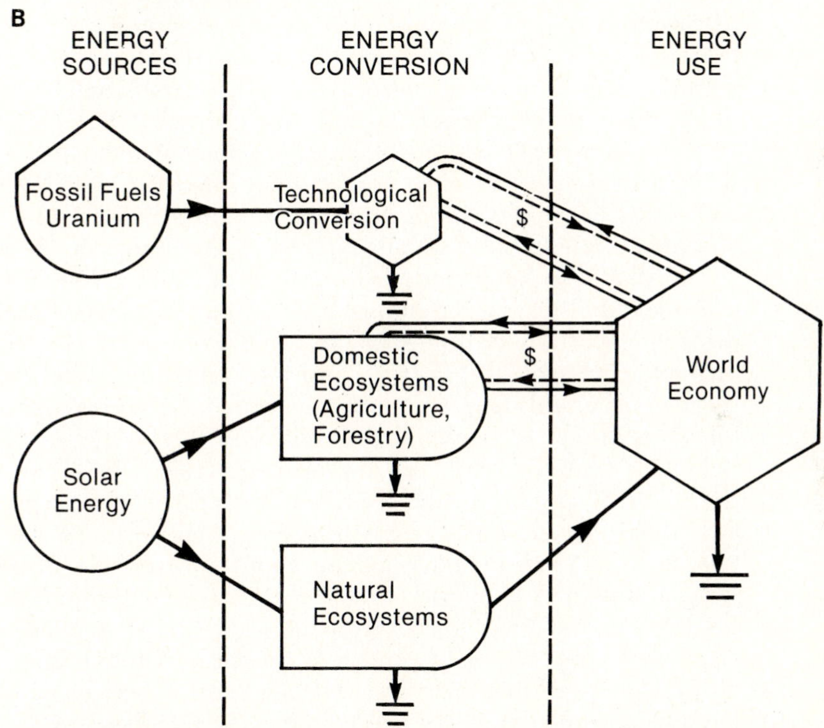

Figure 3–23 *A*. In conventional economics, money is not involved until fish are caught; the work of the estuary to produce fish is given no value. The total value of the estuary in terms of useful work for humans is at least ten times the value of harvested products. (Solid arrows represent energy flows; broken arrows represent money flows.) (See Gosselink, Odum, and Pope, 1974.) *B*. Energy support system for humans. Money flows ($) accompany energy flows from human-made and domesticated ecosystems but not from natural ecosystems. (Diagrams by H. T. Odum.)

been discussed for at least two decades, little has been put into practice. Manufactured goods and human services continue to be given very high values, while equally important goods and services of nature continue to be left out of consideration or, at best, grossly undervalued. As Brown (1979) points out, the global economy ultimately depends on certain basic ecosystems such as the seas, forests, and agriculture. As these resources are depleted or stressed, the world economy suffers accordingly; goods and services of all kinds become scarcer and more costly to produce or preserve, resulting in worldwide inflation.

The current failure of money to track all the energy flows vital to humans is shown in Figure 3–23B. The two basic energy sources and three basic conversion systems vital for humankind are shown. Money circulates to pay for, so to speak, the technological work of the urban-industrial system and for goods and services of the agroecosystem, but not for the equally vital input of goods and services from natural ecosystems. Economists and ecologists, although often differing in their perception of the urgency of market failures and the means to correct them, generally agree that economic theory coupled with properly understood energy theory provides the potential to include the work of nature as an economic value, not as a "free" good, and thus upgrade the economic system to the ecosystem level.

Mishan's *Technology and Growth* (1970), Ellul's *The Technological Society* (1967), and Schumacher's *Small is Beautiful* (1973) are three hard-hitting critiques of the shortcomings of economics in technological society. For more positive approaches that attempt to bridge the "credibility gap" between economic and ecological thinking (in addition to works already cited in this section), see H. T. Odum (1972, 1973), Daly (1973), Westman (1977), Anderson (1977), Ayres (1978), and Smith and Krutilla (1979). An especially promising approach is based on using embodied energy (see page 147 for definition) as a common denominator for goods and services of both humans and nature. Costanza (1980) recently used input-output analyses to calculate total (direct and indirect) i.e., embodied, energy required to produce goods and services in the United States' economy and showed that there was a strong relation between embodied energy and dollar value in many sections of the economy. A discussion of the need for new approaches in economics is continued in Chapter 8, Section 7, and in the Epilogue.

4 Biogeochemical Cycles

1. Patterns and Basic Types of Biogeochemical Cycles

Statement

The chemical elements, including all the essential elements of protoplasm, tend to circulate in the biosphere in characteristic paths from environment to organisms and back to the environment. These more or less circular paths are known as **biogeochemical cycles.** The movement of those elements and inorganic compounds that are essential to life can be conveniently designated as **nutrient cycling.** Each cycle can also be conveniently divided into two compartments or pools: (1) the **reservoir pool**, the large, slow-moving, generally nonbiological component, and (2) the **labile** or **cycling pool**, a smaller but more active portion that is exchanging (i.e., moving back and forth) rapidly between organisms and their immediate environment. From the standpoint of the biosphere as a whole, biogeochemical cycles fall into two basic groups: (1) **gaseous types**, in which the reservoir is in the atmosphere or hydrosphere (ocean), and (2) **sedimentary types**, in which the reservoir is in the earth's crust.

Explanation

As was emphasized in Chapter 2, Section 2, it is profitable in ecology to study not only organisms and their environmental relations but also the basic nonliving environment in relation to organisms. We have seen how the two ecosystem divisions, the biotic and the abiotic, coevolve and influence the behavior of each other (Chapter 2, Section 4). Of the 90-odd elements known to occur in nature, between 30 and 40 are known to be required by living organisms. Some elements, such as carbon, hydrogen, oxygen, and nitrogen, are needed in large quantities; others are needed in small, or even minute, quantities. Whatever the need may be, essential elements exhibit definite biogeochemical cycles. The nonessential elements (i.e., elements not required for life), although less closely coupled with organisms, also cycle, often flowing along with essential elements because of chemical affinity with them. (Examples will be discussed in Section 6.)

"Bio" refers to living organisms and "geo" to the rocks, air, and water of the earth. Geochemistry is concerned with the chemical composition of the earth and with the exchange of elements between different parts of the earth's crust, its atmosphere, and its oceans, rivers, and other bodies of water. Fortescue (1980) has reviewed geochemistry from an ecological and holistic approach in terms of **landscape geochemistry.** This concept is credited to the Russian Polynov (1937) and is defined as the role of chemical elements in the synthesis and decomposition of all kinds of materials, with special emphasis on weathering. **Biogeochemistry**, a term probably first coined by another Russian, Vernadskii (1926), but made prominent by G. E. Hutchinson's early monographs (1943, 1944, 1950), thus becomes the study of the exchange (that is, the back-and-forth movement) of materials between living and nonliving components of the biosphere. Excerpts from key papers in the field of biogeochemistry are presented by Pomeroy (1974).

In Figure 4–1, a biogeochemical cycle is superimposed on a simplified energy-flow diagram to show how the one-way flow of energy drives the cycle of matter. Elements in nature are almost never homogeneously distributed, nor are they present in the same chemical form throughout the ecosystem. In Figure 4–1, the reservoir pool, that portion that is chemically or physically remote from organisms, is indicated by the box labeled "nutrient pool," whereas the cycling portion is designated by the stippled circle going from autotrophs to heterotrophs and back again. Sometimes, the reservoir portion is called the "unavailable" pool and the active cycling pool the "available" or "exchangeable" pool. For example, agronomists

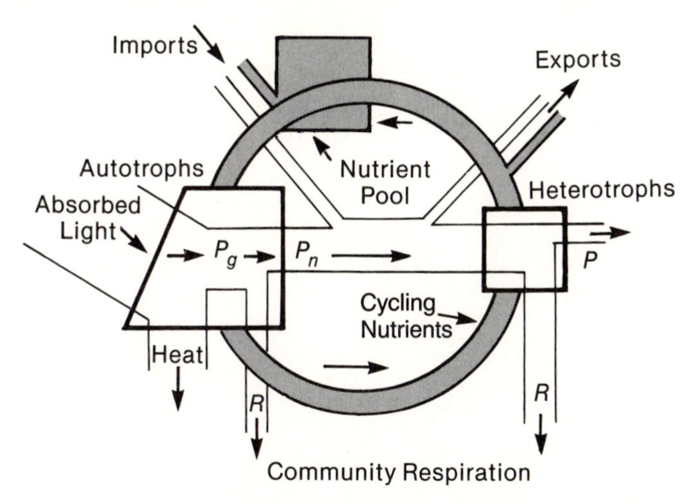

Figure 4–1 A biogeochemical cycle (shaded circle) superimposed upon a simplified energy-flow diagram, contrasting the cycling of material with the one-way flow of energy. P_g = gross production; P_n = net primary production, which may be consumed within the system by heterotrophs or exported from the system; P = secondary production; R = respiration. (After E. P. Odum, 1963.)

routinely measure fertility of soil by estimating the concentration of exchangeable nutrients, that usually small part of total soil nutrient content that is quickly available to plants. Such designations are permissible, provided one clearly understands that the terms are relative. An atom in the reservoir is not necessarily permanently unavailable to organisms, because there are slow fluxes between available and unavailable components. Methods used to estimate exchangeable nutrients in soil testing (usually extraction with weak acids and bases) are at best only rough or approximate indicators. The relative size of reservoir pools is important when one assesses the effect of human activity on biogeochemical cycles. Generally, the smallest pools will be the first affected by changes in fluxes.

The rationale for classifying biogeochemical cycles into gaseous types and sedimentary types is that some cycles, such as those involving carbon, nitrogen, or oxygen, self-adjust rather quickly to perturbations because of the large atmospheric or oceanic reservoirs, or both. Local increases in CO_2 production by oxidation or combustion, for example, tend to be quickly dissipated by air movement and the increased output compensated for by increased plant uptake and carbonate formation in the sea. Gaseous-type cycles, with large atmospheric reservoirs, can be considered "well buffered" globally because of a large capacity to adjust to change. However, there are definite limits to the self-adjustment capacity of even so large a reservoir as the atmosphere. Sedimentary cycles, which involve elements such as phosphorus or iron, tend to be much less cybernetically controlled and more easily disrupted by local perturbations because the great bulk of material is in a relatively inactive and immobile reservoir in the earth's crust. Consequently, some portion of the exchangeable material tends to get lost for long periods of time when "downhill" movement is more rapid than "uphill" return. Return or recycle mechanisms in many cases are chiefly biotic.

Hutchinson, in a classic essay (1948a), points out that human beings are unique not only in requiring the 40 essential elements but also in using nearly all the other elements and the newer synthetic ones as well. We have so speeded up the movement of many materials that the cycles tend to become imperfect, or the process becomes "acyclic," resulting in the paradoxical situation of too little here and too much there. For example, we mine and process phosphate rock with such careless abandon that severe local pollution results near mines and phosphate mills. Then, with equally acute myopia, we increase the input of phosphate fertilizers in agricultural systems without controlling in any way the inevitable increase in runoff that severely stresses our waterways and reduces water quality. The concept of "man as the mighty geological agent" was introduced in Chapter 2 (see page 53).

The aim of conservation of natural resources in the broadest sense is to make acyclic processes become more cyclic. The concept of "recycle" must become a major goal for society. Recycling of water is a good start, because if the hydrologic cycle can be maintained and repaired, there is a better chance of controlling nutrients that move along with the water.

Examples

Three examples will illustrate the principle of cycling. The nitrogen cycle (Figure 4–2) is an example of a very complex and well-buffered gaseous-type cycle; the phosphorus cycle (Figure 4–4) is an example of a simpler, less well-regulated sedimentary type. Both these elements are often very important factors, limiting or controlling the abundance of organisms, and hence have received much attention and study. The sulfur cycle (Figure 4–5) is a good one to illustrate the links between air, water, and the earth's crust, since there is active cycling within and between each of these pools. Both the nitrogen and the sulfur cycles illustrate the key role played by microorganisms and the complications caused by industrial air pollution.

The Nitrogen Cycle Figure 4–2 shows three different ways to picture the complexities of the nitrogen cycle; each diagram illustrates a major overall feature or driving force. Figure 4–2A brings out the circularity of flows and the kinds of microorganisms required for the basic exchanges between organisms and environment. The nitrogen in protoplasm is broken down from organic to inorganic form by a series of decomposer bacteria, each specialized for a particular part of the job. Some of this nitrogen ends up as ammonia and nitrate, the forms most readily used by green plants. The air, which contains 80 percent nitrogen, is the greatest reservoir and safety valve of the system. Nitrogen is continually entering the air by the action of denitrifying bacteria and continually returning to the cycle through the action of nitrogen-fixing bacteria or algae (biofixation) and through the action of lightning and other physical fixation.

Figure 4–2B emphasizes the processes, namely fixation, assimilation, nitrification, denitrification, decomposition, leaching, runoff, rainout, and so on. Also shown are some estimates (numbers in parentheses) of annual global fluxes in teragrams (1 Tg = 10^6 metric tons), including estimates for magnitude of the two flows directly related to human activities, namely emissions into the atmosphere and industrial fixation that is largely added to farmlands in the form of nitrogen fertilizers. The latter is an appreciable amount, about equal to natural atmospheric fixation but less than biofixation, which, of course, is also enhanced by the use of legume crops. Because the atmospheric N_2 reservoir has not recently changed, flow in and out of the atmospheric reservoir (denitrification versus fixation) is thought to be generally balanced, with perhaps a small excess of fixation.

In Figure 4–2C, the components of the nitrogen cycle are shown in terms of the energy necessary for operation of the cycle. The steps from proteins down to nitrates provide energy for organisms that accomplish the breakdown, whereas the return steps require energy from other sources, such as organic matter or sunlight. For example, the chemosynthetic bacteria *Nitrosomonas* (which converts ammonia to nitrite) and *Nitrobacter* (which converts nitrite to nitrate) obtain energy from the breakdown, whereas denitrifying and nitrogen-fixing bacteria require energy from other sources to accomplish their respective transformations.

Until about 1950, the capacity to fix atmospheric nitrogen was thought to be limited to these few, but abundant, kinds of microorganisms:

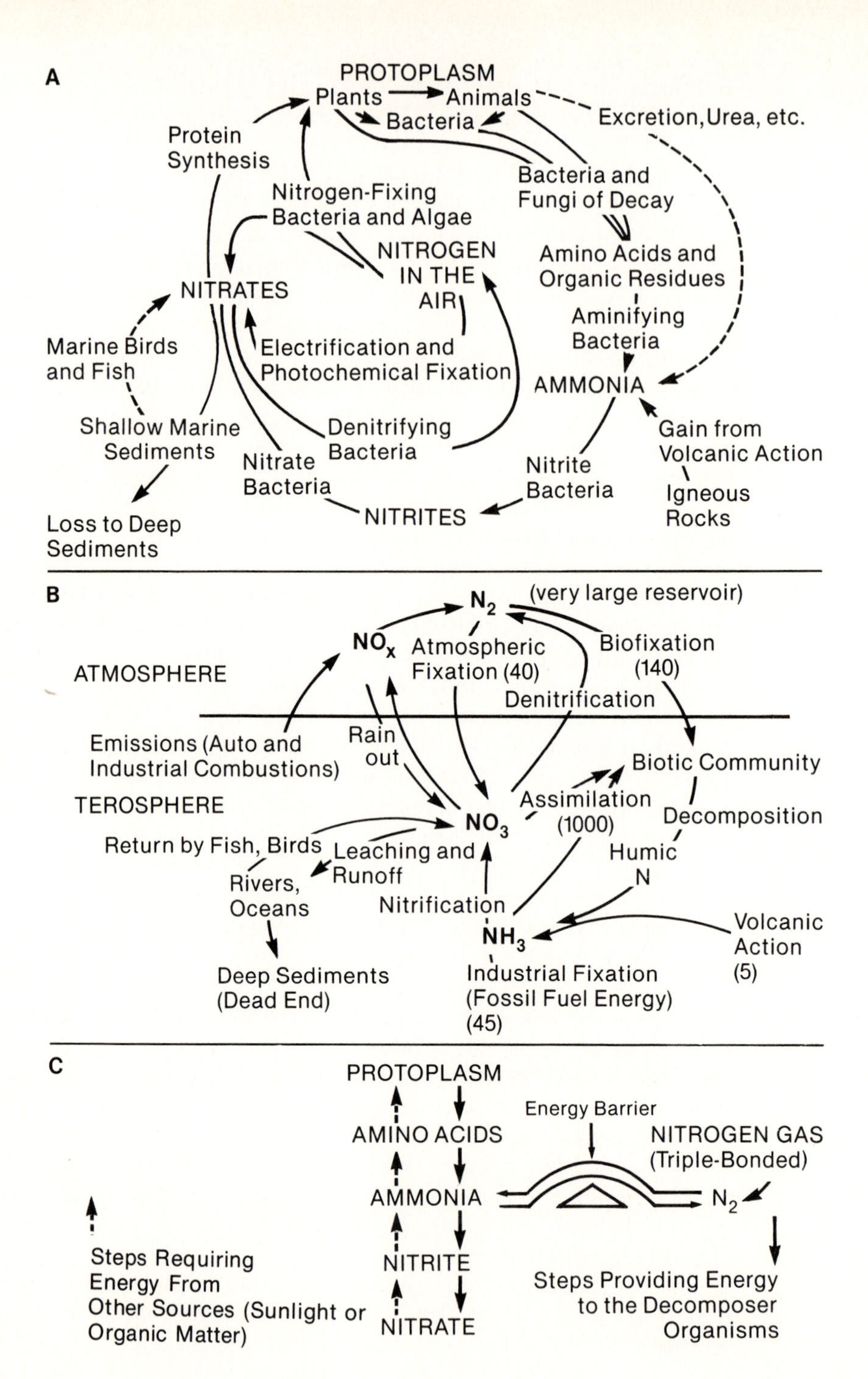
A
PROTOPLASM
Plants
Animals
Bacteria
Excretion, Urea, etc.
Protein Synthesis
Bacteria and Fungi of Decay
Nitrogen-Fixing Bacteria and Algae
NITROGEN IN THE AIR
Amino Acids and Organic Residues
NITRATES
Aminifying Bacteria
Marine Birds and Fish
Electrification and Photochemical Fixation
AMMONIA
Shallow Marine Sediments
Denitrifying Bacteria
Gain from Volcanic Action
Nitrate Bacteria
Nitrite Bacteria
Igneous Rocks
NITRITES
Loss to Deep Sediments
B
N_2 (very large reservoir)
NO_x
Atmospheric Fixation (40)
Biofixation (140)
ATMOSPHERE
Denitrification
Emissions (Auto and Industrial Combustions)
Rain out
Biotic Community
TEROSPHERE
Assimilation (1000)
NO_3
Decomposition
Return by Fish, Birds
Leaching and Runoff
Humic N
Rivers, Oceans
Nitrification
NH_3
Volcanic Action (5)
Deep Sediments (Dead End)
Industrial Fixation (Fossil Fuel Energy) (45)
C
PROTOPLASM
AMINO ACIDS
Energy Barrier
NITROGEN GAS (Triple-Bonded)
AMMONIA
N_2
NITRITE
NITRATE
Steps Requiring Energy From Other Sources (Sunlight or Organic Matter)
Steps Providing Energy to the Decomposer Organisms

1. Free-living bacteria—*Azotobacter* (aerobic) and *Clostridium* (anaerobic).
2. Symbiotic nodule bacteria on legume plants—*Rhizobium* (see Figure 4–3).
3. Blue-green algae, also known as cyanobacteria—*Anabaena*, *Nostoc*, and other members of the order Nostocales.

It was then discovered that the purple bacterium *Rhodospirillum* and other representatives of the photosynthetic bacteria are nitrogen fixers (see Kamen and Gest, 1949; Kamen, 1953) and that a variety of *Pseudomonas*-like soil bacteria also have this capacity (see Anderson, 1955). Later it was discovered that actinomycetes (a kind of primitive fungus) in root nodules of alders (*Alnus*) and certain other non-leguminous woody plants fix nitrogen as efficiently as do *Rhizobium* bacteria in legume nodules, even though the actinorhizal nodules are less complex and evolutionally less advanced (see Tjephema and Winship, 1980, for comparison of N-fixation by the two kinds of nodules). So far, 160 species in five genera in eight families of dicots have been shown to possess actinomycete-induced nodules, including (in addition to *Alnus*) *Ceanothus*, *Comptonia*, *Elaeagnus*, *Myrica*, *Casuarina*, *Coriaria*, *Araucaria*, and *Ginkgo* (see Torrey, 1978). Unlike legumes, which are largely tropical in origin, these nitrogen fixers originate in the temperate zone. Most species are adapted to poor sandy or boggy soils where available nitrogen is scarce. Some species such as alders have the potential of increasing forest yields when interplanted with timber species.

Nitrogen fixation by blue-green algae or cyanobacteria may take place in free-living forms or be symbiotic with fungi as in certain lichens or with mosses, ferns, and at least one seed plant. The fronds of the small floating aquatic fern *Azolla* contain small pores filled with symbiotic blue-green algae (*Anabaena*) that actively fix nitrogen (see Moore, 1969). For centuries, this fern has played an important role in paddy rice culture in the Orient; before the rice seedlings are planted, the flooded paddies are covered with the aquatic ferns,

◀Figure 4–2 Three ways of picturing the nitrogen biogeochemical cycle, an example of a relatively well-buffered self-regulating cycle with a large gaseous reservoir. In *A* the circulation of nitrogen between organisms and environment is depicted along with microorganisms responsible for key steps. In *B* the basic processes are emphasized, and estimates of the magnitude of key flows are shown. Numbers in parentheses are teragrams (1 Tg = 10^6 metric tons) per year. In *C* the basic steps are arranged in an ascending-descending series, with the high energy forms on top to distinguish steps that require energy from those that release energy.

Figure 4–3 *A*. Root nodules on a legume, the location of nitrogen-fixing bacteria of the symbiotic or mutualistic type. The legume shown is blue lupine, a cultivated variety used in the southeastern United States. (U.S. Soil Conservation Service Photo.) *B*. Nodules induced by nitrogen-fixing actinomycetes on woody nonleguminous plants. Nodules shown are on roots of *Camptonia* ("sweet fern," a myrtle). (Courtesy of Dr. John G. Torrey.)

which fix enough nitrogen to supply the crop as it matures. This practice, as well as encouragement of free-living blue-green algae, permits rice to be grown season after season in the same paddy without the addition of fertilizer. As in the case of legume nodule bacteria, symbiotic blue-green algae are more efficient than free-living ones (for a review of N-fixation by blue-green algae, see Peters, 1978).

The key to biofixation is the enzyme nitrogenase that catalyzes the splitting of N_2. This enzyme also can reduce acetylene to ethylene, thereby providing a convenient way to measure nitrogen fixation in nodules, soils, water, or wherever one suspects that fixation is occurring. The acetylene reduction method, together with use of the isotopic tracer ^{15}N, has resulted in a "measurement breakthrough," which is revealing that the ability to fix nitrogen is widespread among photosynthetic, chemosynthetic, and heterotrophic microorganisms. There is even evidence that algae and bacteria growing on leaves and epiphytes in humid tropical forests fix appreciable quantities of atmospheric nitrogen, some of which may be used by the trees themselves. In short, it appears that biological nitrogen fixation goes on in both the autotrophic and heterotrophic strata of ecosystems and in both aerobic and anaerobic zones of soils and aquatic sediments.

Nitrogen fixation is especially energy expensive because much energy is required to break the triple bond of molecular N_2 ($N \equiv N$) so that it can be converted (with addition of hydrogen from water) to two molecules of ammonia (NH_3). For biofixation by legume nodule bacteria, some 10 gm of glucose (about 40 kcal) of plant photosynthate is required to fix 1 gm of nitrogen (efficiency, 10 percent). Free-living fixers are less efficient and may require up to 100 gm of glucose to fix 1 gm of nitrogen (efficiency, 1 percent). See Gutschick (1978) for a review of the energy cost of N-fixation. Similarly, a lot of fossil fuel energy has to be expended in industrial fixation, which is why, pound for pound, nitrogen fertilizer is more expensive than most other fertilizers.

In summary, only the prokaryotes, the nonnucleated and most primitive microorganisms, can convert biologically useless nitrogen gas into the nitrogen forms required to build and maintain living protoplasm. When these microorganisms form mutually beneficial partnerships with higher plants, nitrogen fixation is greatly enhanced. The plant provides a congenial home (i.e., the root nodule), protects the microbes from too much O_2, which inhibits fixation, and furnishes the microbes with the high-quality energy required. In return, the plant gets a readily assimilable supply of fixed nitrogen.

Cooperation for mutual benefit, a survival strategy very common in natural systems, is one that humanity needs to emulate. In fact, a dream of the new breed of genetic engineers is to induce nodule formation in corn, wheat, rice, and other major food crops. If they succeed, money and energy will be saved in cultivation of self-fertilizing crops. In the meantime, the widespread occurrence of already evolved N-fixation systems can be more efficiently used. For example, several varieties of beans could be better utilized in agriculture, especially in the tropics. Nitrogen-fixers work hardest when the nitrogen supply in their environment is low; putting nitrogen fertilizer on a legume crop pretty well shuts down biofixation.

As shown in Figure 4–2B, of the estimated 10^9 metric tons (1000 teragrams) of nitrogen assimilated each year by the global biotic community, about 80 percent is recycled from the land and water stratum (terosphere), and only about 20 percent of the necessary input is "new" nitrogen coming from the atmosphere by fixation and rainout. In contrast, very little of the nitrogen in fertilizers applied to agricultural fields is recycled; most is lost in harvest removal, leaching (runoff), and denitrification. Reduction of the latter loss by use of new synthetic chemicals that inhibit nitrifying bacteria is a possibility now being researched (see Huber et al., 1977). If nitrification (i.e., the ammonia-to-nitrate step) is reduced, then ammonia fertilizers that are relatively immobile compared with nitrate will be retained in the soil. As always, the effect of new chemicals on other organisms and the other processes in the N-cycle, as well as the effects of runoff into aquatic ecosystems, must be carefully checked before the chemicals are recommended for widespread use. According to Burris (1978), use of nitrogen fertilizer in the United States has increased 12 times since 1950 (from 1 million to 12 million tons). Since crop yields have increased no more than twofold, a lot of this massive increase in use of mineral fertilizers is wasted. Furthermore, too much nitrate in food and water can be bad for humans (see the 1978 report of the National Academy of Science). Wastage of nitrogen and energy is avoided when grain crops are rotated or interplanted with legumes (see Heichel, 1976). For more on N-fixation, see Nutman (1976) and Brill (1979).

The self-regulating feedback mechanisms, shown very simply by the arrows in Figure 4–2, make the nitrogen cycle a relatively perfect one, when a large area or the biosphere as a whole is considered. Some nitrogen from heavily populated regions of land, fresh water, and shallow seas is lost to the deep ocean sediments and thus gets out of circulation, at least for a while (a few million years perhaps). This loss is compensated for by nitrogen entering the air from volcanic gases (and also from our "industrial volcanos"). Thus, volcanic action is not to be entirely lamented; it has some use after all. Capping the world's volcanos, even if technically possible, might very well cause the death by starvation of more people than it could save from damage by eruptions.

The Phosphorus Cycle The phosphorus cycle appears somewhat simpler than the nitrogen cycle, because phosphorus occurs in few chemical forms. As shown in Figure 4–4, phosphorus, an important and necessary constituent of protoplasm, tends to circulate the organic compounds being broken down eventually to phosphates, which are again available to plants. The great reservoir of phosphorus is not the air, however, but the rocks and other deposits formed in past geological ages. These deposits are gradually eroding, releasing phosphates to ecosystems, but much phosphate escapes into the sea, where part

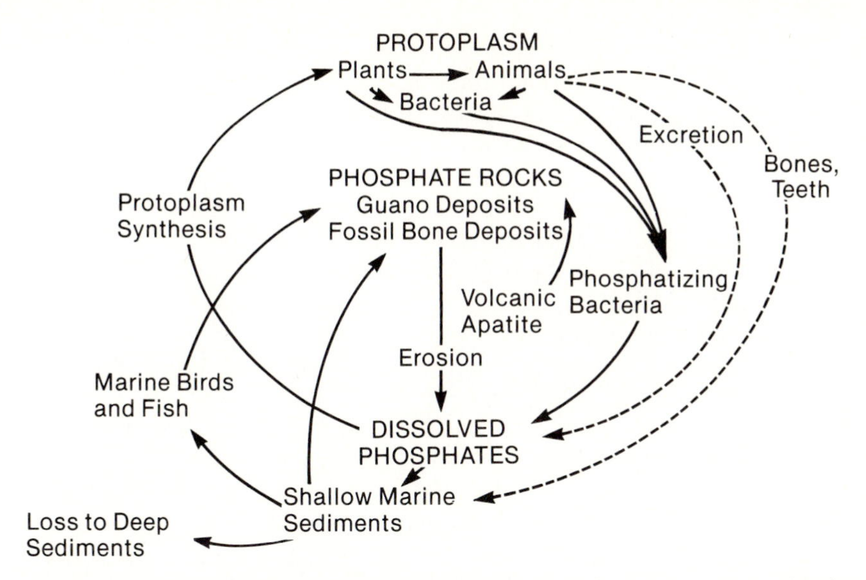

Figure 4–4 The phosphorus cycle. Phosphorus is a rare element compared with nitrogen. Its ratio to nitrogen in natural waters is about 1 to 23 (Hutchinson, 1944). Chemical erosion in the United States has been estimated at 34 metric tons per square kilometer per year. Fifty-year cultivation of virgin soils of the Middle West reduced the P_2O_5 content by 36 percent (Clarke, 1924). As shown in the diagram, the evidence indicates that return of phosphorus to the land has not been keeping up with loss to the ocean.

of it is deposited in the shallow sediments and part of it is lost to the deep sediments. The means of returning phosphorus to the cycle may be inadequate to compensate for the loss. In some parts of the world there is no extensive uplifting of sediments at present, and transport of fish from sea to land is not adequate to replace phosphorus flowing from land to sea. Sea birds have apparently played an important role in returning phosphorus to the cycle (witness the fabulous guano deposits on the coast of Peru). This transfer of phosphorus and other materials by birds from the sea to land is continuing, but apparently not at the rate at which it occurred in the past.

Unfortunately, human activities appear to hasten the rate of loss of phosphorus and thus to make the phosphorus cycle less "cyclic." Although a lot of marine fish are harvested, Hutchinson estimates that only about 60,000 tons of elementary phosphorus per year is returned in this manner, compared with 1 or 2 million tons of phosphate rock that is mined and used for fertilizer, much of which is washed away and lost. There is no immediate cause for concern, since the known reserves of phosphate rock are large. However, mining and processing of phosphate for fertilizer creates severe local pollution problems, as is evident in the Tampa Bay area of Florida, where there are very large deposits. The excess of dissolved phosphate in the waterways resulting from increased input of urban-industrial and agricultural waste water is a great concern at present. Ultimately, phosphorus will have to be recycled on a large scale to avoid famine. Of course, a few geological upheavals raising the "lost sediments" might accomplish it for us. Who knows? One experimental procedure for recycling phosphorus "uphill" involves spraying waste water on upland vegetation or passing it through natural wetlands (marshes and swamps) instead of piping it directly into streams and rivers. (For more on this, see Bouwer, 1968, and Woodwell, 1977.)

At any rate, take a good look at the diagram of the phosphorus cycle. Its importance will loom large in the future, because, of all the macronutrients (vital elements required in large amounts by life), phosphorus is one of the most scarce in terms of its relative abundance in available pools on the earth's surface. Further discussion of phosphorus as a limiting factor appears in Chapter 5.

The Sulfur Cycle A comprehensive diagram of the sulfur cycle is shown in Figure 4–5. Many of the main features of biogeochemical cycling are illustrated by this diagram; for example:

1. The large reservoir in soil and sediments and a smaller reservoir in the atmosphere.
2. The key role in the rapidly fluxing pool (the center "wheel" in Figure 4–5) played by specialized microorganisms that function like a relay team, each carrying out a particular chemical oxidation or reduction (see legend, Figure 4–5).
3. Microbial recovery from deep sediments resulting from an upward movement of a gaseous phase (H_2S), as was discussed on page 36.
4. The interaction of geochemical and meteorological processes (erosion, sedimentation, leaching, rain, adsorption-desorption, and so on) and biological processes (production and decomposition).
5. The interdependence of air, water, and soil in regulation of the cycle at the global level.

Sulfate (SO_4), like nitrate and phosphate, is the principal available form that is reduced by autotrophs and incorporated into proteins, sulfur being an essential constituent of certain amino acids. Not as much sulfur is required by the ecosystem as nitrogen and phos-

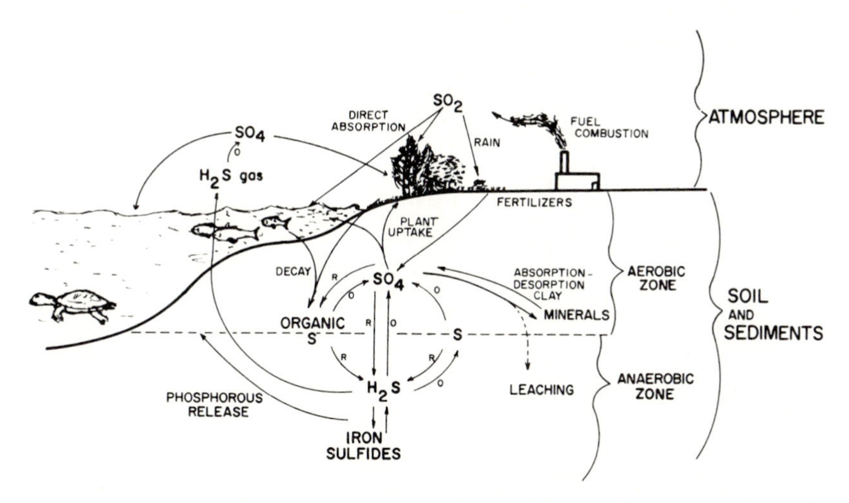

Figure 4–5 The sulfur cycle linking air, water, and soil. The center wheel-like diagram shows oxidation (O) and reductions (R) that bring about key exchanges between the available sulfate (SO_4) pool and the reservoir iron sulfide pool deep in soils and sediments. Specialized microorganisms are largely responsible for the following transformations: $H_2S \rightarrow S \rightarrow SO_4$, colorless, green, and purple sulfur bacteria; $SO_4 \rightarrow H_2S$ (anaerobic sulfate reduction), desulfovibrio bacteria; $H_2S \rightarrow SO_4$ (aerobic sulfide oxidizers), thiobacilli bacteria; organic $S \rightarrow SO_4$ and H_2S, aerobic and anaerobic heterotrophic microorganisms, respectively. The metabolism of these various sulfur bacteria is described in Chapter 2. Primary production, of course, accounts for the incorporation of sulfate into organic matter, while animal excretion is a source of recycled sulfate. Sulfur oxides (SO_2) released into the atmosphere on burning of fossil fuels, especially coal, are becoming increasingly bothersome components of industrial air pollution.

phorus, nor is sulfur as often limiting to the growth of plants and animals. Nevertheless, the sulfur cycle is a key one in the general pattern of production and decomposition, as was referred to in Chapter 2, Section 3. For example, when iron sulfides are formed in the sediments, phosphorus is converted from insoluble to soluble form, as shown by the "phosphorus release" arrow in Figure 4–5, and thus enters the pool available to living organisms. Here is an excellent illustration of how one cycle regulates another. Recovery of phosphorus as part of the sulfur cycle is most pronounced in anaerobic sediments of wetlands (see Patrick et al., 1973, 1974), which are also important sites for recycling of nitrogen and carbon. The interesting metabolism of the several kinds of sulfur bacteria has already been discussed in Chapter 2 (pages 31, 34), and the importance of the reduced-sulfur food chain was noted in Chapter 3 (Figure 3–11B).

Effect of Air Pollution Both the nitrogen and the sulfur cycles are increasingly being affected by industrial air pollution. The oxides of nitrogen (N_2O and NO_2) and sulfur (SO_2), unlike nitrates and sulfates, are toxic to varying degrees. Normally, they are only transitory steps in their respective cycles and are present in most environments in very low concentrations. The combustion of fossil fuels, however, has greatly increased the concentrations of these volatile oxides in the air, especially in urban areas, to the point where they adversely affect important biotic components of ecosystems. When plants, fish, birds, or microbes are poisoned, humans eventually are also affected adversely. These oxides constitute about one third of the industrial air pollutants discharged into the air over the United States. Fortunately, the passage of the Clean Air Act of 1977 and the tightening of emission standards have reduced the volume, so that the Council on Environmental Quality in its 1979 report could state that "Overall, the nation's air quality is improving." This is a trend that must continue, since the present situation is not "good," only less "bad" than it was.

The burning of coal is a major source of SO_2, and automobile exhaust, along with other industrial combustions, is a major source of NO_2. In a sense, such combustions "fix" nitrogen, but in a more poisonous form than natural biofixation. Sulfur dioxide is damaging to photosynthesis, as was discovered in the early 1950s when leafy vegetables, fruit trees, and forests showed signs of stress in the Los Angeles Basin. The destruction of vegetation around copper smelters, described on pages 27–29, is largely caused by this pollutant. Furthermore, SO_2 interacts with water vapor to produce dilute sulfuric acid (H_2SO_4) droplets that fall to earth as **acid rain**, a truly alarming development that is receiving more attention from the public and from researchers. Acid rain is no longer just a local problem of urban areas; its impact has spread into pristine areas of the Adirondack and Appalachian mountains. It has also become a major problem in Scandinavia and northern Europe. In many ways, the building of tall smoke stacks for coal-burning power plants so that local air pollution could be reduced has aggravated the problem, since the longer the oxides remain in cloud layers, the more acid is formed. This is a good example of a short-term "quick fix" that produces a more severe long-term problem. The solution is to remove the sulfur from the fuel or from the emissions. Acid rain has its greatest impact on soft-water lakes or streams and already acid soils that lack pH buffers (such as carbonates, calcium, salts, and other bases). The increase in acidity (decrease in pH) in some Adirondack lakes has rendered them incapable of supporting fish.

The oxides of nitrogen are also threatening the quality of life. They irritate the respiratory membranes of higher animals and humans. Furthermore, chemical reactions with other pollutants produce a synergism (total effect of the interaction exceeds the sum of the effects of each substance) that increases the danger. For example, in the presence of ultraviolet radiation in sunlight, NO_2 reacts with unburned hydrocarbons (both produced in large quantities by automobiles) to produce a photochemical smog, which not only makes one's eyes tear but also, like smoking cigarettes, is dangerous to one's health.

For more on the N-, P-, and S-cycles, see the Swedish Scope report edited by Svenson and Söderlund (1976). For a review of the effect of air pollution on vegetation, see Kozlowski (1980).

2. Quantitative Study of Biogeochemical Cycles

Statement

The rates of exchange or transfers from one place to another are more important in determining the structure and function of an ecosystem than the amounts present at any one time in any one place. To understand and thereby better control the human role in the cycles of materials, we must quantify **cycling rates** as well as **standing stocks**. During the past 25 years, improvements in tracers, mass chemistry, monitoring, and remote sensing techniques have permitted the measurement of cycling rates in sizable units such as lakes and forests and the starting of the important task of quantifying biogeochemical cycles at the global level. Tracers also opened the way to compartment analysis ("box" models; see Figure 1–2) and to sophisticated mathematical modeling.

Examples

Diagrams such as Figures 4–2, 4–4, and 4–5 show only the broad outlines of biogeochemical cycles. Quantitative relations—that is, how much material passes along the routes shown by the arrows, and how fast it moves—are still poorly known, especially for large systems. The numbers for some of the flows in the nitrogen cycle (Figure 4–2B) are, at most, "first approximations"; for many of the major flows, we cannot even guess about the global scale. Radioactive isotopes, which have been generally available since 1946, are providing a tremendous stimulus for cycling studies, since these isotopes can be used as tracers or tags to follow the movement of materials. Tracer studies in ecosystems, as in organisms, are designed so that the amount of radioactive element introduced is extremely small compared with the amount of nonradioactive element already in the system. Therefore, neither the radioactivity nor the extra ions disturb the system. What happens to the tracer (which can be detected in extremely small amounts by the telltale radiations it emits) simply reflects what is normally happening to that particular material in the system.

Ponds and lakes are especially good sites for study because their nutrient cycles are relatively self-contained over short periods. Since the pioneer experiments of Coffin, Hayes, Jodrey, and Whiteway (1949) and Hutchinson and Bowen (1948, 1950), many researchers have reported the results of the use of radiophosphorus (^{32}P) and other sophisticated techniques in studies of phosphorus circulation in lakes. Hutchinson (1957) and Pomeroy (1970) have summarized studies on the rate of cycling of phosphorus and other vital elements.

Phosphorus generally does not move evenly and smoothly from organism to environment and back to organism as one might think from looking at the diagram in Figure 4–3, even though a long-term equilibrium tends to be established. At any one time, most of the phosphorus is tied up either in organisms or in solids (i.e., organic detritus and inorganic particles that make up the sediments). In lakes, 10 percent is the maximum amount likely to be in a soluble form at any one time. Rapid back-and-forth movement or exchange occurs all the time, but extensive movement between solid and dissolved states is often irregular or "jerky." Periods of net release from the sediments are followed by periods of net uptake by organisms or sediments, depending on seasonal temperature conditions and activities of organisms. Generally, uptake rate is more rapid than release rate. Plants readily take up phosphorus in the dark or under other conditions when they cannot use it. When producers exhibit rapid growth, which often occurs in the spring, all the available phosphorus may become tied up in producers and consumers, which may take up more nutrient than they need at the moment (aquatic biologists speak of this as luxury uptake). Accordingly, the concentration of phosphorus in the water at any one time may bear little relation to productivity of the ecosystem. A low level of dissolved phosphate could mean an impoverished system or a system that is very active metabolically; only by measuring the flux rate can the true situation be determined. According to Pomeroy (1960), "a rapid flux of phosphate is typical of highly productive systems, and the flux rate is more impor-

Table 4–1 **Estimates of the Turnover Time of Phosphorus in Water and Sediments of Three Lakes as Determined with the Use of ^{32}P***

Lake	Area (km^2)	Depth (M)	Turnover Time in Days		Ratio Mobile P to Total P in Water
			Water	Sediments	
Bluff	0.4	7	5.4	39	6.4
Punchbowl	0.3	6	7.6	37	4.7
Crecy	2.04	3.8	17.0	176	8.7

* After Hutchinson, 1957.

tant than the concentration in maintaining high rates of organic production."

The concept of turnover, as first introduced in Chapter 2 (page 60), is useful for comparing exchange rates between different components of an ecosystem. After equilibrium has been established, the **turnover rate** is the fraction of the total amount of a substance in a component that is released (or that enters) in a given length of time, whereas **turnover time** is the reciprocal of this, that is, the time required to replace a quantity of substance equal to the amount in the component. For example, if 1000 units are present in the component and 10 go out or enter each hour, the turnover rate is 10/1000 or 0.01 or 1 percent per hour. Turnover time would then be 1000/10 or 100 hours. **Residence time**, a term widely used in the geochemical literature, is a concept similar to turnover time. It refers to the time a given amount of substance remains in a designated compartment of a system. Data on turnover time for two large components, the water and the sediments, of three lakes are given in Table 4–1. The smaller lakes have a shorter turnover time, presumably because the ratio of bottom "mud" surface to water volume is greater. In general, the turnover time for the water of small or shallow lakes is a matter of days or weeks; for large lakes, it may be a matter of months.

Studies with ^{32}P-tagged fertilizers in land ecosystems have revealed similar patterns; much of the phosphorus is "locked up" and unavailable to plants at any given time (see Comar, 1957, for a summary of some of these experiments).

One very practical result of intensive studies of nutrient cycles has been repeatedly demonstrated: overfertilization can be just as "bad," from the standpoint of human interest, as underfertilization. When more materials are added than can be used by the organisms active at the time, the excess is often quickly tied up in soil or sediments or is lost completely (as by leaching) and is unavailable when increased growth is most desired. Too many people mistakenly believe that if 1 pound of nutrient (or pesticide) is recommended for a given area of their garden or pond, 2 pounds will be twice as good. These believers in "more is always better" need to understand the subsidy-stress principle as diagrammed in Figure 3–5. Subsidies inherently tend to turn into stresses, if one is not prudent. The dumping of fertilizers in ecosystems such as fish ponds is not only wasteful insofar as the desired results are concerned but is also likely to create unanticipated changes in the system as well as "downstream" pollution. Since different organisms are adapted to specific levels of materials, continued excess fertilization may result in a change in the kinds of organisms, perhaps discouraging the ones we want and encouraging the kinds we do not want. The destruction of an oyster industry caused by a change in phytoplankton populations resulting from increased fertilization by phosphorus and nitrogen materials is described in Chapter 5 (see pages 228, 229).

During the 1950s and 1960s, an active interdisciplinary field of **radiation ecology** developed. It is concerned not only with studying the effects of ionizing radiation but also with developing and using the new tracer tools. The interaction of physicists, chemists, biologists, and mathematicians produced a sort of "hybrid vigor" in this field that profoundly influenced the theory and practice of ecology, especially its biogeochemical aspects. (For a brief historical review of radiation ecology and some of its accomplishments, see E. P. Odum, 1980a.) First came compartment analysis, in which the major structural or "standing state" components of an ecosystem are depicted as compartments connected with flows of nutrients and/or energy (as in the simple "box model" of Figure 2–2), many of which can be verified by use of tracers. Mathematical equations are then "fitted" to observed rates of flow. The rapid development of computers during this period made it possible to run and integrate many equations simultaneously.

An early version of a compartment model of the phosphorus exchange within a Georgia salt marsh is shown in Figure 4–6. In this model, seasonal pulses are built into the circuits, since, as already indicated, movement of materials is neither continuous nor a linear function of time. In the marsh, a major pulse in the growth of the marsh grass occurs during the warmer months when phosphorus is "pumped" up through grass roots that penetrate deeply into the anaerobic sediments. This "recovery" process was experimentally verified by injecting ^{32}P into deep sediments. Two major pulses of decomposition release phosphorus into the water, one in the hottest

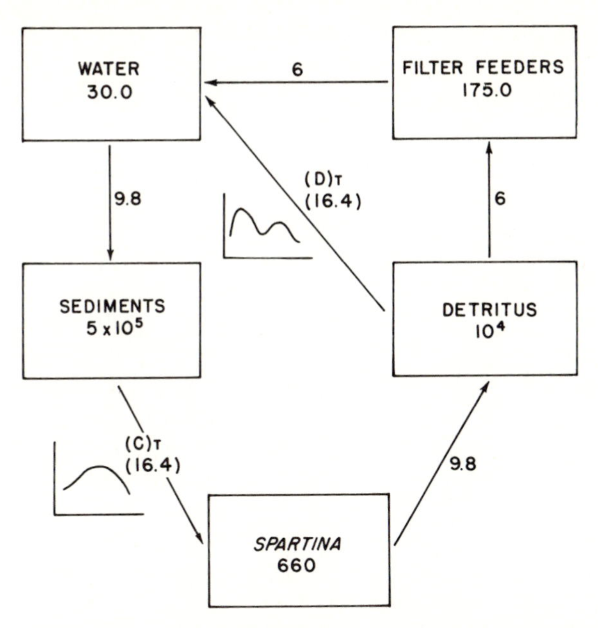

Figure 4–6 Compartmental flux diagram of phosphorus in a Georgia salt marsh ecosystem (see Pomeroy, Johannes, Odum, and Roffman, 1969). Two large reservoirs of phosphorus (sediments and detritus) and the three most active compartments (water, *Spartina* or marsh grass, and detritus-feeding animals) are shown. Quantities within compartments are standing stocks in mg P/m^2; transfer fluxes are expressed in mg P/m^3/day. This is a linear model with two variable coefficients, (D)t and (C)t, that mimic seasonal pulses in release of phosphorus from decomposing detritus and uptake of phosphorus by marsh grass, respectively. A graph of each seasonal pulse is shown. Quantities in parentheses are integrated means of the variable transfers.

part of the summer and another in the fall when large quantities of dead grass are washed out of the marsh by high seasonal tides. When systems of differential equations were set up to describe how the content of phosphorus in one compartment affects that of adjoining parts of the system, and when the whole model was then tested with an analog computer, it was found that several "throughputs" had to be adjusted to keep the "contents" of the smaller compartments stable. Once the computer was adjusted so that the compartments neither overfilled nor drained out, a simplified working model of the behavior of phosphorus in a very large natural system was produced. These procedures illustrate the two points emphasized in Chapter 1, Section 3: (1) a realistic though greatly simplified model can be constructed from a relatively small amount of field data, and (2) such models can be "tuned" by computer manipulations to determine what transfer coefficients will make the model a better mimic of the real world.

This particular model reflects the importance of filter-feeder and detritus complexes in recycling phosphorus in this estuarine system, as was determined by field observations and experiments. For example, Kuenzler (1961) had shown that the population of filter-feeding mussels (*Modiolus demissus*) alone "recycles" from the water every 2½ days a quantity of particulate phosphorus equivalent to the amount present in the water (i.e., a turnover time of only 2½ days for particulate phosphorus in the water). Kuenzler (1961a) also measured the energy flow of the population and concluded that the mussel population is more important to the ecosystem as a biogeochemical agent than as a transformer of energy (i.e., as a potential source of food for other animals or humans). This example illustrates the fact that a species does not have to be a link in our food chain to be valuable to us. Many species are valuable in indirect ways that are not apparent on superficial examination.

What good are all those species in nature that humans cannot eat or sell? This question will be discussed again in later chapters. As observations and sophisticated mathematical modeling are now showing, interactions of cause and effect among many species control transfers in the compartments that have direct value to humanity.

3. Watershed Biogeochemistry

Statement

Bodies of water are not closed systems but need to be considered as parts of larger drainage basins or watershed systems (see Chapter 2, Section 7). The watershed systems provide a sort of minimum ecosystem unit insofar as practical management is concerned. Long-term studies (10 years or more of year-round research) on experimental, instrumented watersheds (outdoor laboratories, as it were), such as those ongoing at Hubbard Brook in New Hampshire, Coweeta in western North Carolina, and many other sites in the United States and elsewhere in the world, have greatly advanced our understanding of the basic biogeochemical processes as they occur in relatively undisturbed headwater ecosystems. Such studies in turn have provided a basis for comparison with agricultural, urban, and other more "domesticated" watersheds where most people live. These comparisons reveal just how careless and unnecessarily wasteful many human activities are, and point to means of reducing downhill losses and restoring the cyclic behavior of our vital nutrients, all of which, of course, also conserve energy.

Examples

A quantitative model of the calcium cycle for mountainous, forested watersheds of the Hubbard Brook study area in New Hampshire is presented in Figure 4–7. The data are based on studies of six watersheds ranging from 12 to 48 hectares (Bormann and Likens, 1967, 1979; Likens et al., 1977). Precipitation, which averaged 123 cm (58 inches) per year, was measured by a network of gauging stations, and the amount of water leaving the watershed in the drainage stream of each watershed unit was measured by a V-notched weir similar to that shown in Figure 2–4. From the concentration of calcium and other minerals in the input and output water and in the biotic and soil pools, a watershed input-output "budget" can be calculated, as shown in a simplified manner in Figure 4–7A.

Retention by and recycling within the undisturbed but rapidly growing forest proved so effective that the estimated loss from the ecosystem was only 8 kg calcium/ha/year (and equally small amounts of other nutrients). Since 3 kg of the calcium was replaced in rain, an input of only 5 kg/ha would be needed to achieve a balance. This amount is thought to be easily supplied by the normal rate of weathering from the underlying rock that constitutes the reservoir pool. Experiments with the radionuclide ^{45}Ca to measure turnover on Oak Ridge, Tennessee, watersheds have demonstrated how understory trees, such as dogwood, act as calcium "pumps," which counter the downward movement in the soil and thus circulate calcium between organisms and the active upper layers of litter and soil (see Thomas, 1969).

On one of the Hubbard Brook experimental watersheds, all the vegetation was felled, and regrowth the next three seasons was suppressed by aerial application of herbicides. Even though the soil was little disturbed and no organic matter was removed by this procedure, the loss of mineral nutrients in stream outflow increased three to 15 times over losses from the undisturbed control watersheds. The sixfold increase in loss of calcium and 15-fold increase in nitrogen loss are shown in Figure 4–7B. Increased stream flow out of the cut-over ecosystem resulted primarily from the elimination of plant transpiration, and it was the additional stream flow that carried out additional minerals. To some extent, outputs are related to what

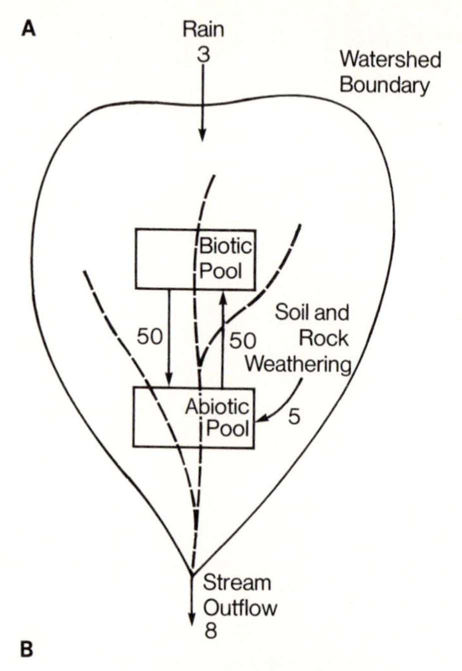

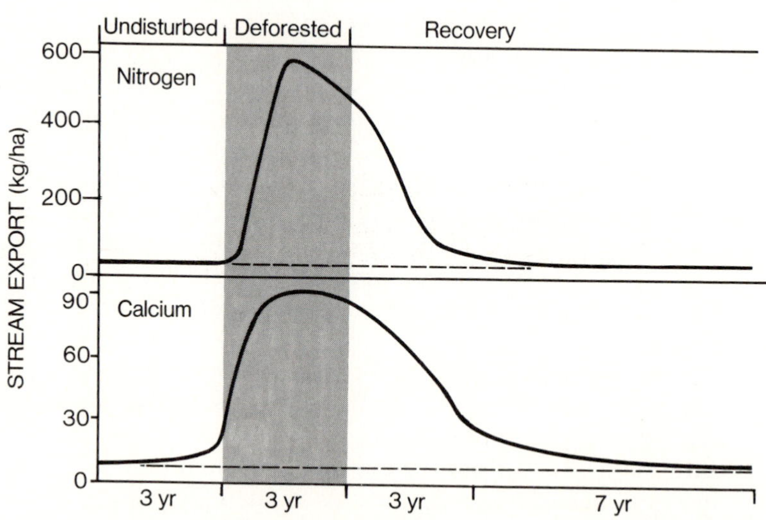

Figure 4–7 *A*. The balanced calcium budget of a forested watershed, Hubbard Brook Experimental Watershed, New Hampshire. Figures are calcium flows in kilograms per hectare per year. The inputs and outputs are small compared with exchanges between biotic and abiotic pools within the watershed ecosystem. *B*. Effect of deforestation and natural reforestation (recovery) on stream output of nitrogen and calcium. (After Bormann and Likens, 1967, 1979.)

geochemists call "relative mobility." Potassium and nitrogen are very mobile (i.e., easily removed by leaching); calcium, however, is more tightly held in the soil.

When vegetation was allowed to recover (no further application of herbicides), nutrient losses declined rapidly, with something like a "balanced budget" being restored in 3 to 5 years, although 10 to 20 years were required for all nutrients to return to the baseline output conditions of an undisturbed forested watershed (Figure 4–7B). Rapid recovery of nutrient retention, long before the species composition and biomass of the original forest can be restored, is aided by a number of biocontrol adaptations such as what Marks (1974) calls the "buried seed strategy." Seeds of rapidly growing pioneer trees such as pin cherry remain viable for years when buried in the soil. When the forest is removed, these seeds sprout, and the fast-growing cherry trees quickly form a sort of temporary forest that stabilizes water and nutrient fluxes and reduces soil loss from the watershed. Such quick recovery adaptations have evolved, of course, in response to natural perturbations such as storms and fire. In fact, forests (and other ecosystems as well) subjected to (and therefore adapted to) natural periodic disruptions are more resilient and recover more quickly after human disturbances than do forests in benign physical environments not so subject to severe natural perturbations. Accordingly, inherent resiliency is an ecosystem-level property that needs to be considered when a harvest procedure or other management practice is being decided.

Studies at the Coweeta site in North Carolina reveal the true costs to society as a whole of increasing water flow downstream to meet the demands of a wasteful industrial society. Removing forests from the hills will increase water "yield" to the valleys (see Hibbert, 1967), but at the expense of water quality as well as wood production and air-regenerating capacity of the watershed.

Nutrient losses from undisturbed forested watersheds along headwater streams are small and mostly replaced by inputs from rain and weathering. The picture downstream where human activities are more intense is quite different. Figure 4–8 shows how concentration of nitrogen and phosphorus in waters of streams and rivers increases sharply as watersheds are increasingly "domesticated," that is, as percent of area in agricultural and urban use increases. Concentrations in water flowing out of a 100 percent urban-agricultural landscape are sevenfold higher than in streams draining a completely forested watershed. It is interesting that N and P are about equally affected, since the N/P ratio remains the same, about 28 : 1, compared with an N/P ratio in biomass of 16 : 1. Eighty percent of phosphorus

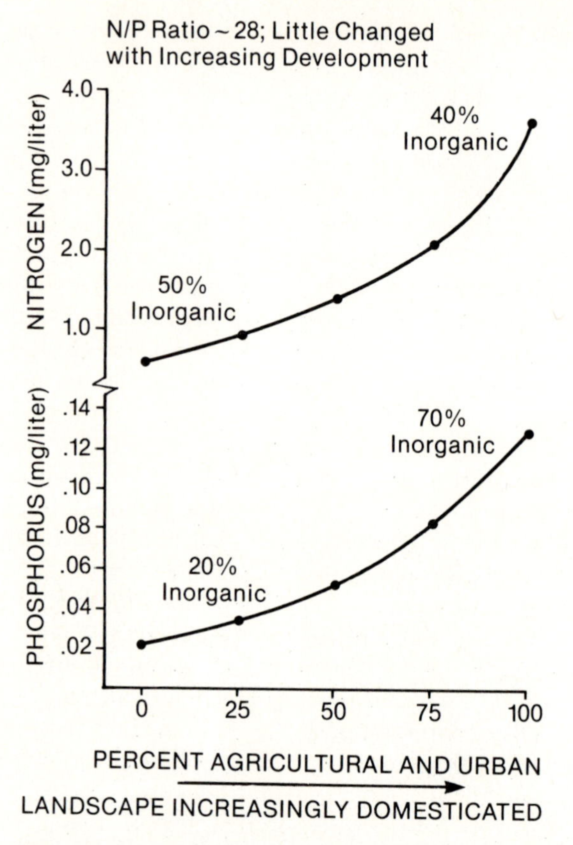

Figure 4–8 Nitrogen and phosphorus in stream water as a function of the percent of agricultural plus industrial development on the watershed. Although the N/P ratio does not change in this gradient, a purely urban landscape tends to release relatively more nitrogen, whereas more phosphorus tends to run off from a predominantly agricultural watershed. Zero is natural vegetation. (Based on data of Omernik, 1977.)

output from agricultural and urban landscapes is inorganic (phosphate), whereas organic phosphorus predominates in runoff from watersheds completely occupied by forest or other natural vegetation.

Though Figure 4–8 depicts only nitrogen and phosphorus, most other nutrients and many other chemicals as well (including the toxic ones) show a similar pattern of increasing runoff with increasing intensity of land and energy use by humans. The large outputs of nutrients and other chemical materials from "domesticated" and especially "industrialized" landscapes are, of course, a more or less direct result of large inputs of agricultural and industrial chemicals and food, the mineral part of which is released by decomposition and sewage treatment. Very little of these large inputs is recycled since so much of it runs off or, as in the case of toxic wastes, builds up in soil and groundwater reservoirs. Technology can facilitate recycling and reduce wasteful "once-through" flow, but there is little incentive to implement such know-how until (1) inputs become scarce, or expensive, or both, and (2) outputs cause serious declines in water quality or threaten human health. Then it "pays" to allocate the necessary money and energy to do the job. The example of paper recycling (see Figure 4–15) illustrates this process of decision making in the real world of economics and politics.

4. The Global Cycling of Carbon and Water

Statement

At the global level, the CO_2 cycle and the hydrologic cycle are probably the two most important biogeochemical cycles as far as humanity is concerned. Both are characterized by small but very active atmospheric pools that are vulnerable to man-made perturbations, which, in turn, can change weather and climates. A network of worldwide measurements has been established to detect significant changes in the CO_2 and H_2O cycles that can quite literally affect our future on earth.

Explanation

For the CO_2 cycle (Figure 4–9A), the atmospheric pool is very small compared with that for carbon in the oceans and in fossil fuels and other storages in the earth's crust. Until the onset of the industrial age, it is believed, flows between atmosphere and continents and oceans were balanced, as shown by the solid lines in Figure 4–9A. During the past century, however, the CO_2 content has been rising because of new anthropogenic inputs, as shown by the dotted lines in Figure 4–9A. Although the burning of fossil fuels is believed to be the main source of the new input, agriculture and deforestation also contribute.

A net loss of CO_2 in agriculture (i.e., addition of more CO_2 into the atmosphere than is removed) may seem surprising, but it results because the CO_2 fixed by crops (many of which are active for only a part of the year) does not compensate for the CO_2 released from the

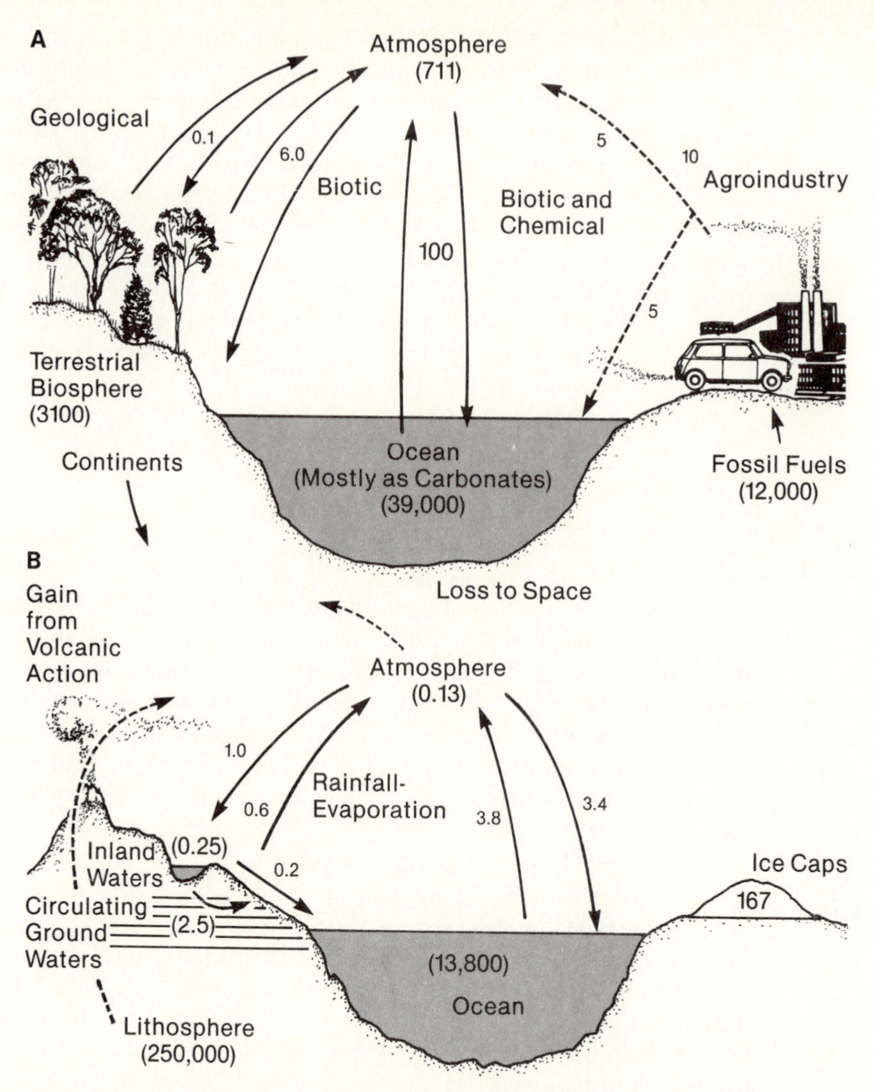

Figure 4–9 *A*. The CO_2 cycle. Figures are 10^9 tons CO_2 in major biosphere compartments and circulating between compartments (arrows). (Data from 1981 report of the Council on Environmental Quality.) *B*. The hydrologic cycle. Figures are geograms (10^{20} grams) H_2O in major biosphere compartments and circulating between compartments (arrows). (Data from Hutchinson, 1957.)

soil, especially that resulting from frequent plowing. Forest removal, of course, may release the carbon stored in wood, especially if the wood is immediately burned, and this is followed by oxidation of humus if the land is used for agriculture or urban development.

Few agree on the "CO_2 problem" and the relative contribution of different human activities to atmospheric enrichment. At one extreme, Woodwell et al. (1978) believe that destruction of biotic pools is as important as the burning of fossil fuels. Broecker et al. (1979), however, conclude that these sources are minor compared with the combustion of fossil fuels. Bolin (1977) takes an intermediate position. All agree that forests are important carbon "sinks," since forest biomass is estimated to contain 1.5 times and forest humus 4 times the amount of carbon in the atmosphere.

The rapid oxidation of humus and the release of gaseous CO_2 normally held in the soil has other, more subtle effects that are just now being recognized, including effects on the cycling of other nutrients. For example, Nelson (1967) used clam shells to demonstrate that deforestation and agriculture have caused a decline in the amount of certain trace elements in soil water runoff. He found that clam shells from Indian middens 1000 to 2000 years old contained 50 to 100 percent more manganese and barium than contemporary shells. By a process of elimination, Nelson concluded that reduced flow of CO_2-charged acidic water percolating deep in the soil has decreased the rate of dissolution of these elements from the underlying rocks. In other words, water now tends to run off rapidly over the surface instead of filtering down through humus layers in the soil. In ecological terms, the flux between the reservoir and the exchangeable pool has been fundamentally altered by our present management of the landscape. If we recognize what has happened and learn to compensate, such changes need not be detrimental. Agronomists are discovering that they now must add trace minerals to fertilizers to maintain yields in many areas, because agroecosystems do not regenerate these nutrients as well as do natural systems.

Recall how the earth's atmosphere came to have its present very low CO_2 content and very high O_2 content. The evolution of the atmosphere was briefly outlined in Chapter 2, Section 4, in connection with the Gaia hypothesis (see also Figure 8–11). When life began on earth more than 2 billion years ago, the atmosphere, like that of the planet Jupiter today, was composed of volcanic gases (atmospheric formation by crustal outgassing, as a geologist would phrase it). It contained quantities of CO_2 but little if any oxygen, and the first life was anaerobic. The buildup of oxygen and decline of CO_2 over geological time have resulted because P (production) has, on average, exceeded R (respiration). Also, geological and purely chemical processes, such as the release of O_2 from iron oxides or the formation of reduced nitrogen compounds and the splitting of oxygen from water by ultraviolet action, have probably contributed to oxygen accumulation (see Cloud, 1978). Both the low concentration of CO_2 and the high concentration of O_2 are now limiting to photosynthesis: most plants increase their rate of photosynthesis if either the CO_2 concentration is increased or the O_2 concentration is decreased experimentally. This makes green plants very responsive regulators.

Even though the earth's photosynthetic green belt and the carbonate system of the sea tend to keep CO_2 in the atmosphere stable, the spiraling increase in consumption of fossil fuels (consider the tremendous amount of CO_2 that would be released if even half of the large fossil fuel pool, as shown in Figure 4–9A, was burned) coupled with the decrease in the "removal capacity" of the green belt is beginning to exceed cybernetic control, so the CO_2 content of the atmosphere is now rising gradually. Recall from the previous discussion that the content of the small, active compartments is most affected by changes in fluxes or "throughputs." At the beginning of the Industrial Revolution (about 1800), it is believed, the CO_2 concentration in our air was about 290 ppm (parts per million) = .29 percent. In 1958, when accurate measurements were first made, the concentration was 315; in 1980, it had risen to 335. Should the concentration double the preindustrial level, which could happen by the middle of the next century if the present rate of increase continues, a warming of the global climate is likely to occur, with a mean rise in temperature of 1.5 to 4.5°C, along with a rise in sea levels (as polar ice melts) and changes in rainfall patterns that could disrupt agricultural production. As recently documented (Gornitz et al., 1982; Etkins and Epstein, 1982), mean sea level has already started to rise, about 12 cm in this century. These threats (climate change and flooding of coastal regions) should become a factor in planning national and international energy policy. For reviews of the "CO_2 problem," see Baes et al. (1977) and panel reports of the Council on Environmental Quality (1981) and the National Academy of Science (1979).

In the next century, a new but uncertain balance will exist between increasing CO_2 (which tends to warm the earth) and increasing dust or particulate pollution (which reflects incoming radiant energy and cools the earth). Any significant net change in the heat budget, either way, will affect climates (see Bryson, 1974, for a good review of possibilities).

In addition to CO_2, two other forms of carbon are present in the atmosphere in small amounts: carbon monoxide (CO) at about 0.1 ppm and methane (CH_4) at about 1.6 ppm. As with CO_2, these flux rapidly and thus have short residence times, about 0.1 year for CO, 3.6 years for CH_4, and 4 years for CO_2. Both CO and CH_4 arise from incomplete or anaerobic decomposition of organic matter; in the atmosphere both are oxidized to CO_2. An amount of CO equal to that from natural decomposition is now injected into the air by incomplete burning of fossil fuels, especially automobile exhaust. Carbon monoxide, a deadly poison to humans, is not a global threat but becomes a worrisome pollutant in urban areas when the air is stagnant. Concentrations of up to 100 ppm are not uncommon in areas of heavy automobile traffic (pack-a-day cigarette smokers receive up to 400 ppm, which reduces their oxyhemoglobin 3 percent, a stress that can lead to anemia and other circulatory and oxygen-related sicknesses).

Methane is believed to have a beneficial function in maintaining the stability of the upper atmospheric ozone layer, which blocks out lethal ultraviolet solar radiation (see page 95). An important function of the world's wetlands and coastal seas is the production of methane.

For a good review of the carbon cycle as a whole, see Chapter 6 in Garrels, Mackenzie, and Hunt (1975).

As shown in the diagram of the hydrologic cycle (Figure 4–9B), the H_2O atmospheric compartment is small, and it has a more rapid turnover rate and shorter residence time in the atmosphere than CO_2. The water cycle, like the CO_2 cycle, is beginning to be affected by human activities on the global scale. Though worldwide monitoring of rainfall and streamflows has been maintained, we urgently need to monitor more completely all of the major fluxes.

In Figure 4–10 the hydrologic cycle is shown in terms of energy, with an "uphill" loop driven by the sun and a downhill loop that releases energy usable by ecosystems and for generating hydroelectric power. As was shown in Table 3–3, about one third of all solar energy is dissipated in driving the hydrologic cycle. Again, we depend on solar energy as a "non-market" service. Too often we do not appreciate the service because we do not pay money for it. If we disrupt this service, however, we will indeed pay dearly!

Two aspects of the H_2O cycle need special emphasis:

1. More water evaporates from the sea than returns by rainfall, and vice versa for the land. In other words, a considerable part of the rainfall that supports land ecosystems, including most food production, comes from water evaporated over the sea. In many areas (the Mississippi Valley, for example) as much as 90 percent of the rainfall is estimated to come from the sea (Benton, Blackburn, and Snead, 1950).

2. Since an estimated 0.25 geogram (1 geogram = 10^{20} grams or 10^{14} metric tons) of water is in freshwater lakes and rivers, and 0.2 geogram runs off each year, the turnover time is about 1 year. The difference between the annual rainfall (1.0 geogram) and the runoff (0.2 geogram), or 0.8, is an estimate of the annual recharge rate of ground

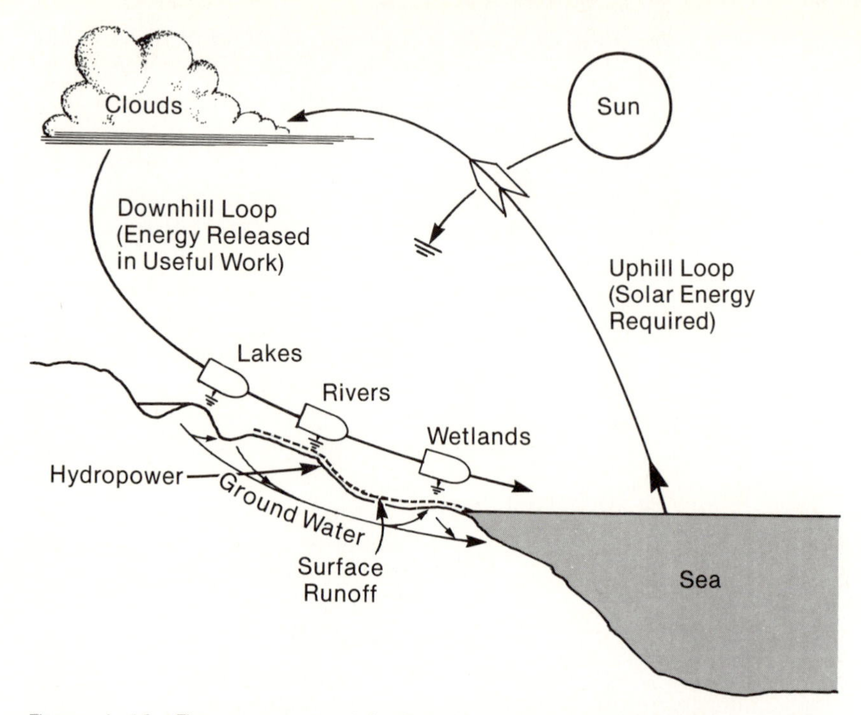

Figure 4–10 The energetics of the hydrologic cycle viewed as two loops: (1) the uphill loop driven by solar energy and (2) the downhill loop that releases energy to lakes, rivers, and wetlands and performs useful work of direct benefit to humans (such as hydropower). The surface runoff recharges and is recharged by groundwater reservoirs, although many in dry regions are not currently being recharged as fast as they are being pumped out for human use.

waters. As already indicated, human activities tend to increase the rate of runoff (by paving over the earth, ditching and diking rivers, compacting agricultural soils, deforestation, and so on), which reduces the recharge of the very important ground water compartment. In the United States, about half the drinking water, most irrigation water, and, in many sections, a large part of industrial-use water comes from ground water. In dry areas such as the western Great Plains, water in the underground aquifers is essentially "fossil" water stored during earlier, wetter geological periods that is not now being recharged. Consequently, it is a nonrenewable resource like oil. A case in point is the heavily irrigated grain region of western Nebraska, Oklahoma, Texas, and Kansas, where the principal aquifer, called the Ogallala, will be pumped out within 30 to 40 years.* Land

* See feature story in *Time*, May 10, 1982.

use will then have to revert to grazing and dry land farming unless water in huge quantities can be piped from large Mississippi Valley rivers—a very expensive and energy-demanding "public works" project that would place a burden on all taxpayers of the nation. As of 1982, it is not possible to predict what will be decided, but the political controversy will certainly be bitter, and many people will be caught in economic collapses that always come when nonrenewable resources are exploited without regard for tomorrow.

Figure 4–11 is a graphic model of the downhill loop in the H_2O cycle showing how biotic communities adjust to the changing conditions in what has been called the river continuum (i.e., the gradient from small to large streams; Vannote et al., 1980). Headwater streams are small and often completely shaded so that little light is available to the aquatic community. Consumers depend largely on leaf and other organic detritus entering from the watershed. Large particles of organic matter such as leaf fragments predominate, as do aquatic insects and other primary consumers belonging to a class called "shredders" by stream ecologists. The headwaters ecosystem is heterotrophic with a *P/R* ratio of much less than one.

In contrast, midsections of rivers are wider, are no longer shaded, and depend less on imported organic matter from watersheds, since autotrophic algae and aquatic macrophytes provide primary produc-

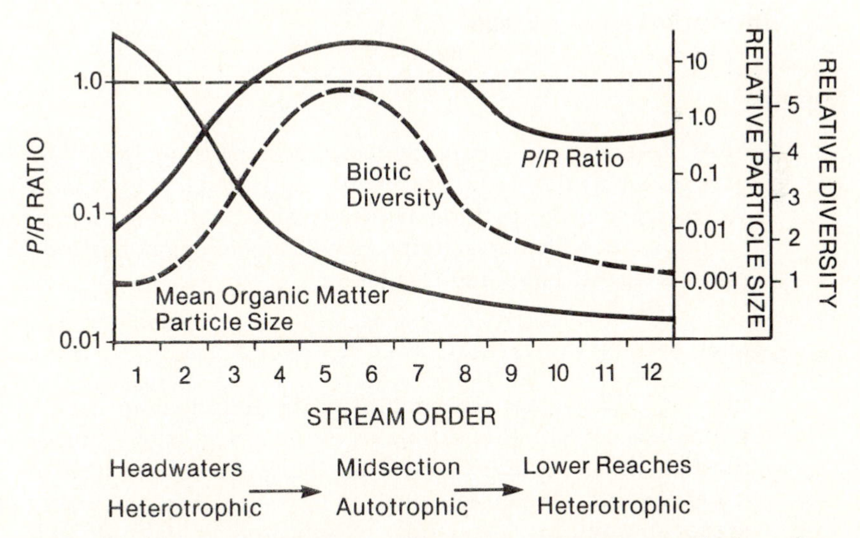

Figure 4–11 The river continuum; changes in community metabolism, diversity, and particle size from headwater streams to large rivers. (After Vannote et al., 1980.)

tion. Finely divided particulate organic matter and filter-feeders with "collector" adaptations (equipped with nets and strainers) predominate. Community metabolism is autotrophic with a P/R ratio of one or more (Figure 4–11). Species diversity and diurnal range of temperature generally peak in the stream midsection. In the lower reaches of large rivers, current is reduced and water is usually muddy, thereby decreasing light penetration and aquatic photosynthesis. The stream then again becomes heterotrophic with a reduced variety of species in most trophic levels.

As elsewhere in the biosphere, organisms are not just passive adapters to a gradient of changes in the physical environment. The concerted action of stream animals, for example, acts to recycle and reduce downstream loss of nutrients to the ocean. Aquatic insects, fish, and other organisms collect particulate and dissolved materials that are held and cycled through the food chain and often moved upstream or out on the watershed as part of the life cycle of the more mobile species. This process has been called material spiralling by limnologists (see Elwood and Nelson, 1975).

For excellent summaries of the hydrologic cycle, see Chapter 4 in Hutchinson, *A Treatise on Limnology* (1957), and Chapter 5 in Garrels, Mackenzie, and Hunt, *Chemical Cycles and the Global Environment* (1975).

5. The Sedimentary Cycle

Statement

Most elements and compounds are more earthbound than nitrogen, oxygen, carbon dioxide, and water, and their cycles follow a basic sedimentary cycle pattern in that erosion, sedimentation, mountain building, and volcanic activity, as well as biological transport, are the primary agents affecting circulation.

Explanation

Figure 4–12 is a general picture of the sedimentary cycle of earthbound elements. Some estimates of the amounts of material passing through the cycle are marked on the arrows. Very little is known about the flow of materials deep in the earth. The movement of solid matter through the air as dust is indicated as "fallout," which may come to earth as rain or as dry fallout. To the natural fallout (as from volcanos, dust storms, or forest fires) humans are adding more mate-

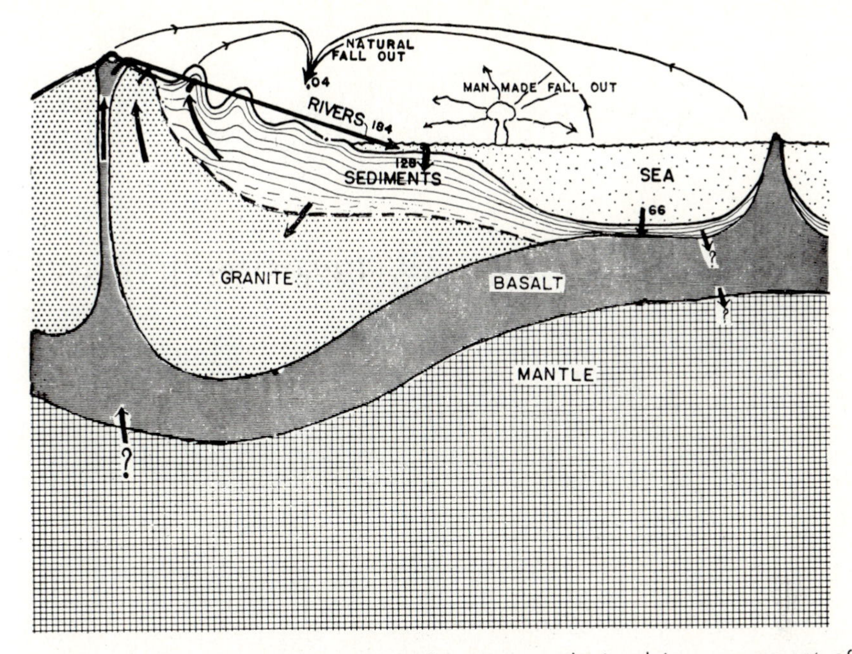

Figure 4–12 A diagram of the sedimentary cycle involving movement of the more "earthbound" elements. Where estimates are possible, the amounts of material are estimated in geograms per million year (1 geogram = 10^{20} grams). The continents are sediment-covered blocks of granite floating like corks on a layer of basalt that underlies the oceans. Below the black basalt is the mantle layer, which extends 2900 km down to the core of the earth. Granite is the light-colored, very resistant rock often used for tombstones; basalt is the black rock found in volcanoes. (Diagram prepared by H. T. Odum.)

rials, relatively small in amount but significant biologically because of their poisonous effects (as in radioactive fallout) or their action in blocking the incoming solar radiation, which, as already indicated, can cool the earth and bring about climatic change the opposite of that caused by an increase in CO_2.

The chemical elements available to the communities of the biosphere are those that by their geochemical nature tend to be enclosed within the types of rocks that come to the surface. Elements abundant in the mantle are scarce at the surface. As already indicated, phosphorus is one of the elements whose scarcity on the earth's surface often limits plant growth.

The general "downhill" tendency of the sedimentary cycle is well shown in Figure 4–12. Estimated annual flow of sediments from each major continent to the oceans is given in Table 4–2. It is significant that Asia, the continent with the oldest civilizations and most

Table 4–2 Estimated Annual Sediment Flow from Continents to Oceans*

Continent	Drainage Area (10^6 mi²)	Sediment Discharge Tons/mi²†	Sediment Discharge Total, 10^9 tons
North America	8.0	245	1.96
South America	7.5	160	1.20
Africa	7.7	70	0.54
Australia	2.0	115	0.23
Europe	3.6	90	0.32
Asia	10.4	1530	15.91
Total	39.2	—	20.16

* After Holeman, 1968.

† Conversion to volume: 1330 tons = 1 acre-foot.

intensive human pressure, loses the most soil. Although the rate varies, the lowlands and the oceans tend to gain soluble or usable mineral nutrients at the expense of the uplands during periods of minimum geological activity. Under such conditions, local biological recycling mechanisms are extremely important in keeping the downhill loss from exceeding the regeneration of new materials from underlying rocks, as was emphasized in the discussion of the watershed calcium cycle. In other words, the longer vital elements can be kept within an area and used over and over again by successive generations of organisms, the less new material will be needed from the outside.

For upstream ecosystems, the sources of sedimentary materials are limited. Unfortunately, as in the case of phosphorus, we tend to disrupt this homeostasis, often unwittingly, by not understanding the symbiosis between life and matter, which may have taken thousands of years to evolve. The material spiralling adaptation of stream biota mentioned in the previous section is an example of such symbiosis. The stopping of salmon runs by dams can cause a decline not only of salmon but also of nonmigratory fish, game, and even timber production in high-altitude regions. When salmon spawn and die in the uplands, they deposit a load of valuable nutrients recovered from the sea. The removal of large volumes of timber without the return of the contained minerals to soil (as would normally occur during the decay of logs) undoubtedly also contributes to the impoverishment of uplands where the pool of nutrients was already limited by ages of geological leaching and erosion. One can readily visualize how the destruction of biological recycling mechanisms could impoverish the whole ecosystem for many years, since reestablishing a circulating pool of minerals would take a long time. In such a case, devising the means of returning limiting materials (and keeping them *in situ*) would be far more effective than stocking fish or planting tree seedlings. The sudden rush of materials into the lowlands, resulting from human-accelerated erosion, does not necessarily benefit the lowland ecosystems, which may not get a chance to assimilate the nutrients before they pass into the sea beyond the range of light and, thus, out of biological circulation (at least for a time). Equally likely, organisms may be smothered by the onslaught of silt, mud, and sand or may be poisoned by toxins.

6. Cycling of Nonessential Elements

Statement

Although nonessential elements may have no known value to any organism, they often pass back and forth between organisms and environment in the same general manner as do the essential elements. Many of them are involved in the general sedimentary cycle, and some find their way into the atmosphere. Many nonnutrients become concentrated in certain tissues, sometimes because of chemical similarity to specific vital elements, even though such concentration may be poisonous. Chiefly because of human activities, the ecologist must now be concerned with the cycling of many of the nonessential elements, and we all must be concerned with the increasing volume of toxic wastes that are discharged or that inadvertently escape into the environment and contaminate the basic cycles of vital elements.

Explanation

Most nonessential elements have little effect in the concentrations normally found in most natural ecosystems, probably because organisms have become adapted to their presence. Therefore, their movement would be of little interest were it not for the byproducts of the mining, manufacturing, chemical, and agricultural industries that contain high concentrations of heavy metals, toxic organic compounds, and other potentially dangerous materials that too often find their way into the environment. Consequently, the cycling of just about everything is important. Even a very rare element, if in the form of a highly toxic metallic compound or a radioactive isotope, can become of biological concern, because a small amount of material (from the biogeochemical standpoint) can have a marked biological effect.

Examples

Strontium is a good example of a previously almost unknown element that now must receive special attention, because radioactive strontium is particularly dangerous to humans and other vertebrates. Strontium behaves like calcium, with the result that radioactive strontium gets into close contact with blood-making tissue in our bones. About 7 percent of the total sedimentary material flowing down rivers is calcium. For every 1000 atoms of calcium, 2.4 atoms of strontium move along with the calcium to the sea. When uranium is fissioned in the preparation and testing of nuclear weapons and in atomic power plants, radioactive strontium-90 is a waste product, one of a number of fission products that decay very slowly. Strontium-90 is new material added to the biosphere; it did not exist in nature before the atom was split. Tiny amounts of radioactive strontium released in fallout from weapons testing and escaping from nuclear reactors have now followed calcium from soil and water into vegetation, animals, human food, and human bones. In 1970, enough strontium was present in the bones of people to have carcinogenic effects, according to some medical scientists. With the international ban on weapons-testing in the atmosphere, the threat has been, for a time, reduced. Radioactive cesium-137, another dangerous fission product, behaves like potassium and therefore cycles rapidly through food chains. Large amounts of radioactive fission products are now stored in tanks at atomic energy facilities. Not knowing what to do about these wastes has, as already noted, limited the peaceful uses of atomic energy. The problem of hazardous wastes will be considered in more detail in Chapter 5.

Mercury is another example of a natural element that had little impact on life before the industrial age because of its low concentration and low mobility. Mining and manufacturing have changed all that, so mercury and other heavy metals (such as cadmium, copper, zinc) are now a problem. Figure 4–13 compares the present cycling of mercury with the situation as it is presumed to have existed before human beings. Industrial activities have introduced two new flows, mining and emissions to the atmosphere (dotted lines in Figure 4–12), which have increased the amount of mercury moving into the soil and rivers, thereby increasing the potential contact with living

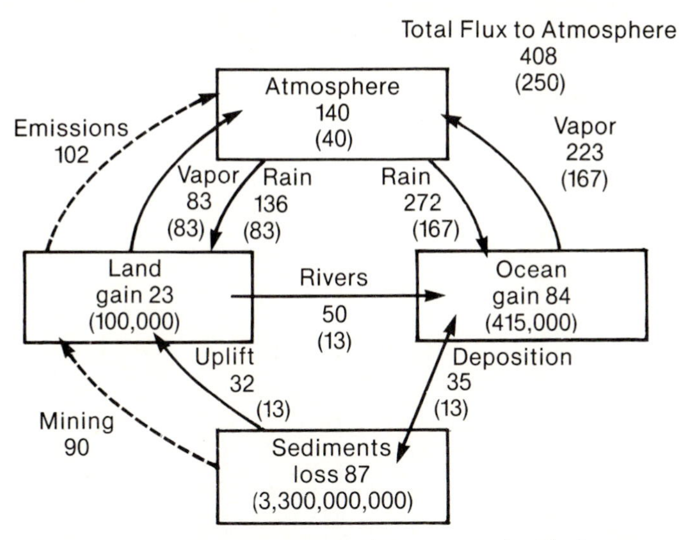

Figure 4–13 Present mercury cycle compared with the mercury cycle before human beings. Reservoir (boxes) amounts in units of 10^8 gm; fluxes (arrows) in 10^8 gm/year. Estimated prehuman amounts and fluxes shown in parentheses. Human activities have established two new flows (mining and emissions), shown by dashed lines, and have increased flux to the atmosphere by 60 percent, with a corresponding increase in rainout. (After Wollast, Billen, and Mackenzie, 1975.)

organisms. As with so many substances, microorganisms play an important role. In this case, insoluble forms of mercury are transformed into soluble and very poisonous methyl-mercury that is quite mobile.

7. Nutrient Cycling in the Tropics

Statement

The pattern of nutrient cycling in the tropics, especially the wet tropics, is, in several important ways, different from that in the north temperate zone. In cold regions a large portion of the organic matter and available nutrients is in the soil or sediment at all times; in the tropics a much larger percentage is in the biomass and is recycled within the organic structure of the system, aided by a number of nutrient-conserving, biological adaptations that include mutualistic symbiosis between microorganisms and plants. When this evolved and well-organized biotic structure is removed, nutrients are rapidly lost by leaching under conditions of high temperatures and heavy

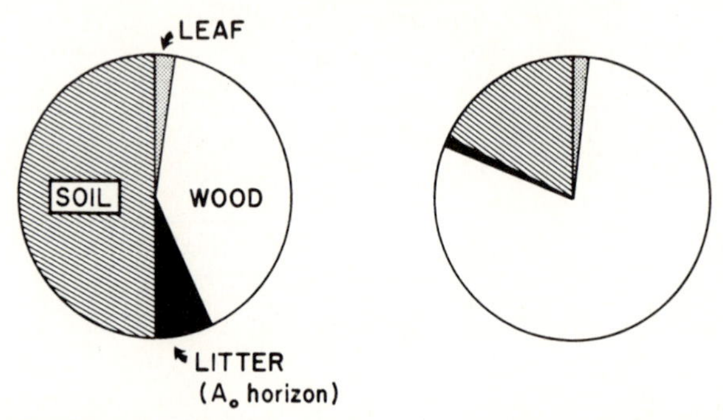

Figure 4–14 Distribution of organic carbon accumulated in abiotic (soil, litter) and biotic (wood, leaves) compartments of a northern and a tropical forest ecosystem. A much larger percentage of total organic matter is in the biomass in the tropical forest. (Redrawn from Kira and Shidei, 1967.)

rainfall, especially on sites that are initially poor in nutrients. For this reason, agricultural strategies of the temperate zone, involving the monoculture of short-lived annual plants, are quite inappropriate for tropical regions. An ecological reevaluation of tropical agriculture in particular, and environmental management in general, is urgent if past mistakes are to be corrected and if future ecodisasters are to be avoided. At the same time, the rich species diversity of the tropics must somehow be preserved.

Explanation

Figure 4–14 compares the distribution of organic matter in a northern and a tropical forest. Interestingly, in this comparison both ecosystems contain about the same amount of organic carbon, but over half is in litter and soil in the northern forest, whereas more than three fourths is in vegetation in the tropical forest. Northern and tropical forests are further compared in Table 4–3. About 58 percent of total nitrogen is in the biomass of a tropical forest—44 percent of it above ground—compared with 6 percent and 3 percent, respectively, in a pine forest in Britain.

When a forest in the temperate zone is removed, the soil retains nutrients and structure and can be farmed for many years in the "conventional" manner, which means plowing one or more times a year, planting annual species, and applying inorganic fertilizers.

Table 4–3 Distribution of Nitrogen in gm/m^2 Within a Temperate and a Tropical Forest Ecosystem*

	British 55-year Pine Forest	Tropical Gallery Forest
Leaves	12.4	52.6
Aboveground wood†	18.5	41.2
Roots	18.4	28.2
Litter	40.9	3.9
Soil	730.8	85.3
% N above ground	3.0	44.0
% in biomass	6.0	57.8
Root/shoot‡ ratio	0.60	0.30
Leaf/wood ratio	0.34	0.76

* After Ovington, 1962.

† Not including leaves.

‡ Including leaves.

During the winter, freezing temperatures help hold in nutrients and control pests and parasites. In the humid tropics, however, forest removal takes away the land's ability to hold and recycle nutrients (as well as to combat pests) in the face of high year-round temperatures and periods of leaching rainfall. Too often, crop productivity declines rapidly and the land is abandoned, creating the pattern of "shifting agriculture" about which so much has been written. Nutrient cycling in particular and community control in general, then, tend to be more physical in the north and more biological in the south. This brief account, of course, oversimplifies complex situations, but the contrast underlies what now appears to be the basic ecological reason why sites in the subtropics or tropics that support luxurious and highly productive forests or other vegetation yield so poorly under a northern method of crop management.

Jordan and Herrera (1981) point out that the degree to which tropical forests "invest," as it were, in nutrient-conserving recycling mechanisms depends on the geology and basic fertility of the site. Large areas of tropical forests, such as over most of the eastern and central Amazon basin, are on ancient, highly leached Precambrian soils or nutrient-poor sand deposits. These oligotrophic sites support forests as luxurious and productive as those found on the more eutrophic (i.e., fertile) sites in the mountains of Puerto Rico, Costa Rica, and the Andean foothills. Tropical rain forests on oligotrophic sites are ecologically similar to the luxurious coral reefs in the nutrient-poor waters of the central Pacific (see pages 5, 6). Intricate symbiosis between autotrophs and heterotrophs involving special microorganism intermediaries is the key to success in both types of ecosystem.

Among the mechanisms that are especially well developed in rain forest ecosystems on oligotrophic sites, Jordan and Herrera list the following:

1. Root mats consisting of many fine feeder roots penetrating the surface litter quickly recover nutrients from leaf fall and rain before they can be leached away. Root mats apparently also inhibit activities of denitrifying bacteria, thus blocking loss of nitrogen to the air.
2. Mycorrhizal fungi (see page 126) associated with root systems act as nutrient traps and greatly facilitate recovery of nutrients and retention within the biomass. (This symbiosis for mutual benefit is widespread on oligotrophic sites in the temperate zone as well.)
3. Evergreen leaves with thick, waxy cuticles retard loss of water and nutrients from trees and also resist herbivores and parasites.
4. "Drip-tips" on leaves (long, pointed leaf tips) drain off rain water, thereby reducing leaching of leaf nutrients.
5. Algae and lichens that cover surfaces of many leaves scavenge nutrients from rainfall, some of which are immediately available for uptake by the leaves; lichens also may fix nitrogen.
6. Thick bark inhibits diffusion of nutrients out from the phloem and subsequent loss by stem flow (i.e., rain running down trunks).

In summary, the nutrient-poor tropical ecosystem is able to maintain high productivity under natural conditions through a variety of nutrient-conserving mechanisms that, in the words of Went and Stark (1968), produce "direct cycling" from plant back to plant, more or less bypassing the soil. When such forests are cleared for agriculture or tree plantations, these mechanisms are destroyed and productivity declines very rapidly, as do crop yields. When abandoned, the forest recovers slowly, if at all. In contrast, forests on eutrophic sites are more resilient. Shifting cultivation on these sites may not be such a bad idea after all.

Development and testing of crop plants with well-developed mycorrhizal and N-fixing root systems and greater use of perennial plants are ecologically sound goals for warm temperature (such as southeastern United States) and tropical climates. Paddy rice culture is successful in the tropics because of the special nutrient-retention features of this ancient type of agriculture. Rice paddies have been cultivated on the same site for more than 1000 years in the Philippines (Sears, 1957), a record of success that few agricultural systems in use today can claim. These rice terraces are interspersed with patches of forests that were originally preserved by religious taboos. To avoid what Hutchinson (1967a) has called "the technological quick-fix with an ecological backlash," we should first find out whether the intermixture of forest and paddies has something to do with longevity of the food-producing system before rushing in and recommending that the forests be bulldozed so that more rice can be planted.

One thing is certain: *Industrialized agrotechnology of the temperate zone cannot be transferred unmodified to tropical regions.*

8. Recycle Pathways: The Recycle Index

Statement

It is instructive to review the subject of biogeochemistry in terms of recycle pathways, since recycling of water and nutrients is a vital process in ecosystems and is becoming an important concern for humankind. Five major recycling pathways can be distinguished: (1) by microbial decomposition and the detritus complex; (2) by animal excretions; (3) direct recycling from plant to plant through microbial symbionts (as described in the previous section); (4) by physical means involving direct action of solar energy; and (5) by use of fuel energy as in industrial fixation of nitrogen. Recycling requires dissipation of energy from some source such as organic matter (Paths 1, 2, 3), solar radiation (Path 4), or fuel (Path 5). The relative amount of recycling in different ecosystems can be compared by calculating a cycling index based on the ratio of the sum of the amounts cycling between compartments within the system and total throughflow.

Explanation

It is appropriate now to focus on the cycling of nutrients within the biologically active portion of the ecosystem. Recall that the same approach was used for energy in Chapter 3: the total energy environment was considered first, and then attention was focused on the fate of that small portion involved in the food chain. Also, a final discussion of biological regeneration is relevant because, as already emphasized, "recycle" must increasingly become a major goal for human society.

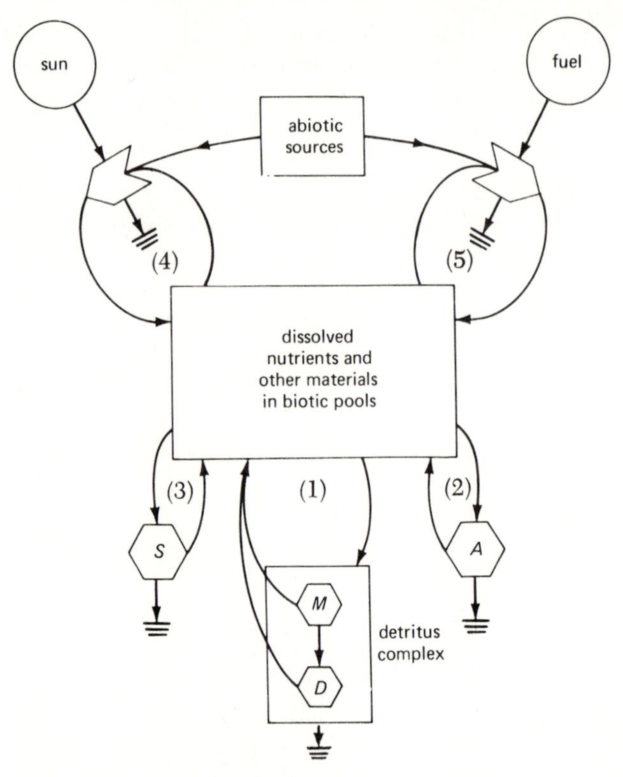

Figure 4–15 Five major recycle pathways (1), (2), (3), (4), (5). A = animals; M = free-living microorganisms; S = symbiotic microorganisms; D = detritus consumers. Energy for Paths 1 to 3 comes from organic matter and for Paths 4 and 5 from solar or fuel energy. See text for further explanation.

Basic recycle pathways are diagrammed in Figure 4–15. The classic assumption has been that bacteria and fungi are the major agents in nutrient regeneration, and recycle Path 1 in Figure 4–15 is certainly dominant in soils of the temperate zone where the regeneration process has been most studied. The complex interaction between microbes and small animal detritivores was described in some detail in Chapter 2. Where small plants such as grass or phytoplankton are heavily grazed by animals, recycling by way of animal excretion may be important (Path 2, Figure 4–15). Johannes (1964) postulated that in the water column of the sea, nitrogen and phosphorus regenerate mainly because of animal excretion; very small animals or "microzooplankton," which are too small to be collected in plankton nets (and thus overlooked in the early studies of marine communities), play an especially important role. Measurements of turnover rates indicate that zooplankton release during their life many times the amount of soluble nutrients as would be released by microbial decomposition of their bodies after their death (Harris, 1959; Rigler, 1961; Pomeroy et al., 1963; and many others). These excretions include dissolved inorganic and organic compounds of phosphorus, nitrogen, and CO_2, which are directly usable by producers without any further chemical breakdown by bacteria. It may be presumed that other vital nutrients are regenerated in this manner.

Direct recycling by symbiotic microorganisms, as described in some detail in the previous section, is designated Path 3 in Figure 4–15. This path can be expected to be important in nutrient-poor or oligotrophic environments. Water, as we have seen (Section 4), is largely recycled by direct action of solar energy and by the weathering and erosional processes associated with downhill flows of water that bring the sedimentary elements into biotic cycles from abiotic reservoirs, as shown in Path 4 of Figure 4–15. Human beings enter the recycle picture when they expend fuel energy to desalinate water from the sea, produce fertilizers, or recycle aluminum or other metals.

Finally, nutrients may be released from the dead bodies of plants and animals and from fecal pellets, without being attacked by microorganisms, as can be demonstrated by placing such materials under sterile conditions. This possibility for recycling can be designated "autolysis" (= self-dissolving). In aquatic or moist environments, especially where the bodies or dead particles are small (thus exposing a large surface-to-volume ratio), 25 to 75 percent of the nutrients may be released by autolysis before microbial attack begins, according to Johannes' (1968) review of the literature. Autolysis can be considered a sixth major recycle pathway, one that does not involve metabolic energy. As already emphasized in Chapter 3, recycle work accomplished mechanically or physically can provide an energy subsidy for the system as a whole. In the design of disposal systems for human and industrial wastes it is frequently profitable to provide an input of mechanical energy to pulverize organic matter and thus hasten its decomposition. Physical breakdown by the activities of large animals is also undoubtedly important in the release of nutrients from large, resistant pieces of detritus such as leaves or logs.

Cycling is not a "free" service; there is almost always an energy cost. When sunlight and organic matter are the sources for recycle work, human beings do not need to pay directly for the service by using expensive fuels. Without being disrupted or poisoned, natural recycle mechanisms can do most of the work of recycling water and nutrients. Industrial materials (such as metals) involved in manufacturing are quite another matter; recycling is costly in fuel and money, but there is little choice when supplies become limited.

The Cycling Index Cycling within ecosystems may be compared in terms of the proportion of incoming material that flows around from one compartment to another before exiting from the system. Jordan, Kline, and Sasscer (1972), theorizing that wood was the chief recycling site in forests, used flow through wood as an estimate of the amount of calcium and manganese recycled in forest ecosystems. As Finn (1976, 1978) has pointed out, such an estimate would be a minimum one at best because there are many other sites and processes, especially for complex cycles such as nitrogen. Finn suggested that the recycled fraction should be the sum of amounts cycled through each compartment and proposed the following:

$$CI = \frac{TST_c}{TST}$$

where CI is cycling index, TST_c is total system throughflow that is cycled, and TST is total throughflow. Throughflow is defined as the sum of all inputs minus change in storage within the system if it is negative or, alternatively, all output flows plus change in storage if it is positive. To calculate flows and cycling indices, Finn resorted to a matrix analysis similar to that used by Hannon (1973) for economic input-output analysis.

Using two different compartment models, Finn (1978) calculated the cycling index for calcium in the Hubbard Brook watersheds (see Figure 4–7) to be between 0.76 and 0.80. This means that about 80 percent of the calcium throughflow is recycled. Cycling indices were even higher for potassium and nitrogen. Nutrients in this watershed appear to recycle in the following order (from highest to lowest CI): $K > Na > N > Ca > P > Mg > S$. This ordering relates to input of each element from outside the system, mobility of the element, and biological requirements of biota. Indices are generally low for nonessential elements such as lead or for essential elements required in very small amounts relative to availability, such as copper. Elements that society considers valuable, such as platinum and gold, are 90 percent or more recycled. As would be expected, the cycling index for energy (calorie flow) is zero, since, as emphasized many times in this book, energy passes straight through and does not cycle.

The cycling index tells nothing about cycling rate, that is, the speed with which materials move around the circuits. The cycling rate is much more rapid in a tropical forest or in a warm ocean than in the tundra or a cold lake, but a satisfactory means of quantifying such rates has not yet been devised.

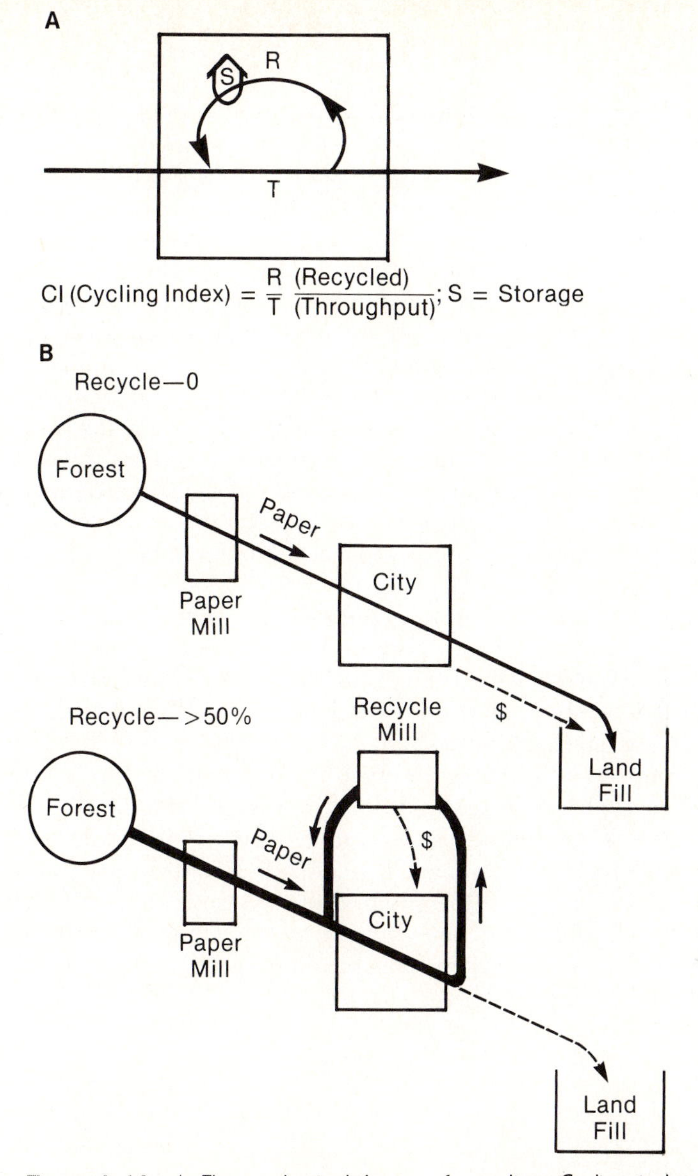

Figure 4–16 *A*. The ecological theory of recycling. Cycling index (CI) is low (0 to 10 percent) (1) during early stages of ecosystem development (succession), (2) when resources are abundant, and (3) for nonessential elements. CI is high (>50 percent) (1) during mature stages of ecosystem development, (2) when resources are scarce, and (3) for essential elements. The key consideration is that energy (which cannot be recycled) is required to drive the recycle loops. *B*. Conditions not conducive (upper diagram) and conducive (lower diagram) to recycling of paper. Benefits of recycling for the public as a whole are that harmful environmental impacts (on forests, streams, and land) are reduced and taxes for city services are reduced. Links include (1) citizen participation, (2) collection system (modified garbage pick-up), (3) sorting and packaging warehouse, (4) recycle mill, (5) transportation to mill, (6) market for used paper, and (7) profit for city or cost less than that of land fill.

Recycling of Paper Paper provides a good example of how recycling develops in urban-industrial systems in a parallel manner to recycling of important materials in natural systems. As shown in Figure 4–16A, recycling in natural ecosystems, as might be measured by a cycling index, increases as biotic components of ecosystems become larger and more complex, or as resources in the input environment become scarce, or as waste products pile up in the output environment to the detriment of the life within the ecosystem, or as all three occur.

As long as there are plenty of trees, paper mills, and vacant land for disposal of waste paper, there is little incentive to invest in facilities and energy to recycle some of the paper flowing through the city (Figure 4–16B). As environs of the city become congested, land values rise and it becomes increasingly difficult and expensive to maintain land fills or disposal sites. Pressure may also come from input environments if pulpwood supplies or mill production falls short of demand. In both cases, it "pays" to consider recycling. For recycling to be successful, a market must exist for used newspaper and cardboard, i.e., a recycling mill. Such a mill would represent an energy-saving recycling mechanism or dissipative structure in a natural ecosystem, as is the case for forests and coral reefs. In other words, something new has to be added to the total system, an emergent property as it were, if recycling conservation is to be efficient.

Too often, because of political inertia and fragmentation of governments (local, county, state, and federal levels), cities and towns wait too long to initiate recycling and suffer unnecessary conflicts and costs of continuing with outdated procedures. You as a citizen can help by checking into the situation in your own town to see what plans are being formulated for recycling of paper, metals, and other materials.

5
Limiting Factors and the Physical Environment

1. Concept of Limiting Factors; Liebig's "Law" of the Minimum

Statement

The presence and success of an organism or a group of organisms depends upon a complex of conditions. Any condition that approaches or exceeds the limits of tolerance is said to be a limiting condition or a limiting factor. Under steady-state conditions, the essential material available in amounts most closely approaching the minimum need tends to be the limiting one, a concept that has come to be known as Liebig's "law" of the minimum. The concept is less applicable under "transient-state" conditions when the amounts, and hence the effects, of many constituents are rapidly changing.

Explanation

The idea that an organism is no stronger than the weakest link in its ecological chain of requirements was first clearly expressed by Justus

Liebig in 1840. Liebig was a pioneer in studying the effect of various factors on the growth of plants. He found, as do agriculturists today, that the yield of crops was often limited not by nutrients needed in large quantities, such as carbon dioxide and water, since these were often abundant in the environment, but by some raw material, such as zinc, for example, needed in minute quantities but very scarce in soil. His statement that "growth of a plant is dependent on the amount of foodstuff which is presented to it in minimum quantity" has come to be known as Liebig's "law" of the minimum.

Extensive work since the time of Liebig has shown that two subsidiary principles must be added to the concept if it is to be useful in practice. The first is a constraint that Liebig's law is **strictly applicable only under steady-state conditions**, that is, when inflows balance outflows of energy and materials. To illustrate, suppose that carbon dioxide was the major limiting factor in a lake, and productivity was therefore in equilibrium with the rate of supply of carbon dioxide coming from the decay of organic matter. Assume that light, nitrogen, phosphorus, and other vital elements were available in excess of use in this steady-state equilibrium (and hence not limiting factors for the moment). If a storm brought more carbon dioxide into the lake, the rate of production would change and depend upon other factors as well. While the rate is changing, there is no steady state and no minimum constituent. Instead, the reaction depends on the concentration of *all* constituents present, which in this transitional period differs from the rate at which the least plentiful is being added. The rate of production would change rapidly as various constituents were used up, until some constituent, perhaps carbon dioxide again, became limiting. The lake system would once again be operating at the rate controlled by the law of the minimum.

The second important consideration is **factor interaction.** Thus, high concentration or availability of some substance, or the action of some factor other than the minimum one, may modify the rate of utilization of the latter. Sometimes organisms can substitute, at least in part, a chemically closely related substance for one that is deficient in the environment. Thus, where strontium is abundant, mollusks can substitute strontium for calcium to a partial extent in their shells. Some plants have been shown to require less zinc when growing in the shade than when growing in full sunlight; therefore, a low concentration of zinc in the soil would less likely be limiting to plants in the shade than under the same conditions in sunlight.

Limits of Tolerance Concept Not only may too little of something be a limiting factor, as proposed by Liebig, but also too much, as in the case of such factors as heat, light, and water. Thus, organisms have an ecological minimum and maximum; a range in between represents the **limits of tolerance.** The concept of the limiting effect of maximum as well as minimum was incorporated into the "law" of tolerance by V. E. Shelford in 1913. Since about 1910, much work has been done in "toleration ecology," so that the limits within which various plants and animals can exist are known. Especially useful are what can be termed "stress tests," carried out in the laboratory or field, in which organisms are subjected to an experimental range of conditions (see Hart, 1952). Such a physiological approach has helped ecologists to understand the distribution of organisms in nature; however, it is only part of the story. All physical requirements may be well within the limits of tolerance for an organism, and the organism may still fail because of biological interrelations. Studies in the intact ecosystem must accompany experimental laboratory studies, which, of necessity, isolate individuals from their populations and communities. We come back again to the concept of emergent properties as discussed in Chapter 1.

Some subsidiary principles to the "law" of tolerance may be stated as follows:

1. Organisms may have a wide range of tolerance for one factor and a narrow range for another.
2. Organisms with wide ranges of tolerance for all factors are likely to be most widely distributed.
3. When conditions are not optimum for a species with respect to one ecological factor, the limits of tolerance may be reduced for other ecological factors. For example, Penman (1956) reports that when soil nitrogen is limiting, the resistance of grass to drought is reduced. In other words, he found that more water was required to prevent wilting at low nitrogen levels than at high levels.
4. Frequently, it is discovered that organisms in nature are not actually living at the optimum range (as determined experimentally) of a particular physical factor. In such cases, some other factor or factors are found to have greater importance. Certain tropical orchids, for example, actually grow better in full sunlight than in shade, provided they are kept cool (see Went, 1957). In nature, they grow only in the shade because they cannot tolerate the heat of direct sunlight. In many cases, population interactions (such as competition, predators, parasites, and so on), as will be discussed in detail in Chapter 7, prevent organisms from taking advantage of optimum physical conditions.
5. Reproduction is usually a critical period when environmental factors are most likely to be limiting. The limits of tolerance for

reproductive individuals, seeds, eggs, embryos, seedlings, and larvae are usually narrower than for nonreproducing adult plants or animals. Thus, an adult cypress tree will grow continually submerged in water or on dry upland, but it cannot reproduce unless there is moist, unflooded ground for seedling development. Adult blue crabs and many other marine animals can tolerate brackish water or fresh water that has a high chloride content and thus are often found for some distance up rivers. The larvae, however, cannot live in such waters; therefore, the species cannot reproduce in the river environment and never become established permanently. The geographical range of game birds is often determined by the impact of climate on eggs or young rather than on the adults. One can cite hundreds of other examples.

For the relative degree of tolerance, a series of terms have come into general use in ecology that use the prefixes *steno-*, meaning "narrow," and *eury-*, meaning "wide." Thus,

stenothermal–eurythermal	refers to temperature
stenohydric–euryhydric	refers to water
stenohaline–euryhaline	refers to salinity
stenophagic–euryphagic	refers to food
stenoecious–euryecious	refers to habitat selection

The concept of limiting factors is valuable because it gives the ecologist an "entering wedge" into the study of complex situations. Environmental relations of organisms are apt to be complex, but fortunately, all possible factors are not equally important in a given situation or for a given organism. Studying a particular situation, the ecologist can usually discover the probable weak links and focus attention, initially at least, on those environmental conditions most likely to be critical or limiting. If an organism has a wide limit of tolerance for a relatively constant factor in moderate quantity in the environment, that factor is not likely to be limiting. Conversely, if an organism is known to have definite limits of tolerance for a factor that also is variable in the environment, then that factor merits careful study, since it might be limiting. For example, oxygen is so abundant, constant, and readily available in the terrestrial environment that it is rarely limiting to land organisms, except to parasites or those living in soil or at high altitudes. On the other hand, oxygen is relatively scarce and often extremely variable in water and is thus often an important limiting factor to aquatic organisms, especially animals. Therefore, the aquatic ecologist has an oxygen determination apparatus in readiness and measures the gas's concentration as one of the procedures in studying an unknown situation. The terrestrial ecologist, on the other hand, would less often need to measure oxygen concentration, even though, of course, it is just as vital a physiological requirement on land as in water.

To sum up, primary attention should be given to factors that are "operationally significant" to the organism at some time during its life cycle. The aim of environmental analysis, for example, preparing an environmental impact statement, is not to make long, uncritical lists of possible "factors," but rather to achieve these more significant objectives: (1) to discover, by observation, analysis, and experiment, which factors are "operationally significant" and (2) to determine how these factors affect the individual, population, or community, as the case may be. In this manner, the effect of disturbances or proposed environmental alterations can be predicted with reasonable accuracy.

The actual range of tolerance in nature is almost always narrower than the potential range of activity. This might be indicated by noting short-term behavioral response in the laboratory, because the metabolic cost of physiological regulation at extreme conditions reduces the limits of tolerance under field conditions at both upper and lower limits. If a fish in a pond receiving heated water from an industry or power plant has to devote all or most of its metabolic energy to coping with the elevated temperature stress, it will have insufficient energy for food-getting and reproductive activities required for survival under nonlaboratory conditions. Adaptation becomes more costly as extreme conditions are approached, and the organism becomes increasingly susceptible to other factors such as disease or predation.

Examples

For an example of limiting factor concepts, compare the conditions under which brook trout (*Savelinus*) eggs and leopard frog (*Rana pipiens*) eggs develop and hatch. Trout eggs develop between 0° and 12°C, with optimum at about 4°C. Frog eggs develop between 0° and 30°C, with optimum at about 22°C. Thus, trout eggs are stenothermal and low-temperature tolerant, compared with frog eggs, which are eurythermal and high-temperature tolerant. Trout, both eggs and adults, are in general relatively stenothermal, but some species are more eurythermal than the brook trout. Likewise, of course, species of frogs differ. These concepts, and the use of terms in regard to temperature, are illustrated in Figure 5–1. In a way, the evolution of narrow limits of tolerance might be considered a form of specialization (as discussed in the ecosystem chapter) that results in greater

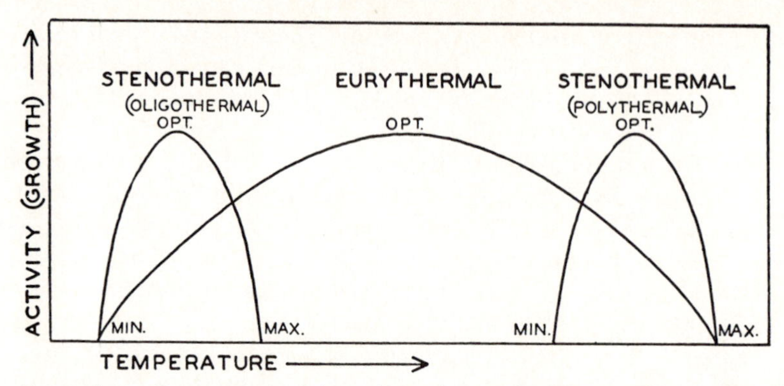

Figure 5–1 Comparison of the relative limits of tolerance of stenothermal and eurythermal organisms. Minimum, optimum, and maximum lie close together for a stenothermal species, so that a small difference in temperature, which might have little effect on a eurythermal species, is often critical. Note that stenothermal organisms may be either low-temperature tolerant (oligothermal), high-temperature tolerant (polythermal), or in between. (After Ruttner, 1963.)

efficiency at the expense of adaptability and contributes to increased diversity in the community as a whole.

The Antarctic fish *Trematomus bernacchi* and the desert pupfish *Cyprinodon macularius* provide an extreme contrast in limits of tolerance related to the very different environments in which they live. *Trematomus bernacchi* has a limit of temperature tolerance of less than 4 degrees Centigrade in the range of −2° to +2°C and is thus extremely stenothermally adapted to cold. As the temperature rises to 0°C, the rate of metabolism increases but then declines as the temperature of the water rises to +1.9°C. At this point, the fish become immobile with heat prostration (see Wohlschlag, 1960). In contrast, the desert fish is eurythermal and also euryhaline, tolerating temperatures between 10° and 40°C and salinity ranging from that of fresh water to that greater than sea water. The ecological performance, of course, is not equal throughout such a range; food conversion, for example, is greatest at 20°C and 15 percent salinity (Lowe and Heath, 1969).

An example of a predictive model based on singling out several operationally significant limiting factors is shown in Figure 5–2. After extensive field research in the rich fishing grounds of Georges Bank off New England's coast in the 1940s, Gorden Riley and co-workers found they could predict the amount and seasonal distribution of plankton by means of a formula based on key limiting factors in the

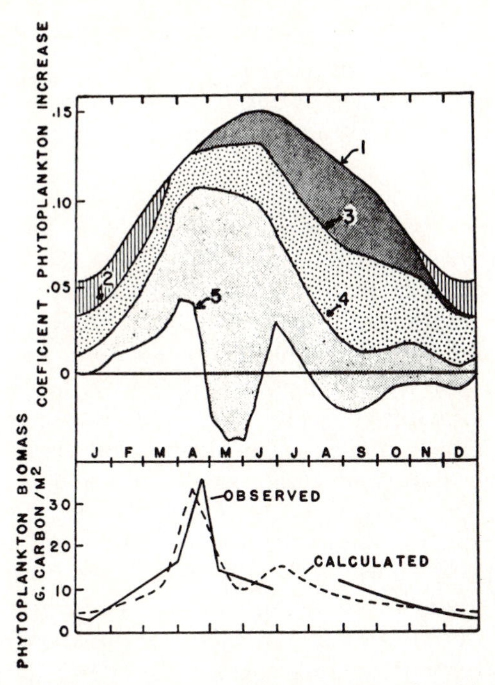

Figure 5–2 Theoretical operation of six limiting factors on phytoplankton (upper figure), and observed and calculated density during one annual cycle in the waters at Georges Bank off the coast of New England (lower figure). The limiting factors are (1) light and temperature; (2) turbulence (which carries cells below photic zone); (3) phosphate depletion; (4) phytoplankton respiration; and (5) zooplankton grazing. Only during spring and late summer are conditions favorable for rapid growth of the population. (After Riley, 1952.)

environment together with physiological coefficients determined from laboratory experiments. As shown in Figure 5–2, the limiting factors found to be important for phytoplankton are light, temperature, turbulence, phosphate, respiration, and grazing by zooplankton. The theoretical seasonal pattern in the intensity of operation of the limiting factors is diagrammed in the upper part of Figure 5–2; the observed and calculated biomass of phytoplankton for a specific locality is plotted in the lower diagram. Only in the spring and to a lesser extent in late summer do limitations "relax" sufficiently to allow rapid growth of the phytoplankton populations. In general, the observed was within 25 percent of the calculated, remarkably close considering the complexity of the situation. Riley's pioneer model had great significance, since it showed that useful, predictive models of complex situations could be developed by judiciously selecting only a few of the many factors involved. This is how the concept of limiting factors can be usefully applied.

The following briefly outlined examples further demonstrate the importance of the concept of limiting factors and the limitations of the concept itself.

1. Ecosystems developing on unusual geological formations often provide instructive sites for analysis of limiting factors, since one or more of the important chemical elements may be unusually scarce or unusually abundant. Such a situation is provided by serpentine soils (derived from magnesium-iron-silicate rocks), which are low in major nutrients (Ca, P, N) and high in magnesium, chromium, and nickel, with concentrations of the latter two approaching toxic levels for organisms. Vegetation growing on such soils has a characteristically stunted appearance, which contrasts sharply with adjacent vegetation on nonserpentine soils, and comprises an unusual flora with many endemic species and varieties (i.e., found only on serpentine soils). Despite the twin limitations of scarce major nutrients and abundant toxic metals, a biotic community has evolved over geological time that can tolerate the conditions, but at a reduced level of community structure and productivity. For a comparison of ecosystem function on serpentine and nonserpentine soils, see McNaughton (1968).

2. Great South Bay on Long Island, New York, dramatically shows how too much of a good thing can completely change an ecosystem, to the detriment of a seafood industry in this case. This story, which might be entitled "The Ducks vs. the Oysters," has been well documented, and the cause-and-effect relations have been verified by experiment (Ryther, 1954). The establishment of large duck farms along the tributaries leading into the bay resulted in extensive fertilization of the waters by duck manure and a consequent great increase in phytoplankton density. The low circulating rate in the bay allowed the nutrients to accumulate there rather than be flushed out to sea. The increase in primary productivity might have been beneficial had not the organic form of the added nutrients and the low nitrogen-phosphorus ratio completely changed the type of producers. The normal mixed phytoplankton of the area consisting of diatoms, green flagellates, and dinoflagellates was almost completely replaced by very small, little-known green flagellates of genera *Nannochloris* and *Stichococcus*. (The most common species was so little known to marine botanists that it had to be described as a new species.) The famous blue-point oysters, which had been thriving for years on a diet of the normal phytoplankton and supporting a profitable industry, could not utilize the newcomers as food and gradually disap-

peared. Oysters were found starving to death with intestines full of undigested green flagellates. Other shellfish were also eliminated, and all attempts to reintroduce them failed. Culture experiments demonstrated that the green flagellates grow well when nitrogen is in the form of urea, uric acid, and ammonia, whereas the diatom *Nitzschia*, a "normal" phytoplankter, requires inorganic nitrogen (nitrate). The flagellates could "short-circuit" the nitrogen cycle; that is, they did not have to wait for organic material to be reduced to nitrate (see Figure 4–2). This case is a good example of how a normally rare "specialist" in the usual fluctuating environment takes over when unusual conditions are stabilized.

This example also points up the frequent experience of laboratory biologists, who find that the common species of unpolluted nature are often hard to culture in the laboratory under conditions of constant temperature and enriched media because they have adapted to the opposite, i.e., low nutrients and variable conditions. On the other hand, the "weed" species, normally rare or transitory in nature, are easy to culture because they are stenotrophic and thrive on enriched (i.e., "polluted") conditions. A good example of such a weed species is *Chlorella*, an alga often used in laboratory study of algal physiology and photosynthesis. As already noted in Chapter 2, *Chlorella* has been highly touted for life support in space travel and for solving the world food problem, but so far it has proved disappointing on both counts.

3. The marine hydroid (Coelenterate) *Cordylophora caspa* is apparently an example of a euryhaline organism that does not actually live in waters of an optimum salinity for its growth. Kinne (1956) studied this species under laboratory conditions of controlled salinity and temperature and found that a salinity of 16 parts per thousand resulted in the best growth. Yet the organism was never found at this salinity in nature, but always at a much lower salinity; obviously some condition, perhaps a biotic one, present in its natural habitat but not in the laboratory cultures, is limiting.

Field observation and analysis should always be combined with laboratory experimentation. Probably, no situation in nature can be really understood from either field observation or experimentation alone, since each approach has obvious limitations. It is especially easy to be fooled or to make a premature conclusion because of a limited observation of a single situation. Hunters, fishermen, amateur naturalists, and laypersons who are much interested in nature's complexes, and who are often keen observers, nevertheless are too often guilty of "jumping to conclusions" about limiting factors. Thus, a sportsman may see an osprey catch a fish or a hawk catch a quail and conclude that predators are the principal limiting factors in fish and quail populations. Actually, when the situation is well studied, more basic but less spectacular factors are generally found to be more important than large predators. Unfortunately, much time and money is wasted on predator control without the real limiting factors ever being discovered or the situation improved from the standpoint of increased yield. However, where organisms are confined to limited space, predation can be a major limiting factor as described in the seventh example of this series.

4. In the 1950s, Andrewartha and Birch (1954) started a lively discussion in the ecological literature when they suggested that distribution and abundance are controlled by the same factors. Accordingly, study at range margins should be a good way to single out which factors are limiting. However, many ecologists think that quite different factors may limit abundance in the center of ranges and distribution at the margins, especially since Carson (1958) and other geneticists have reported that individuals in marginal populations may have different gene arrangements from central populations. In any event, the biogeographical approach becomes especially interesting when one or more regional environmental factors suddenly or drastically change. Then, a natural experiment is set up that is often superior to a laboratory experiment, because factors other than the one being considered continue to vary normally instead of being "controlled" in an abnormal, constant manner.

Certain birds that have extended their breeding ranges within the last 50 to 100 years provide examples of fortuitous field tests that help determine limiting factors. For example, when song birds that thrive in domesticated habitats (towns, suburbs, or farms), such as the American robin (*Turdus migratorius*), the song sparrow (*Melospiza melodia*), and the house wren (*Troglodytes aedon*), recently extended their range southward, it seemed likely that alteration of the vegetation by humans was the cause, since temperature and other climatic factors had not changed appreciably at the time of the invasion (Odum and Burleigh, 1946; Odum and Johnston, 1951). Elton (1958) has extensively reviewed the animal invasions and the limiting factors involved.

5. To determine the optimal region for crops, one must consider not only average yields but also variation from year to year. In attempting to determine which state in the Middle West was optimal for growing barley, Klages (1942) considered not only the average yield over a period of years but also the coefficients of variation of the yields. The region with the highest average yield and the lowest coefficient of variation (hence the fewest crop failures) was judged to be the opti-

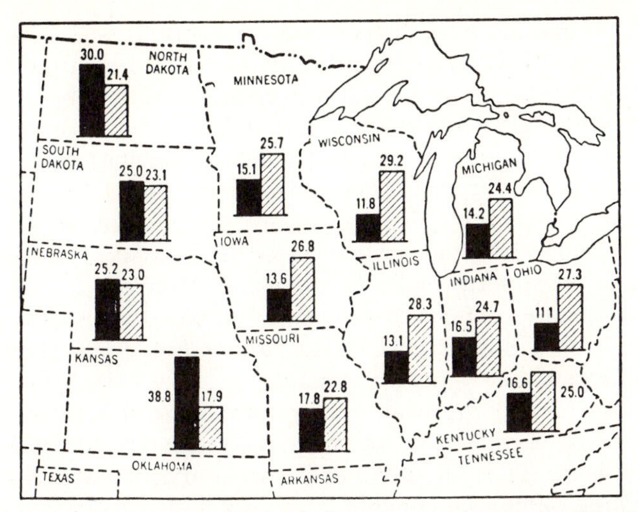

Figure 5–3 A method of determining the optimal regions for agricultural crops by comparing average yields and variability in yields from year to year. Cross-hatched columns indicate average yields in bushels per acre, and solid columns, the coefficient of variation in the yields of barley in the states of the upper Mississippi Valley. (After Klages, 1942.)

mal region. As shown in Figure 5–3, Wisconsin and Ohio proved to be the optimal states for barley, as determined by this method.

6. Ever since the time of Liebig, what might be called the "artificial enrichment experiment" has been a widely used approach to determine which mineral nutrients might be limiting. This broad category includes the very unquantitative trial and error fertilization experiments that characterize the early development of agriculture, the unplanned cultural eutrophication previously mentioned, as well as more carefully designed experiments. As emphasized earlier, enrichment experiments create a transient, or unsteady, state that may make interpretation of results difficult. Nevertheless, if background knowledge about the ecosystem is adequate and if accessory factors are considered, the enrichment approach can be useful and quantitative. Experiments of Menzel and Ryther (1961) and Menzel, Hulbert, and Ryther (1963) can be cited as examples. These investigators were interested in finding out what nutrients limit phytoplankton productivity in the Sargasso Sea, which is a kind of "marine desert." Their experiments brought out the importance of the time factor. Experiments lasting 1 hour, 24 hours, and several days often gave different results because the species composition sometimes changed during

the longer experiments in response to the enrichment. The investigators concluded that experiments should not be much longer than the generation or turnover time of the organisms if the objective is to determine what factor limits the original populations. On the other hand, if the experiment is too short, the conclusions may also be misleading. For example, iron enrichment produced increased carbon uptake by the phytoplankton during the first 24 hours, but to sustain the increased production rate for several days, the levels of nitrogen and phosphorus had to be increased. Menzel and co-workers concluded that the chief value of enrichment experiments was to determine which populations could evolve into more productive ones in the presence of more nutrients.

7. An experimental approach to determining biotic limiting factors involves adding or removing species populations. The intertidal zone on rocky seashores is a good habitat for such experiments. Extensive work by Paine (1966), Dayton (1971), Connell (1977), and others has shown that intertidal communities tend to have strong dominants, that is, species capable of excluding others in the same trophic position. With space in the narrow intertidal zones as always potentially limiting, the main factors in preventing monopolization by a single species are predation (for animals) and grazing (for plants) (see Dayton, 1975). Biological regulation at the population level will be considered in greater detail in Chapter 6.

8. A classic study of multiple factor limitation is that of McLeese (1956) on the American lobster. He determined experimentally the limits of tolerance of lobsters to water temperature, salinity, and oxygen concentration as single varying factors and in combination. When salinity is optimal, the lobster can survive at a higher temperature than it can when exposed to a lower than optimal salinity, and similarly for lower than optimal oxygen concentration. Accordingly, tolerable (i.e., sublethal) conditions for a given factor can become lethal if another interacting factor is not optimal.

9. Another example of interaction of multiple factors is modeled in Figure 5–4, which shows the percentage of mortality of larval (zoea) fiddler crabs for different combinations of temperature and salinity. As previously noted, the limits of tolerance of the young are likely to be lower than those of adults, so survival of a species over the long term may be determined by the ability of reproductive stages to tolerate a given set of conditions. Figure 5–4A shows that larval mortality is lowest when the temperature is 25 ± 5°C and salinity 30 ± 5 ‰. When a very small amount of cadmium, a substance especially toxic to crustacea, is added to the environment, the pattern of temperature-salinity interaction is changed (Figure 5–4B). The opti-

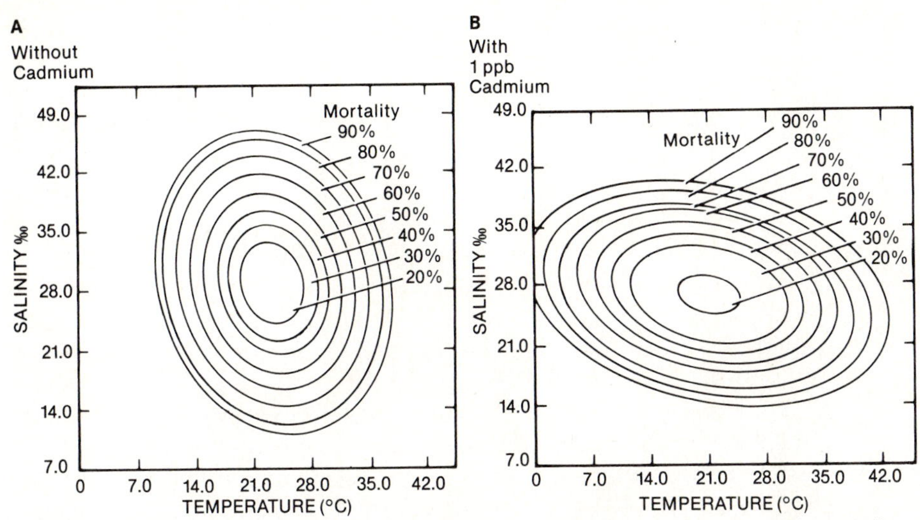

Figure 5–4 Interaction of temperature, salinity, and cadmium (a toxic substance) on crab larvae mortality. Estimation of percentage mortality of first-stage fiddler crab (*Uca pugilator*) zoea based on response surface fitted to observed mortality under 13 combinations of salinity and temperature (*A*) and with the addition of 1 ppb cadmium (*B*). (After Vernberg et al., 1974.)

mal zone (inner circle of lowest mortality) is reduced, and the tolerance to salinity is narrowed, but interestingly, the range of temperature tolerance seems to increase somewhat (note that the ring of 50 percent mortality extends several degrees lower in B than in A).

2. Factor Compensation and Ecotypes

Statement

Organisms are not just "slaves" to the physical environment; they adapt themselves and modify the physical environment to reduce the limiting effects of temperature, light, water, and other physical conditions of existence. Such **factor compensation** is particularly effective at the community level of organization but also occurs within the species. Species with wide geographical ranges almost always develop locally adapted populations called **ecotypes** that have optima and limits of tolerance adjusted to local conditions. Compensation along gradients of temperature, light, or other factors may involve genetic races (with or without morphological manifestations) or physiological acclimation without a genetic change.

Explanation

Species that range widely along a gradient of temperature or other conditions often differ physiologically and sometimes morphologically in different parts of their range. Such locally adapted populations are known as ecotypes. Reciprocal transplants provide a convenient method of determining to what extent genetic fixation is involved in ecotypes. McMillan (1956), for example, found that prairie grasses of the same species (and to all appearances identical) transplanted into experimental gardens from different parts of their range responded quite differently to light. In each case, the timing of growth and reproduction was adapted to the area from which the grasses were transplanted. The possibility of genetic fixation in local strains has often been overlooked in applied ecology; restocking or transplanting of plants and animals may fail because individuals from remote regions were used instead of locally adapted stock. Factor compensation in local or seasonal gradients can also be accomplished by physiological adjustments in organ functions or by shifts in enzyme-substrate relationships at the cellular level. Somero (1969), for example, points out that immediate temperature compensation is promoted by an inverse relationship between temperature and enzyme-substrate affinity, whereas longer-term evolutionary adaptation is more likely to involve changes in enzyme-substrate affinity itself. Animals, especially larger ones with well-developed powers of movement, compensate through adapted behavior that avoids the extremes in local environmental gradients. Lizards, for example, can move out into the sun during the day and under warm rocks or into burrows at night, a behavior regulation that can be just as effective in maintaining an optimum body temperature as internal physiological regulation of warm-blooded vertebrates. At the community level, factor compensation is more frequently accomplished by replacement of species in environmental gradients, both latitudinal and seasonal; examples will be noted in Chapter 6.

Examples

Figure 5–5 illustrates two instances of temperature compensation, one at the species level and one at the community level. As shown in Figure 5–5A, northern jellyfish can swim actively at low temperatures that would completely inhibit individuals from the southern populations. Both populations are adapted to swim at about the same rate, and both function, to a remarkable extent, independently of the temperature variations in their particular environment. Figure 5–5B

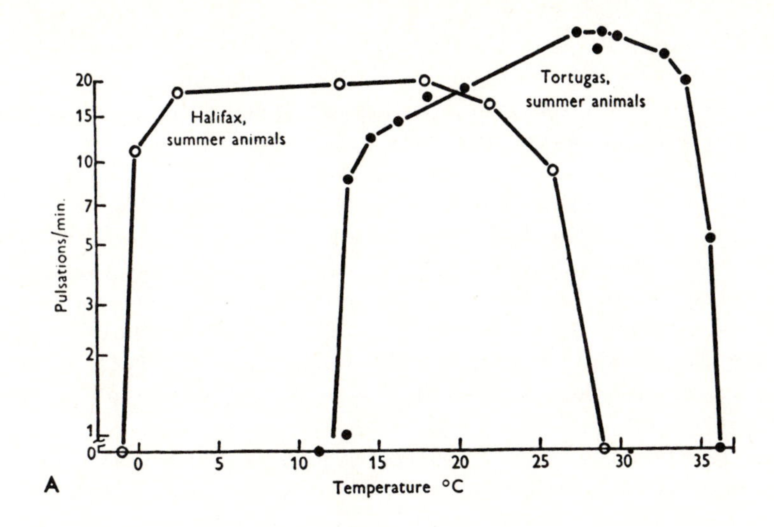

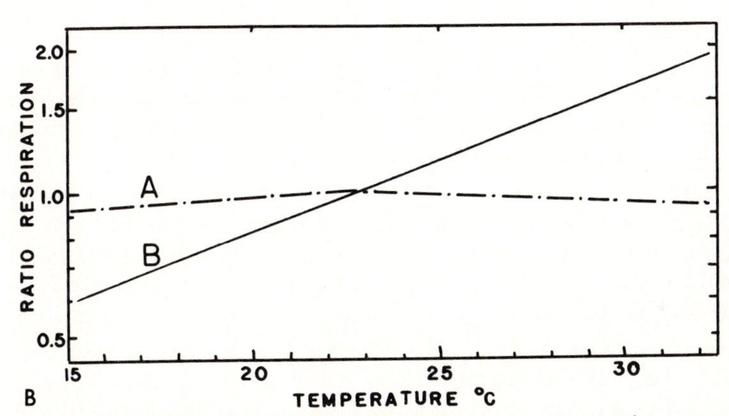

Figure 5–5 Temperature compensation at the species and community levels. *A*. The relation of temperature to swimming movement in northern (Halifax) and southern (Tortugas) individuals of the same species of jellyfish, *Aurelia aurita*. The habitat temperatures were 14° and 29°C, respectively. Note that each population is acclimated to swim at a maximum rate at the temperature of its local environment. The cold-adapted form shows an especially high degree of temperature independence. (From Bullock, 1955, after Mayer.) *B*. The effect of temperature on respiration of a balanced laboratory microcosm community (A) and a single species component, *Daphnia* (a small crustacean) (B). The relative change in the rate of CO_2 production is plotted as a ratio of the rate at 23°C, the temperature to which the microcosm was adapted. (Redrawn from Beyers, 1962.)

shows that the rate of respiration of the whole community in a balanced microcosm is less affected by temperature than is the respiratory rate of one species (*Daphnia*). In the community, many species with different optima and temperature responses develop reciprocal adjustments and acclimations, enabling the whole to compensate for ups and downs in temperature. In the example shown (Figure 5–5B), temperatures 8 to 10°C above or below the temperature to which the microcosm was acclimated did result in a slight decrease in respiration; however, the effect was negligible compared with the more than twofold effect of this range of temperature on *Daphnia*. In general, then, metabolic rate-temperature curves will be flatter for ecosystems than for species, another example, of course, of community homeostasis.

Compensation occurs along seasonal as well as geographical gradients. A striking and well-studied example is that of the creosote bush, *Larrea*, which dominates low-altitude, hot deserts of the southwest United States. Although *Larrea* is a C_3 plant, a photosynthetic mode not especially adapted to hot and dry conditions (see page 32), it can shift its thermal optimal temperature upward from winter to summer by acclimation. High photosynthetic rates are maintained by an additional acclimation to drought stress as measured by leaf water potential. For details on these acclimations, see Mooney et al. (1976).

In nutrient-poor environments, efficient recycling between autotrophs and heterotrophs often compensates for nutrient scarcity. Coral reefs and rain forests are examples cited previously. McCarthy and Goldman (1979) report that nitrogenous nutrients in the waters of the North Atlantic are so low that they cannot be detected by standard instruments. Yet phytoplankton photosynthesis occurs at a high rate. Rapid and efficient uptake of nutrients released in zooplankton excretion and bacterial action compensates for an overall scarcity of nitrogen.

For reviews of the physiological basis for factor compensation, see Bullock (1955), Fry (1958), and Prosser (1967).

3. Conditions of Existence as Regulatory Factors

Statement

Light, temperature, and water (rainfall) are ecologically important environmental factors on land; light, temperature, and salinity are the important ones in the sea. In fresh water, other factors such as oxygen may be of major importance. In all environments, the chemical nature and cycling rates of basic mineral nutrients are noteworthy. All these physical conditions of existence may be not only limiting factors in the detrimental sense but also regulatory factors in the beneficial sense: adapted organisms respond to these factors so that the community of organisms "mitigates," as it were, deleterious effects and achieves the maximum performance and homeostasis possible under the existing conditions.

Explanation and Examples

Organisms not only adapt to the physical environment in the sense of tolerating it, but also use the natural periodicities in the physical environment to time their activities and to "program" their life histories so they can benefit from favorable conditions. When one adds interactions between organisms and reciprocal natural selection between species (coevolution; see page 482), the whole community becomes programmed to respond to seasonal and other rhythms. The biological literature reports many examples of adaptive responses. Usually, the responses of a particular group of organisms (for example, *Environmental Control of Plant Growth*, edited by Evans, 1963) or of a particular habitat are described (for example, *Adaptations of Intertidal Organisms*, edited by Lent, 1969). Detailed consideration of regulatory adaptations is beyond the scope of this text, but perhaps two examples will suffice to emphasize the points of special ecological interest.

A dependable cue by which organisms time their activities in temperate zones is the day-length period or **photoperiod.** In contrast to most other seasonal factors, day length is always the same for a given season and locality. The amplitude in the annual cycle increases with increasing latitude, thus providing latitudinal as well as seasonal cues. At Winnipeg, Canada, the maximum photoperiod is 16.5 hours (in June) and the minimum is 8 hours (in late December). In Miami, Florida, the range is only 13.5 to 10.5 hours. The photoperiod has been shown to be the timer or trigger that sets off physiological sequences that cause the growth and flowering of many plants, molting, fat deposition, migration and breeding in birds and mammals, and the onset of diapause (resting stage) in insects. Photoperiodicity is coupled with what is now widely known as the organism's **biological clock** to create a timing mechanism of great versatility. The biological clock is a physiological mechanism for measuring time. The most common and perhaps basic manifestation is the circadian rhythm (*circa* = about; *dies* = day) or the ability to time and repeat functions at about 24-hour intervals even without conspicuous environmental cues such as daylight. The biological clock couples environmental and physiological rhythms and enables organisms to anticipate daily, seasonal, tidal, and other periodicities. (For a review of biological clock theories, see Brown, Hastings, and Palmer, 1970; for a review of ecological aspects of endogenous rhythmicity, see Enright, 1970.)

Day length acts through a sensory receptor, such as an eye in animals or a special pigment in the leaves of a plant, which, in turn, activates one or more back-to-back hormone and enzyme systems that bring about the physiological or behavioral response. It is not known just where in this sequence time is actually measured. Although the higher plants and animals are widely divergent in morphology, the linkage with environmental photoperiodicity is similar.

Among the higher plants, some species bloom on increasing day length and are called long-day plants; others that bloom on short days (less than 12 hours) are known as short-day plants. Animals likewise may respond to either long or short days. In many, but by no means all, photoperiod-sensitive organisms, the timing can be altered by experimental or artificial manipulation of the photoperiod. As shown in Figure 5–6, an artificially speeded-up light regimen can bring brook trout into breeding condition four months early. Florists can often force flowers to bloom out of season by altering the photoperiod. In migratory birds, there are several months after the fall migration when the birds are refractory to photoperiod stimulation. The short days of fall are apparently necessary to "reset" the biological clock, as it were, and prepare the endocrine system for a response to long days.

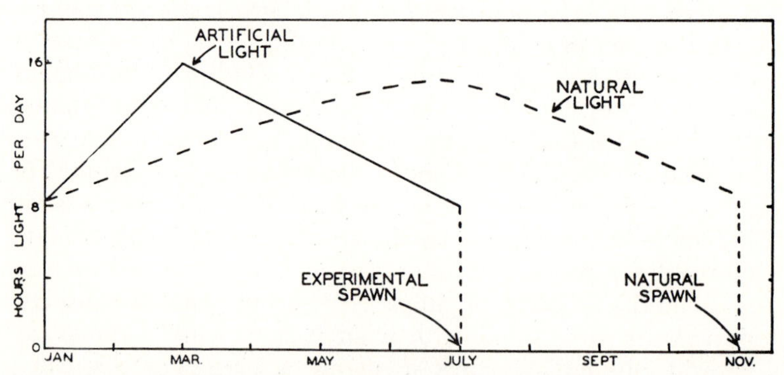

Figure 5–6 Control of the breeding season of the brook trout by artificial manipulation of the photoperiod. Trout, which normally breed in the autumn, spawn in summer when day length is increased artificially in the spring and then decreased in the summer to simulate autumn conditions. (Redrawn from Hazard and Eddy, 1950.)

Anytime after late December, an artificial increase in day length will bring on the sequence of molting, fat deposition, migratory restlessness, and gonad enlargement that normally occurs in the spring. The physiology of this response in birds is now fairly well known (see reviews by Farner, 1964, 1964a), but it is uncertain whether the fall migration is initiated by the direct stimulus of short days or is timed by the biological clock that was set by the long photoperiods of the spring.

Photoperiodism in certain insects is noteworthy because it provides a sort of "birth control." Long days of late spring and early summer stimulate the "brain" (actually a nerve cord ganglion) to produce a neurohormone that starts producing a diapause or resting egg that will not hatch until next spring no matter how favorable the temperatures, food, and other conditions are (see Beck, 1960). Thus, population growth is halted before, rather than after, the food supply becomes critical.

Some researchers have shown that the number of underground nitrogen-fixing root nodules on legumes (see Figure 4–3) is controlled by photoperiod acting through the leaves of the plant. Since nitrogen-fixing bacteria in the nodules require food energy manufactured by the plant leaves to do their work, the more light and chlorophyll, the more food that becomes available to the bacteria. Maximum coordination between the plant and its microbial partners is thus enhanced by the photoperiod regulator.

In striking contrast to day length, rainfall in a desert is highly unpredictable; yet desert annuals, which constitute the largest number of species in many desert floras, use this factor as a regulator. The seeds of many such species contain a germination inhibitor that must be washed out by a certain minimum of a rain shower (for example, a half inch or more). This shower provides all the water necessary to complete the life cycle back to seeds again. If such seeds are placed in moist soil in the greenhouse, they fail to germinate, but do so quickly when treated with a simulated shower of the necessary magnitude (see Went, 1957). Seeds may remain viable in the soil for many years, "waiting," as it were, for the adequate shower, which explains why deserts bloom—become quickly covered by flowers—a short time after heavy rainfall.

4. Brief Review of Physical Factors of Importance as Limiting Factors

The broad concept of limiting factors is not restricted to physical factors, since biological interrelations (co-actions or biological factors) are just as important in controlling the actual distribution and abundance of organisms in nature. However, the latter will be considered in Chapters 6 and 7 dealing with populations and communities; the physical and chemical aspects of the environment will be reviewed in this section. To present all that is known in this field would require a book in itself and would be beyond the scope of the present outline of ecological principles. Therefore, only a few items that ecologists have found important and worth studying will be noted.

Temperature

Compared with the range of thousands of degrees known to occur in the universe, life, as we know it, can exist only within a tiny range of about 300 degrees Centigrade—from about −200° to 100°C. Actually, most species and most activity are restricted to an even narrower band of temperatures. Some organisms, especially in a resting stage, can exist at very low temperatures at least briefly, whereas a few microorganisms, chiefly bacteria and algae, can live and reproduce in hot springs where the temperature is close to the boiling point. The upper temperature tolerance for hot-spring bacteria is about 88°C and for blue-green algae about 80°, compared with 50°C for the most tolerant fish and insects (see Brock, 1967). In general, the upper limits are more quickly critical than the lower limits, though many organisms appear to function more efficiently toward the upper limits of their ranges of tolerance. The range of temperature variation tends to be less in water than on land, and aquatic organisms generally have narrower ranges of tolerance to temperature than do equivalent land animals. Temperature, therefore, is universally important and is often a limiting factor. Temperature rhythms, along with rhythms of light, moisture, and tides, largely control the seasonal and daily activities of plants and animals. Temperature is often responsible for the zonation and stratification that occur in both water and land environments. It is also one of the easiest of environmental factors to measure. The mercury thermometer, one of the first and most widely used precision scientific instruments, has more recently been supplemented by electrical "sensing" devices, such as platinum resistance thermometers, thermocouples (bimetallic junctions), and thermistors (metallic oxide thermally sensitive resistors), which permit not only measurement in "hard-to-get-at" places but also the continuous and automatic recording of measurements. Furthermore, advances in the technology of telemetry now make it feasible to radiotransmit temperature information from the body of a lizard deep in its burrow or from a migratory bird flying high in the atmosphere.

Variability of temperature is extremely important ecologically. A temperature fluctuating between 10° and 20°C and averaging 15°C does not necessarily have the same effect on organisms as a constant temperature of 15°C. *Organisms normally subjected to variable temperatures in nature* (as in most temperate regions) *tend to be depressed, inhibited, or slowed down by constant temperature.* Thus, to give the results of one pioneer study, Shelford (1929) found that eggs and larval or pupal stages of the codling moth developed 7 to 8 percent faster under conditions of variable temperature than under a constant temperature having the same mean. In another experiment (Parker, 1930), grasshopper eggs kept at a variable temperature showed an average accelerated development of 38.6 percent and nymphs a development of 12 percent when compared with development at constant temperature.

Whether variation in itself is responsible for the accelerating effect or whether the higher temperature causes more growth than is balanced by the low temperature is uncertain. In any event, the stimulating effect of variable temperature, in the temperate zone at least, may be accepted as a well-defined ecological principle, and one that might be emphasized, since the tendency has been to conduct experimental work in the laboratory under conditions of constant temperature.

Because organisms are sensitive to temperature changes, and because temperature is so easy to measure, it is sometimes overrated as a limiting factor. One must guard against assuming that temperature is limiting when other, unmeasured factors may be more important. Plants, animals, and especially communities can often compensate or acclimate to temperature. The beginning ecologist who studies a particular organism or problem should by all means consider temperature, but he or she must not stop there.

Radiation: Light

As aptly expressed by Pearse (1939), light puts organisms on the horns of a dilemma: direct exposure of protoplasm to light causes death, yet light is the ultimate source of energy, without which life could not exist. Consequently, many of the structural and behavioral characteristics of organisms are concerned with solving this problem. In fact, as noted in the discussion of the Gaia Hypothesis (Chapter 2, Section 4) and again in Chapter 8, the evolution of the biosphere as a whole has chiefly involved the taming of incoming solar radiation so that its useful wavelengths could be exploited and its dangerous ones mitigated or shielded out. Light, therefore, is not only a vital factor but also a limiting one, at both the maximum and minimum levels. There is not another factor of greater interest to ecologists!

The total radiation environment and something of its spectral distribution was considered in Chapter 3, as was the primary role of solar radiation in ecosystem energetics. Consequently, this chapter discusses light as a limiting and controlling factor. Radiation consists of electromagnetic waves of a wide range in length. As shown in Figure 5–7, two bands of wavelengths readily penetrate the earth's atmosphere: the visible band, together with some parts of adjacent bands, and the low-frequency radio band, having wavelengths greater than 1 cm. Whether the long radio waves are ecologically significant is unknown, despite researchers who assert positive effects on migrating birds or other organisms. As shown in Figure 3–2, the solar radiations that penetrate the upper atmosphere and reach the earth's surface consist of electromagnetic waves ranging in length from about 0.3 micron to 10 microns (μ); this equals from 300 to 10,000 mμ, or 3000 to 100,000 Å.* To the human eye, visible light lies in the range of 3900 to 7600 Å (390 to 760 mμ), as shown in Figure 5–7. This figure also shows the energy-matter interaction of different bands and the kind of sensors used to detect and measure them. The role of ultraviolet (below 3900 Å) and infrared light (above 7600 Å) has been considered in Chapter 3. The role that high-energy, very short wave gamma radiation, as well as other types of ionizing radiation, may play as ecological limiting factors in the atomic age involves many special and complex considerations and will be briefly reviewed in the next section.

Ecologically, the quality (wavelength or color), the intensity (actual energy measured in gram-calories or foot candles), and the duration (length of day) of light are known to be important. Both animals and plants respond to different wavelengths of light. Color vision in animals sporadically occurs in different taxonomic groups, apparently being well developed in certain species of arthropods, fish, birds, and mammals but not in other species of the same groups (among mammals, for example, color vision is well developed only in primates). The rate of photosynthesis varies somewhat with different wavelengths. In terrestrial ecosystems, the quality of sunlight does not vary enough to have an important differential effect on the rate of photosynthesis, but as light penetrates water, the reds and blues are

* A micron (μ) is one thousandth of a millimeter (10^{-3} mm); a millimicron (mμ) is one millionth of a millimeter (10^{-6} mm); an angstrom (Å) is one tenth of a millimicron (10^{-7} mm).

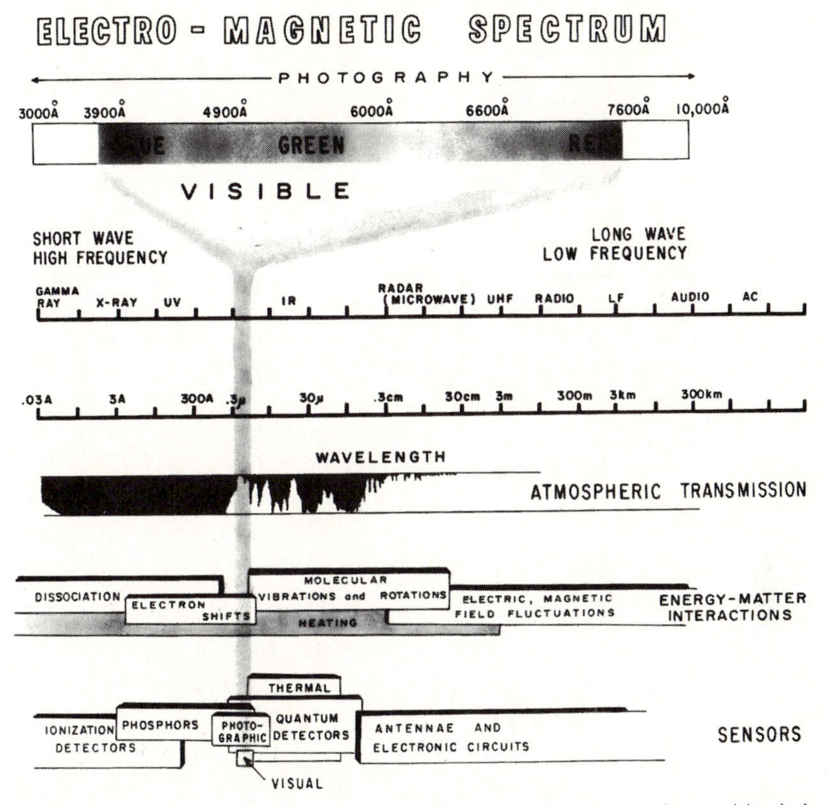

Figure 5–7 The electromagnetic spectrum in relation to the visible light, with an indication of atmospheric transmission, energetics, and methods of detection of different wave-frequency bands. Å = angstrom = 0.1 millimicron (mμ) = 0.0001 micron (μ). (Modified from Colwell et al., 1963.)

filtered out, and the resultant greenish light is poorly absorbed by chlorophyll. The marine red algae (Rhodophyta), however, have supplementary pigments (phycoerythrins) enabling them to utilize this energy and to live at greater depths than would be possible for the green algae.

As was discussed in Chapter 3, the intensity of light (i.e., the energy input) impinging on the autotrophic layer controls the entire ecosystem through its influence on primary production. The relationship of intensity to photosynthesis in both terrestrial and aquatic plants follows the same general pattern of linear increase up to an optimum or **light saturation** level, followed in many instances by a decrease at the high intensities of full sunlight. As noted in Chapter

2, Section 5, plants with the C_4 type of photosynthesis reach light saturation at high intensities and are not inhibited by full sunlight (see Figure 2–6).

As would be expected, factor compensation occurs, since individual plants as well as communities adapt to different light intensities by becoming "shade-adapted" (i.e., reach saturation at low intensities) or "sun-adapted" (see Figure 3–8). Diatoms that live in beach sand or on intertidal mudflats are remarkable in that they reach a maximum rate of photosynthesis when light intensity is less than 5 percent full sunlight, and they can maintain a net production at less than 1 percent (Taylor, 1964). Yet these diatoms are only slightly inhibited by high intensities. Phytoplankton, in contrast, are shade-adapted and are very much inhibited by high intensities, which accounts for the fact that peak production in the sea usually occurs below, rather than at, the surface (see Figure 3–6).

The role of light duration, or photoperiodicity, has been considered.

Ionizing Radiations

Very high-energy radiations that can remove electrons from atoms and attach them to other atoms, thereby producing positive and negative **ion pairs**, are known as ionizing radiations. Light and most solar radiation do not have this ionizing effect. It is believed that ionization is the chief cause of injury to protoplasm and that the damage is proportional to the number of ion pairs produced in the absorbing material. Ionizing radiations are sent out from radioactive materials on earth and are also received from space. Isotopes of elements that emit ionizing radiations are called **radionuclides** or **radioisotopes.**

Ionizing radiation in the environment has been increased appreciably by our efforts to utilize atomic energy. Nuclear weapons tests have injected into the atmosphere radionuclides that then come to earth all over the globe as **fallout.** About 10 percent of the energy of a nuclear weapon is in residual radiation (Glasstone, 1957). Nuclear power plants (including fuel processing and disposal of wastes at other sites), medical research, and other peaceful uses of atomic energy produce local "hot spots" and wastes that often escape into the environment while being transported or stored. Failure (so far) to avoid accidental releases and to solve the radioactive waste problem is the main reason why atomic energy has not lived up to its potential as an energy source for human societies. Because of atomic energy's importance in the future, this factor will be reviewed in some detail.

Of the three ionizing radiations of primary ecological concern, two are corpuscular (alpha and beta) and one is electromagnetic (gamma radiation and the related x-radiation). Corpuscular radiation consists of streams of atomic or subatomic particles that transfer their energy to whatever they strike. **Alpha particles** are parts of helium atoms and are huge on the atomic scale of things. They travel only a few centimeters in air and may be stopped by a sheet of paper or the dead layer of human skin, but when stopped they produce a large amount of ionization locally. **Beta particles** are high-speed electrons—much smaller particles that may travel several feet in air or up to a couple of centimeters in tissue and give up their energy over a longer path. **Ionizing electromagnetic radiations**, on the other hand, are like light, only of much shorter wavelength (see Figure 5–7). They travel great distances and readily penetrate matter, releasing their energy over long paths (the ionization is dispersed). For example, **gamma rays** penetrate biological materials easily; a given "ray" may go right through an organism without having any effect, or it may produce ionization over a long path. The effect of gamma rays depends on the number and energy of the rays and the distance of the organism from the source, since intensity decreases exponentially with distance. Important features of alpha, beta, and gamma radiation are diagrammed in Figure 5–8. Thus, the alpha, beta, gamma series is

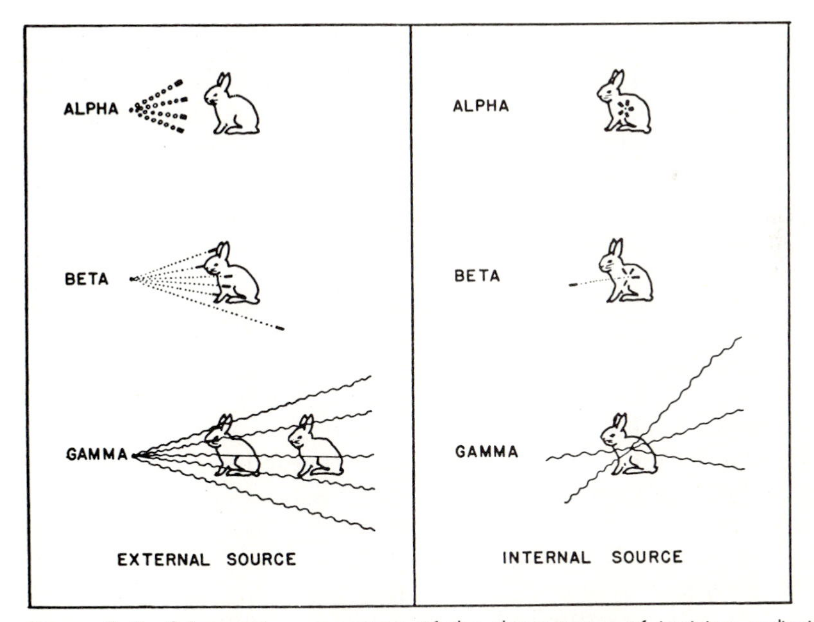

Figure 5–8 Schematic comparison of the three types of ionizing radiations of greatest ecological interest, showing relative penetration and specific ionization effect. The diagram is not intended to be quantitative.

one of increasing penetration but decreasing concentration of ionization and local damage. Therefore, biologists often class radioactive substances that emit alpha or beta particles as "internal emitters," because their effect is likely to be greatest when absorbed, ingested, or otherwise deposited in or near living tissue. Conversely, radioactive substances that are primarily gamma emitters are classed as "external emitters," because they are penetrating and can produce their effect without being taken inside.

Other types of radiation are of at least indirect interest to the ecologist. **Neutrons** are large, uncharged particles that do not cause ionization themselves but wreak local havoc by bumping atoms out of their stable states. Neutrons thus induce radioactivity in nonradioactive materials or tissues through which they pass. For a given amount of absorbed energy, "fast" neutrons may do ten times, and "slow" neutrons five times, the local damage of gamma rays. Neutrons are restricted to the vicinity of reactors or atomic explosions, but, as indicated previously, they are of primary importance in producing radioactive substances that can and do become widely distributed in nature. **X-rays** are electromagnetic radiations similar to gamma rays but originate from the outer electron shell rather than from the nucleus of the atom and are not sent out from radioactive substances dispersed in the environment. Since they and gamma rays have similar effects, and since they are easily obtained from an x-ray machine, x-rays are used in experimental studies on individuals, populations, or even small ecosystems. **Cosmic rays** are radiations from outer space that are mixtures of corpuscular and electromagnetic components. The intensity of cosmic rays in the biosphere is low, but they are a major hazard in space travel. Cosmic rays and ionizing radiation from natural radioactive substances in soil and water produce what is known as **background radiation** to which the present biota is adapted. In fact, the biota may depend on this background radiation for maintaining genetic fluidity. Background varies three- to fourfold in various parts of the biosphere; it is lowest at or below the surface of the sea and highest at high altitudes on granitic mountains. Cosmic rays increase in intensity with increasing altitude, and granitic rocks have more naturally occurring radionuclides than do sedimentary rocks.

A study of radiation phenomena requires two types of measurement: (1) a measure of the number of disintegrations occurring in an amount of radioactive substance and (2) a measure of radiation dose in terms of the energy absorbed, which can cause ionization and damage.

The basic unit of the quantity of a radioactive substance is the **curie** (Ci) and is defined as the amount of material in which 3.7×10^{10}

atoms disintegrate each second, or 2.2×10^{12} disintegrations per minute (dpm). The actual weight of material making up a curie is very different for a long-lived, slowly decaying isotope compared with that of a rapidly decaying one. Approximately one gram of radium, for example, is a curie; very much less (about 10^{-7} gram) of newly formed radiosodium would emit 3.7×10^{10} disintegrations a second. Since a curie represents a rather large amount of radioactivity from the biological standpoint, smaller units are widely used: **millicurie** (mCi) = 10^{-3} Ci; **microcurie** (μCi) = 10^{-6} Ci; **nanocurie** (nCi) (formerly called a millimicrocurie, mμc) = 10^{-9} Ci; **picocurie** (pCi) (formerly called a micromicrocurie, $\mu\mu$c) = 10^{-12} Ci. The possible range of activity is so tremendous that one must be careful about the position of the decimal point. The curie indicates how many alpha or beta particles or gamma rays are emitted from a radioactive source, but this information does not tell how the radiation might affect organisms in the line of fire.

Radiation dose, the other important aspect of radiation, has been measured with several scales. The most convenient unit for all types of radiation is the **rad**, which is defined as the absorbed dose of 100 ergs of energy per gram of tissue. The **roentgen** (R) is an older unit, which strictly speaking is to be used only for gamma and x-rays. Actually, for effects on living organisms, the rad and the roentgen are nearly the same. A unit 1/1000 smaller, namely the milliroentgen (mR) or the millirad (mrad), conveniently quantifies the kind of radiation levels often encountered in the environment. The roentgen or rad is a unit of total dose. The **dose rate** is the amount received per unit time. Thus, if an organism is receiving 10 mR per hour, the total dose in a 24-hour period would be 240 mR or 0.240 R. The time over which a given dose is received is a very important consideration.

Instruments that measure ionizing radiation consist of two basic parts: a detector and a rate meter or electronic counter (scaler). Gaseous detectors, such as geiger tubes, are often used to measure beta radiation; solid or liquid scintillation detectors (substances that convert the invisible radiation to visible light recorded by a photoelectric system) are widely used to measure gamma and other types of radiation.

In general, the higher, more complex organisms are more easily damaged or killed by ionizing radiation; human beings are about the most sensitive of all.

Comparative sensitivity of three diverse groups of organisms to single doses of x- or gamma radiation is shown in Figure 5–9. Large, single doses delivered at short time intervals (minutes or hours) are known as **acute doses**, in contrast to **chronic doses** of sublethal radia-

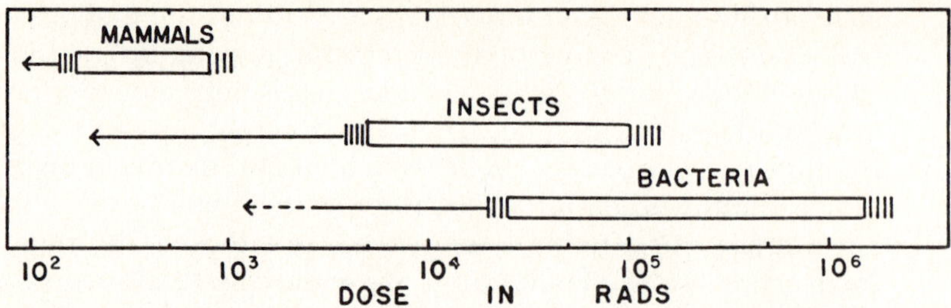

Figure 5–9 Comparative radiosensitivity of three groups of organisms to single acute doses of x- or gamma radiation. See text for explanation.

tion that might be experienced continuously over a whole life cycle. The left ends of the bars indicate levels at which severe effects on reproduction (temporary or permanent sterilization, for example) may be expected in the more sensitive species of the group, and the right ends of the bars indicate levels at which a large portion (50 percent or more) of the more resistant species would be killed outright. The arrows to the left indicate the lower range of doses that would kill or damage sensitive life-history stages such as embryos. Thus, a dose of 200 rads will kill some insect embryos in the cleavage stage, 5000 rads will sterilize some species of insects, but 100,000 rads may be required to kill all adult individuals of the more resistant species. In general, mammals are the most sensitive and microorganisms the most resistant of organisms. Seed plants and lower vertebrates fall somewhere between insects and mammals. Most studies have shown that rapidly dividing cells are most sensitive (which explains why sensitivity decreases with age). Thus, any component—whether a part of an organism, a whole organism, or a population—undergoing rapid growth is likely to be affected by comparatively low levels of radiation regardless of taxonomic relationships.

The effects on low-level chronic doses are more difficult to measure, since long-term genetic as well as somatic effects may be involved. For growth response, Sparrow (1962) has reported that a chronic dose of 1 R per day continued for ten years (total dose 25,000 R) produces about the same amount of growth reduction in pine trees (which are relatively radiosensitive) as an acute dose of 60 R. Any increase in the ionizing radiation environment above background level, or even a high natural background, can increase the rate of production of deleterious mutations (as can many chemicals and food additives in the human diet and environment).

In higher plants, sensitivity to ionizing radiation has been shown to be directly proportional to the size of the cell nucleus or, more specifically, to chromosome volume or DNA content (Sparrow and Woodwell, 1962; Sparrow et al., 1963). Sensitivity varies almost three orders of magnitude as does chromosome volume when plants are irradiated in the laboratory. In the field, other considerations such as shielding of sensitive growing or regeneration parts (as when underground) would determine relative sensitivity.

Among higher animals, no simple or direct relationship between sensitivity and cellular structure has been found; effects on specific organ systems are more critical. Thus, mammals are very sensitive to low doses because rapidly dividing, blood-making tissue in the bone marrow is especially vulnerable. The digestive tract is also sensitive, but brain damage occurs only at quite high levels. Even very low levels of chronic ionizing radiation can cause cancerous growth in bone and other vulnerable tissue that may not appear until many years after exposure (as is now being documented in the cases of soldiers exposed to the early atomic bomb tests). Whether there is a threshold on which to base a "permissible level" or whether any increase above background is detrimental is a question that is unsettled but much debated as we seek to assess the risks and benefits of atomic energy.

Between 1950 and 1970, effects of gamma radiation on whole communities and ecosystems were studied at several sites. Gamma sources, usually either cobalt-60 or cesium-137, of 10,000 Ci or more have been placed in fields and forests at the Brookhaven National Laboratory on Long Island (see Woodwell, 1962, 1965), in a tropical rain forest of Puerto Rico (see H. T. Odum and Pigeon, 1970), and in a desert in Nevada (see French, 1965). The effects of unshielded reactors (which emit neutrons as well as gamma radiation) on fields and forests have been studied in Georgia (see Platt, 1965) and at the Oak Ridge National Laboratory in Tennessee (see Witherspoon, 1965, 1969). A portable gamma source has been used to study short-term effects on a wide variety of communities at the Savannah River Ecology Laboratory in South Carolina (see McCormick and Golley, 1966; Monk, 1966b; McCormick, 1969). A lake bed community subjected to low-level chronic radiation from atomic wastes has been under study at the Oak Ridge Laboratory for many years.

No higher plant or animal survives when close to these powerful sources. Growth inhibition in plants and a reduced diversity of animal species were noted at levels as low as 2 to 5 rads per day. Although resistant forest trees or shrubs (in the case of the desert) persisted at rather high dose rates (10 to 40 rads per day), the vegetation was stressed and became vulnerable to insects and disease. In the second year of the experiment at Brookhaven, for example, an outbreak of oak leaf aphids occurred in the zone receiving about 10 rads per day. In this zone, aphids were more than 200 times as abundant as in the normal, unradiated oak forest.

When radionuclides are released into the environment, they often become dispersed and diluted, but they may also become concentrated in living organisms and during food-chain transfers by various means, which are categorized under the general heading of "biological magnification." Radioactive substances may also simply accumulate in water soils, sediments, or air if the input exceeds the rate of natural radioactive decay; thus, an apparently innocuous amount of radioactivity can soon become lethal.

The ratio of a radionuclide in the organism to that in the environment is often called the **concentration factor.** The chemical behavior of a radioactive isotope is essentially the same as that of the nonradioactive isotope of the same element. Therefore, the observed concentration by the organism is not the result of the radioactivity, but merely demonstrates in a measurable way the difference between the density of the element in the environment and in the organism. Thus, radioactive iodine (^{131}I) concentrates in the thyroid just as nonradioactive iodine does. Also, some of the synthetic radionuclides become concentrated because of chemical affinity with nutrients that are naturally concentrated by organisms.

Two examples will sufficiently illustrate the concentrative tendencies of two of the most troublesome, long-lived radionuclides that are byproducts of the fission of uranium (hence called fission products). Strontium-90 (^{90}Sr) tends to cycle like calcium; cesium-137 (^{137}Cs) behaves like potassium. Concentration factors for ^{90}Sr in various parts of a food web in a lake receiving low-level wastes are diagrammed in Figure 5–10. Since, as already noted, the blood-making bone marrow tissue is especially sensitive to the beta radiation of ^{90}Sr, the 3000 or 4000× concentration in bone is significant. In assessing the impact of releases of radioactive material into the environment, one must allow for ecological concentration.

Concentration factors are likely to be greater in nutrient-poor than in nutrient-rich soils and water. Concentration is also greater in "thin" vegetation such as lichen-covered rocks or the arctic tundra. Unfortunately, the Laplanders and Eskimos eating reindeer or caribou meat ingest more fallout radionuclides than do those of us who eat from the Iowa corn-hog food chain.

Table 5–1 shows the concentration of fallout cesium-137 (determined by whole body count) in deer to be much higher on the sandy, low-lying coastal plain than in the adjacent Piedmont region where soils are well drained and have a high clay content. Since average rainfall is the same for both regions, the input of fallout from the atmosphere to the soil is probably also the same.

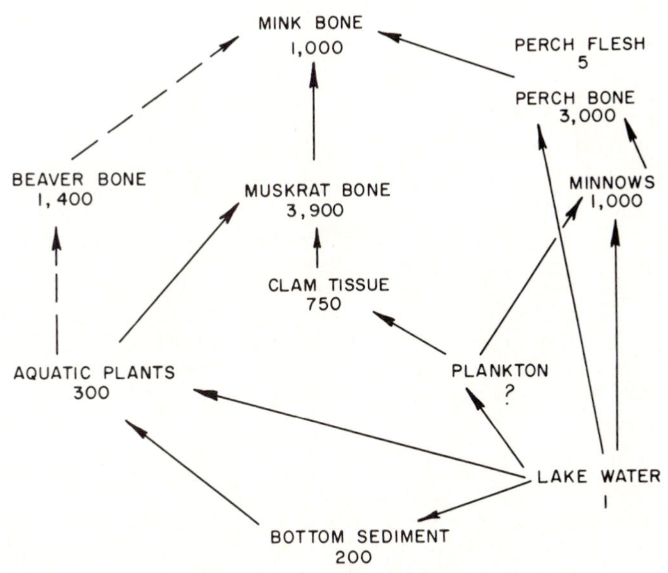

Figure 5–10 Concentration of strontium-90 in various parts of the food web of a small Canadian lake receiving low-level atomic wastes. Average concentration factors are shown in terms of lake water = 1. (After Ophel, 1963; used by permission of Biology and Health Physics Division, Atomic Energy of Canada Limited, Chalk River, Ontario.)

Table 5–1 **Comparison of the Concentration of Cesium-137 (Resulting from Fallout) in White-tailed Deer in Coastal Plain and Piedmont Regions of Georgia and South Carolina***

Region	Number of Deer	Cesium-137 (pCi/kg Wet Weight)	
		Mean and Standard Error†	Range
Lower coastal plain	25	18,039 ± 2359	2076–54,818
Piedmont	25	3007 ± 968	250–19,821

* Data from Jenkins and Fendley (1968).

† Difference between regions highly significant at 99 percent level.

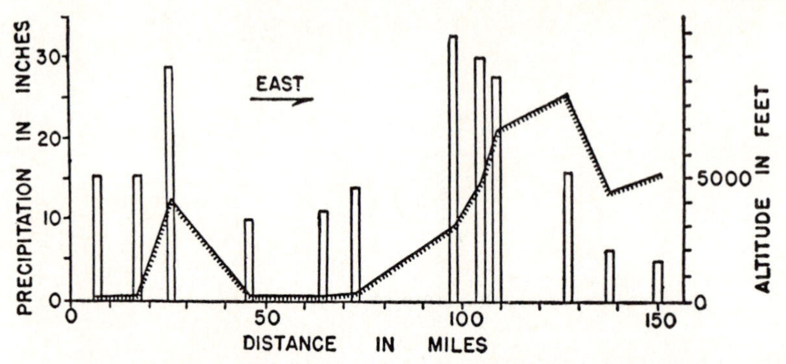

Figure 5–11 Mean annual rainfall (vertical columns) in relation to altitude (ornamented line) at a series of stations extending from Palo Alto on the Pacific Coast eastward across the Coast Range and the Sierra Nevada to Oasis Ranch in the Nevada desert. The diagram shows (1) the approach effect on the west edge of the Sierra, (2) the zone of maximum rainfall on the middle western slope of the Sierra, and (3) rain shadows to the landward of the two mountain ranges.

Water

Water, a physiological necessity for all protoplasm, is from the ecological viewpoint chiefly a limiting factor in land environments or in water environments in which the amount can fluctuate greatly, or where high salinity fosters water loss from organisms by osmosis. Rainfall, humidity, the evaporating power of the air, and the available supply of surface water are the principal factors measured. Each of these aspects is briefly described.

Rainfall is determined largely by geography and the pattern of large air movements or weather systems. A relatively simple example is shown in Figure 5–11. Moisture-laden winds blowing off the ocean deposit most of their moisture on the ocean-facing slopes; the resulting "rain shadow" produces a desert on the other side. The higher the mountains, the greater the effect, in general. As the air continues beyond the mountains, some moisture is picked up, and rainfall may again increase somewhat. Thus, deserts are usually found "behind" high mountain ranges or along the coast where winds blow from large, dry interior land areas rather than off the ocean. Distribution of rainfall over the year is an extremely important limiting factor for organisms. The situation provided by a 35-inch rainfall evenly distributed is entirely different from that provided by 35 inches of rain that falls largely during a restricted part of the year. In the latter case, plants and animals must be able to survive long droughts. Rainfall generally tends to be unevenly distributed over the seasons in the tropics and subtropics, often with well-defined wet and dry seasons resulting. In the tropics, this seasonal rhythm of moisture regulates the seasonal activities (especially reproduction) of organisms, much as the seasonal rhythm of temperature and light regulates temperate zone organisms. In temperate climates, rainfall tends to be more evenly distributed throughout the year, with many exceptions. The following tabulation gives a rough approximation of the climax biotic communities that may be expected with different annual amounts of rainfall evenly distributed in temperate latitudes:

0–10 inches per year—desert
10–30 inches per year—grassland, savanna,* or open woodland
30–50 inches per year—dry forest
More than 50 inches per year—wet forest

Actually, the biotic situation is determined not by rainfall alone but by the balance between rainfall and potential evapotranspiration, which is the loss of water by evaporation from the ecosystem.

Humidity represents the amount of water vapor in the air. Absolute humidity is the actual amount of water in the air expressed as weight of water per unit of air (grams per kilograms of air, for example). Since the amount of water vapor that air can hold (at saturation) varies with the temperature and pressure, **relative humidity** represents the percentage of vapor actually present compared with saturation under existing temperature-pressure conditions. Relative humidity is usually measured by noting the difference between a wet and a dry bulb thermometer mounted on an instrument called a psychrometer. If both thermometers read the same, the relative humidity is 100 percent; if the wet bulb thermometer reads less than the dry bulb one, as is usually the case, the relative humidity is less than 100 percent. The exact value is determined by consulting prepared tables. Relative humidity may also be conveniently measured by a hygrograph, which provides a continuous record. Human hair, especially long blond hair, expands and contracts in proportion to relative humidity, and strands of it can thus be made to operate a lever writing on a moving drum. As with measurement of temperature, various electrical sensing devices are coming into widespread use. One such

* A savanna is a grassland with scattered trees or scattered clumps of trees, a type of community intermediate between grassland and forest (see Figure A–7 in the Appendix).

device uses the ability of a film of lithium chloride to change its electrical resistance in proportion to changes in relative humidity. Other hydroscopic materials are under experimentation.

In general, relative humidity has been the measurement most used in ecological work, although the converse of relative humidity, **vapor pressure deficit** (the difference between partial pressure of water vapor at saturation and the actual vapor pressure), is often preferred as a measure of moisture relations because evaporation tends to be proportional to vapor pressure deficit rather than to relative humidity.

Because of the daily rhythm of humidity in nature (high at night, low during the day, for example), as well as vertical and horizontal differences, humidity along with temperature and light helps regulate the activities of organisms and limit their distribution. Humidity is especially important in modifying the effects of temperature, as will be noted in the next section.

The evaporative power of the air is important ecologically, especially for land plants, and is usually measured by evaporimeters, which measure evaporation from the surface of a porous bulb filled with water. Animals may often regulate their activities to avoid dehydration by moving to protected places or becoming active at night; plants, however, cannot move. From 97 to 99 percent of water that enters plants from the soil is lost by evaporation from the leaves. This evaporation, called **transpiration**, is a unique feature of the energetics of terrestrial ecosystems. When water and nutrients are nonlimiting, growth of land plants is closely proportional to the total energy supply at the ground surface. Since most of the energy is heat, and since the fraction providing latent heat for transpiration is nearly constant, growth is also proportional to transpiration (Penman, 1956).

Transpiration has positive aspects, also. Evaporation cools the leaves and is one of the several processes that aid cycling of nutrients. Other processes include transport of ions through the soil to the roots, transfer of ions across root boundaries, translocation within the plant, and foliar leaching (see Kozlowski, 1964, 1968). Several of these processes require the expenditure of metabolic energy, which may exert rate-limiting control on both water and mineral transport (see Fried and Broeshart, 1967). Thus, transpiration is not a simple function of exposed physical surface; a forest does not necessarily lose more water than a grassland. Transpiration as an energy subsidy in moist forests was discussed in Chapter 3. If the air is too moist (approaching 100 percent relative humidity), as in certain tropical cloud forests, trees are stunted and much of the vegetation is epiphytic, presumably because of the lack of "transpiration pull" (see H. T. Odum and Pigeon, 1970).

Despite the many biological and physical complications, total evapotranspiration is broadly correlated with the rate of productivity. For example, Rosenzweig (1968) found that evapotranspiration was a highly significant predictor of the annual aboveground net primary production in mature or climax terrestrial communities of all kinds (deserts, tundras, grasslands, and forests); however, the relationship was not reliable in unstable or developmental vegetation. He presents the following regression equation (including 5 percent confidence interval for slope and intercept):

$$\log_{10} P_n = (1.66 \pm 0.27) \log_{10} AE - (1.66 \pm 0.07)$$

where P_n is net aboveground production in grams per square meter, and AE is the annual actual evapotranspiration in millimeters. Knowing the latitude and mean monthly temperatures and precipitation (basic weather record), one can estimate AE from meteorological tables (see Thornthwaite and Mather, 1957) and then, using the equation, predict what a well-adjusted, mature natural community should be able to produce. Rosenzweig hypothesizes that the relationship between AE and P_n is due to the fact that AE measures the simultaneous availability of water and solar energy, the most important rate-limiting resources in terrestrial photosynthesis. The poor correlation between AE and P_n in developmental communities is logical, since such communities have not yet reached equilibrium conditions with their energy and water environment.

The ratio of growth (net production) and water transpired is called the **transpiration efficiency** and is usually expressed as grams of dry matter produced per 1000 grams of water transpired. Most species of agricultural crops, as well as a wide range of noncultivated species, have a transpiration efficiency of 2 or less: 500 grams or more of water are lost for every gram of dry matter produced (Norman, 1957). Drought-resistant crops, such as sorghum and millet, may have efficiencies of 4. Strangely enough, desert plants can do little, if any, better. Their unique adaptation involves not the ability to grow without transpiration but the ability to become dormant when water is not available (instead of wilting and dying as would be the case in non-desert plants). Desert plants that lose their leaves and expose only green buds or stems during dry periods do show a high transpiration efficiency (Lange et al., 1969). Cacti that employ the CAM type of photosynthesis reduce water loss by keeping their stomates closed during the day (see Section 5 of Chapter 2).

The available surface water supply is, of course, related to the rainfall of the area, but there are often great discrepancies. Owing to underground sources or supplies from nearby regions, animals and plants may have access to more water than that which falls as rain. Likewise, rainwater may quickly become unavailable to organisms. Wells (1928) has spoken of the North Carolina sandhills as "deserts in the rain," because the abundant rain of the region sinks so quickly through the porous soil that plants, especially herbaceous ones, find very little water available in the surface layer. The plants and small animals of such areas resemble those of much dryer regions. Other soils in the western plains of the United States retain water so tenaciously that crops can be raised without a single drop of rain falling during the growing season. The plants can use the water stored from winter rains.

The general nature of the hydrologic cycle has been considered in detail in Chapter 4; the important and still poorly understood relationships between surface water and ground water and between rainfall and the atmospheric and ocean pools were discussed in that chapter. Ecologists generally agree that we need to know more about water resources and to do a better job of managing them before we seriously consider manipulation of rainfall, i.e., "rain-making," which is now technically possible in some dry regions. Too severe a removal of vegetative cover and poor land-use practices, which destroy soil texture and increase erosion, have already increased runoff to such an extent that local deserts are produced in regions of adequate rainfall.

On the more positive side, irrigation and artificial impoundment of streams have helped increase local water supplies. However, these mechanical engineering devices, useful though they often are, should never be regarded as substitutes for sound agricultural and forestry land-use practices, which trap the water at or near its sources for maximum usefulness. The ecological viewpoint—water as a cyclic commodity within the whole ecosystem—is very important. People who think all floods and erosion and water-use problems can be solved by building big dams or any other mechanical device alone may know engineering, but they do not know their ecology. Despite millions spent on flood control and other attempts to tame the Mississippi River, the cost of flood damage has increased. The more the river is constricted by dikes and the more the watershed is urbanized, the higher the water rises and the worse the flood is when water does break through or rises over the barriers. For a well-documented account of attempts to control the Mississippi, see Belt, 1975.

Dew may contribute appreciably and, in areas of low rainfall, vitally to precipitation. Dew and ground fog are especially important in coastal forests and in deserts. Oberlander (1956) and Azevedo and Morgan (1974) report that fog on the West Coast may account for as much as two to three times more water than the annual precipitation and that some tall trees in position to intercept coastal fog as it moves inland may get as much as 150 cm of "rainfall" dripping down from the limbs.

Ground Water

For humankind, ground water is one of the most important resources, because, in contrast to most other organisms, we do have access in many regions to a great deal more water than falls as rain. Cities and irrigated agriculture located in deserts and other dry regions are made possible by this accessibility. Unfortunately, much of this underground water was stored in past ages, and reservoirs in arid regions are either not being replenished at all or are being replenished at a slower rate than they are being pumped out. Arid-region ground water, like oil, is nonrenewable.

Ground water provides 25 percent of the fresh water used for all purposes in the United States and about 50 percent of drinking water. In 1975, about 70 percent of ground water went for irrigation, and this use has increased sharply into the 1980s. Fifteen percent goes to industry and a like percent to urban and rural domestic water use. As the quality of surface water has deteriorated, use of ground water has increased. As with other abundant common property or "free" goods of the earth, ground water tends to be taken for granted and was very little studied until depletion and pollution showed plainly that limiting factors were involved.

The largest stores of ground water are in **aquifers**, porous underground strata, often limestone, sand, or gravel, bounded by impervious rock or clay that holds the water in like a giant pipe or elongated tank. Water enters where the permeable strata are close to the surface or otherwise intersect the surface water table; water may leave the aquifer by way of springs or other discharges at or near the surface. Where aquifers slope seaward from higher ground recharge areas, water in the deeper aquifers is under pressure and will rise above the surface like a fountain when a well is drilled into it (the so-called artesian well). The geographical distribution of aquifers and other substantial stores of ground water is mapped in Figure 5–12.

According to the U.S. Water Resources Council's 1978 report, "The Nation's Water Resources," the volume of ground water in the United States is about 50 times the annual flow of surface water and

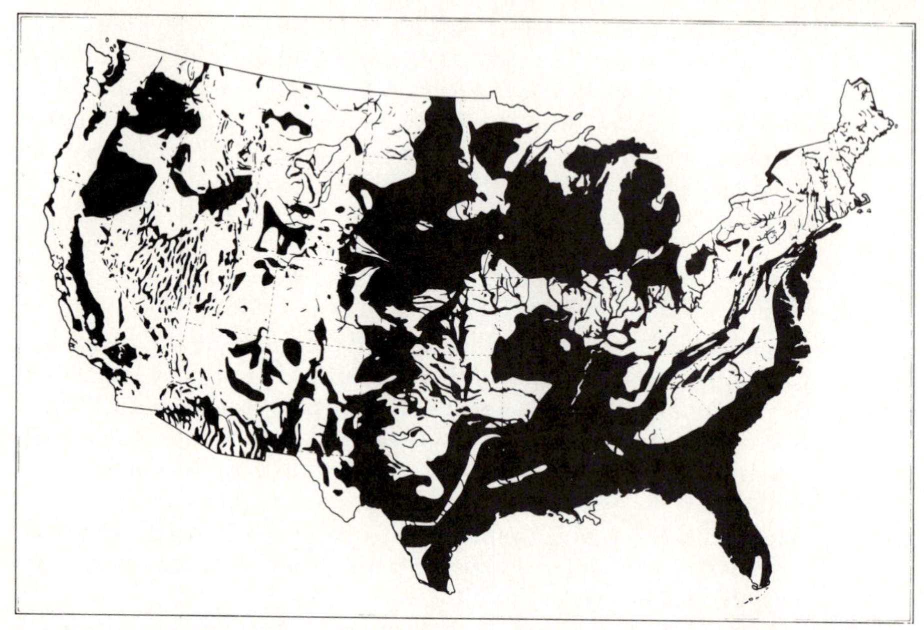

Figure 5–12 Ground water resources of the United States. About half of the country is underlain by aquifers capable of yielding very large volumes of water. Aquifers in the mid-continent and western region, which are poorly recharged, are being "overdrawn" or "mined" in many areas where withdrawals are the source of irrigation water. (Courtesy of U.S. Water Resources Council.)

four times the volume of the Great Lakes. The annual input (rain and snowbelt recharge) and output (water returned to the hydrologic cycle of rivers, oceans, and the atmosphere) for this huge reservoir are estimated to be about 1 part in 120 parts of total volume. Although our withdrawals total only about one tenth of the recharge volume, some of the most heavily used aquifers are located in regions of low or no recharge, as already noted. In 1975, about one fourth of all withdrawals were overdrafts (exceeding recharge), mostly in agricultural regions of the West. An example is the Ogallala aquifer of the high plains of Texas, Kansas, Oklahoma, Nebraska, and eastern Colorado where irrigated grain production provides an important part of the export market that this nation counts on to balance payments for imported oil. Fossil water and fossil fuel (to pump the water) have combined to make a billion-dollar economy in this region. Within the next couple of decades, it is predicted, this aquifer will, for all practical purposes, be "pumped out." The fossil water will be gone before fossil fuel is exhausted, but the latter becomes useless without water. Then the region will be faced with severe economic depression and depopulation, and the nation will have to find some other place to grow grain—unless, of course, it is feasible to build an aqueduct from the Mississippi River system! For more on ground water, see Freeze and Cherry (1979), Sheridan (1981), and the U.S. Water Resources Council's report (1978).

Depletion is not the only threat to ground water. Contamination with toxic chemicals may be an even greater threat. At least the problem of toxic wastes does have technological solutions if societies are willing and able to pay the cost to protect a water resource that in the long run is more precious than oil or gold. In fact, one could argue for the proposition that usable fresh water is potentially a greater limiting factor for civilization than is energy. Water problems vary with the region, as shown in Figure 5–13, but no region is without a "water problem" of some sort. Since water is, at least in part, a non-market commodity, public opinion and political intervention are necessary to prevent both wasteful allocation of this resource and complete depletion.

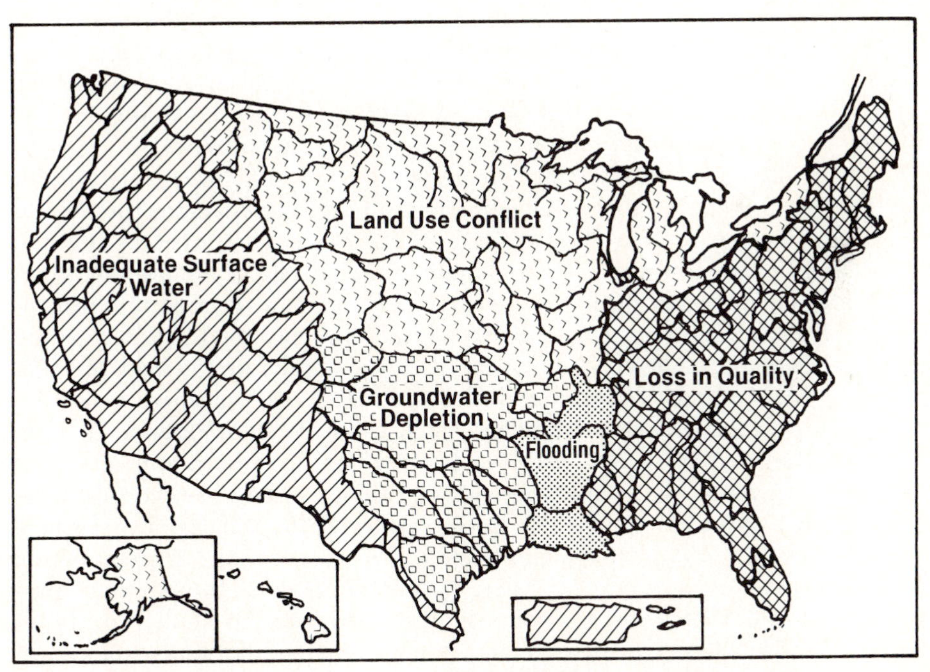

Figure 5–13 Regional view of water problems in the United States.

Temperature and Moisture Acting Together

By considering the ecosystem concept first among ecological concepts, we have avoided creating the impression that environmental factors operate independently of one another. This chapter attempts to show that consideration of individual factors is a means of approaching complex ecological problems but is not the ultimate objective of ecological study, which is to evaluate *the relative importance of various factors as they operate together in actual ecosystems.* Temperature and moisture are so generally important in terrestrial environments and so closely interacting that they are usually conceded to be the most important part of climate. Thus, it may be well to consider them together before proceeding to other factors.

The interaction of temperature and moisture, as in the case of the interaction of most factors, depends on the relative as well as the absolute values of each factor. Thus, temperature exerts a more severe limiting effect on organisms when there is either very much moisture or very little than when there are moderate conditions. Likewise, moisture is critical in the extremes of temperature. In a sense, this is another aspect of the principle of factor interaction, which was discussed earlier in the chapter. For example, the boll weevil can tolerate higher temperatures when the humidity is low or moderate than when it is very high. Hot, dry weather in the cotton belt is a signal for the cotton farmers to look for an increase in the weevil population. Hot, humid weather is less favorable for the weevil but, unfortunately, not so good for the cotton plant.

Large bodies of water greatly moderate land climates because of the high latent heat of evaporation and melting characteristic of water, which is to say that many heat calories are required to melt ice and evaporate water. In fact, there are two basic types of climate: (1) the continental climates are characterized by extremes of temperature and moisture, and (2) the marine climates are characterized by less extreme fluctuation because of the moderating effect of large bodies of water (large lakes thus produce local "marine climates").

Classifications of climate, such as those of Köppen or Thornthwaite (1931, 1948), are based largely on quantitative measures of temperature and moisture, taking into consideration the effectiveness of precipitation and temperature as determined by seasonal distribution and the mean values. The comparison of precipitation of potential evapotranspiration (which depends on temperature) provides a particularly accurate picture of climates, as shown in Figure 5–14, which contrasts climates of three distinctly different biological regions or biomes. The period of soil moisture utilization represents

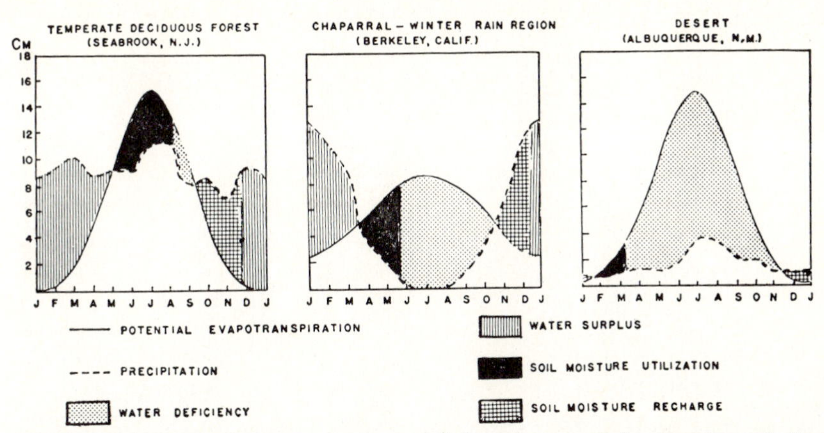

Figure 5–14 Relationship between rainfall and potential evapotranspiration (evaporation from soil plus transpiration from vegetation) in three localities representing three distinctly different ecological regions. The dotted area in the charts ("water deficiency") indicates the season during which water may be expected to be a limiting factor, whereas the vertical extent of this area indicates the relative severity of this limitation. (After Thornthwaite, 1955.)

the principal period of primary production for the community as a whole and thus determines the supply of food available to the consumers and decomposers for the entire annual cycle. In the deciduous forest region, water is likely to be severely limiting only in late summer, more so in the southern than in the northern portion of the region. Native vegetation is adapted to withstand the periodic summer droughts, but some agricultural crops grown in the region are not. After bitter experiences with many late summer crop failures, farmers in the southern United States, for example, are finally beginning to provide for irrigation in the late summer. In the winter rain region, the main season of production is late winter and spring; in the desert, the effective growing season is much reduced.

Climographs, or charts in which one major climatic factor is plotted against another, are another useful method of graphically representing temperature and moisture in combination. In temperature-rainfall or temperature-humidity charts, mean monthly values are plotted with the temperature scale on the vertical axis and either humidity or rainfall on the horizontal axis. The resulting polygon shows the temperature-moisture conditions and makes possible the graphic comparison of one year with another, or comparison of climate of one biotic region with that of another as shown in Figure 5–15. Climographs have been useful in determining the suitability of

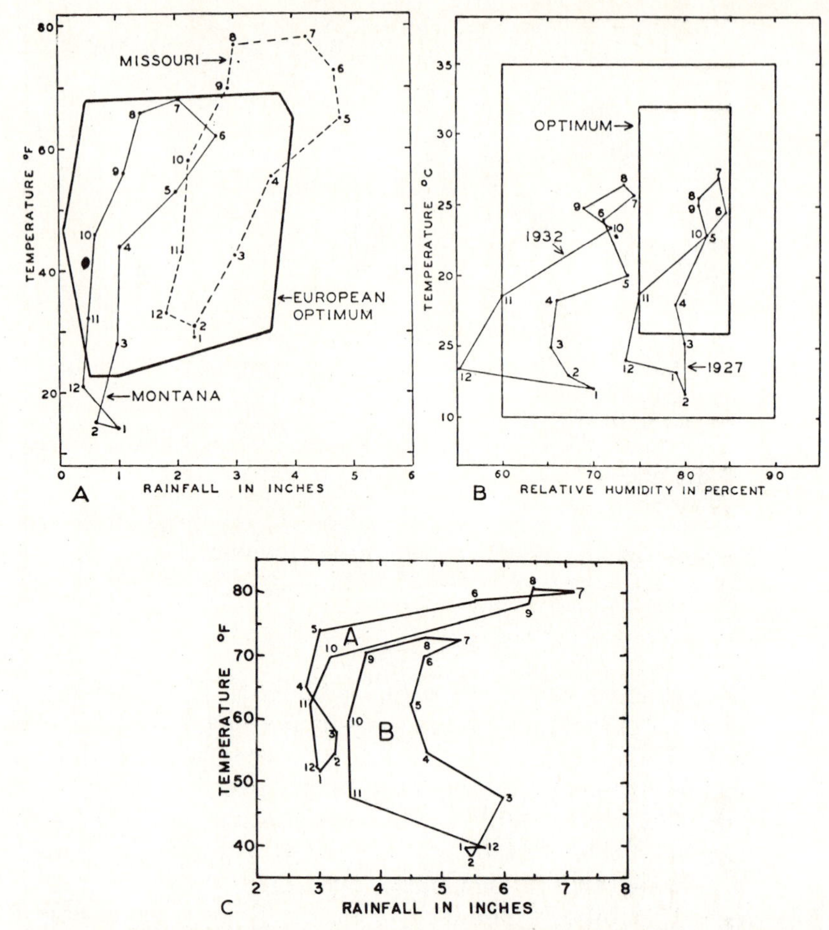

Figure 5–15 Temperature-moisture climographs. *A*. Temperature-rainfall monthly averages for Havre, Montana, where the Hungarian partridge has been successfully introduced, and Columbia, Missouri, where it has failed, compared with average conditions in the European breeding range. (Redrawn from Twomey, 1936.) *B*. Temperature-humidity conditions at Tel Aviv, Israel, for two different years, compared with optimum (inner rectangle) and favorable (outer rectangle) conditions for Mediterranean fruit fly. Damage to oranges was much greater in 1927. (After Bodenheimer, 1937.) *C*. Temperature-rainfall climographs for coastal Georgia (sea level) (A) and northern Georgia (altitude 2000 to 3000 feet) (B). In A, a pronounced wet and dry season is the conspicuous seasonal variant as is characteristic of a subtropical climate, while in B, seasonal differences in temperature are more pronounced than seasonal changes in rainfall. The climatic climax vegetation (see page 469 for definition of this term) in the coastal locality is a broad-leaved evergreen forest, while in the more northern area it is a temperate deciduous forest.

temperature-moisture combinations for proposed introductions of plants or game animals. Plots of other pairs of factors, such as plots of temperature and salinity in marine environments, may also be instructive. For more on climographs, see Smith (1940).

Climate chambers, which provide another useful approach to the study of combinations of physical factors, vary from simple temperature-humidity cabinets in use in most laboratories to large controlled greenhouses, such as the "phytotron" in which any desired combination of temperature, moisture, and light can be maintained. These chambers are often designed to control environmental conditions so that the investigator can study the genetics and physiology of cultivated or domesticated species. However, the chambers can be useful for ecological studies, especially when natural rhythms of temperature and humidity can be simulated. Experiments of this sort help single out factors that may be "operationally significant," but they can reveal only part of the story, since many significant aspects of the ecosystem cannot be duplicated indoors but must be experimented with outdoors.

Atmospheric Gases

Except for the large variations in water vapor already discussed, the atmosphere of the major part of the biosphere is remarkably homeostatic. Interestingly, the present concentration of carbon dioxide (0.03 percent by volume) and oxygen (21 percent by volume) is somewhat limiting to many higher plants. It is well known that photosynthesis in many plants can be increased by moderately increasing CO_2 concentration, but it is not so well known that decreasing the oxygen concentration experimentally can also increase photosynthesis. Beans, for example, increase their rate of photosynthesis by as much as 50 percent when the oxygen concentration around their leaves is lowered to 5 percent (Björkman, 1966). C_4 plants are not inhibited by high O_2 concentration as are C_3 plants. (See Chapter 2, Section 5.) Thus, C_4 grasses, including corn and sugar cane, do not show oxygen inhibition. The reason for inhibition in C_3 broad-leaved plants may be that they evolved when the CO_2 concentration was higher and the O_2 concentration lower than it is now.

Deeper in soils and sediments and in the bodies of large animals (the rumen of cattle being an anaerobic system) oxygen becomes limiting to aerobes, and CO_2 increases. This results in a slowdown in the rate and/or a change in the end-products of decomposition (the importance of which was discussed in Chapter 2). Anthropogenic impact on the CO_2 cycles was discussed in Chapter 4.

The situation in aquatic environments differs from that in the atmospheric environment because amounts of oxygen, carbon dioxide, and other atmospheric gases dissolved in water and thus available to organisms vary from time to time and place to place. Oxygen is a prime limiting factor, especially in lakes and in waters with a heavy load of organic material. Though oxygen is more soluble in water than is nitrogen, the actual quantity of oxygen that water can hold under the most favorable conditions is much less than that constantly present in the atmosphere. Thus, if 21 percent by volume of a liter of air is oxygen, there will be 210 cc of oxygen per liter. By contrast, the amount of oxygen per liter of water does not exceed 10 cc. Temperature and dissolved salts greatly affect the ability of water to hold oxygen; the solubility of oxygen is increased by low temperatures and decreased by high salinities. The oxygen supply in water comes chiefly from two sources: by diffusion from the air and from photosynthesis by aquatic plants. Oxygen diffuses into water very slowly unless helped along by wind and water movements; light penetration is an all-important factor in the photosynthetic production of oxygen. Therefore, important daily seasonal and spatial variations may be expected in the oxygen concentration of aquatic environments.

Carbon dioxide, like oxygen, may be present in water in highly variable amounts, but its behavior in water is rather different and its ecology is not as well known. It is therefore difficult to make general statements about carbon dioxide's role as a limiting factor. Although present in low concentrations in the air, carbon dioxide is extremely soluble in water, which also obtains large supplies from respiration, decay, and soil or underground sources. Thus, the "minimum" is less likely to be important than is the case with oxygen. Furthermore, unlike oxygen, carbon dioxide chemically combines with water to form H_2CO_3, which in turn reacts with available limestones to form carbonates ($—CO_3$) and bicarbonates ($—HCO_3$). A major reservoir pool of biospheric CO_2 is the carbonate system of the oceans (see Figure 4–8A). These compounds not only provide a source of nutrients but also act as buffers, helping to keep the hydrogen ion concentration of aquatic environments near the neutral point. Moderate increases in CO_2 in water seem to speed up photosynthesis and the developmental processes of many organisms. CO_2 enrichment, along with increased nitrogen and phosphorus, may be a key to cultural eutrophication (Lange, 1967; Kuentzel, 1969). High CO_2 concentrations may be definitely limiting to animals, especially since such high concentrations of carbon dioxide are associated with low concentrations of oxygen. Fishes respond vigorously to high concentrations and may be killed if the water is too heavily charged with unbound CO_2.

Hydrogen ion concentration, or pH, is closely related to the carbon dioxide complex, and being relatively easy to measure, it has been much studied in natural aquatic environments. Unless values are extreme, communities compensate for differences in pH by mechanisms already described in this chapter and show a wide tolerance for the naturally occurring range. However, when the total alkalinity is constant, pH change is proportional to CO_2 change and therefore is a useful indicator of the rate or rates of total community metabolism (photosynthesis and respiration). Soils and waters of low pH (i.e., acidic) are frequently deficient in nutrients and low in productivity.

Industrial pollution (smog, acid rain, toxic chemicals, etc.) will be discussed in Section 5 of this chapter.

Biogenic Salts: Macronutrients and Micronutrients

Dissolved salts vital to life may be conveniently termed **biogenic salts.** About half of the 54 elements in the periodic table have now been shown to be essential to either plants or animals or, in most cases, both. As already indicated, nitrogen and phosphorus salts are of major importance, and the ecologist may do well to consider these first as a matter of routine. Hutchinson (1957) states the case for phosphorus as the prime limiting factor:

> Of all the elements present in living organisms, phosphorus is likely to be the most important ecologically, because the ratio of phosphorus to other elements in organisms tends to be considerably greater than the ratio in the primary sources of the biological elements. A deficiency of phosphorus is therefore more likely to limit the productivity of any region of the earth's surface than is a deficiency of any other material except water.

Agricultural and industrial demands create conditions in which too much of such nutrients as nitrogen and phosphorus, as well as too little, becomes limiting (discussed in Chapter 4).

The interaction of nitrogen and phosphorus is especially worthy of attention. The N/P ratio in "average" biomass is about 16 : 1 and in streams and rivers about 28 : 1 (see Figure 4–8). Schindler (1977) reports experiments in which fertilizers with different N/P ratios were added to a whole lake. When the N/P ratio was reduced to 5, nitrogen-fixing blue-green algae dominated the phytoplankton and fixed enough nitrogen to raise the ratio to within the range of many natural lakes. Schindler hypothesizes that lake ecosystems have evolved natural mechanisms to compensate for deficiencies of nitrogen and carbon but not for phosphorus, which does not have a gaseous phase. Accordingly, primary production is very often correlated with available phosphorus (reaffirming Hutchinson's statement).

After nitrogen and phosphorus, potassium, calcium, sulfur, and magnesium merit consideration. Mollusks and vertebrates need calcium in especially large quantities, and magnesium is a necessary constituent of chlorophyll, without which no ecosystem could operate. Elements and their compounds needed in relatively large amounts are often known as **macronutrients.**

In recent years, great interest has developed in the study of elements and their compounds that are necessary for the operation of living systems but that are required only in extremely minute quantities, often as components of vital enzymes. These elements are generally called **trace elements** or **micronutrients.** Since minute requirements seem to be associated with an equal or even greater minuteness in environmental occurrence, the micronutrients are important as limiting factors. The development of modern methods of microchemistry, spectrography, x-ray diffraction, and biological assay has greatly increased our ability to measure even the smallest amounts. Also, the availability of radioisotopes of many trace elements has greatly stimulated experimental studies. Deficiency diseases due to the absence of trace elements have been known to exist for a long time. Pathological symptoms have been observed in laboratory, domestic, and wild plants and animals. Under natural conditions, deficiency symptoms of this sort are sometimes associated with peculiar geological history and sometimes with a deteriorated environment of some sort, often a direct result of poor management by human beings. An example of peculiar geologic history is found in southern Florida. The potentially productive organic soils of this region did not meet expectation (for crops and cattle) until it was discovered that this sedimentary region lacked copper and cobalt, which are present in most areas. A possible case of micronutrient deficiency resulting from changes in land management was discussed on page 201 in connection with the discussion of the CO_2 cycle.

Ten micronutrients are especially important to plants: iron (Fe), manganese (Mn), copper (Cu), zinc (Zn), boron (B), silicon (Si), molybdenum (Mo), chlorine (Cl), vanadium (V), and cobalt (Co).

These elements can be arranged by function into three groups: (1) those required for photosynthesis—Mn, Fe, Cl, Zn, and V; (2) those required for nitrogen metabolism—Mo, B, Co, Fe; and (3) those required for other metabolic functions—Mn, B, Co, Cu, and Si. All these elements except boron are essential for animals, which also may require selenium, chromium, nickel, fluorine, iodine, tin, and perhaps even arsenic (Mertz, 1981). The dividing line between macro- and micronutrients, of course, is neither sharp nor the same for all groups of organisms; sodium and chlorine, for example, are needed in larger amounts by vertebrates than by plants. Sodium, in fact, is often added to the preceding list as a micronutrient for plants. Many micronutrients resemble vitamins, because they act as catalysts. The trace metals often combine with organic compounds to form "metallo-activators"; cobalt, for example, is a vital constituent of vitamin B_{12}. Goldman (1960) documented a case in which molybdenum is limiting to a whole ecosystem when he found that addition of 100 parts per billion to the water of a mountain lake increased the rate of photosynthesis. He also found that in this particular lake, concentration of cobalt was high enough to be inhibitory to the phytoplankton. As with macronutrients, too much can be as limiting as too little. For a nicely illustrated article on the essential elements, see Frieden (1972).

Currents and Pressures

The atmospheric and hydrospheric media in which organisms live are not often completely still for any period of time. Currents in water not only greatly influence the concentration of gases and nutrients but also act directly both as limiting factors at the species level and often as energy subsidies that increase productivity at the community level. Thus, the differences between the species composition of a stream and that of a small pond community are due largely to the big difference in the current factor. Many stream plants and animals are morphologically and physiologically adapted to maintaining their position in the current and are known to have very definite limits of tolerance to this specific factor. On the other hand, water flow that acts as an energy subsidy is a key to the productivity of wetland ecosystems as shown in Figure 5–16.

On land, wind exerts a limiting effect on the activities and even the distribution of organisms and also sometimes acts beneficially. Birds, for example, remain quiet in protected places on windy days, which are therefore poor days for the ecologist to attempt a bird census. Plants may be modified structurally by the wind, especially

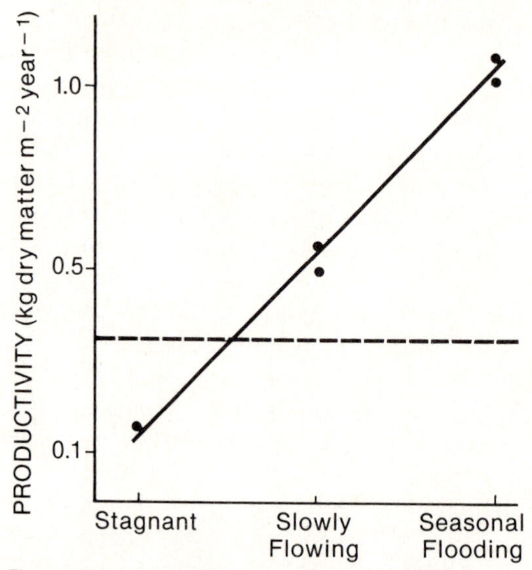

Figure 5–16 Productivity of swamp forests in relation to water flow regimes. The horizontal dotted line represents an average level of regional productivity for upland forests not subjected to flooding. Stagnant conditions thus stress the ecosystem, while flowing and fluctuating water conditions act as subsidies that enhance productivity. The five data points are from the study of Conner and Day, 1976.

when other factors are also limiting, as in alpine regions. Whitehead (1957) demonstrated experimentally that wind limits the growth of plants in exposed locations on mountains. When he erected a wall to protect the vegetation from wind, the height of the plants increased.

On the other hand, air movement can enhance productivity in the same manner as water flow, as is apparently the case for certain tropical rain forests (see Chapter 3, Section 3). Storms are important, even though they may be only local. Hurricanes as well as ordinary winds transport animals and plants for great distances and, when these storms strike land, the winds may change the composition of the forest communities for many years to come. In a recent study of New England forests, Oliver and Stephens (1977) reported that the effects of two hurricanes that occurred before 1803 could still be seen in the structure of vegetation. It has been observed that insects spread faster in the direction of the prevailing winds than in other directions to areas that seem to offer equal opportunity for the establishment of the species. In dry regions, wind is an especially important limiting factor for plants, since it increases the rate of water loss by transpiration, and as already noted, desert plants have many special adaptations to mitigate this limitation.

Barometric pressure has not been shown to be an important direct limiting factor for organisms, although some animals appear able to detect differences, and, of course, barometric pressure has much to do with weather and climate, which are directly limiting to organisms. In the ocean, however, hydrostatic pressure is important because of the tremendous gradient from the surface to the depths. In water, the pressure increases 1 atmosphere for every 10 meters. In the deepest part of the ocean, the pressure reaches 1000 atmospheres. Many animals can tolerate wide changes in pressure, especially if the body does not contain free air or gas. When the body does, gas embolisms may develop. In general, the great pressures found in the depth of the ocean exert a depressing effect, so that the pace of life is slower.

Soil

It is sometimes convenient to think of the biosphere as comprising the atmosphere, the hydrosphere, and the pedosphere, the latter being the soil. Each of these divisions owes many of its characteristic features to the ecological reactions and co-actions of organisms and to the interplay of ecosystems and basic cycles between them. Each is composed of a living and a nonliving component more easily separated theoretically than practically. Biotic and abiotic components are especially intimate in soil, which by definition consists of the weathered layer of the earth's crust with living organisms and products of their decay intermingled. Without life, the earth would have a crust and might have air and water, but the air and water, and especially the soil, would be entirely different from these components as they are now. Thus, soil not only is a "factor" of the environment of organisms but also is produced by them. In general, soil is the net result of the action of climate and organisms, especially vegetation, on the parent material of the earth's surface. Soil thus is composed of a parent material, the underlying geologic or mineral substrate, and an organic increment in which organisms and their products are intermingled with the finely divided and modified parent material. Spaces between the particles are filled with gases and water. The texture and porosity of the soil are highly important characteristics and largely determine the availability of nutrients to plants and soil animals.

The cut edge of a bank or a trench (Figure 5–17) shows that soil is composed of distinct layers, which often differ in color. These

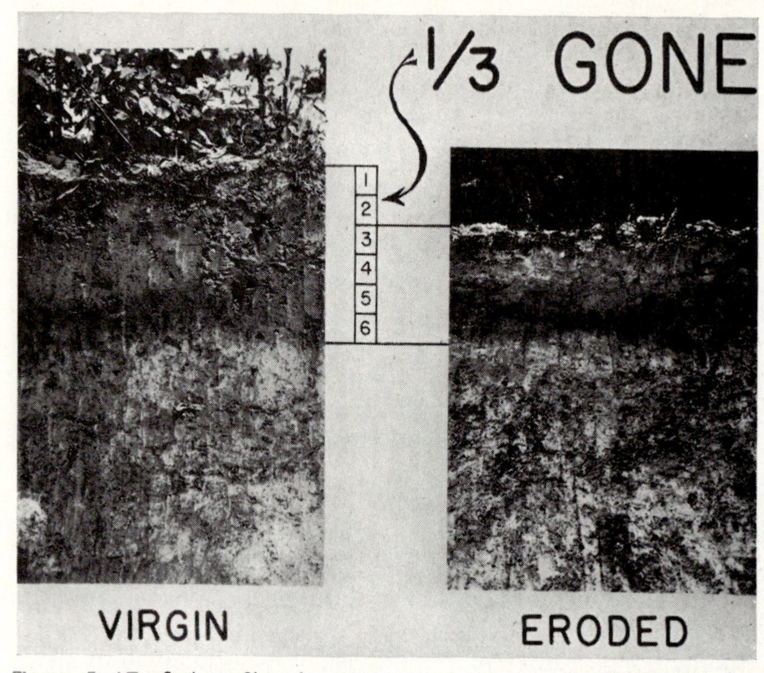

Figure 5–17 Soil profile of a virgin area compared with that of an eroded area in the deciduous forest region. In the left picture, 1–2 represents the A-1 horizon, 3–4 the A-2 horizon, and 5–6 the B-1 layer (accumulation of leached material). Compare with Figure 5–18. (U.S. Soil Conservation Service Photo.)

layers are called soil horizons, and the sequence of horizons from the surface down is called a soil profile. The upper horizon, or **A horizon** ("top soil"), is composed of the bodies of plants and animals that are being reduced to finely divided organic material by **humification** (described in Chapter 2, Section 5). In a mature soil, this horizon is usually subdivided into distinct layers representing progressive stages of humification. These layers (Figures 5–17 and 5–18) are designated (from the surface downward) as litter (A-0), humus (A-1), and leached (light-colored) zone (A-2). The A-0 layer is sometimes subdivided as A-1 (litter proper), A-2 (duff), and A-3 (leaf-mold). The litter (A-0) horizon represents the detritus component and can be considered a sort of ecological subsystem in which microorganisms (bacteria and fungi) work in partnership with small arthropods (soil mites and collembola) to decompose the organic material. When the latter are removed, decomposition is markedly reduced (as was shown in Figure 2–9). The annual input into the litter subsystem from leaf fall in forests increases from north to south (Figure 5–19).

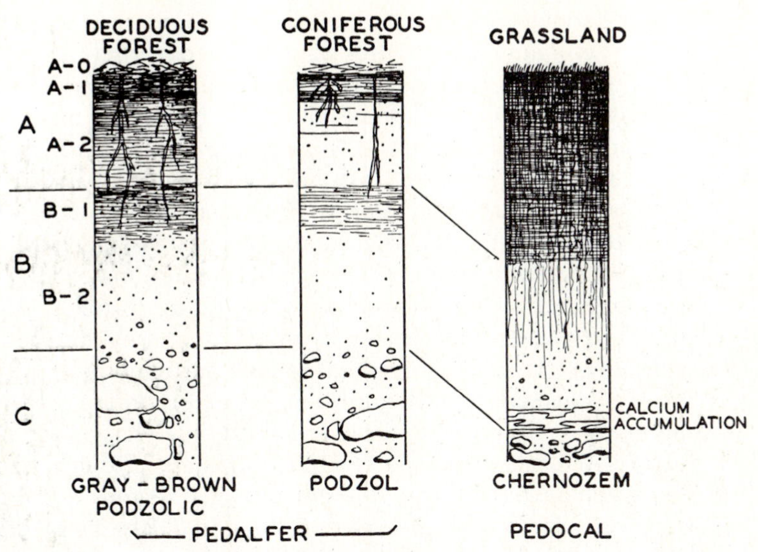

Figure 5–18 Simplified diagrams of three major soil types that are characteristic of three major biotic regions. See legend to Figure 5–22.

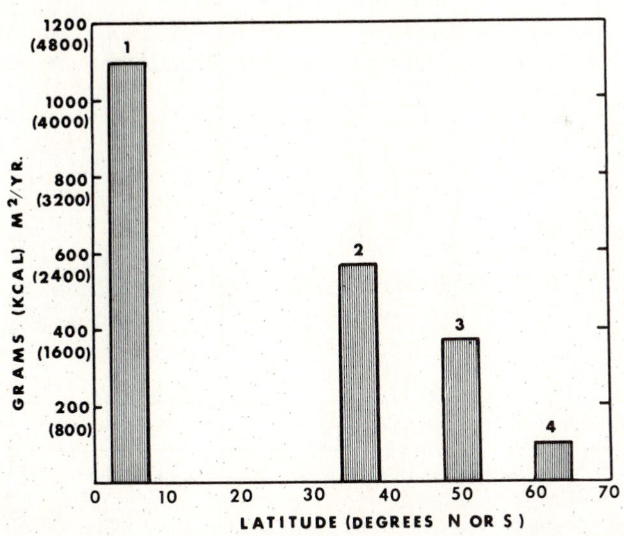

Figure 5–19 Annual litter fall in forests in relation to latitude. 1. Equatorial forests. 2. Warm temperate forests. 3. Cool temperate forests. 4. Arctic-alpine forests. (Bar graph drawn from data compiled by Bray and Gorham, 1964.)

The next major horizon, or **B horizon**, is composed of mineral soil in which the organic compounds have been converted by decomposers into inorganic compounds by **mineralization** and thoroughly mixed with finely divided parent material. The soluble materials of B horizon are often formed in the A horizon and deposited, or leached by downward flow of water, in B horizon. The dark band in Figure 5–17 represents the upper part of B horizon where materials have accumulated. The third horizon, or **C horizon**, is the more or less unmodified parent material. This parent material may represent the original mineral formation that is disintegrating in place, or it may have been transported to the site by gravity (colluvial deposit), water (alluvial deposit), glaciers (glacial deposit), or wind (eolian deposit, or loess). Transported soils are often extremely fertile (witness the deep loess soils of Iowa and the rich soils of the deltas of large rivers).

The soil profile and the relative thickness of the horizons are generally characteristic for different climatic regions and for topographical situations (Figures 5–18, 5–20, and 5–21). Thus, grassland soils differ from forest soils in that humification is rapid, but mineralization is slow. Since the entire grass plant, including roots, is short-lived, with each year are added large amounts of organic material, which decays rapidly, leaving little litter or duff but much humus. In the forest, litter and roots decay slowly, and since mineralization is rapid, the humus layer remains narrow (Figure 5–18). The average humus content of grassland soil, for example, is 600 tons per acre, compared with 50 tons per acre for forest soils (Daubenmire, 1974). In a forest-grassland buffer zone in Illinois, one can easily tell

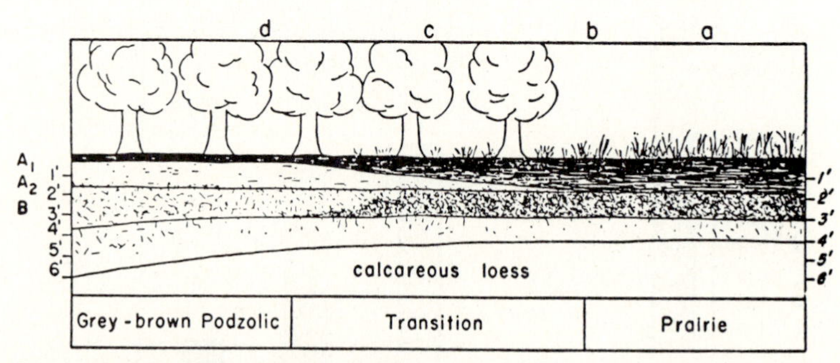

Figure 5–20 Soil-vegetation relationships in a prairie-forest transition zone. Distinctly different soils develop from the same parent material (calcareous loess, or wind-transported C horizon in this case) under influence of different vegetation and climate. The decrease in organic matter, the development of a podzolic A horizon (with narrow humus layer; see Figure 5–18), and the increased structural development of the B horizon are the main features differentiating forest soils from prairie soils. (After Crocker, 1952.)

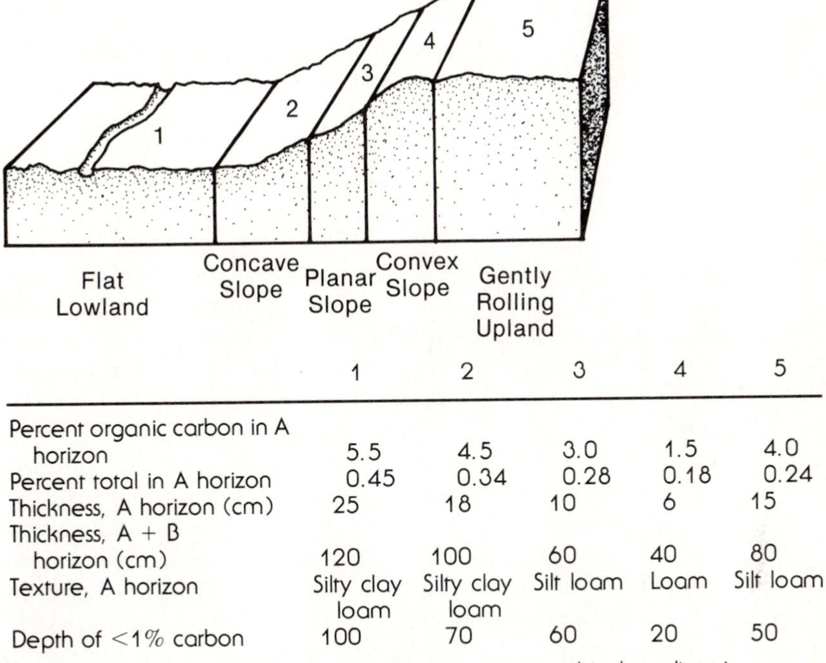

	1	2	3	4	5
Percent organic carbon in A horizon	5.5	4.5	3.0	1.5	4.0
Percent total in A horizon	0.45	0.34	0.28	0.18	0.24
Thickness, A horizon (cm)	25	18	10	6	15
Thickness, A + B horizon (cm)	120	100	60	40	80
Texture, A horizon	Silty clay loam	Silty clay loam	Silt loam	Loam	Silt loam
Depth of <1% carbon	100	70	60	20	50

Figure 5–21 Variation of soil properties in a topographical gradient in a grassland region. (After Coleman et al., 1983.)

by the color of the soil which cornfield was once prairie and which was forest. The prairie soil is much blacker owing to its high humus content. Given adequate rainfall, it is no accident that the "granaries of the world" are located in grassland regions.

Topographical conditions greatly influence the soil profile within a given climatic region. Steep slopes, especially if misused by humans, tend to have thin A and B horizons owing to erosion. Figure 5–21 shows how soil characteristics vary in a gradient from lowlands to uplands in a grassland region. Flat and gently sloping lands have deeper, more mature (well-developed profile), and more productive soils than do steeply sloping lands.

Sometimes on poorly drained land, water may leach materials rapidly into the deeper layers, forming a mineral "hardpan" through which plant roots, animals, and water cannot penetrate. Figure 8–10 illustrates an extreme case of hardpan condition that supports a stunted, "pigmy" forest in a region where "normal" soils support giant redwoods. Poorly drained areas such as bogs also favor the accumulation of humus, since poor aeration slows decay.

Classification of soil types has become a highly empirical subject. The soil scientist may recognize dozens of soil types as occurring within a county or state. Local soil maps are widely available from county and state soil conservation agencies and from state universities. Such maps and the soil descriptions that accompany them provide useful background for studies of terrestrial ecosystems. The ecologist, of course, should do more than merely name the soil on his or her study area. At the very minimum, three important attributes should be measured in at least the A and B horizons: (1) texture—the percent of sand, silt, and clay (or more detailed determination of particle size); (2) percent organic matter; and (3) exchange capacity—an estimate of the amount of exchangeable nutrients. The "available" minerals rather than the total amount determine potential fertility, other conditions being favorable (as emphasized in Chapter 4).

Major soil types of the United States are listed in Table 5–2, arranged in order of area occupied. Alfisols and mollisols make the best agriculture soils, but these constitute only about 22 percent of the land area. Huge areas of our country are unsuitable for intensive crop production unless soils are heavily amended with fertilizers and water. The same can be said for the earth as a whole.

Since soil is the product of climate and vegetation, a map of major soil types of the world, as shown in Figure 5–22, becomes a

Table 5–2 **Soils of the United States***

	Land area ($km^2 \times 10^6$)	Percent
Mountain soils	25.6	20.0
Aridisol (desert soils)	24.4	19.1
Alfisol (moderately weathered forest soils)	17.1	13.4
Inceptisol (weakly developed soils)	11.5	9.0
Mollisol (grassland soils)	11.1	8.7
Oxisol (oxidized, tropical soils)	11.1	8.7
Entisol (recent soils; profile undeveloped)	10.8	8.4
Ultisol (highly weathered forest soils)	7.2	5.6
Spodosol (sandy, poorly developed soils)	5.6	4.4
Vertisol (expandable clay soils)	2.3	1.8
Histosol (organic soils; some wetlands)	1.2	0.9
Total	127.9	100.0

* Listed in order of land area occupied. Estimated by J. E. Witty, Soil Conservation Service, U.S. Department of Agriculture; see Bartholomew and Alexander, 1981.

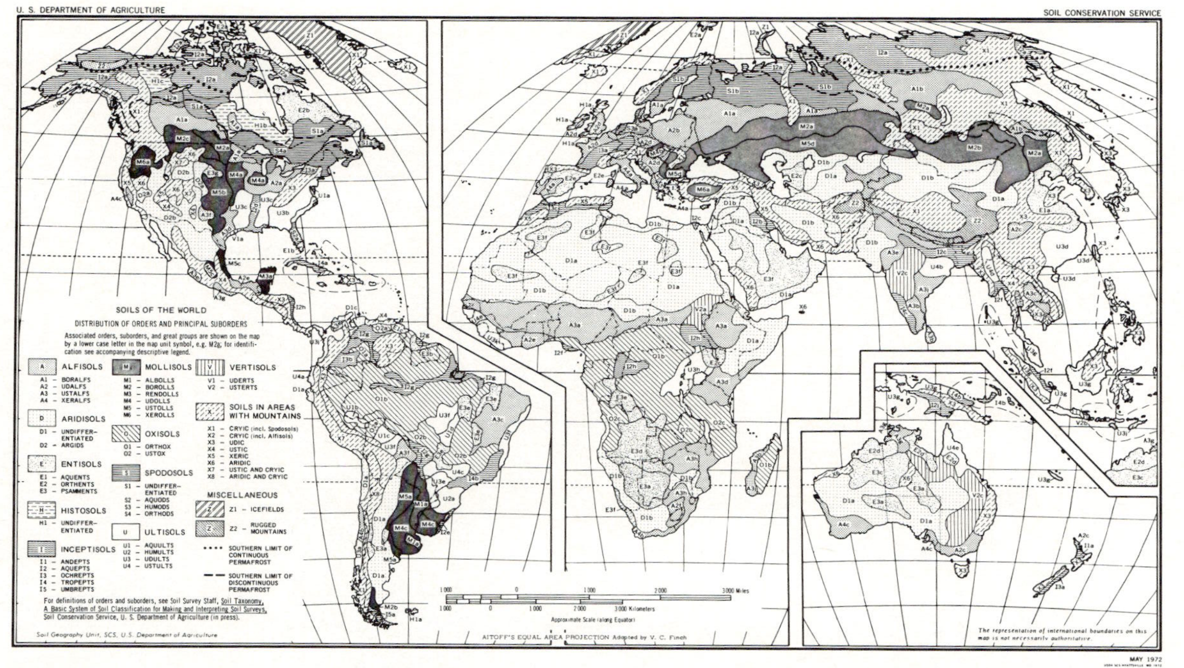

Figure 5–22 Soils of the world: distribution of orders and principal suborders. *A*. Alfisols: soils with subsurface horizons of clay accumulation and medium to high base supply; either usually moist or moist for 90 consecutive days when temperature is suitable for plant growth. *D*. Aridisols: soils with pedogenic horizons, usually dry in all horizons and never moist as long as 90 consecutive days when temperature is suitable for plant growth. *E*. Entisols: soils without pedogenic horizons; either usually wet, usually moist, or usually dry. *H*. Histosols: organic soils. *I*. Inceptisols: soils with pedogenic horizons of alteration or concentration but without accumulations of translocated materials other than carbonates or silica; usually moist or moist for 90 consecutive days when temperature is suitable for plant growth. *M*. Mollisols: soils with nearly black, organic-rich surface horizons and high base supply; either usually moist or usually dry. *O*. Oxisols: soils with pedogenic horizons that are mixtures principally of kaolin, hydrated oxides, and quartz and are low in weatherable minerals. *S*. Spodosols: soils with accumulation of amorphous materials in subsurface horizons; usually moist or wet. *U*. Ultisols: soils with subsurface horizons of clay accumulation and low base supply; usually moist or moist for 90 consecutive days when temperature is suitable for plant growth. *V*. Vertisols: soils with high content of swelling clays; deep, wide cracks develop during dry periods. *Z*. Miscellaneous.

composite map of climate and vegetation. Given a favorable parent material and not too steep a topography, the action of organisms and climate will tend to build up a soil characteristic of the region. From a broad ecological viewpoint, the soils of a given region may be lumped into two groups: mature soils on level or gently rolling topography, largely controlled by climate and vegetation of the region, and immature (in terms of profile development) soils, largely controlled by local conditions of topography, water level, or unusual type of parent material. The degree of soil maturity (the extent to which equilibrium between soil, climate, and vegetation has been reached) varies greatly with the region. Wolfanger (1930), for example, estimated that 83 percent of the soils in Marshall County, Iowa, were mature, compared with only 15 percent of the soils in Bertie County, North Carolina, which is located on the sandy "geologically young" Coastal Plain. The role the soil type plays in ecosystem function depends on the stage in geological and ecological development. Highly recommended readings are the small books by Kellogg, 1975 (soils from the agricultural viewpoint), and Richards, 1974 (soil as an ecosystem).

Soil Erosion In the 1930s, the Soil Conservation Service (SCS) was established by the United States government to combat soil erosion that was ruining thousands of acres of farm and forest land. The "dust bowl" was also taking its toll on the western plains at about this time. The program that was developed to save soil is an excellent example of how government should work in a democracy. A close linkage was established between Washington, state governments, land-grant state universities, and counties where county agents worked directly with land owners. Washington provided funds and universities contributed the knowledge, but decisions were made locally. Terracing, contour plowing, farm ponds, cover crops, buffer strips of permanent vegetation, crop rotation, and other measures, together with improvements in economic status of farmers, reversed the tide of soil loss, and a soil conservation ethic became generally accepted by farmers. Perhaps partly because of its success, the SCS had so much support in Congress and in the states that it became increasingly bureaucratic (i.e., less responsive to real needs) and extended its activities into other areas such as channeling streams and building large dams that often had questionable value in soil preservation. Then, suddenly, in the 1970s, soil erosion, per se, again became an urgent national problem because of two new trends:

1. Industrialization of farming, emphasizing "cash crops" that are treated not so much as food but as commodities for sale, especially on

the overseas market. Unfortunately, when farms are operated as businesses, often by corporations or other absentee owners, crop yield on the short term is maximized rather than maintenance of long-term fertility and productivity.

2. Urban sprawl, with roads and housing developments mushrooming into the rural countryside with little or no concern about soil loss. (SCS failed to set up "city agents" to help the developer avoid erosion!) Erosion, as related to land use, is illustrated by bar graphs in Figure 5–23, which shows just how severe erosion can be in urban and farming regions.

Although very severe soil losses from urban and suburban construction are usually of short duration, appreciable erosion from urban landscape and farmland tends to continue for years unless measures are taken to reduce it. The Soil Conservation Service considers an annual loss from row crop fields of 5 tons/acre as the maximum "tolerable" level for good, deep soils, and 2 tons/acre for poorer, thinner soils. A recent survey cited in the Council on Environmental Quality (CEQ) report "Global Future—Time to Act" (1981) indicates

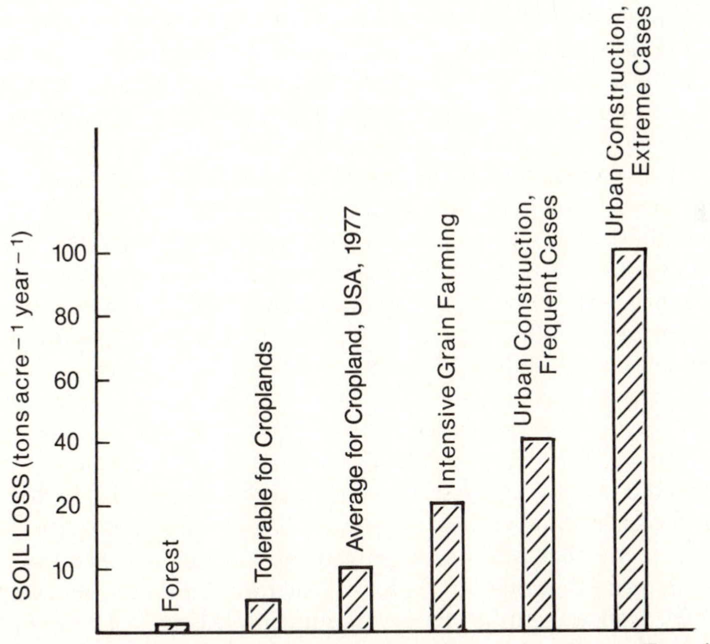

Figure 5–23 Soil losses (soil erosion) in relation to land use. (From U.S. Department of Agriculture Service Report, 1978.)

that half of prime farmland in Iowa and Illinois is now losing 10 to 20 tons/acre/year. Another estimate (Batie and Healy, 1980) suggests that 25 percent of all farmland in the United States is losing soil at a rate greater than the "tolerable" 5 tons/acre/year. To put all of this in perspective, consider that an acre of good topsoil 6 inches deep (about plow depth) weighs about 1000 tons, so 1 inch equals about 167 tons. An annual loss of 10 tons/acre results in a loss of 1 inch of topsoil about every 17 years. At this rate, a corn yield costs 3+ tons of soil lost, given that a bushel of corn weighs 60 lbs and 100 bushels per acre is a good yield. Furthermore, Langdale et al. (1979) estimate that for every inch of topsoil lost, a yield reduction of 6 bu or about 10 percent (of annual crop) occurs. If all these estimates are realistic and the degradation continues, it is difficult to see how needs and demands for more food on fewer available acres can be met.

Erosion, of course, is not the only problem that threatens the capacity of soil to produce food and fiber for humans. Soil compaction resulting from intensive cultivation with ever larger and heavier farm machinery definitely reduces yields; this factor can be mitigated by limited-till farming practices (i.e., less plowing). According to the CEQ report cited previously, about half of the world's irrigated lands are "damaged" to some extent by salination (salt accumulation) or alkalinization (alkali accumulation). So far, yields have been maintained despite declines in soil quality by pouring on more fertilizer and more water. This method works at least for as long as these subsidies are cheap, which will be less and less the case in the near future.

Ultimately, the fate of the soil system depends on societies' willingness to intervene in the marketplace to forgo some short-term benefits so that soils are preserved for long-term benefits. The short-term economic costs of soil conservation can be greatly reduced by designing more efficient and more harmonious agroecosystems (discussed in Chapter 2, Section 7). The real problem is political and economic, not ecological or technical.

Fire

Research during the last 45 years has necessitated a rather drastic reorientation of our ideas about fire as an ecological factor. Fire is a major factor that is almost a part of the normal "climate" in shaping the history of vegetation in most of the terrestrial environments of the world. Consequently, biotic communities adapt and compensate for this factor just as they do for temperature or water. As with most environmental factors, human beings have greatly modified its effect, increasing its influence in many cases and decreasing it in others. Failure to recognize that ecosystems may be "fire adapted" has resulted in a great deal of "mismanagement" of our natural resources. Properly used, fire can be an ecological tool of great value. It is thus an extremely important limiting factor if for no other reason than that the control of fire is far more possible than the control of many other limiting factors. For a discussion of fire and early man, see Sauer (1950, 1963).

Fire is an important factor in forest and grassland regions of temperate zones and in tropical areas with dry seasons. In most parts of the United States, especially southern and western states, it is difficult to find a sizable area that does not give evidence of fire having occurred on it during the last 50 years at least. In many sections, fires are started naturally by lightning. Early man, the North American Indian for example, regularly burned woods and prairies for practical reasons. Fire was a factor in natural ecosystems long before modern times. Accordingly, it should be considered an important ecological factor along with such other factors as temperature, rainfall, and soil and should be studied with an open mind.

As an ecological factor, fire is of different types with different effects. Two extreme types are shown in Figure 5–24. For example, **crown fires** or "**wildfires**" (i.e., very intense and out of control) often destroy all the vegetation and soil organic matter as well, whereas **surface fires** have entirely different effects. Crown fires are limiting to most organisms; the biotic community must start to develop all over again, more or less from scratch, and it may be many years before the area is productive. Surface fires, on the other hand, exert a selective effect; they are more limiting to some organisms than to others and thus favor the development of organisms with high tolerance to fire. Also, light surface fires supplement bacterial action in breaking down the bodies of plants and in making mineral nutrients more quickly available to new plant growth. Nitrogen-fixing legumes often thrive after a light burn. In regions especially subject to fire, regular light surface fires greatly reduce the danger of severe crown fires by keeping the combustible litter to a minimum. In examining an area in regions where fire is a factor, the ecologist usually finds some evidence of the past influence of fire. Whether fire should be excluded in the future (assuming that it is practical) or should be used as a management tool will depend entirely on the type of community that is desired or seems best from the standpoint of regional land use.

Several examples taken from well-studied situations illustrate how fire acts as a limiting factor and how fire is not necessarily "bad" from the human standpoint. On the coastal plain of the southeastern

Figure 5–24 See legend on opposite page.

United States, the long-leaf pine is more resistant to fire than any other tree species, and pines in general are more resistant than hardwoods. The terminal bud of seedling long-leaf pines is well protected by a bunch of long fire-resistant needles (Figure 5–24, inset). Thus, ground fires selectively favor this species. In the complete absence of fire, scrub hardwoods grow rapidly and choke out the long-leaf pines. Grasses and legumes are also eliminated, and the bobwhite and other animals dependent on legumes do not thrive in the complete absence of fire in forested lands. Ecologists generally agree that the magnificent virgin, open stands of pine of the coastal plain and the abundant game associated with them are part of a fire-controlled, or a "fire climax," ecosystem.

Figure 5–24 The two extremes of fire. *A*. Result of a severe crown fire in Idaho with subsequent severe erosion of the watershed. *B*. A controlled burning operation of a long-leaf pine forest in southwest Georgia that removes hardwood competition, stimulates growth of legumes, and improves reproduction of valuable pine timber. Burning is done under damp conditions late in the afternoon (fire is stopped at night by dewfall). Note that the smoke is white (indicating little loss of nutrients) and that the thin line of fire can be stepped over at many points. Ants, soil insects, and small mammals are not harmed by such light surface fires. *C*. A mature long-leaf pine forest that results from controlled burning. Shown in the picture is E. V. Komarek, pioneer fire ecologist lecturing to students. *Inset:* Long-leaf pine seedling showing terminal bud well protected from fire by long needles. (*A*, U.S. Forest Service Photo; *B*, photo by Leon Neel, Tall Timbers Research Station; *C*, photos by E. P. Odum.)

A good place to observe the long-term effects of intelligent use of fire is the Tall Timbers Research Station in northern Florida and the adjacent plantations of southwestern Georgia, where for many years more than a million acres have been managed according to principles developed by the late Herbert Stoddard and E. V. and Roy Komarek, who began studying the relation of fire to the entire ecological complex in the 1930s. Stoddard (1936) was one of the first to advocate the use of controlled or "prescribed" burning for increasing both timber and game production, at a time when professional foresters believed that all fire was bad. For years, high densities of both quail and wild turkeys have been maintained on land devoted to highly profitable timber crops through the use of a system of "spot" burning aided by a diversification in the land use. Between 1963 and 1978, an annual "fire ecology conference" has been held at the Tall Timbers Station. The Proceedings from these conferences review not only local experience but also fire-soil-vegetation-climate interrelationships all over the world.

Fire is especially important in grassland. Under moist conditions (as in tall grass prairies of the Middle West), fire favors grass over trees, and under dry conditions (as in the southwest United States), fire is often necessary to maintain grassland against the invasion of desert shrubs. The main growth centers and energy storages of grasses are underground, so they sprout quickly and luxuriously after the dry aboveground parts burn, which also releases nutrients to soil surface. A close coupling of fire and grazing has been shown to be the key to maintaining the incredible diversity of antelope and other large herbivores and their predators on the east African savannas. Perhaps the most studied type of fire in the ecosystem is the chaparral vegetation of coastal California, the Mediterranean region, and other areas with a winter-rain, dry-summer climate. Here, fire interacts with plant-produced antibiotics or "allelopathics" to produce a unique cyclic climax as described in Chapter 8. (See also Figure A–8 in the Appendix.)

An example of the use of fire in management of game on the British heather moors is depicted in Figure 5–25. Extensive experimentation over the years has shown that burning in patches or strips of about 1 hectare each, with about six such patches per square kilometer, results in highest grouse populations and game yields. The grouse, which are herbivores that feed on buds, require mature (unburned) heather for nesting and protection against enemies but find more nutritious food in the regrowth on burned patches. This example of a compromise between maturity and youth in an ecosystem is very relevant to human beings, as will be discussed in Chapter 8.

Figure 5–25 British heather moor burned in strips and patches (the light-colored areas of about 1 hectare each) to increase game production. This photo illustrates a desirable combination of young and mature vegetation (as discussed in detail in Chapter 7, Section 3) and also the principle of the "edge effect" (see Chapter 7). (After Picozzi, 1968; reproduced with the author's permission.)

As would be expected, plants have evolved special adaptations to fire just as they have for other limiting factors. Fire-dependent and fire-tolerant species can be divided into two basic types: (1) "resprout species" that put more energy into underground storage organs and less into reproductive structure (inconspicuous flowers, little nectar, few seeds), and thus can quickly regenerate after fire has killed exposed parts, and (2) "mature-die" species that do just the opposite, i.e., produce abundant, resistant seeds ready to germinate just after fire.

The question of whether to burn or not to burn can certainly be confusing to the citizen, since seasonal timing and intensity are so critical to determining the consequences of burning. Human carelessness tends to increase "wildfire"; therefore, it is necessary to have a strong campaign for fire protection in forests and recreation areas. The individual citizen should never start or cause fires anywhere in nature, but should recognize that the use of fire as a tool by trained persons is part of good land management. A recommended review of the ecology of fire is the volume edited by Kozlowski and Ahlgren (1974), as well as the earlier review papers by Ahlgren and Ahlgren (1960) and Cooper (1961).

5. Anthropogenic Stress and Toxic Waste as a Limiting Factor for Industrial Societies

Statement

Natural ecosystems exhibit considerable resistance, or resilience, or both to periodic severe or acute disturbance, probably because they are naturally adapted to it. Many organisms, in fact, require stochastic (random) disturbance, such as fire or storms, for long-term persistence. Accordingly, ecosystems may recover rather well from many periodic anthropogenic disturbances such as a pollution episode or harvest removal. Chronic (persistent or continued) disturbance, however, may have pronounced and prolonged effects, especially in the case of industrial chemicals new to the environment. In such cases, organisms have no evolutionary history of adaptation. Unless the highly toxic wastes that are the current byproduct of high-energy, industrialized societies are reduced, contained, or otherwise isolated from the global life-support systems, toxic wastes will directly threaten health and be a major limiting factor for humankind.

Explanation

Although somewhat arbitrary, as is any classification, it may be instructive to consider anthropogenic stress on ecosystems under two categories: (1) acute stress characterized by sudden onset of disturbance, sharp rise in intensity, and short duration; and (2) chronic stress involving long duration or frequent recurrence but not high intensity—a "constantly vexing" disturbance, as one dictionary puts it. Natural ecosystems exhibit considerable ability to deal with or recover from acute stress (discussed in Chapter 2, Section 6). The buried seed strategy, described in Chapter 4, Section 3, is an example of a quick recovery mechanism that facilitates forest regrowth after clearcut. The effects of chronic stress are more difficult to assess, since responses will not be so dramatic. It may be years before the full effects are known, just as it took many years to understand the link between cancer and smoking or the relationship between cancer and chronic low-level ionizing radiation. Whether environmental stress and environmental cancer (disorderly growth at population or community level) provides an analogous situation remains to be seen. It is quite certain, however, that a good deal of human cancer is linked to pollutants in food, water, and the environment (see Epstein, 1974; Reif, 1981).

Of special concern are industrial wastes that contain potential stressors that are new chemical creations and hence are environmental factors to which living organisms and ecosystems have not yet had a period of evolutional history for adaptation or accommodation. Chronic exposure to such anthropogenic factors can be expected to result in basic changes in structure and function of biotic communities as acclimation and genetic adaptation occur. During the transition or adaptation period, some evidence indicates that organisms may be especially vulnerable to secondary factors, such as a disease that can have catastrophic results.

The increasing volume of toxic waste that affects human health either because of direct contact or through contamination of food and drinking water is approaching crisis proportions. In 1978, the Council on Environmental Quality estimated that 50 or more million tons of toxic wastes were being produced annually. Under a heading of "The Poisoning of America," a 1980 news magazine* reviewed the situation as follows:

> Of all of man's interventions in the natural order, none is accelerating quite so alarmingly as creation of chemical compounds. Through their genius, modern alchemists brew as many as 1000 new concoctions each year in the U.S. alone. At last count, nearly 50,000 chemicals were on the market. Many have been an undeniable boon to mankind—but almost 35,000 of these used in the U.S. are classified by the federal EPA as being either definitely or potentially hazardous to human health.

Perhaps the greatest danger and potential disaster is contamination of ground water and the deep aquifers that provide a large percentage of water for cities, industry, and agriculture. Unlike surface water, ground water is almost impossible to purify once it has become polluted, since it is not exposed to sunlight, strong flow, and other

* *Time*, September 22, 1980.

natural purification processes that cleanse surface water. Already, cities in the industrial heartlands can no longer use local ground water for drinking because of contamination; they must pipe in water at great expense.

As with radioactive wastes, the handling of toxic wastes before 1980 was considered a business "externality" not worthy of serious attention. The unwanted material was just dumped somewhere until several local disasters came to public attention. The Love Canal in New York, where a residential area built on top of a waste dump had to be abandoned, has received wide press coverage, as has the Kepone that poisoned a large section of the James River in Virginia (as well as workers in the plant that made the insecticide). These and other incidents aroused public concern and government action. Now, somewhat belatedly, both industry and government are trying to work together to establish special waste management centers capable of handling wastes from all industries in the state or region served. The first order of business is to contain safely the most poisonous material so that detoxification (where possible), incineration, immobilization in glass or ceramic materials, and other technology can be developed (in other words, put the same genius that created the problem to work to reduce the hazard). The practice of illegal "midnight dumping" must be replaced by some kind of secure containment (Figure 5–26). The next logical step would be to find substitutes for the most toxic chemicals so that output of materials requiring special care might be reduced. Toxic-waste treatment facilities should be located where there are no aquifers, rivers, or residential areas, and they should be well buffered by unoccupied greenbelts. Properly managed, such centers can become industries in themselves, providing jobs and contributions to private sectors of the economy.

Hazardous wastes in water, soil, and food are not the only concerns, since millions of tons of stressful substances are entering the atmosphere yearly. Accumulation of carbon dioxide and acid rain, two anthropogenic stresses of special concern, were already discussed (Section 2 of Chapter 4) so only the magnitude of air pollution as a whole will be discussed now.

Estimates of the amount of the five major air pollutants expelled into the atmosphere over the continental United States in selected years are shown in Table 5–3. Transportation is the source of about 60

Figure 5–26 The choice for disposal of toxic wastes. The problem can no longer be considered an economic externality. *A.* Very bad—insecure landfill. *B.* Better—one design for a secure landfill. (Courtesy of *Chemical and Engineering News.*) ▶

MIDNIGHT HAZARDOUS WASTE DISPOSAL
NO DUMPING

A

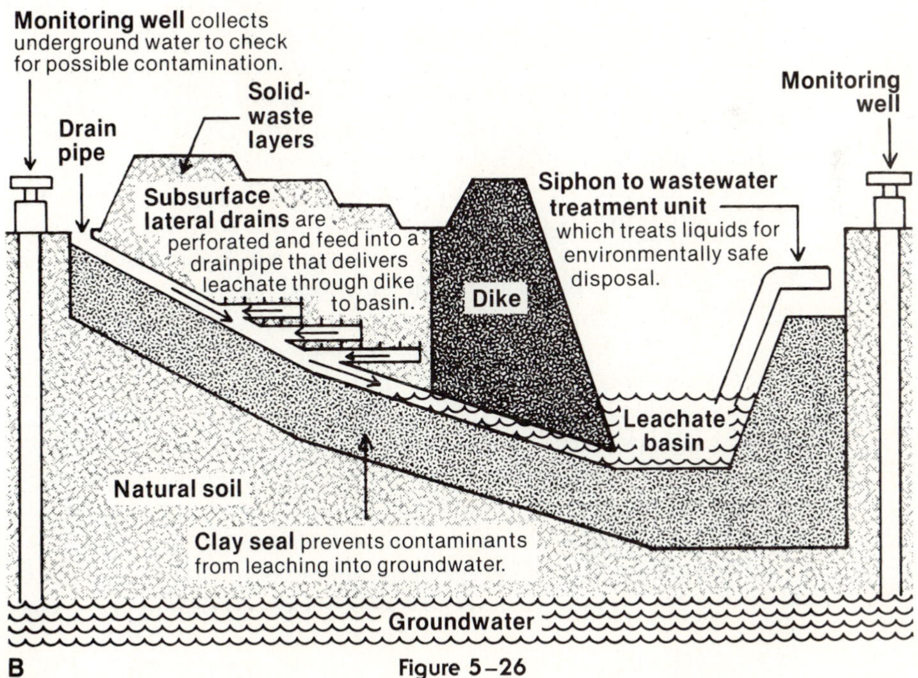
Monitoring well collects underground water to check for possible contamination.
Drain pipe
Solid-waste layers
Subsurface lateral drains are perforated and feed into a drainpipe that delivers leachate through dike to basin.
Dike
Siphon to wastewater treatment unit which treats liquids for environmentally safe disposal.
Monitoring well
Leachate basin
Natural soil
Clay seal prevents contaminants from leaching into groundwater.
Groundwater

B

Figure 5–26

Table 5–3 Relative Magnitude of Air Pollution in the United States in Million Metric Tons per Year and Trends 1966–1978*

	TSP†	SO_2	NO_x	VOC‡	CO	Total
1966	12	23	8	15	65	112
1970	23	30	20	28	103	204
1973	20	30	22	28	104	204
1978	13	27	23	28	102	193
Percentage change 1970–1978	−43	−10	+13	0	0	−5

* 1970–1978 data from CEQ Environmental Quality, Annual Report 1980; 1966 data from NAS Waste Management and Control Report, 1966.

† Total suspended particles.

‡ Volatile organic compounds (including hydrocarbons).

percent, with industry contributing 20 percent, generation of electricity 12 percent, and space heating 8 percent. Although high concentrations that build up over such cities as Tokyo, Los Angeles, and New York during temperature inversions (air trapped under a warm upper layer of air that prevents a vertical rise of pollutants) pose the greatest direct threat to human health, countrywide and global effects are also serious. As shown in Table 5–3, the volume of pollutants peaked early in the 1970s and declined about 5 percent toward the end of that decade, with a 43 percent decline in particulates. As an indication of an improvement in the nation's air quality, the Council on Environmental Quality's 1980 report notes that the number of unhealthful or hazardous days (as determined by a somewhat arbitrary air quality standard) in 23 urban areas decreased 18 percent between 1974 and 1978. Increased fuel efficiency, reduced energy consumption, and federally mandated air pollution control devices have apparently combined to at least halt the increase in air pollution. A similar leveling-off has occurred in Europe.

Air pollution provides the negative feedback signal that may well save industrialized society from extinction because (1) it provides a clear danger signal easily perceived by everyone, and (2) everyone contributes to it (by driving a car, using electricity, buying a product, and so on) and suffers from it, so it cannot be blamed on a convenient villain. A holistic solution must evolve, since attempts to deal with any one pollutant as a separate problem (the one problem–one solution approach) is not only ineffectual but also usually just shifts the problem from one place or environment to another.

Air pollution also provides an excellent example of an augmentative synergism in that combinations of pollutants react in the environment to produce additional pollution, which greatly aggravates the total problem (in other words, total effect is greater than the sum of the individual effects). For example, two components of automobile exhaust combine in the presence of sunlight to produce new and even more toxic substances, known as photochemical smog:

$$\text{Nitrogen oxides} + \text{Hydrocarbons} \xrightarrow[\text{in sunlight}]{\text{Ultraviolet radiation}} \text{Peroxyacetyl nitrate (PAN) and Ozone } (O_3)$$

Both secondary substances not only cause eye-watering and respiratory distress in humans but also are extremely toxic to plants. Ozone increases respiration of leaves, killing the plant by depleting its food. Peroxyacetyl nitrate blocks the "Hill reaction" in photosynthesis, thus killing the plant by shutting down food production (see Taylor et al., 1961, and Dugger et al., 1966). The tender varieties of cultivated plants become early victims, so that certain types of agriculture and horticulture are no longer possible near big cities. Other photochemical pollutants that go under the general heading of polynuclear aromatic hydrocarbons (PAH) are known carcinogens. A. J. Haagen-Smit was a co-recipient of the first Tyler Ecology Award for his pioneer work on Los Angeles smog in the early 1950s (Haagen-Smit et al., 1952; Haagen-Smit, 1963). For a review of the effect of air pollution on forests, see W. H. Smith (1981).

If stringent efforts are not continued during the 1980s to halt the deterioration of environmental quality, toxic substances rather than scarcity of resources as such may well be the limiting factor for industrial societies.

Examples

Thermal pollution is becoming a commonplace example of chronic stress, since low-utility heat is a byproduct of any conversion of energy from one form to another, as dictated by the second law of thermodynamics. Power plants and other large energy converters release great quantities of heat to both air and water, with atomic power plants requiring especially large volumes of cooling water. To generate 1 kilowatt hour of electricity, the waste heat released to the atmosphere and to cooling water is 1600 and 5300 BTU, respectively, for a fossil-fueled power plant and 500 and 7600 BTU, respectively, for a modern nuclear power plant. Using the conversion factors in Table 3–1, we find that an average-sized nuclear power plant that produces 3000 megawatts of electricity also produces waste heat at the rate of more than 20×10^9 BTU/hr.

The surface cooling capacity of water ranges from about 1.5 to 7.5 BTU/hr/ft²/°F difference between air and water, depending on wind and water temperatures. Consequently, a lot of water surface is required to disperse heat, something on the order of 1.5 acres/megawatt in a temperate locality, or 45 acres for a 3000-megawatt power station.

The use of powered cooling devices such as cooling towers can reduce the space and water volume needed, of course, but at a considerable cost, since expensive fuels replace solar energy. Also, cooling towers can cause other impacts if chlorine or other chemicals are used to keep surfaces free of algae.

In general, the effects of increasing the water temperature in ponds, lakes, or streams follows the subsidy-stress gradient shown in Figure 3–5 in that both positive and negative responses result. Moderate increases often act as subsidies in that productivity of the aquatic community and growth of fish may be increased, but in time or with increasing heat loading, stress effects begin to enter the picture. Local detrimental effects of thermal pollution on aquatic ecosystems that have been observed can be listed as follows: (1) A rise in water temperature often increases the susceptibility of organisms to toxic materials (which will undoubtedly be present in waste water) and to fungal and other diseases. (2) Critical stenothermal periods of life histories may be exceeded (see page 233). (3) Elevated temperatures tend to foster replacement of normal algal populations by less desirable blue-green algae, especially if there is also nutrient enrichment. (4) As water temperature rises, animals need more oxygen, yet warm water holds less (see page 264). For details on all these effects, see Krenkel and Parker, 1969.

The National Environmental Research Park at the Department of Energy Savannah River Plant is an excellent place to observe the long-term effects of thermal pollution. The Savannah River Ecology Laboratory located on the site has focused on the study of thermal effects since the establishment of atomic energy facilities there in the 1950s and has sponsored two large symposia that combine work and ideas from other study sites as well (Gibbons and Sharitz, 1974; Esch and McFarlane, 1975). A large artificial lake constructed as a cooling pond is especially interesting because it has a "warm" arm (receiving heated water) and a "cool" arm that receives no heated water and is at ambient temperature (normal for the region). Furthermore, since the reactors are periodically turned "off" and "on," one can observe the effect from one temperature state to the other. Turtles and bass grow faster and achieve larger sizes when water temperature is elevated a few degrees, and the active season for alligators is prolonged into the winter months. Thus, the first observed effects were generally subsidies, but after a few years, definite stress effects began to appear, such as debilitating diseases that shorten the life span and increase mortality. As shown in Figure 5–27, the percent of bass infected with red-sore disease rises and falls with the seasons but is consistently higher in the thermally enriched zones of the lake. Also, after ten years or more of elevated temperatures, there is evidence of genetic change in populations of fish and also in cattails (*Typha*) that grow along the shore of the warm "arm." For a brief review of all these studies, see Gibbons and Sharitz (1981).

These illustrations emphasize the importance of looking for secondary or delayed responses when assessing the effect of chronic anthropogenic perturbation.

Pesticides Increasingly heavy applications of insecticides and other pesticides in agriculture have resulted in contamination of soil and water by yet another class of toxic substances. This threat to the health of ecosystems and of humans may soon be reduced for the simple rea-

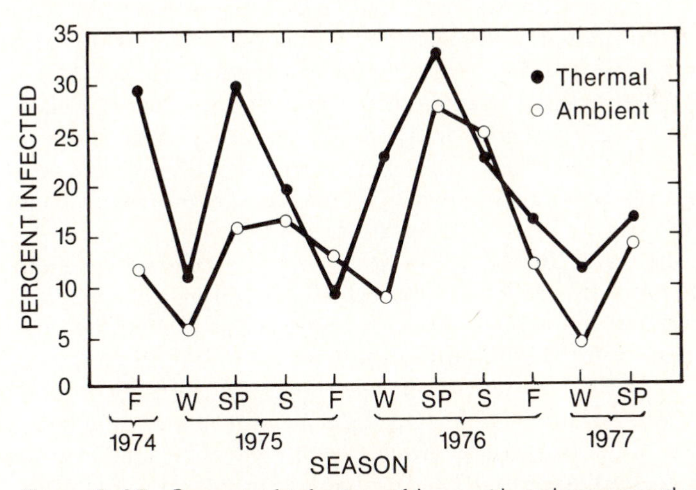

Figure 5–27 Percent of infection of bass with red-sore mouth disease varies with the season (higher in summer) but is consistently higher in "thermal" waters where temperature is elevated by chronic thermal pollution from nuclear reactors at the Savannah River Plant. (After Esch and Hazen, 1978.)

son that exclusive dependence on chemical poisons fails to achieve long-term control, but rather produces booms and busts in crop yields. Alternative systems of pest control are being developed that may soon reduce the need for massive applications of what are in reality very dangerous poisons.

Paradoxically, the resilience and adaptability of nature is the root cause of the failure of broad-spectrum insecticides such as the organochlorines (the DDT generation of insecticides) and the organophosphates. Too often, pests develop immunity or become even more abundant after the pesticide has been dissipated or detoxified because their natural enemies were destroyed by the treatment. Also, a pest species that is successfully killed off is sometimes replaced by other species that are more resistant, less well known, and therefore even more difficult to deal with.

Efforts to control cotton's insect pest provide a good example of the boom-and-bust syndrome. Cotton has been one of the crops most heavily treated with insecticides (as much as 50 percent of all insecticide used in United States agriculture is sprayed on cotton). In the 1950s massive aerial spraying of the chlorinated hydrocarbons in the Canete Valley of Peru, made possible by foreign aid funds from the United States, resulted in a doubling of yield for about six years. There followed, however, a complete crop failure as pests became resistant and other species of insects moved in. I have described this well-documented case history elsewhere (Odum, 1975). The same thing happened in the 1960s in Texas, a major cotton-growing state, as is documented in detail in a recent paper by Adkisson et al. (1982). In both cases yields were restored by adoption of what entomologists call integrated pest management. The strategy involves using short-season varieties (which mature before pest populations can build up) and cultural practices (tilling, irrigation, fertilizing) that discourage pests and encourage their natural enemies, combined with judicious use of several kinds of insecticides, including some old-fashioned ones such as sulfur dust. Careful timing of application to catch insects in the early stages of their life cycle, when they are most vulnerable, is an important aspect. The new control system verifies the age-old common-sense wisdom that it never pays to put all your eggs in one basket. Nature's diversity and resilience must be met with diverse technological innovations that must be continually updated as conditions change and nature reacts. In other words, our "war" with insects and other competitors probably can never actually be "won," but involves a continuous effort that is one of the costs of "pumping out disorder" necessary to maintain a large and complex civilization.

There is currently considerable optimism that what Carroll Williams (1967) has called the third generation of pesticides will soon be added to the arsenal available for integrated systems. (The first generation, according to Williams' classification, were the botanicals and inorganic salts; the second generation were members of the DDT family.) The third generation are the biochemicals, the hormones and pheromones (sex attractants) that direct behavior and are species-specific. For reviews, see Blum (1969) and Silverstein (1981).

For more on integrated control management, see Smith and van den Bosch (1967), Kennedy (1968), Flint and van den Bosch (1981), Batra (1982), and the National Academy and CEQ reviews.

6 Population Dynamics

The more purely biological aspects of ecology, that is, the interaction of organisms with organisms as they function in ecosystems, are the subject of this chapter and the next. In Chapters 2 through 5, the great physical and chemical forces that act as primary forcing functions were discussed; however, organisms do not just passively adapt to these forces but actively modify, change, and regulate the physical environment within limits imposed by the natural laws that determine the transformation of energy and cycling of materials. In other words, human beings are not the only population that modifies and attempts to control the environment. Referring back to the levels-of-organization chart of Figure 1–1, this chapter and the next one focus on the biotic levels of populations and communities. Interaction at these levels between genetic systems and environment determines the course of natural selection and thereby not only how individual organisms optimize their survival but also how ecosystems as a whole have changed and are changing over evolutionary time.

1. Properties of the Population Group

Statement

A **population** may be defined as any group of organisms of the same species (or other groups within which individuals may exchange genetic information) occupying a particular space and functioning as a part of a **biotic community**, which, in turn, is defined as an assemblage of populations that function as an integrative unit through coevolved metabolic transformations in a prescribed area of physical habitat. A population has various characteristics, which, although best expressed as statistical functions, are the unique possession of the group and are not characteristic of the individuals in the group. Some of these properties are density, natality (birth rate), mortality (death rate), age distribution, biotic potential, dispersion, and growth form. Populations also possess genetic characteristics directly related to their ecology, namely, adaptiveness, reproductive (Darwinian) fitness, and persistence (i.e., probability of leaving descendants over long periods of time).

Explanation

As was well expressed by pioneer population ecologist Thomas Park (in Allee et al., 1949), a population has characteristics or "biological attributes" that it shares with its component organisms, and it has characteristics or "group attributes" unique to the group. Among the population's biological attributes is its life history: it grows, differentiates, and maintains itself as does the organism. It has a definite organization and structure that can be described. By contrast, group attributes, such as birth rate, death rate, age ratio, and genetic fitness, apply only to the population. Thus, an individual is born, dies, and ages, but it does not have a birth rate, a death rate, or an age ratio. These latter attributes are meaningful only at the group level.

Definitions and brief resumés of basic population attributes follow:

Density Population density is population size in relation to some unit of space. It is generally assayed and expressed as the number of individuals, or the population biomass, per unit area or volume—for example, 200 trees per acre, 5 million diatoms per cubic meter of water, or 200 pounds of fish per acre of water surface. Sometimes, it is important to distinguish between **crude density**, the number (or biomass) per unit total space, and **specific** or **ecological density**, the number (or biomass) per unit of habitat space (available area or volume that can actually be colonized by the population). Often, it is more important to know whether a population is changing (increasing or decreasing) than to know its size at any one moment. In such cases, indices of **relative abundance** are useful; these may be time-relative, as, for example, the number of birds seen per hour. Another useful index is **frequency** of occurrence, as, for example, the percentage of sample plots occupied by a species. In descriptive studies of vegetation, density and frequency are often combined to provide an **importance value** index for each species.

Figure 6–1 and Table 6–1 illustrate how densities encountered in populations of mammals and fish are related to trophic (energy) level and to size of individual. Though density of mammals as a class may range over nearly five orders of magnitude, the range for any given species or trophic group is much less (Figure 6–1). The lower the trophic level, the higher the density, and within a given level, the larger the individuals, the larger the biomass. Since large organisms

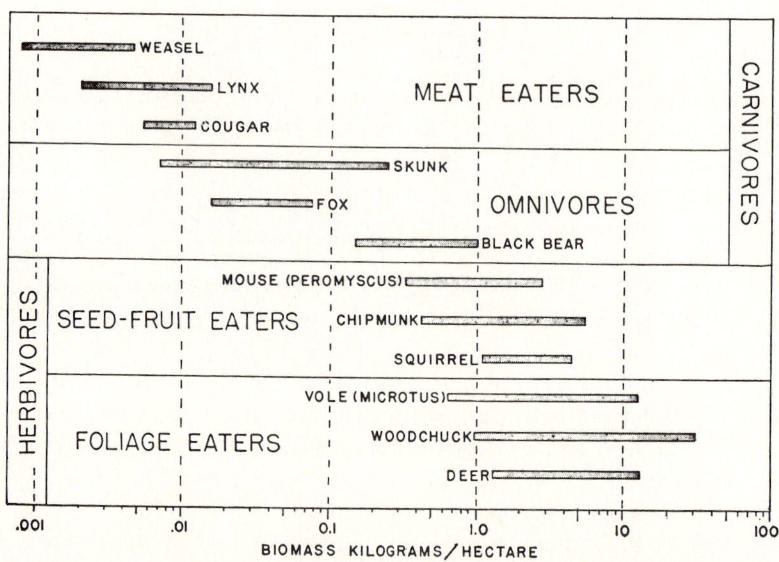

Figure 6–1 The range of population density (as biomass per hectare) of various species of mammals as reported from preferred habitat of the species in localities where humans are not unduly restrictive. Species are arranged according to trophic levels and according to individual size within the four levels to illustrate the limits imposed by trophic position and size of organism on the expected standing crop. (Graph prepared from data collected by Mohr, 1940, plus results of later studies.)

Table 6–1 Density in Fish Populations Illustrating the Effect of Trophic Level on Biomass Density (A) and the Relationships Between Numbers and Biomass in a Population with a Rapidly Changing Age and Size Structure (B)

A. Mixed Populations, Biomass per Unit Area:

Fish in artificial ponds in Illinois. (Data from Thompson and Bennett, 1939.) Fish groups arranged in approximate order of food chain relations with "rough fish" occupying the lowest trophic level and "game fish" the highest.

	Fish in Pounds per Acre		
	Pond No. 1	Pond No. 2	Pond No. 3
Game and pan fish (bass, bluegills, etc.)	232	46	9
Catfish (bullheads and channel cats)	0	40	62
Forage fish (shiners, gizzard shad, etc.)	0	236	3
Rough fish (suckers, carp, etc.)	0	87	1143
Totals	232	409	1217

B. Comparison of Individual and Biomass Density Where Size of Organism Undergoes Pronounced Change with Age:

Fingerling sockeye salmon in a British Columbia lake. The salmon hatch in streams and in April enter the lake, where they remain until mature. Note that between May and October the fish grew rapidly in size, with the result that biomass increased three times, even though the number of fish was greatly reduced. From October to the next April very little growth occurred, and continued death of fish reduced the total biomass. (Data from Ricker and Forester, 1948.)

	May	October	April
Individuals, thousands in the lake	4000	500	250
Biomass, metric tons in the lake	1.0	3.3	2.0

have lower rates of metabolism per unit weight than small organisms, a larger population biomass can be maintained on a given energy base.

When the size and metabolic rate of individuals in the population are relatively uniform, density expressed in terms of number of individuals is quite satisfactory as a measure, but so often the situation exists as shown in Table 6–1B. The relative merits of numbers, biomass, and energy flow parameters as indices were discussed in Chapter 3 (see especially Table 3–15). Recall the following statement from that chapter: "Numbers overemphasize the importance of small organisms, and biomass overemphasizes the importance of large organisms." Components of energy flow "provide a more suitable index for comparing any and all populations in an ecosystem."

Many special measures and terms in wide use apply only to specific populations or groups of populations. Forest ecologists, for example, often use "basal area" (= total cross-section area of trunks) as a measure of tree density. Foresters, however, determine "board feet per acre" as a measure of the commercially usable part of the tree. These, and many others, are density measures as the concept has been broadly defined, since they all express in some manner the size of the "standing crop" per unit area.

As already indicated, relative abundance is often a useful measure when one must know how the population is changing, or when conditions are such that the absolute density cannot be determined. The terms "abundant," "common," "rare," and so forth are most useful when tied to something that is measured or estimated in a manner that makes comparison meaningful.

As might be imagined, relative abundance "indices" are widely used with populations of larger animals and terrestrial plants, where it is imperative that a measure applicable to large areas be obtained without excessive expenditure of time and money. For example, administrators charged with setting up annual hunting regulations for migratory waterfowl must know whether the populations are smaller than, larger than, or the same as in the previous year if they are to adjust the hunting regulations to the best interest of both the birds and the hunters. To do this, these officials must rely on relative abundance indices obtained from field checks, hunter surveys, questionnaires, and nesting censuses. Such information is often summarized in terms of number observed or killed per unit time effort. Percentage indices are widely used in the study of vegetation, and specially defined terms have come into general use. For example, frequency equals percent of sample plots in which the species occurs, abundance equals percent of individuals in a sample, and cover equals percent of ground surface covered as determined by projection of areal parts. One should be careful not to confuse these indices with true density measures, which are always in terms of a definite amount of space.

The contrast between **crude density** and **ecological density** can be illustrated by Kahl's study (1964) of the wood stork in the Florida Everglades. As shown in Figure 6–2, the density of small fish in the area as a whole goes down as the water levels drop during the winter dry season, but the ecological density in the contracting pools of water increases as fish are crowded into smaller and smaller water areas. The stork times its egg laying so that young will be hatched

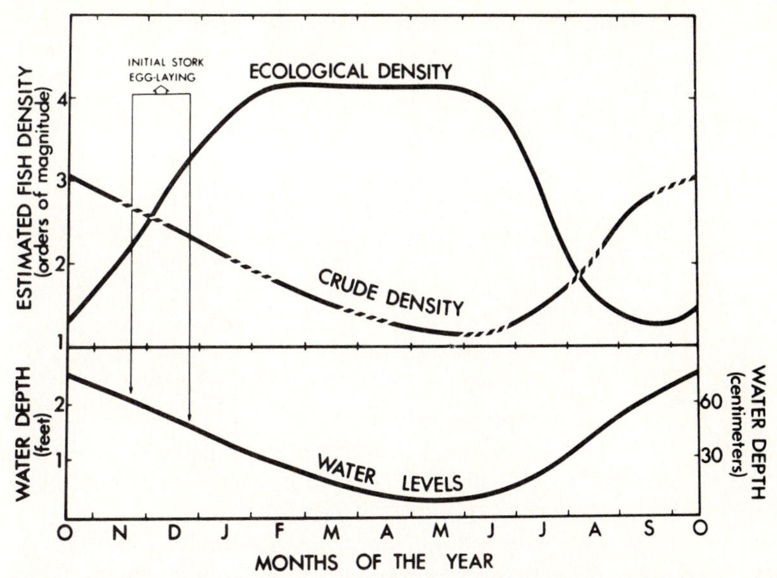

Figure 6–2 Ecological and crude density of fish prey in relation to the breeding of the stork predator. As water levels fall during the dry season in southern Florida, the crude density of small fish declines (i.e., the number of fish per square mile of total area declines because size and number of ponds are reduced), but the ecological density (i.e., number per square meter of remaining water surface) increases as fish are crowded into smaller water areas. Nesting of the stork is timed so that maximum food availability coincides with greatest food demand by the growing nestlings. (After Kahl, 1964.)

when ecological density of fish is at a peak and it is easy for the adults to catch the fish that provide the chief food for the young. In the wood stork's world, ecological density of its food, not crude density, is what counts. In the human's world, it is much the same: the concentrated (high quality) energy and concentrated production of food on the 15 to 20 percent of land capable of high yields count the most, not widely dispersed, hard-to-collect resources.

Many different techniques for measuring population density have been tried, and methodology is an important field of research in itself. There is little point in going into detail on methods here, since methods are better learned by consulting field and methods manuals or, best of all, by consulting with an experienced investigator who has reviewed the original literature and then has modified and improved existing methods to fit a specific field situation. There is no substitute for experience when it comes to field censusing.

Methods fall into several broad categories:

1. Total counts, sometimes possible with large or conspicuous organisms (for example, bison on open plains) or with organisms that aggregate into large breeding colonies (such as sea birds or seals).
2. Quadrat sampling that involves counting and weighing of organisms in plots or transects of appropriate size and number to get an estimate of density in the area sampled.
3. Marking-recapture methods (for mobile animals), in which a sample of population is captured, marked, and released, and a proportion of marked individuals in a later sample is used to determine total populations (see Zippin, 1958).
4. Removal sampling, in which the number of organisms removed from an area in successive samples is plotted on the y-axis of a graph and the number previously removed is plotted on the x-axis. If the probability of capture remains reasonably constant, the points will fall on a straight line that can be extended to the zero point (x-axis), which would indicate theoretical 100 percent removal from the area (see Menhinick, 1963).
5. Plotless methods (applicable to sessile organisms such as trees). The point-quarter method is an example: from a series of random points, the distance to the nearest individual is measured in each of four quarters. The density per unit area can be estimated from the mean distance (see Phillips, 1959).

Natality Natality is the ability of a population to increase. Natality rate is equivalent to the birth rate in the terminology of human population study (demography). In fact, it is simply a broader term covering the production of new individuals of any organism whether such new individuals are born, are hatched, are germinated, or arise by division. **Maximum** (sometimes called absolute or physiological) **natality** is the theoretical maximum production of new individuals under ideal conditions (i.e., no ecological limiting factors, reproduction being limited only by physiological factors) and is a constant for a given population. **Ecological** or **realized natality** (or just plain "natality," without a qualifying adjective) refers to population increase under an actual or specific environmental condition. It is not a constant for a population but may vary with the size and age composition of the population and the physical environmental conditions. Natality is generally expressed as a rate determined by dividing the number of new individuals produced by time (the absolute or crude natality rate) or as the number of new individuals per unit of time per unit of population (the specific natality rate).

The difference between crude and specific natality or birth rate can be illustrated this way: suppose a population of 50 protozoa in a pool increases by division to 150 in an hour. The crude natality is 100/hr, and the specific natality (average rate of change per unit-population) is 2 per hour per individual of the original 50. Or suppose there were 400 births in a town of 10,000; the crude birth rate is 400/yr, and the specific birth rate is 0.04 per capita (4 per 100, or 4 percent). In human demography it is customary to express specific birth rates in terms of number of females of reproductive age, rather than in terms of total population.

The contrast between maximum and realized natality can be illustrated with data obtained by Laskey (1939) for a wild population of bluebirds provided with artificial nesting boxes in a Nashville, Tennessee, city park and with data on flour beetles cultivated in Park's laboratory. Female bluebirds laid 510 eggs (about 15 per female in three successive broods); that number represents maximum natality. Since only 265 eggs survived to fledged young, the ecological or realized natality was 52 percent of maximum (about eight per female or four per pair of birds). In contrast, a population of flour beetles in a large laboratory container produced in one experiment about 12,000 eggs (see Park, Ginsburg, and Horwitz, 1945), of which only 773 or 6 percent survived to larval stage. In general, species that do not protect or take special care of eggs or young have high potential and low realized natality. Other considerations that affect natality will be discussed in subsequent sections.

Mortality Mortality refers to death of individuals in the population. It is more or less the antithesis of natality. Mortality is equivalent to death rate in human demography. Like natality, mortality may be expressed as the number of individuals dying in a given period (deaths per time), or as a specific rate in terms of units of the total population or any part thereof. **Ecological** or **realized mortality**—the loss of individuals under a given environmental condition—is, like ecological natality, not a constant but varies with population and environmental conditions. A theoretical **minimum mortality**, a constant for a population, represents the loss under ideal or nonlimiting conditions. Even under the best conditions, individuals would die of "old age" determined by their **physiological longevity**, which, of course, is often far greater than the average **ecological longevity.** Often, the **survival rate** is of greater interest than the death rate. If the death rate is expressed as a fraction, M, then survival rate is $1 - M$.

Since mortality varies greatly with age, as does natality, especially in the higher organisms, specific mortalities at as many different ages of life history stages as possible are of great interest inasmuch as they enable ecologists to determine the forces underlying the crude, overall population mortality. A complete picture of mortality in a population is given systematically by the **life table**, a statistical device developed by students of human populations. Raymond Pearl first introduced the life table into general biology when he applied it to data obtained from laboratory studies of the fruit fly, *Drosophila* (Pearl and Parker, 1921). Deevey (1947, 1950) has assembled data for the construction of life tables for several natural populations, ranging from rotifers to mountain sheep. Since Deevey's reviews, numerous life tables have been published for various natural and experimental populations. The life tables for an Alaskan population of wild mountain sheep, a laboratory population of rice weevils, and a hypothetical population are displayed in Tables 6–2 and 6–3. In Table 6–3, age-specific natality rates are also included. The age of the sheep was determined from the horns (the older the sheep, the more bony rings). When a sheep is killed by a wolf or dies for any other reason, its horns remain preserved for a long period. For several years, Adolph Murie studied the relation between wolves and mountain sheep in Mt. McKinley National Park, Alaska. He collected many horns, thus providing admirable data on the age at which sheep die in an environment subject to all the natural hazards, including wolf predation (but not including predation by humans, as sheep were not hunted in Mt. McKinley National Park).

The life table consists of several columns, headed by standard notations, giving (l_x), the number of individuals out of a given population (1000 or any other convenient number) that survive after regular time intervals (day, month, year, and so on, given in column x); (d_x) the number dying during successive time intervals; (q_x), the death rate or mortality during successive intervals (in terms of initial population at beginning of period); and (e_x), the life expectancy at the end of each interval. As shown by Table 6–2, the average age of mountain sheep was better than 7 years, and if a sheep survived the first year or so, its chances of survival were good until relative old age, despite the abundance of wolves and the other vicissitudes of the environment.

Curves plotted from life-table data may be very instructive. When data from column l_x are plotted with the time interval on the horizontal coordinate and the number of survivors on the vertical coordinate, the resulting curve is called a **survivorship curve.** If a semilogarithmic plot is used, with the time interval on the horizontal coordinate expressed as a percentage of the mean length of life (see column x', Table 6–2) or as a percentage of the total life span, species

Table 6–2 **Life Table for the Dall Mountain Sheep (*Ovis d. dalli*)***

Age (years)	Age as Percent Deviation from Mean Length of Life	Number Dying in Age Interval out of 1000 Born	Number Surviving at Beginning of Age Interval out of 1000 Born	Mortality per Thousand Alive at Beginning of Age Interval	Expectation of Life, or Mean Life-Time Remaining to Those Attaining Age Interval (years)
x	x'	d_x	l_x	$1000\ q_x$	e_x
0–0.5	−100	54	1000	54.0	7.06
0.5–1	−93.0	145	946	153.0	—
1–2	−85.9	12	801	15.0	7.7
2–3	−71.8	13	789	16.5	6.8
3–4	−57.7	12	776	15.5	5.9
4–5	−43.5	30	764	39.3	5.0
5–6	−29.5	46	734	62.6	4.2
6–7	−15.4	48	688	69.9	3.4
7–8	− 1.1	69	640	108.0	2.6
8–9	+13.0	132	571	231.0	1.9
9–10	+27.0	187	439	426.0	1.3
10–11	+41.0	156	252	619.0	0.9
11–12	+55.0	90	96	937.0	0.6
12–13	+69.0	3	6	500.0	1.2
13–14	+84.0	3	3	1000	0.7

* From Deevey (1947); data from Murie (1944). Based on known age at death of 608 sheep dying before 1937 (both sexes combined). Mean length of life, 7.09 years. A small number of skulls without horns, but judged by their osteology to belong to sheep 9 years old or older, have been apportioned *pro rata* among the older age classes.

of widely different life spans may be compared. Furthermore, a straight line on a semilogarithmic plot indicates a constant specific rate of survival.

Survivorship curves are of three general types, as shown in Figure 6–3. A highly convex curve (A in Figure 6–3) is characteristic of species such as the Dall sheep, in which the population death rate is low until near the end of the life span. (Plot column l_x, Table 6–3, on semilog graph paper and compare the shape of the curve with the models in Figure 6–3.) Many species of large animals and, of course, humans, exhibit this type of survivorship. At the other extreme, a highly concave curve (C in Figure 6–3) results when mortality is high

Table 6–3 Life Tables That Include Both Age-specific Survival and Natality

A. Life Table for a Hypothetical Population with a Simple Life History

Age	Age-specific Survival Rate (as Fractions)	Age-specific Death Rate	Age-specific Natality Rate (Offspring per Female Aged x)	
x	l_x	d_x	m_x	$l_x m_x$*
0	1.00	0.20	0	0.00
1	0.80	0.20	0	0.00
2	0.60	0.20	1	0.60
3	0.40	0.20	2	0.80
4	0.20	0.10	2	0.40
5	0.10	0.05	1	0.10
6	0.05	0.05	0	0.00
7	0.00			

$\Sigma l_x m_x$ = net reproductive rate (R_0) = 1.90

during the young stages. Oysters or other shellfish and oak trees are good examples; mortality is extremely high during the free-swimming larval or the acorn-seedling stages, but once an individual is well established on a favorable substrate, life expectancy improves considerably. Intermediate are patterns in which the age-specific survival is more nearly constant so that curves approach a diagonal straight line on a semilog plot (B in Figure 6–3). A "stairstep" type of survivorship curve may be expected if survival differs greatly in successive stages of life history as is often the case in holometabolous insects (insects with complete metamorphosis such as butterflies). In the model shown (B_1 in Figure 6–3), the steep segments represent egg, pupation, and short-lived adult stages; the flatter segments rep-

Table 6–3 (*Continued*)

B. Life Table for Laboratory Population of Rice Weevils (*Calandra oryzae*) under Optimum Conditions (29°C and 14% Moisture Content of Rice), Sex Ratio Equal†

Pivotal Age in Weeks	Age-specific Survival (as Fractions)	Age-specific Natality (Female Offspring per Female Aged x)	
x	l_x	m_x	$l_x m_x$
4.5	0.87	20.0	17.400
5.5	0.83	23.0	19.090
6.5	0.81	15.0	12.150
7.5	0.80	12.5	10.000
8.5	0.79	12.5	9.875
9.5	0.77	14.0	10.780
10.5	0.74	12.5	9.250
11.5	0.66	14.5	9.570
12.5	0.59	11.0	6.490
13.5	0.52	9.5	4.940
14.5	0.45	2.5	1.125
15.5	0.36	2.5	0.900
16.5	0.29	2.5	0.800
17.5	0.25	4.0	1.000
18.5	0.19	1.0	0.190
			R_0 = 113.560

* Based on l_x at the start of the age period.

† After Birch, 1948.

resent larval and pupal stages that suffer less mortality (see Itô, 1959). Probably no population in the real world has a constant age-specific survival rate throughout the whole life span (B_2, Figure 6–3), but a slightly concave or sigmoid curve (B_3, Figure 6–3) is characteristic of many birds, mice, and rabbits. In these cases, the mortality is high in the young but lower and more nearly constant in the adult (1 year or older).

The shape of the survivorship curve is related to the degree of parental care or other protection given to the young. Thus, survivorship curves for honeybees and robins (which protect their young) are

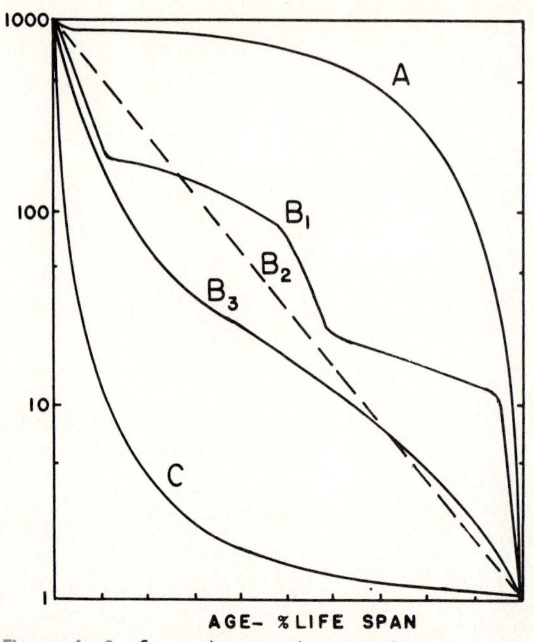

Figure 6–3 Several types of survivorship curves plotted on the basis of survivors per thousand log scale (vertical coordinate) and age as a percent of the life span (horizontal coordinate). Curve A is of the convex type, in which most of the mortality occurs toward the end of the life span. B_1 is a stairstep type of curve, in which survival rate undergoes sharp changes in transition from one life history stage to another. B_2 is a theoretical curve (straight line), in which age-specific survival remains constant. B_3 is a slightly sigmoid type of curve that approaches B_2. Curve C is of the concave type, in which mortality is very high during the young stages.

much less concave than are those for grasshoppers or sardine fish (which do not protect their young). The latter species, of course, compensate by laying many more eggs (ratio of maximum to realized natality is high, as noted in the previous section).

The shape of the survivorship curve very often also varies with the density of the population. Survivorship curves for two deer populations living in the chaparral region of California are shown in Figure 6–4; the survivorship curve of the denser population is quite concave. In other words, deer living in the managed area, where the food supply was increased by controlled burning, have a shorter life expectancy than deer in the unmanaged area, presumably because of increased hunting pressure, intraspecific competition, and so on. From the viewpoint of the hunter, the managed area is most favora-

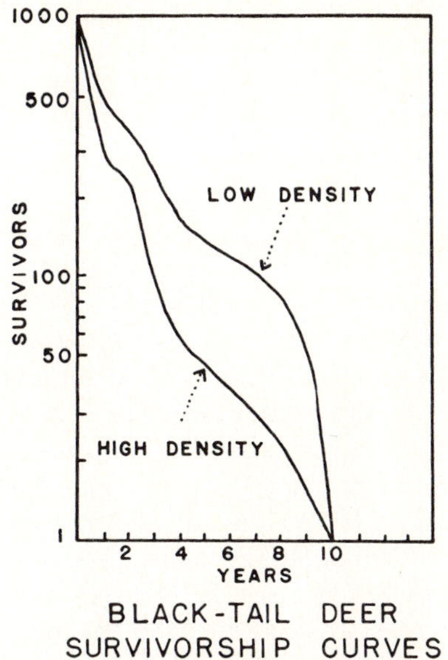

BLACK-TAIL DEER
SURVIVORSHIP CURVES

Figure 6–4 Survivorship curves for two stable deer populations living in the chaparral region of California. The high-density population (about 64 deer per square mile) is in a managed area where an open shrub and herbaceous cover is maintained by controlled burning, thus providing a greater quantity of browse in the form of new growth. The low-density population (about 27 deer per square mile) is in an unmanaged area of old bushes unburned for 10 years. Recently burned areas may support up to 86 deer per square mile, but the population is unstable and hence survivorship curves cannot be constructed from age distribution data. (After Taber and Dasmann, 1957.)

ble, but from the viewpoint of the individual deer, the less crowded area offers a better chance for a long life. For human populations also, high density has not always been favorable to the individual. Many ecologists believe that rapid growth and high density in human populations is not so much a threat to survival as it is a threat to the quality of life for the individual, even though humans have greatly increased their own "ecological" longevity because of modern medical knowledge, better nutrition, and so forth. The curve denoting a human being's survival approaches the sharp-angled minimum mortality curve. However, human beings have apparently not increased their maximum or "physiological" longevity, since no more people live now to be 100 years old than did in past centuries. These paradoxes are discussed further in Chapter 8.

To prepare the way for mathematical models of population growth to be considered in subsequent sections, it is instructive to add age-specific natality (as offspring per female produced in a unit of time) to the life table, as in Table 6–3 (the m_x column), so that it is not just a "death" table.

If l_x and m_x are multiplied and the sum of the values is obtained for the different age classes, a net reproductive rate (R_0) is calculated. Thus:

$$R_0 \text{ (net reproductive rate)} = \Sigma\, l_x \cdot m_x$$

(In this case l_x refers to females only.) In Table 6–3A, the net reproductive rate of 1.9 indicates that for every one female, 1.9 offspring are produced. If the sex ratio is equal, this means that the population just about replaces itself during the generation. For grain weevils, under the most optimum conditions, the laboratory population multiplies 113.6 times in each generation (i.e., R_0 = 113.6; Table 6–3B). Under stable conditions in nature, R_0 in terms of the total population should be around 1. Paris and Pitelka (1962), using the life table approach and data on l_x and m_x of year classes, calculated R_0 for a sowbug population in a grassland as 1.02, indicating an approximate balance between births and deaths.

The reproductive schedule greatly influences population growth and other population attributes. Natural selection can effect various kinds of change in the life history that will result in adaptive schedules. Thus, selection pressure may change the time when reproduction begins without affecting the total number of offspring produced, or it may affect production or "clutch size" without changing the timing of the reproduction. These and many other aspects can be revealed by life-table analyses.

Population Age Distribution Age distribution, an important characteristic of populations, influences both natality and mortality, as shown by the examples discussed in the preceding section. The ratio of the various age groups in a population determines the current reproductive status of the population and indicates what may be expected in the future. Usually, a rapidly expanding population will contain a large proportion of young individuals, a stationary population will show a more even distribution of age classes, and a declining population will have a large proportion of old individuals, as illustrated by the age pyramids in Figures 6–5 and 6–6. A population may pass through changes in age structure without changing in size. There is evidence that populations have a "normal" or stable age distribution toward which actual age distributions are tending, as first proposed by Lotka (1925) on theoretical grounds. Once a stable age distribution is achieved, unusual increases in natality or mortality result in tempo-

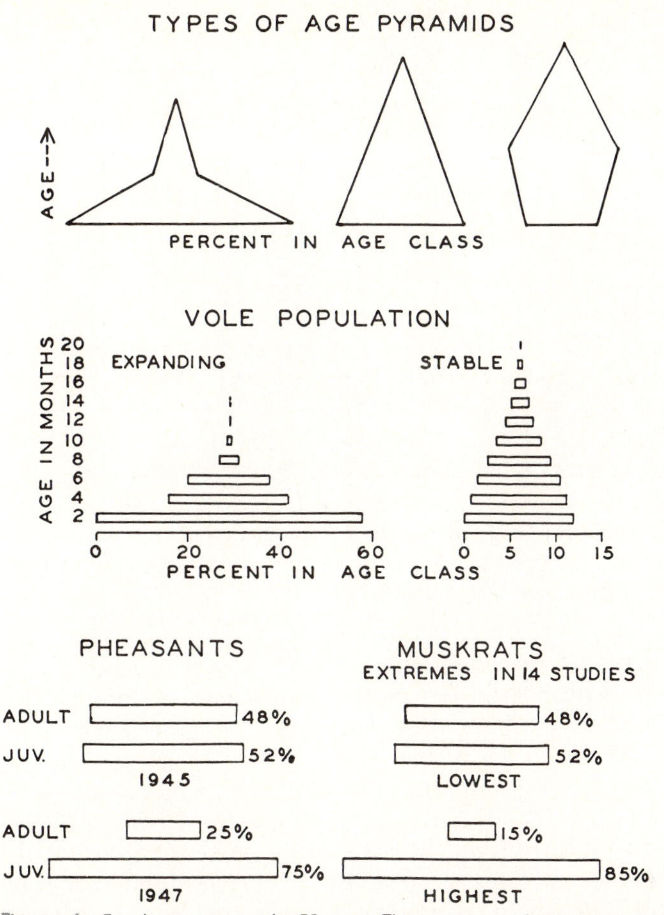

Figure 6–5 Age pyramids. *Upper*. Three types of age pyramids representing a large, moderate, and small percentage of young individuals in the population. *Middle*. Age pyramids for laboratory populations of the vole, *Microtus agrestis*, when expanding at an exponential rate in an unlimited environment (left) and when birth rates and death rates are equal (right) (data from Leslie and Ranson, 1940). *Lower*. Extremes in juvenile-adult ratios in pheasants in North Dakota (data from Kimball, 1948) and in muskrats in eastern United States (data from Petrides, 1950).

rary changes, with spontaneous return to the stable situation. As nations go from pioneer conditions of rapidly expanding densities to mature conditions of stable populations, the percent of individuals in younger age classes decreases as shown in Figure 6–6. This changing age structure with an increasing percentage of older individuals has profound impacts on life styles and economic considerations.

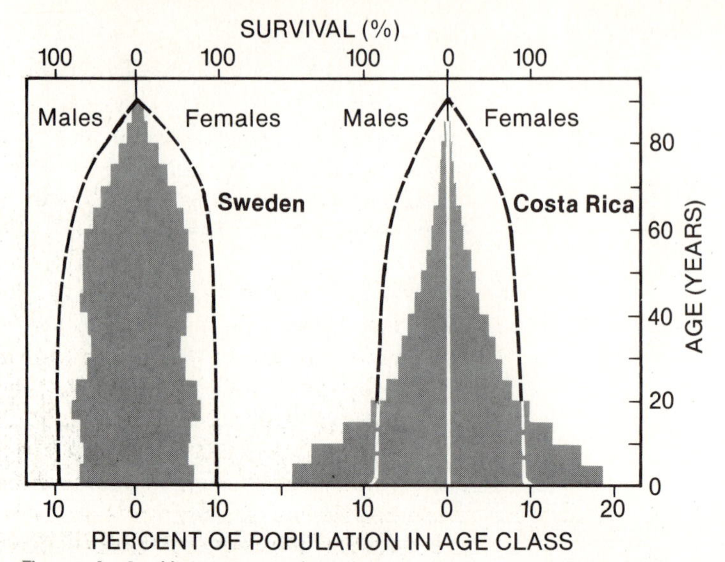

Figure 6–6 Human population age structure calculated separately for males and females in Sweden (1965) and Costa Rica (1963). Because Sweden's population has grown slowly, the age pyramid has a narrow base. A large percent of the population is in the older age classes (40 to 60 years). A baby boom following World War II is responsible for the bulge at ages 15 to 23. Costa Rica's rapid population growth has resulted in a broad-based pyramid with a large percent of the population less than 25 years of age. (After Ricklefs, 1979, from data of Keyfitz and Fieger, 1968.)

In simplistic form, age structure can be expressed in terms of three ecological ages: **prereproductive, reproductive,** and **postreproductive.** The relative duration of these ages in proportion to the life span varies greatly with different organisms. For humans in recent times, the three ages are relatively equal in length, about a third of a life falling in each class. Primitive humans, by comparison, had a much shorter postreproductive period. Many plants and animals have a very long prereproductive period. Some animals, notably insects, have extremely long prereproductive periods, a very short reproductive period, and no postreproductive period. Certain species of mayflies (*Ephemeridae*) and the 17-year cicada are classic examples. Mayflies require from one to several years to develop in the larval stage in the water, and adults emerge to live but a few days. Cicadas have an extremely long developmental history (not necessarily 17 years, however), with adult life lasting less than a single season.

In game birds and fur-bearing mammals, the ratio of first-year animals to older animals, determined during the seaon of harvest (fall or winter) by examining samples from the population taken by hunters or trappers, provides an index to the population trends. In general, a high ratio of juveniles to adults, as shown in the bottom diagrams in Figure 6–5, indicates a highly successful breeding season and the likelihood of a larger population the next year, provided the mortality of juveniles is not excessive. In the muskrat (lower right, Figure 6–5), the highest percentage of juveniles (85 percent) occurred in a population that had been heavily trapped for the previous few years. Reduction of the total population in this manner had apparently resulted in increased natality for those individuals surviving, indicating that muskrat populations are resilient—they recover quickly from acute mortality.

A phenomenon known as the "dominant age class" has been repeatedly observed in fish populations that have a very high potential natality rate. When a large year class occurs because of unusual survival of eggs and larval fish, reproduction is suppressed for the next several years. Hjort's data on herring in the North Sea provide the classic case, as shown in Figure 6–7. Fish of the 1904 year class dominated the catch from 1910 (when this age class was 6 years old and large enough to be caught effectively in commercial fish nets) until 1918 (when, at 14 years of age, they still outnumbered fish of younger age groups). The situation produced something of a cycle in total catch, which was high in 1910 and then declined in subsequent years as the dominant age class declined and before there was replacement from other classes. Something of a compensatory mechanism, in which high survival is followed by a high probability of low survival in subsequent years, is at work here. Fishery biologists have not yet agreed on what environmental conditions result in the unusual survival occurring every now and then.

2. Basic Concepts of Rates

A population is a changing entity. Even when the community and the ecosystem are seemingly unchanging, density, natality, survivorship, age structure, growth rate, and other attributes of component populations are usually in flux as species constantly adjust to seasons, to physical forces, and to one another. Accordingly, ecologists are often more interested in how and at what rate the population is changing than in its size and composition at any one moment. Calculus, the branch of mathematics dealing (in part) with a study of rates, thus becomes an important tool in studying population ecology. A brief

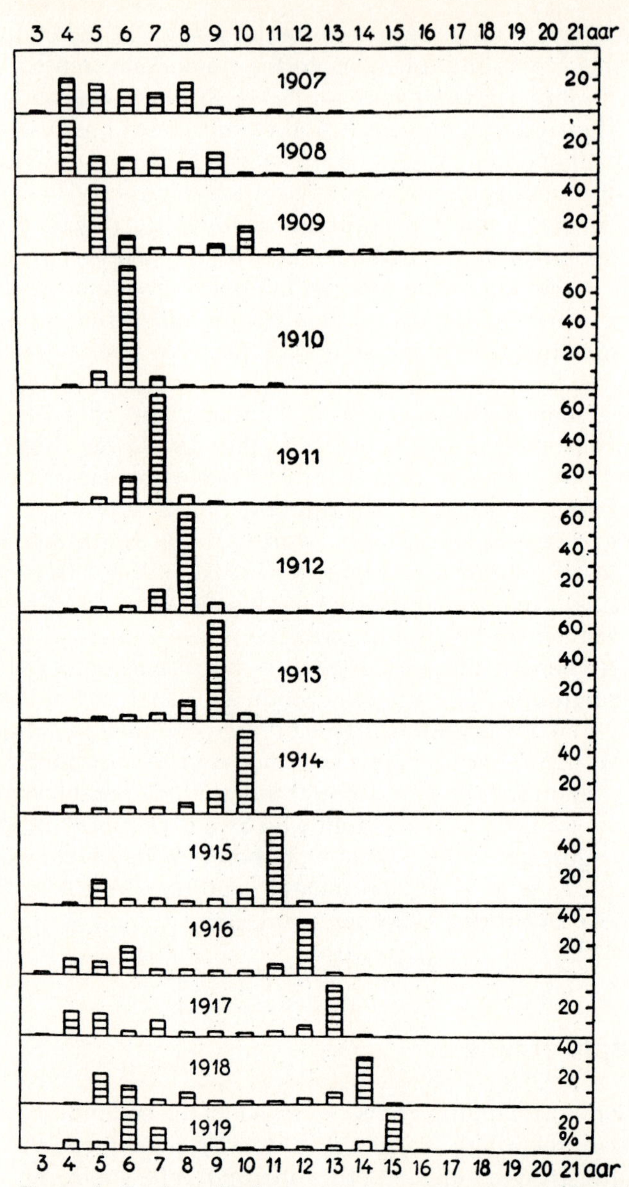

Figure 6–7 Age distribution in the commercial catch of herring in the North Sea between 1907 and 1919, illustrating the dominant age class phenomenon. The 1904 year class was very large and dominated the population for many years. Since fish younger than 5 years are not caught in the nets, the 1904 class did not show up until 1909. Age of the fish was determined by growth rings on scales, which are laid down annually in the same manner as growth rings on trees. (After Hjort, 1926.)

review of concepts concerning rates may help those not yet familiar with this kind of mathematics.

A rate may be obtained by dividing the change in some quantity by the period of time elapsed during the change; such a rate term would indicate the rapidity with which something changes with time. Thus, the number of miles traveled by a car per hour is the speed rate, and the number of births per year is the birth rate. The "per" means "divided by." Remember that, as discussed previously (Chapter 3, Section 3), productivity is a rate, not a standing state.

Customarily, "the change in" something is abbreviated by writing the symbol Δ (delta) in front of the letter representing the thing changing. Thus, if N represents the number of organisms and t the time, then

ΔN = the change in the number of organisms

$\frac{\Delta N}{\Delta t}$ or $\Delta N/\Delta t$ = the average rate of change in the number of organisms per (divided by, or with respect to) time. (This is the growth rate.)

$\frac{\Delta N}{N\Delta t}$ or $\Delta N/(N\Delta t)$ = the average rate of change in the number of organisms per time per organism (the growth rate divided by the number of organisms initially present or, alternatively, by the average number of organisms during the period of time). (This is often called the specific growth rate and is useful when populations of different sizes are to be compared. If multiplied by 100 [i.e., $\Delta N/(N\Delta t) \times 100$], it becomes the percent growth rate.)

Often, we are interested not only in the average rate over a period of time but also in the theoretical instantaneous rate at particular times, in other words, the rate of change when Δt approaches zero. In the language of calculus, the letter d (for derivative) replaces the Δ when instantaneous rates are being considered. In this case the preceding notations become:

dN/dt = the rate of change in the number of organisms per time at a particular instant

$dN/(Ndt)$ = the rate of change in the number of organisms per time per individual at a particular instant

Figure 6–8 shows the difference between a growth curve and a growth rate curve. As will be discussed further in Section 4, s-shaped

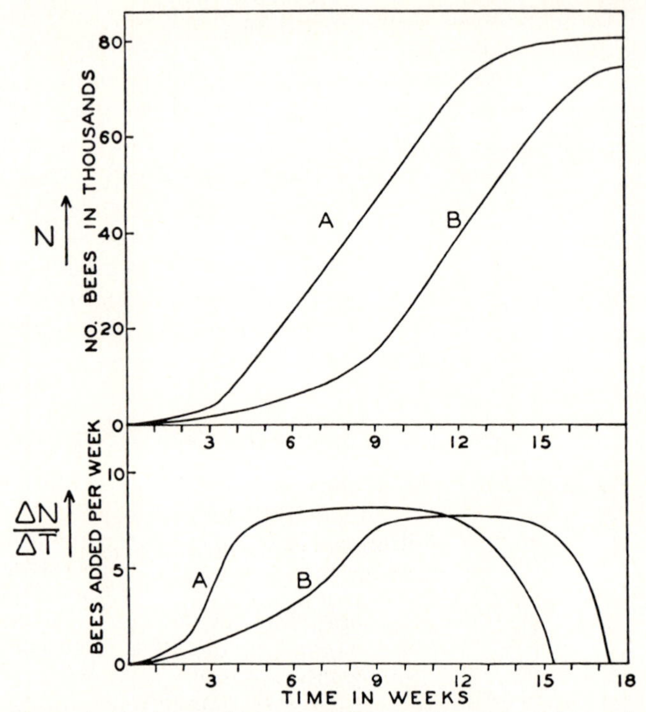

Figure 6–8 A population growth curve (upper) and growth rate curve (lower) for two bee colonies in the same apiary. A. Italian bees. B. Cyprian bees. (Redrawn from Bodenheimer, 1937.)

growth curves and hump-backed growth rate curves are often characteristic of populations in pioneer growth stages.

On the growth curve, the slope (straight line tangent) at any point is the growth rate. Thus, in the case of population A in Figure 6–8, the growth rate was at a maximum between 4 and 11 weeks and zero after 16 weeks. In population B, maximum rate occurred between 9 and 16 weeks and was zero by 18 weeks. The $\Delta N/\Delta t$ notation serves to illustrate the model for the usual purposes of measurement, but the dN/dt notation must be substituted in many types of actual mathematical manipulations of the models.

The instantaneous rate, dN/dt, cannot be measured directly, nor can $dN(Ndt)$ be calculated directly from population counts. The type of population growth curve exhibited by the population would have to be known, and then the instantaneous rate would have to be calculated from equations, as will be explained in Section 4. The instantaneous speed rate of a population cannot be measured as a car's can. The rate can be approximated, of course, by making a census at very short intervals. We can plot whatever census points we have, connect them with lines, and then determine what kind of instantaneous equation best mimics the actual growth curve or, better still, what equation best reflects the factors that control growth of that particular population (thus avoiding a mere "curve fitting" procedure).

3. The Intrinsic Rate of Natural Increase

Statement

When the environment is unlimited (space, food, or other organisms not exerting a limiting effect), the specific growth rate (the population growth rate per individual) becomes constant and maximum for the existing microclimatic conditions. The value of the growth rate under these favorable population conditions characterizes a particular population age structure and is a single index of the inherent power of a population to grow. It may be designated by the symbol r, which is the exponent in the differential equation for population growth in an *unlimited environment* under specified physical conditions:

$$dN/dt = rN;\ r = dN(Ndt) \tag{1}$$

This is the same form as used in Section 2. The parameter r can be thought of as an **instantaneous coefficient of population growth.** The exponential integrated form follows automatically by calculus manipulation:

$$N_t = N_0 e^{rt} \tag{2}$$

where N_0 represents the number at time zero, N_t the number at time t, and e the base of natural logarithms. By taking the natural logarithm of both sides, one converts the equation into a form used in making actual calculations. Thus:

$$\ln N_t = \ln N_0 + \mathrm{rt};\ r = \frac{\ln N_t - \ln N_0}{t} \tag{3}$$

In this manner, the index r can be calculated from two measurements of population size (N_0 and N_t or at any two times during the unlimited growth phase, in which case N_{t1} and N_{t2} may be substituted for N_0 and N_1 and $(t_2 - t_1)$ substituted for t in the previous equations).

The index r is actually the difference between the instantaneous specific natality rate b (i.e., rate per time per individual) and the instantaneous death rate d and may thus be simply expressed:

$$r = b - d \tag{4}$$

The overall growth rate of population under unlimited environmental conditions (r) depends on the age composition and the specific growth rates due to reproduction of component age groups. Thus, there may be several values for a species, depending upon population structure. When a stationary and stable age distribution exists, the specific growth rate is called the **intrinsic rate of natural increase** or r_{max}. The maximum value of r is often called by the less specific but widely used expression **biotic potential**, or reproductive potential. The difference between the maximum r or biotic potential and the rate of increase that occurs in an actual laboratory or field condition is often taken as a measure of the **environmental resistance**, which is the sum total of environmental limiting factors that prevent the biotic potential from being realized.

Explanation

Natality, mortality, and age distribution all are important, but each tells little by itself about how the population is growing as a whole, about what it would do if conditions were different, and about what its best possible performance is, compared with its everyday performances. Chapman (1928) proposed the term **biotic potential** to designate maximum reproductive power. He defined biotic potential as "the inherent property of an organism to reproduce, to survive, i.e., to increase in numbers. It is sort of the algebraic sum of the number of young produced at each reproduction, the number of reproductions in a given period of time, the sex ratio and their general ability to survive under given physical conditions." As one might imagine from the very general preceding definition, biotic or reproductive potential came to mean different things to different people. To some, it meant a nebulous reproductive power lurking in the population, fortunately never allowed to come forth because of the action of the environment (i.e., "if unchecked the descendants of a pair of flies would weigh more than the earth in a few years"). To others, it meant simply and more concretely the maximum number of eggs, seeds, spores, and so forth that the most fecund individual was known to produce, though this would have little meaning in the population sense, since most populations do not contain individuals that are all continually capable of peak production.

Lotka (1925), Dublin and Lotka (1925), Leslie and Ranson (1940), Birch (1948), and others had to translate the rather broad idea of biotic potential into mathematical terms that could be understood in any language (with, sometimes, the help of a good mathematician). Birch (1948) expressed it well when he said: "If the 'biotic potential' of Chapman is to be given quantitative expression in a single index, the parameter *r* would seem to be the best measure to adopt since it gives the intrinsic capacity of the animal to increase in an unlimited environment." The index *r* is also frequently used as a quantitative expression of "reproductive fitness" in the genetic sense, as will be noted later.

For the growth curves discussed in Section 2, *r* is the specific growth rate ($\Delta N/N\Delta t$) when population growth is exponential. Equation 3 in the previous Statement is an equation for a straight line. Therefore, the value of *r* can be obtained graphically. If growth is plotted as logarithms or on semilogarithmic paper, the log of population number plotted against time will give a straight line if growth is exponential; *r* is the slope of this line. The steeper the slope, the higher the intrinsic rate of increase. In Figure 6–9, the same growth curve is plotted in two ways: (1) with numbers (*N*) on an arithmetic scale (left-hand graph) and (2) with *N* on a logarithmic scale (right-hand graph on semilog paper). In this example, a hypothetical population of microorganisms is experiencing six days of exponential growth in which the population increases by a factor of 10 every two

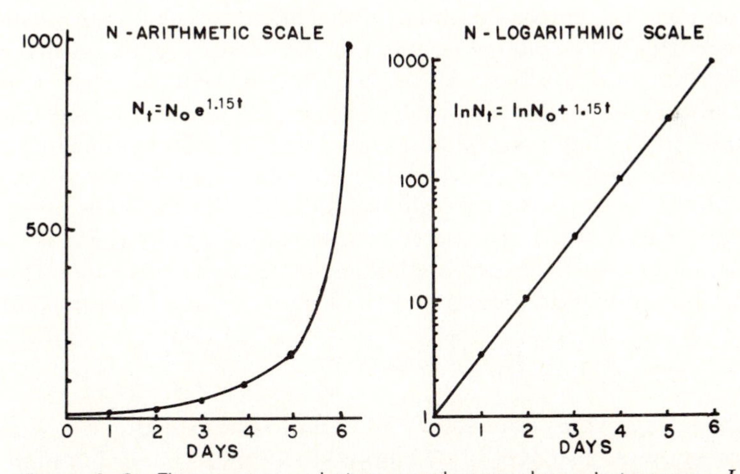

Figure 6–9 The same population growth curve shown in two ways. *Left.* Numbers (*N*) on an arithmetic scale. *Right.* Numbers on a logarithmic scale. In this hypothetical example, a population of microorganisms is experiencing six days of exponential growth during which the population increases tenfold every two days. See text for explanation of equations.

days. The slope of the line on the semilog plot is 1.15, which is the value of r. One can check this by plugging into Equation 3 any two population values, for example, density at Day 2 and Day 4, as follows:

$$r = \frac{\ln N_{t2} - \ln N_{t1}}{t_2 - t_1}$$

$$r = \frac{\ln 100 - \ln 10}{2}$$

From the table of natural logarithms:

$$r = \frac{4.6 - 2.3}{2} = 1.15$$

The extremely wide differences in biotic potential are especially emphasized when expressed as the number of times the population would multiply itself if the exponential rate continued and as the time required to double the population. The two parameters are mathematically derived from the intrinsic rate as follows:

Finite rate of increase: $\lambda = e^r$; $\log_e\lambda = r$; $\lambda = \text{antilog}_e$

Doubling time: $t = \log_e 2/r = 0.6931/r$, as derived from Equation 2 by setting $N_t/N_0 = 2$

Doubling time at the maximum intrinsic rate for flour beetles under optimum laboratory condition is less than a week (see Leslie and Park, 1949). For human beings in 1968, doubling time was 35 years (Ehrlich and Ehrlich, 1970), with population in some countries doubling faster. Smith (1954) suggests that r, and hence doubling time, in the biotic kingdom ranges over six orders of magnitude.

The coefficient of population growth r should not be confused with the net reproductive rate R_0, as discussed in Section 1 (see Table 6–3), which is strictly related to generation time and is not suitable for comparing different populations unless their generation times are similar. However, mean generation time (T) is related to R_0 and r:

$$R_0 = e^{rT}; \text{ therefore, } T = \frac{\log_e R_0}{r}$$

The relationships between T, R_0, and r for various animal populations are graphed on page 52 of Slobodkin's (1962) book. Populations in nature often grow exponentially for short periods when there is ample food and no crowding effects, enemies, and so forth. Under such conditions, the population as a whole is expanding at a terrific rate, even though each organism is reproducing at the same rate as before; the specific growth rate is constant. Plankton blooms, mentioned in previous chapters, pest eruptions, or growth of bacteria in new culture media are examples of situations in which growth may be logarithmic. Many other phenomena such as absorption of light, monomolecular chemical reactions, and compound interest behave similarly. It is obvious that this exponential increase cannot continue for very long; often, it is never realized. Interactions within the population as well as external environmental resistances soon slow down the rate of growth and play a part in shaping population growth form in various ways.

4. Population Growth Form

Statement

Populations have characteristic patterns of increase called population growth forms. For comparison, two basic patterns based on the shapes of arithmetic plots of growth curves can be designated: the **J-shaped growth form** and the **S-shaped** or **sigmoid growth form.** These contrasting types may be combined or modified, or both, in various ways according to the peculiarities of different organisms and environments. In the J-shaped form, density increases rapidly in exponential fashion (as shown in Figure 6–9) and then stops abruptly as environmental resistance or another limit becomes effective more or less suddenly. This form may be represented by the simple model based on the exponential equation considered in the preceding section:

$$\frac{dN}{dt} = rN \text{ with a definite limit on } N$$

In the sigmoid form, the population increases slowly at first (establishment or positive acceleration phase), then more rapidly (perhaps approaching a logarithmic phase), but it soon slows down gradually as the environmental resistance increases in percentage (the negative acceleration phase), until equilibrium is reached and maintained. This form may be represented by the simple logistic model:

$$\frac{dN}{dt} = rN\,\frac{(K - N)}{K}$$

The upper level, beyond which no major increase can occur, as represented by the constant K, is the **upper asymptote** of the sigmoid

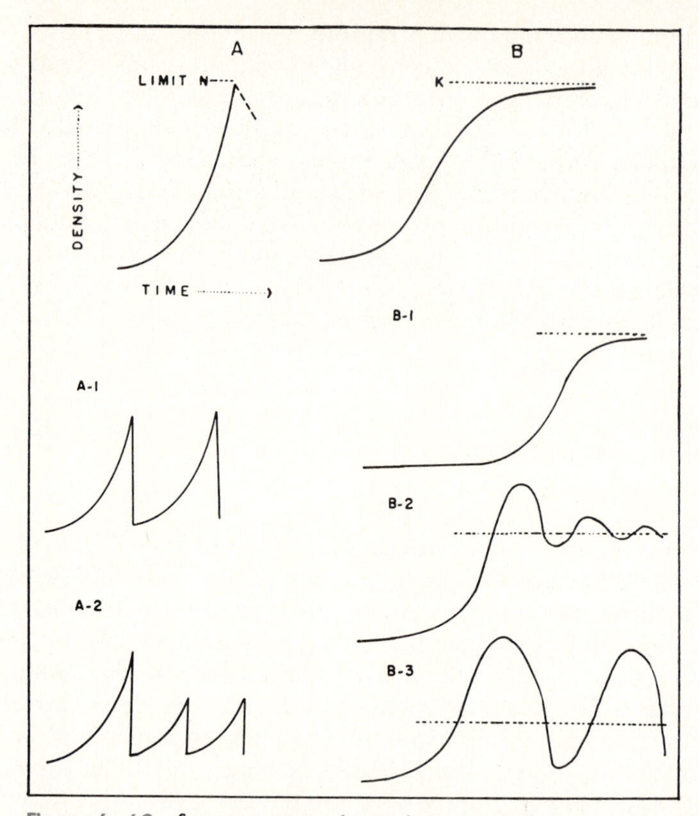

Figure 6–10 Some aspects of population growth form, when plotted on arithmetic scales, showing the J-shaped (exponential) (A) and the S-shaped (sigmoid) (B) forms and some variants. A-1 and A-2 show oscillations that would be inherent in the J-shaped form. B-1, B-2, and B-3 show some possibilities (but by no means all) where there is a delay in density effect, which occurs when elapses between production of young individuals and full influence of the individuals (the case in higher plants and animals). When nutrients or other requisites accumulate prior to population growth, an "overshoot" may occur, as shown in A-2 and B-2. (This explains why new ponds or lakes often provide better fishing than old ones!) (Curves adapted from Nicholson, 1954.)

curve and has been aptly called the **maximum carrying capacity** (see Chapter 3, Section 8). In the J-form there may be no equilibrium level; instead, density may fluctuate according to limitation in resources, physical factors, and interaction with other populations (predation, for example). The two growth forms and certain variants are shown schematically in Figure 6–10.

Explanation

When a few individuals are introduced into or enter an unoccupied area (for example, at the beginning of a season), characteristic patterns of population increase have often been observed. When plotted on an arithmetic scale, the part of the growth curve representing an increase in population often takes the form of an S or a J, as shown in Figure 6–10, curves A and B. It is interesting to note that these two basic growth forms are similar to the two metabolic or growth types that have been described in individual organisms (Bertalanffy, 1957). However, it is not known whether there is a causal relationship between growth of individuals and growth of populations. One can now say only that there are some similarities in patterns. Populations and communities are not "superorganisms" but may have analogous properties (Chapter 2, Section 6).

The equation given previously as a simple model for the J-shaped form is the same as the exponential equation discussed in Section 3, except that a limit is imposed on N. The relatively unrestricted growth is suddenly halted when the population runs out of some resource (such as food or space), or when frost or any other seasonal factor intervenes, or when the reproductive season suddenly terminates (perhaps, for example, because of development of diapause as described in Chapter 5, page 239). When the upper limit of N is reached, the density may remain at this level for a time, or an immediate decline occurs, producing a relaxation-oscillation pattern in density, shown in Figure 6–10, curves A-1 and A-2. This pattern, which Nicholson (1954) has called "density triggered," seems to be characteristic of many populations in nature, such as algal blooms, annual plants, some insects, and perhaps lemmings on the tundra.

A type of growth form also frequently observed follows an S-shaped or sigmoid pattern when density and time are plotted on arithmetic scales. The sigmoid curve is the result of greater and greater action of detrimental factors (environmental resistance) as the density of the population increases, unlike the previous model in which environmental resistance was delayed until near the end of the increase. For this reason, Nicholson (1954) has spoken of the sigmoid type as "density conditioned." A simple case is one in which detrimental factors are linearly proportional to the density. Such a growth form is said to be logistic and conforms to the logistic equation* used as a basis for the model of the sigmoid pattern.

* The logistic equation was first proposed by P. F. Verhulst in 1838; it was extensively used by Lotka and "rediscovered" by Pearl and Reed (1930).

The logistic equation may be written in several ways; three forms, plus the integrated form, follow:

$$\frac{dN}{dt} = rN\frac{(K-N)}{K} \quad or$$
$$= rN - \frac{r}{K}N^2 \quad or$$
$$= rN\left(1 - \frac{N}{K}\right)$$
$$N = \frac{K}{1 + e^{a-rt}}$$

where dN/dt is the rate of population growth (change in number in time), r the specific growth rate or intrinsic rate of increase (discussed in Section 3), N the population size (number), K the maximum population size possible or upper asymptote, e the base of natural logarithm, and a the constant of integration defining the position of the curve relative to the origin. It is the value of $\log_e (K - N)/N$ when $t = 0$.

This equation is the same as the exponential one written in the previous section, with the addition of the expression $(K - N)/K$, $(r/K)N^2$, or $(1 - N/K)$. The latter expressions are three ways of indicating the environmental resistance created by the growing population itself, which brings about an increasing reduction in the potential reproduction rate as population size approaches the carrying capacity. In word form, these equations simply mean the following:

Rate of population increase	equals	Maximum possible rate of increase (unlimited specific growth rate) times numbers in the population	times	Degree of realization of maximum rate
			or	
			minus	Unrealized increase

In summary, then, this simple model is a product of three components: a rate constant (r), a measure of population size (N), and a measure of the portion of available limiting factor not used by the population $(1 - N/K)$.

The logistic equation may also be written in terms of the rate of increase per generation, R, as follows:

$$\frac{dN}{dt} = N \log_e R \left(\frac{K-N}{K}\right)$$

Although the growth of a great variety of populations, representing microorganisms, plants, and animals and including both laboratory and natural populations, has been shown to follow a sigmoid pattern, it does not follow necessarily that such populations increase according to the logistic equation. Many mathematical equations will produce a sigmoid curve. Almost any equation in which the negative factors increase in some manner with density will produce sigmoid curves. As Wiegert (1974) has pointed out, the logistic represents a sort of minimum sigmoid growth form since the limiting effects of both space and resources begin at the very beginning of growth (i.e., the maximum specific rate of increase is achieved only at zero density). In most cases, we would expect less limited growth at first, followed by a slowing down as density increased. Figure 6–11 illustrates this concept of the logistic as the lowest and the exponential as the highest growth form. Most populations would be expected to follow an intermediate pattern.

In populations of higher plants and animals, which have complicated life histories and long periods of individual development, there are likely to be delays in the increase in density and the impact of limiting factors, producing what Nicholson (1954) has called "tardy density conditioned" patterns. In such cases, a more concave growth curve may result (longer period required for natality to become effective), and almost always the population "overshoots" the upper asymptote and undergoes oscillations before settling down at the carrying capacity level (see Figure 6–10, curve B-2). Many modifications of the logistic equation have been suggested to include two

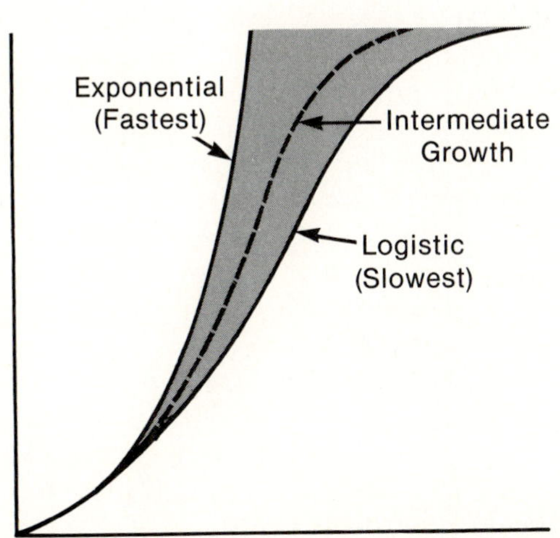

Figure 6–11 Curves showing theoretical upper (exponential) and lower (logistic) growth for any population with identical maximum growth rates and minimum maintenance densities. The hatched area between represents the area within which the growth form of most populations would lie. (After Wiegert, 1974.)

kinds of time lags: (1) the time needed for an organism to start increasing when conditions are favorable and (2) the time required for organisms to react to unfavorable crowding by altering birth and death rates. Let these time lags be $t - t_1$ and $t - t_2$, respectively:

$$\frac{dN(t)}{dt} = rN_{(t-t_1)} \frac{K - N_{(t-t_2)}}{K}$$

When this kind of equation is studied by the use of an analog or microcomputer, the density overshoots and oscillates with decreasing amplitude with time, very much as shown in Figure 6–10, curve B-2.

A major weakness of almost all mathematical models of population growth is that they operate in a closed system, without input or output. Only self-crowding or other internal factors are modeled. As discussed repeatedly in Chapters 2 and 3, the real world consists of open systems in which the input and output environments play major roles in the behavior of the component being considered (see Figure 2–1). This shortcoming is especially apparent when it comes to modeling or trying to predict the growth form of human populations. Edward Deevey (1958) pointed out many years ago that in the absence of "external" controls, such as systematic family planning, growth of human population will likely follow some sort of overshoot-and-oscillation pattern because of time lags involved in appearance of effective self-crowding restraints. Fast-growing cities with their dependence on huge external sources of energy, food, water, and general life-support are especially likely to boom and bust in varying degrees, depending on input factors and the degree to which citizens and governments can anticipate future conditions and plan ahead. Thus, in early stages of urban growth, when economic conditions are favorable (space and resources available and inexpensive) and when the need for services (water, sewage treatment, streets, schools, and so on) is small, the population grows rapidly (with immigration often the major increase factor), as in a J growth pattern. Not until some time later (the time lag) do housing and schools become overcrowded, demands for services increase, taxes rise, and general diseconomies of scale begin to be felt. In the absence of negative feedback, such as is built into the simple logistic, or which can be built into rational city planning, cities will grow too fast for their future good and then suffer decline.

Examples

Figure 6–12 illustrates the sigmoid growth form at its simplest. Figure 6–13 illustrates the J-shaped pattern, in which the thrips (small

Paralleling the dt (instantaneous time) and the Δt (discrete time intervals) notations used in Sections 2, 3, and 4 are the two major mathematical tools used by population and ecosystem modelers: systems of differential equations and systems of difference equations. Differential equations have some advantages for mathematical manipulation, perhaps because they have been so widely used and taught in calculus courses. Difference equations, however, parallel more closely the way data are gathered (at discrete time intervals), and in the case of population growth, they correspond to patterns in species with discrete generations. To compare, let us repeat the differential equations for the J and the S growth forms, and write the analog difference equations for each:

Differential equation for instantaneous exponential growth:

$$dN/dt = rN \tag{1}$$

Difference equation for discrete exponential growth:

$$N_{t+1} = \lambda N_t(r) \quad or \tag{2}$$

$$N_{t+1} = e^r N_t \tag{3}$$

where N_{t+1} is a change during a time interval; λ is a multiplicative growth factor per generation or other specified period (it is a finite rate of increase as previously introduced in Section 3, page 318, and is the analog of "r," the intrinsic rate of increase); $r = \log_e \lambda$, so $\lambda = e^r$, which is substituted for λ in Equation 3 above. For growth, λ must be greater than 1 and r greater than 0; otherwise, the population declines exponentially to extinction.

Differential equations for logistic (density-dependent or sigmoid) growth:

$$dN/dt = rN(1 - N/K) \tag{4}$$

Difference equation for density-dependent growth:

$$N_{t+1} = N_t e^{r(1-N/K)} \tag{5}$$

Several other difference equations have been proposed as discrete analogs of the logistic and other density-dependent equations (see May, 1981; May and Oster, 1976). Note that difference equations are "ready to be used" and do not have to be integrated as do differential equations. Compare Equation 5 with the integrated form of logistic on page 322.

Tools are often chosen because one is interested in describing how something is changing or is more interested in rates of change; in the latter case, the differential equation would be preferred.

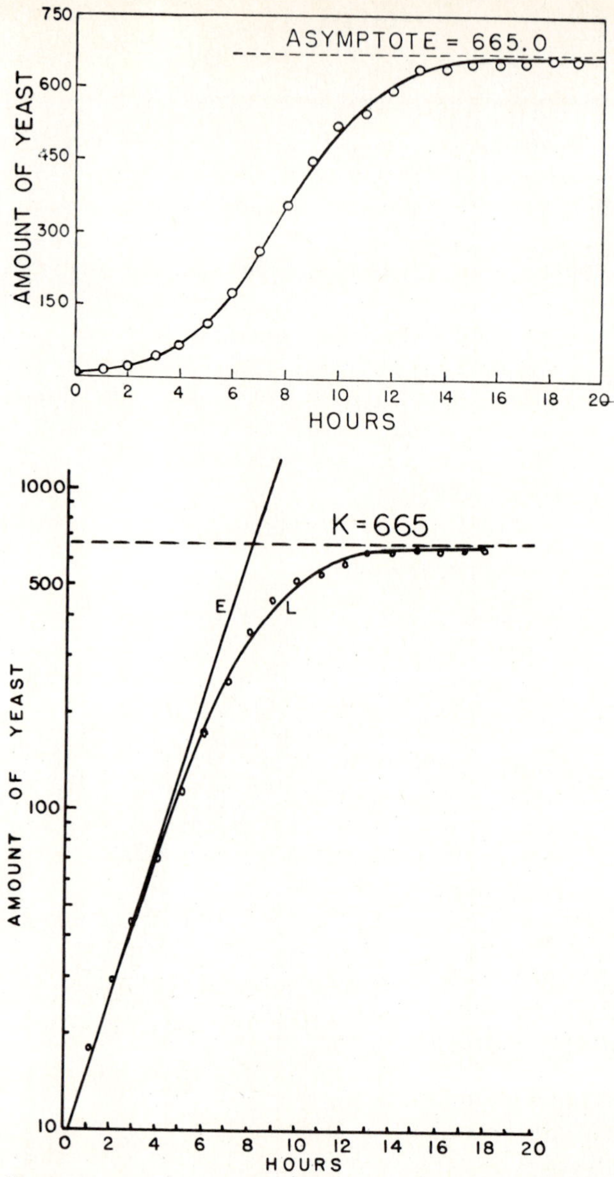

Figure 6–12 Growth of yeast in a culture. A simple case of the sigmoid growth form in which environmental resistance (in this case, detrimental factors produced by organisms themselves) is linearly proportional to the density. Open circles are observed growth values; solid lines are curves drawn from equations. In the upper graph, growth of yeast is plotted on an arithmetic scale and a logistic curve is fitted to the data. In the lower graph, the same data are plotted (L) with the amount of yeast on a logarithmic scale; an exponential curve (E) is included to illustrate what growth would be like without self-limiting constraints. (Drawn from data of Pearl, 1927; upper graph from Allee et al., 1949.)

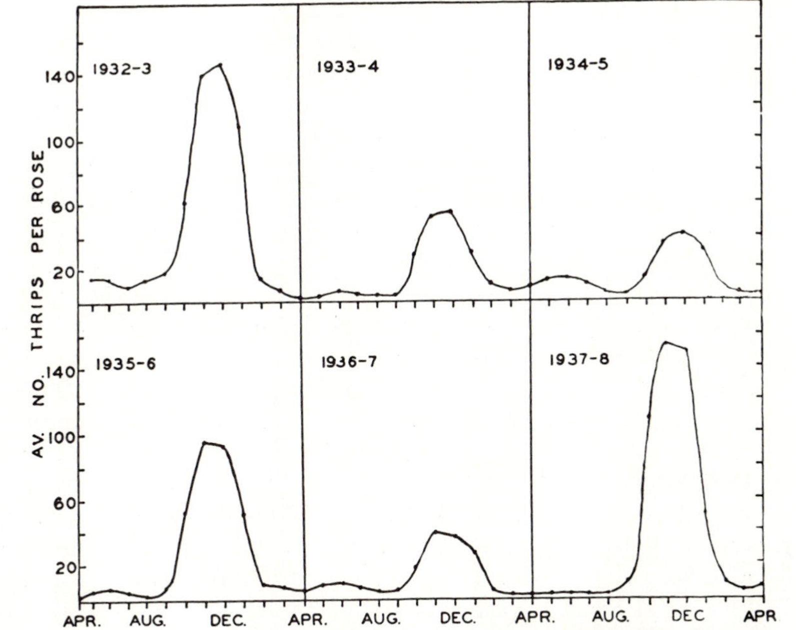

Figure 6–13 Seasonal changes in a population of adult thrips living on roses. (Graph constructed from data by Davidson and Andrewartha, 1948.)

insects) increase rapidly during favorable years until the end of the season calls a halt; then the density just as rapidly declines. In less favorable years, the growth form is more sigmoid. Broadly speaking, the J-shaped form may be considered an incomplete sigmoid curve, since a sudden limiting effect is brought to bear before the self-limiting effects within the population become important.

In Figure 6–12, the growth of the yeast is plotted on both arithmetic (upper graph) and log (lower graph) scales, and a logistic curve is fitted to the data. Note that in the semilog plot, the curve takes the shape of an inverted J rather than an S. In the lower graph, an exponential curve (straight line E) is also shown to illustrate what growth would be like if not limited by the size of the culture vessel and the density of the population. The equations are the integrated forms for the logistic and the exponential (J-shaped growth form) equations as already described; actual values are plugged in for the constants K, a, and r. The actual (observed) growth of the yeast follows very closely the logistic, indicating that effects of self-crowding begin early and are linearly related to density. The area between the two curves in the lower diagram could be taken as a quantitative measure of the environmental resistance. The advantage of the semilog plot is that any deviation from a straight line indicates a change in the growth rate of the population (dN/dt). The more the curve bends, the greater the change.

Even though simple logistic growth form is probably restricted to small organisms or those with simple life histories, a general sigmoid growth pattern may be observed in larger organisms when they are introduced onto previously unoccupied islands, as, for example, the growth of sheep populations on the island of Tasmania (Davidson, 1938) or the growth of a pheasant population introduced on an island in Puget Sound, Washington (Einarsen, 1945).

Recall this book's recurrent theme of open systems. **Population dispersal**, the movement of individuals or their disseminules (seeds, spores, larvae, and so forth) into or out of the population or population area, supplements natality and mortality in shaping the population growth form. **Emigration**—one-way outward movement—affects the local growth form in the same way as mortality; **immigration** acts like natality. **Migration**—periodic departure and return—supplements both natality and mortality seasonally. Dispersal is greatly influenced by barriers and by the inherent power of movement, or vagility, of individuals or disseminules. And, of course, dispersal is the means of colonizing new or depopulated areas. It is also an important component in the flow of genes and in speciation. Dispersal of small organisms and passive propagules generally takes an exponential form in that density decreases by a constant amount of equal multiples of distance from source. Dispersal of large, active animals deviates from this pattern and may take the form of "set distance" dispersal, normally distributed dispersal, or other forms. Stewart's (1952) study of dispersal of banded barn owls is a good example of the interplay of random dispersal and a tendency for southward migration. Within 100 miles, dispersal was nondirectional (about equal chance a banded bird would be recovered at any point in any direction), but beyond this distance, dispersal was definitely directional and southward. For general reviews of dispersal patterns, see Wolfenbarger (1946) and MacArthur and Wilson (1967).

5. Population Fluctuations and "Cyclic" Oscillations

Statement

When populations complete their growth, and $\Delta N/\Delta t$ averages zero over a long time, population density tends to fluctuate above and below the steady-state level even in populations subject to various forms of feedback control. Often, fluctuations result from seasonal or annual changes in availability of resources, or they may be stochastic (random). However, some populations oscillate so regularly that they can be classed as "cyclic."

Explanation

In nature, it is important to distinguish between (1) seasonal changes in population size, largely controlled by life-history adaptations coupled with seasonal changes in environmental factors, and (2) annual fluctuations. For purposes of analysis, annual fluctuations may be considered under two headings: (1) fluctuations controlled primarily by annual differences in extrinsic factors (factors such as temperature and rainfall that are outside the sphere of population interactions) and (2) intrinsic factors, oscillations controlled primarily by population dynamics (biotic factors such as food or energy availability, disease, and so on). In many cases, year-to-year changes in abundance seem clearly correlated with the variation in one or more major extrinsic limiting factors, but some species show such regularity in relative abundance seemingly independent of obvious environmental cues that the term "cycles" seems appropriate (species with such regular variation in population size are often known as "cyclic" species). A summary of theories that have been advanced to explain these cycles will be presented after the following examples.

As has been stressed in earlier chapters, populations modify and compensate for perturbations of physical factors. Therefore, the more highly organized and mature the community, or the more stable the physical environment, or both, the lower will be the amplitude of fluctuations in population density with time.

Examples

We are all familiar with seasonal variations in population size. We expect that at certain times of the year, mosquitoes or gnats will be abundant, the woods will be full of birds, or the fields will be full of ragweed. At other seasons, populations of these organisms may dwindle to the vanishing point. Although it would be difficult to find in nature populations of animals, microorganisms, and herbaceous plants that do not exhibit some seasonal change in size, the most pronounced fluctuations occur with organisms that have limited breeding seasons, especially those with short life cycles and those with pronounced seasonal dispersal patterns (e.g., migration). Figure

6–13, as already mentioned, illustrates not only a J-shaped growth form but also seasonal and annual fluctuations, as revealed by systematic study over the years. A pattern of this sort is probably typical of most insects and of most "annuals" among plants and animals. In both fresh water and the ocean, seasonal cycles in plankton populations have been intensively studied and modeled (an example was described in Chapter 5, Figure 5–2).

An example of a rather irregular variation in population size that appears to be correlated with weather is shown in Figure 6–14. During most years, the heron population in the two areas of Great Britain remains relatively constant; apparently, the local environments provided a rather stable carrying capacity for herons. However, a sharp decrease in density with subsequent recovery occurred after each of three series of severe winters (as indicated along the top of Figure 6–14). The synchronous changes in density in the two areas lend credibility to the correlation with winter mortality. Having more than one study area is a good rule for ecological field study! Incidentally, bird populations have been among the most intensively studied, and results of such studies have contributed much to population theory. For a readable summary of work on bird populations, see Lack (1966).

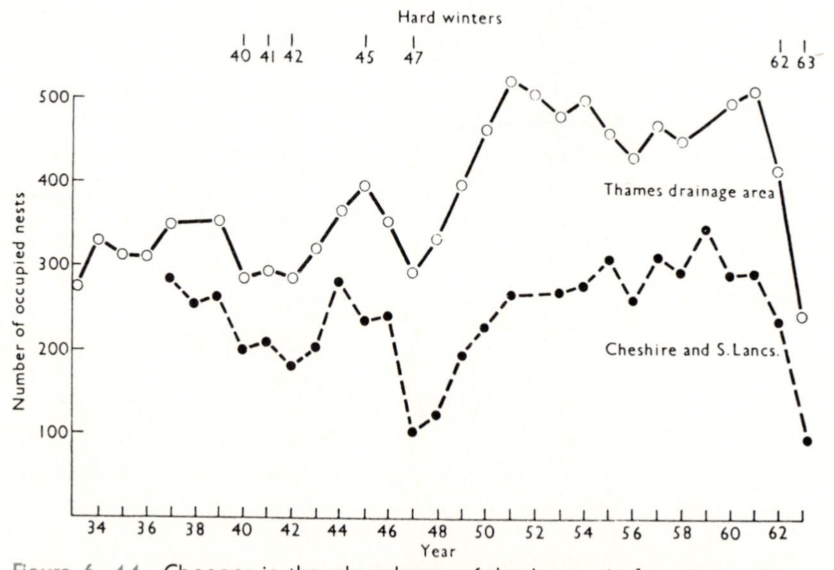

Figure 6–14 Changes in the abundance of the heron *Ardea cinerea* in two areas of Great Britain between 1933 and 1963. A relation between cold winters and a decline in abundance is indicated. (From Lack, 1966.)

Among the best-known examples of "cyclic" oscillations are certain species of northern mammals and birds that exhibit either a nine- to ten-year or a three- to four-year periodicity. A classic example of a nine- to ten-year oscillation is that of the snowshoe hare and the lynx, as shown in Figure 6–15. Since about 1800, the Hudson Bay Company of Canada has kept records of pelts of fur-bearers trapped each year. When plotted, these records show that the lynx, for example, has reached a population peak every nine to ten years, averaging 9.6 years, throughout this time. Peaks of abundance often are followed by "crashes," or rapid declines, and the lynx becomes scarce for several years. The snowshoe hare follows the same cycle: a peak abundance generally precedes that of the lynx by a year or more. Since the lynx largely depends on the hare for food, it is obvious that the cycle of the predator is related to that of the prey. But the two cycles are not strictly a cause-and-effect predator-prey interaction, since the hare's cycle occurs in areas where there are no lynxes. The shorter, or three- to four-year, cycle is characteristic of many northern murids (lemmings, mice, voles) and their predators (especially the snowy owl and foxes). The cycle of the lemming of the tundra and the arctic fox and the snowy owl was first well documented by Elton (1942). Every three or four years, over enormous areas of the northern tundra of two continents, the lemmings (two species in Eurasia and one in North America of the genus *Lemmus*, and one species of *Dicrostonyx* in North America) become abundant only to "crash," often within a single season. Foxes and owls, which increase in numbers as

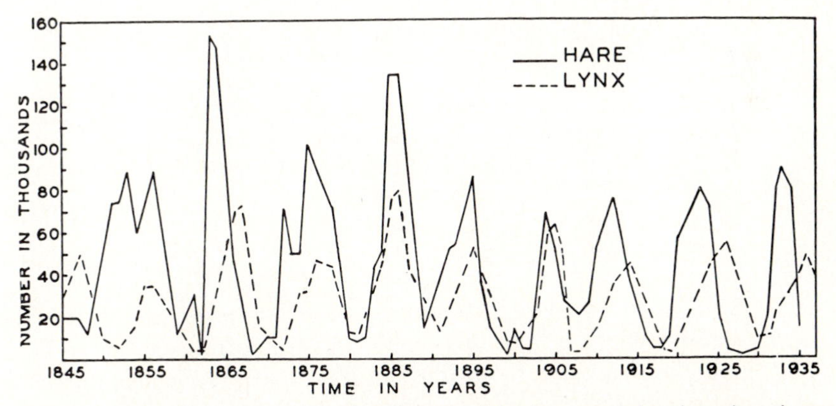

Figure 6–15 Changes in the abundance of the lynx and the showshoe hare, as indicated by the number of pelts received by the Hudson Bay Company. This is a classic case of cyclic oscillation in population density. (Redrawn from MacLulich, 1937.)

Figure 6–16 The showshoe or varying hare, famous in ecological annals for its spectacular cyclic abundance (see Figure 6–15). The individual shown is in its white winter pelage. The change from the brown summer pelage to the white one of winter has been shown to be controlled by photoperiodicity. (U.S. Soil Conservation Service Photo.)

their food increases, decrease very soon afterward. The owls may migrate south into the United States (sometimes as far south as Georgia) in search of food. This eruptive migration of surplus birds is apparently a one-way movement; few if any owls ever return. The owl population thus "crashes" as a result of the dispersal movement. So regular is this oscillation that ornithologists in the United States can count on an invasion of snowy owls every three or four years.

Since the birds are conspicuous and appear everywhere about cities, they attract a lot of attention and get their pictures in the newspaper or their skins mounted in local taxidermy shops. In years between invasions, few or no snowy owls are seen in the United States or southern Canada. Gross (1947) and Shelford (1943) have analyzed the records of the invasions and have shown that they are correlated with the periodic decrease in abundance of the lemming, the owls' chief food.

In Europe, but apparently not in North America, the lemmings themselves become so abundant at the crest of the cycle that they sometimes emigrate from their overcrowded haunts. Elton (1942) vividly describes the famous lemming migrations in Norway. So many of the animals pass through villages that dogs and cats tire of killing them and just ignore the horde. On reaching the sea, many lemmings drown. Thus, similar to the explosive increase in owls, the lemming movement is one-way. These spectacular emigrations do not occur at every four-year peak in density, but only during exceptionally high peaks. Often, the population subsides without the animals leaving the tundra or mountains.

Krebs and Meyers (1974) have reviewed population cycles in small mammals, and Finerty (1980) has applied some new statistical approaches to try to answer the question, "Do cycles exist?" The answer seems to be yes for some species and situations, and no for others.

Two examples from long-term records of violent oscillations in foliage insects in European forests are shown in Figure 6–17. Such pronounced cycles have been reported mostly from northern forests, especially pure stands of conifers. Density may vary over five orders of magnitude (log cycles), from less than one to more than 10,000 per 1000 square meters (upper diagram, Figure 6–17). One can well imagine that with 10,000 potential moths emerging for every 1000 square meters, and several generations in a season, enough caterpillars could be produced to defoliate and even kill the trees, as frequently happens. The cycles of defoliating caterpillars are not so regular as the oscillations of the snowshoe hare, and their periodicity seems to be somewhere between four and ten years (peaks of abundance average 7.8 and 8.8 years in the two examples shown). The cycles of different species are not synchronous.

Periodic outbreaks of the spruce budworm and tent caterpillars are well-known examples of similar patterns in the northern part of North America. Tent caterpillars have been studied by Wellington (1957, 1960) and budworms by MacDonald (1975) and Holling and colleagues (Ludwig et al., 1978; Clark et al., 1979; Holling, 1980).

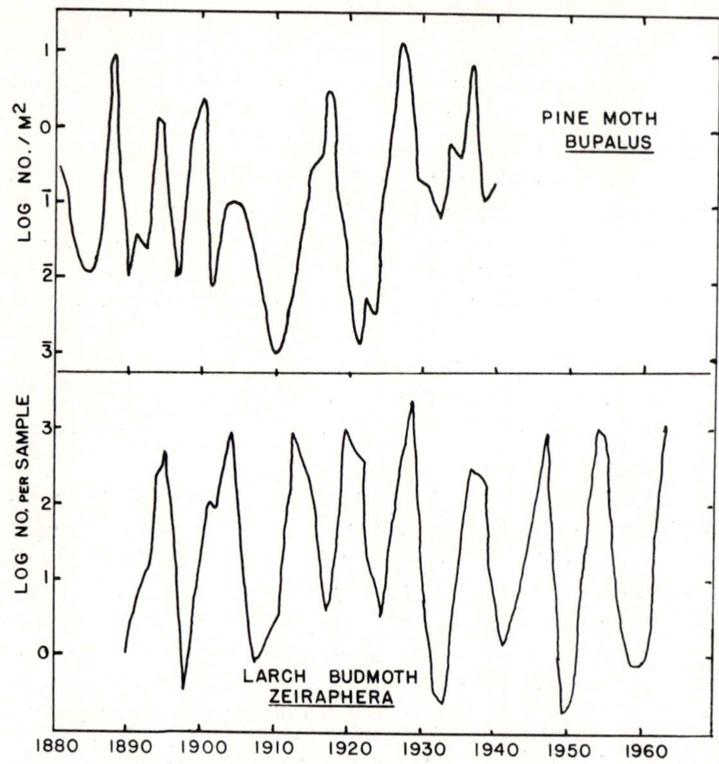

Figure 6–17 Fluctuations in numbers of two species of moths whose larvae feed upon foliage of conifers. *Upper*. The pine moth *Bupalus* in German pine forests at Litzlingen, which are managed as pure stands. *Lower*. The larch budmoth *Zeiraphera griseana* in larch forests in the Swiss valley of Engadin. (Upper graph redrawn from Varley, 1949; lower graph redrawn from Baltensweiler, 1964.)

In the case of tent caterpillars (*Malacosoma*) an alternation of "strong" and "weak" physiological races is thought to have a genetic basis. On the upswing of the cycle, caterpillars build elongated tents, and individuals are active feeders moving far out into the foliage. At peak density, the caterpillars become inactive, build more compact tents, feed less, and become more susceptible to disease. The population thus has evolved a built-in limit-control mechanism that prevents repeated defoliation of trees and thus enhances the long-term survival of both the insect and its host plant.

The budworm cycles are clearly a phenomenon at the ecosystem level because the defoliating insect, its parasites and predators, and the conifers (spruce and balsam fir that often grow in pure stands) are strongly coupled or coevolved. As the forest biomass increases, the large, older trees are vulnerable to a buildup of budworm caterpillars, and many trees are killed by successive defoliation. The death and decomposition of trees and the insect frass and feces return nutrients to the forest floor. Young trees less susceptible to attack are released from shade suppression and grow rapidly, filling in the canopy in a few years. During this time, parasites and bird predators of the insect combine to reduce ecological density of the budworm. In the long view, the budworm is an integrated component that periodically rejuvenates the conifer ecosystem and not the catastrophe it seems if one only observes the dead and dying trees at the peak of the cycle. In fact, after a study of the role of a bark beetle in lodgepole mountain forests, Peterman (1978) concluded that the beetle creates forests that are more useful to humans by thinning overcrowded stands that are of little value for lumber, wildlife, or recreation. Peterman would have us view the beetle as a management tool rather than a pest, and he suggests that in isolated forests it would be better to let the periodic outbreaks run their course. Such a view, of course, contradicts the traditional management of forest pests, which is to attempt to control insects only when density becomes high enough to kill trees outright. An alternate strategy might be simply to harvest old growth trees before the beetles or budworms take them out. This becomes practical now that models have been developed that can predict when an outbreak of insects is likely. Periodic storms perform a function similar to that of the insects: by blowing down old and crowded stands of high mountain forest, storms create a patchwork of young and older forests that are constantly being shifted about on the mountainside (see Chapter 8, Figure 8–3).

In all the cases described so far, the interacting populations of animals and plants have evolved together over long periods and thus have developed strategies not only for survival but also for mutual benefit.

Probably the most famous of all oscillations of an insect population are those involving locusts or grasshoppers. In Eurasia, records of outbreaks of the migratory locust (*Locusta migratoria*) go back to antiquity (Carpenter, 1940). The locusts live in desert or semiarid country and in most years are nonmigratory, eat no crops, and attract no attention. At intervals, however, population density increases to an enormous extent. The locusts actually become morphologically different (for example, they develop longer wings) under effects of crowding* and emigrate into cultivated lands, consuming everything in their path. Uvarov (1957) points out that human activities, such as shifting cultivation and overgrazing by cattle, tend to increase rather than decrease the chance of an outbreak because a patchwork or mosaic of vegetation and bare ground (in which the locusts lay eggs) is favorable for an exponential buildup of population. Here, a population explosion is generated by combined instability and simplicity in the environment. As with the lemming, probably not every population maximum is accompanied by an emigration; therefore, the frequency of the plagues does not necessarily represent true periodicity of the oscillations in density. Even so, outbreaks have been recorded at least once every 40 years between 1695 and 1895. Waloff (1966) presents a more recent historical survey of the upsurges and decrements of the desert locust.

An interesting type of "predator-prey" oscillation involves plants and animals. Seed production in conifers is often cyclic, and seed-eating birds and other animals show corresponding oscillations.

Regular cycles of abundance, such as those of lynx–hare and lemming, which seem almost paradoxical in view of nature's notorious irregularity, remain a puzzle. Two striking features of these oscillations are that (1) they are most pronounced in the less complex ecosystems of northern regions, as is evident from the preceding examples, and (2) although peaks of abundance may occur simultaneously over wide areas, peaks in the same species in different regions do not always coincide by any means. Theories that have been advanced to explain regular cycles range through all the levels-of-organization hierarchy and can be grouped under several headings: (1) meteorological theories, (2) random fluctuations theory, (3) population interaction theories, and (4) trophic-level interaction theories. A very brief summary of each theory will be presented in the next several paragraphs.

Attempts to relate these regular oscillations to climatic factors have so far been unsuccessful, although their synchrony and prominence in northern latitudes would seem to suggest some cyclic event outside the local ecosystem. At one time, the cycle of sunspots, which cause major weather changes, was considered by many an adequate explanation for the cycle of the lynx's abundance and scarcity and for other ten-year cycles. However, MacLulich (1937) and others have demonstrated that there is actually no correlation. So far, no wide-

* Solitary and migratory forms occur in several species of locusts and often were described as separate species before their true relations were known.

spread climatic periodicities that would fit the three- to four-year interval have been demonstrated.

Palmgren (1949) and Cole (1951, 1954) have suggested that what appear to be regular oscillations could result from random variations in the complex biotic and abiotic population environment. If this hypothesis is so, no one factor can be singled out as more important than numerous other factors. However, Keith (1963) made a detailed statistical analysis of the northern bird and mammal cycles and concluded that the ten-year cycle is nonrandom, even though it would be difficult to prove that the shorter cycles might not be due to random fluctuations.

If climatic factors, random or otherwise, should prove not to be the major cause of the violent oscillations, then causes within the populations themselves ("intrinsic factors") would have to be sought. Some evidence exists of possible mechanisms that might operate in conjunction with weather or other changes in physical factors.

Building on Hans Selye's medical theory of stress (the general adaptation syndrome), Christian and co-workers (see Christian, 1950, 1961, and 1963; Christian and Davis, 1964) have amassed considerable evidence from both the field and the laboratory to show that crowding in higher vertebrates results in enlarged adrenal glands. This enlargement is symptomatic of shifts in the neural-endocrine balance that, in turn, cause changes in behavior, in reproductive potential, and in resistance to disease or other stress. Such changes often combine to cause a precipitous decline in population density. For example, snowshoe hares at the peak of density often die suddenly from "shock disease," which has been shown to be associated with enlarged adrenal glands and other evidence of endocrine imbalance.

Chitty (1960, 1967) suggests that genetic shifts account for differences in aggressive behavior and survival that are observed at different phases of the cycle for the vole, a situation similar to that of the tent caterpillar. Such adaptation syndromes certainly dampen oscillation.

A fourth group of theories rely on the idea that cycles of abundance are intrinsic at the ecosystem level rather than at the population level. Certainly, density changes that range over several orders of magnitude must involve not only secondary trophic levels such as predators and prey but also the primary plant-herbivore interactions. Holling's hypothesis on budworms, as described previously, is an example. Another example is the nutrient-recovery hypothesis, proposed to explain microtine cycles in the tundra (Schultz, 1964, 1969; Pitelka, 1964, 1973). According to this hypothesis, which is supported by data from mineral cycling studies, heavy grazing by lemmings during the peak year ties up and reduces the availability of mineral nutrients (especially phosphorus) the following year, so that the food of the lemmings is low in nutritional quality. Growth and survival of the young are thus reduced greatly. During the third or fourth year, the recycling of nutrients is restored, plants recover, and the ecosystem can again support a high density of lemmings.

Large amplitude cycles of abundance are important not because they are particularly common in the world in general, but because a study of them reveals functions and interactions that probably have general application but are not so evident in populations whose density is less variable. The problem of cyclic oscillation in any specific case may well depend on determining whether one to several factors are primarily responsible, or whether causes are so numerous that it would be too difficult to untangle them, even though the total interaction may be understood to be what Cole (1957) calls "secondary simplicity," in that regularity may be "no greater than that encountered in a sequence of random numbers." One or several causes are certainly possible in simple ecosystems, whether experimental or natural; many causes may be more likely in complex ecosystems.

Having considered a number of interesting specifics, we can now consider the more general problem of population regulation.

6. Density-independent and Density-dependent Action in Population Control

Statement

In low-diversity, physically stressed ecosystems, or in those subject to irregular or unpredictable extrinsic perturbations, population size tends to be mainly influenced by physical factors such as weather, water currents, chemical limiting conditions, pollution, and so forth. In high-diversity ecosystems, in benign environments (low probability of periodic physical stress such as storms or fire), populations tend to be biologically controlled, and to some extent, at least, their density is self-regulated. Any factor, whether limiting or favorable (negative or positive) to a population, is (1) **density-independent** if the effect or action is independent of the size of the population, or (2) **density-dependent** if the effect on the population is a function of density. Density-dependent action is usually direct because it intensifies as the upper limit is approached. It may, however, also be inverse (decrease in intensity as density increases). Direct density-dependent factors act like governors on an engine (hence can be termed "density-governing") and for this reason are considered one of the chief agents in preventing overpopulation. Climatic factors often, but by no means always, act in a density-independent manner, whereas biotic factors (competition, parasites, pathogens, and so forth) often, but not always, act in a density-dependent manner.

Explanation

A general theory for regulating population comes logically from the discussion of biotic potential, growth form, and variation around the carrying-capacity level. Thus, the J-shaped growth form tends to occur when density-independent or extrinsic factors determine when growth slows down or stops. The sigmoid growth form, on the other hand, is density-dependent, since self-crowding and other intrinsic effects control population growth.

The behavior of any one population that one might wish to select for study depends on the kind of ecosystem of which that population is a part. Contrasting the "physically controlled" and the "biologically controlled" ecosystem (as we have done) is arbitrary and produces an oversimplified model, but it is a relevant approach, especially since human efforts during most of the industrial revolution have been directed toward replacing self-maintaining ecosystems with monocultures and stressed systems that require a lot of human care. As the cost (in energy as well as in money) of physical and chemical control has risen, as pest resistance to pesticides has increased, and as byproduct toxic chemicals in food, water, and air become more of a threat, biological control is beginning to be seriously reconsidered. Evidence for this is the increasing interest in a new frontier called "integrated pest management," which involves efforts to reestablish natural, density-dependent, ecosystem-level controls in agricultural and forest ecosystems, as was detailed at the end of Chapter 5.

The preceding section showed how physiological and genetic shifts, or the alternation of ecotypes in time, as it were, can dampen oscillations and hasten return of density to lower levels, after it overshoots carrying capacity. However, the question remains about how self-regulation at the population level evolves through natural selection at the individual level, if, indeed, it does. Wynne-Edwards (1962, 1965), whose books and review articles have been widely read and discussed, proposed two mechanisms that can stabilize density at a level lower than saturation: (1) **territoriality**, an exaggerated form of intraspecific competition that limits growth through "land-use" con-

Theories on regulating the population are often correlated with the environment of the theorist. Thus, ecologists working in stressful environments (such as arid regions) or with small organisms (such as insects or plankton, which have short life cycles, high biotic potentials, and high rates of metabolism per gram and, hence, a relatively small standing crop per unit space at any one time) have been impressed with the following: (1) the importance of the period of time when the rate of increase (r) is positive, (2) the importance of density-independent factors such as weather in determining the length of favorable periods, (3) the secondary rather than primary importance of self-limiting forces within the population, and (4) the general lack of stability in the density of any one species even when the ecosystem seems stable. These points have been emphasized in the well-known book by Andrewartha and Birch (1954).

Ecologists working in benign environments (such as English gardens, coral reefs, or tropical forests) or with larger organisms (such as birds, mammals, or forest trees, whose life cycles are longer and whose numbers and biomass more clearly reflect energy flow) have been impressed with the following: (1) the importance of density-dependent factors, especially self-limiting intraspecific competition (as in the sigmoid growth equation) and interspecific checks of various sorts (interspecific competition, parasites, and so forth), (2) the stability or at least consistency in the density patterns, and (3) the general importance of biological control mechanisms. Interestingly, persons working with confined populations of small organisms, such as bacteria or flour beetles in culture, are also impressed with these latter aspects, which is perhaps not surprising, since standing crop per unit space in cultures is greater than would be usual in nature. Likewise, it comes as no surprise that persons working with monoculture crops and forests are unimpressed with the efficiency of biological control.

For an analysis of density-dependent control theories, see the three pioneer papers by Nicholson (1954, 1957, and 1958), Lack (1954), and the Appendix in Lack (1966).

trol (to be discussed more fully in Section 9), and (2) **group behavior**, such as "peck-orders," "sexual dominance," and other behaviors that increase fitness of offspring but reduce their number. These mechanisms tend to enhance the quality of the individual's environment and reduce the probability of extinction that might result from overshooting the availability of resources. The importance of such social and behavioral traits is difficult to test experimentally and is much discussed, as indicated by the very large review book of Cohen et al. (1980). This subject is considered further in Chapter 8, Sections 5 and 6.

To summarize, density-independent factors of the environment tend to cause variations, sometimes drastic, in population density and to cause a shifting of upper asymptotic or carrying-capacity levels. Density-dependent natality and mortality, however, tend to maintain a population in a steady state or to hasten the return to such a level. Density-independent factors of the environment have a greater role in physically stressed ecosystems; density-dependent natality and mortality become more important as extrinsic stress is reduced. As in a smoothly functioning cybernetic system, additional negative feedback control is provided by interactions (both phenotypic and genetic) between populations of different species that are linked together in food chains or by other important ecological relationships.

Examples

Severe storms, sudden drops in temperature, and other drastic changes in physical factors generally provide the most clear-cut examples of density-independent action. In a three-year study of a snail (*Acmaea*) that lives on rocks in the intertidal zone, Frank (1965) found that most changes in the population were density-dependent, except for mortality after severe winter frosts when portions of the rock surface crumble away, removing snails regardless of the number present. A good case of "perfect" density-dependence has already been presented in the growth of yeast (Figure 6–12). As indicated in the discussion, linear density-dependence is unlikely in the open systems of nature. Varley's (1947) intensive study of the knapweed gallfly shows that the action of a major parasitic insect, *Eurytoma curta*, can be density-dependent, since it killed a much greater percentage as well as a larger total number of the host when the population of the host was high (Table 6–4). However, in another study, Varley and Edwards (1957) reported that when the "area of discovery" is low, as was the case with the parasitic wasp *Mormoniella*, the action of the parasite on its dipteran hosts was not necessarily density-dependent. Thus, differences in behavior can be important. Holling (1965, 1966) has incorporated behavioral characteristics into

Table 6–4 **Density-dependent Action of a Parasitic Hymenopteran on Larvae of a Host Gallfly***

Population Level	Larval Population at Beginning of Season (no./m²)	Larvae Killed by Parasite (no./m²)	Percent Population Killed by Parasite
Year of low population	43	6	14
Year of high population	148	66	45

* After Varley, 1947

a series of mathematical models that predict how effectively a given insect parasite will control the insect host at different densities.

The whole subject of population regulation can be effectively summarized by the graphic model of Figure 6–18. This graph is based on the dynamics of a particular species population (an Australian psyllid insect living on eucalyptus trees) but illustrates many of the principles discussed so far. Normally, the population is stabilized at a low level (Level 1, Time I, in Figure 6–18), which is well below the level that is reached when all the food and space resources are exploited. Regulation at a level lower than saturation is accomplished by prevailing weather, density-independent predation and parasitism on the nymphal stages, and density-dependent predation by birds on adults. Occasionally, the stabilizing process fails, usually because of unusually low temperatures that reduce the percentage of parasitism. The density rises above (i.e., "escapes from") the controlled level (Time II in Figure 6–18). A rapid, irruptive, or J-shaped growth then occurs (Time III) because (1) parasitism on nymphs becomes ineffective because of a rapid increase in hyperparasites (parasites that parasitize the psyllid parasites), and (2) bird predators cannot increase as rapidly as the insects (recall the vast difference in the natality of insects and birds). The temporarily unlimited growth is halted at Level 2 as the nymphs run out of food and as the adults run out of oviposition sites. The population then "crashes" as the trees are defoliated and as the predation by birds, ants, and parasites begins to catch up (Time IV). If psyllid numbers are reduced below Level 1,

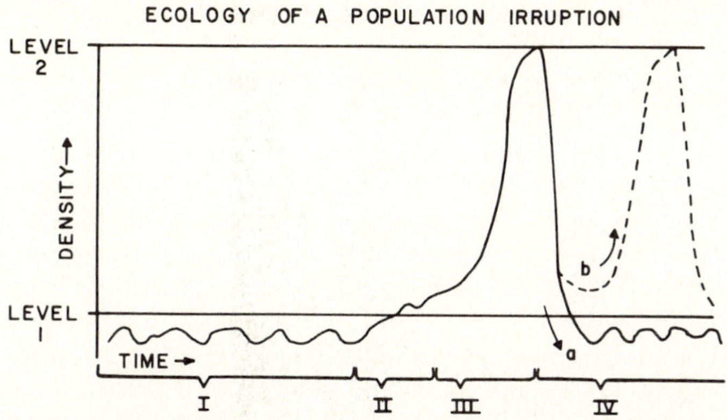

Figure 6–18 Population dynamics of the psyllid insect *Cardiaspina albitextura*, which feeds on *Eucalyptus blakelyi* trees. The population density normally remains stabilized at a low level by the combined density-independent and density-dependent action of weather, parasites, and predators; but occasionally the population "escapes" from natural control and "irrupts" to a high density, resulting in extensive defoliation of trees. Density Level 1 is that below which natural control operates effectively. Level 2 is the density at which food supply and oviposition sites become strongly limiting. (Redrawn from Clark, 1964; see also Clark et al., 1967, page 158.)

the population again comes under control and is likely to remain low for some years, as shown by *a* in Figure 6–18. However, as shown in *b*, another irruption may be generated if density does not fall below the control level.

This model also reiterates some of the difficulties already discussed in the practical control of insect pest irruptions, whose frequency is increased by human disturbances, which act like the "unusual" weather by hampering the natural control mechanisms. Often, the outbreak develops so quickly that it is not detected until the exponential growth is well underway and it is too late for treatment to be effective. Obviously, control measures might help prevent a second outbreak if the treatment was specific for the target insect. Application of a broad-spectrum insecticide during Time IV, however, could do more harm than good, since parasites and predators would also be killed, thus increasing, rather than decreasing, the possibility of another outbreak. In many cases, no treatment at all would be preferable to indiscriminate spraying of insecticide, done without knowledge of the phase of the population cycle or the condition of other populations involved in the natural control mechanism.

7. Population Structure: Internal Distribution Patterns (Dispersion)*

Statement

Individuals in a population may be distributed according to three broad patterns (Figure 6–19): (1) random, (2) uniform (more regular than random), and (3) clumped (irregular, nonrandom). Random distribution occurs where the environment is very uniform and where there is no tendency to aggregate. Uniform distribution may occur where competition between individuals is severe or where there is positive antagonism that promotes even spacing. Clumping of varying degrees represents by far the most common pattern when individuals are considered. However, if individuals of a population tend to form groups of a certain size—for example, pairs in animals, or vegetative clones in plants—the distribution of the *groups* may be more prone to be random or even uniform. Determination of the type of distribution is important in the selection of methods of sampling.

Explanation and Examples

The three patterns of distribution or intrapopulation dispersion are shown simply in Figure 6–19. Each rectangle contains approximately the same number of individuals. In the case of clumped distribution (C), the groups could be of the same or varying size (as shown), and they could be randomly distributed (as shown), uniformly distributed, or themselves aggregated or clumped, with large unoccupied spaces. In other words, there are five types of distribution: (1) uniform, (2) random, (3) random clumped, (4) uniform clumped, and (5) aggregated clumped. All these types are undoubtedly found in nature. From examining Figure 6–19 one may see that a small sample drawn from the three populations could obviously yield very different results. A small sample from a population with a clumped distribution would tend to give either too high or too low a density when the number in the sample is multiplied to obtain the total population. Thus, "clumped" populations require larger and more carefully planned sample techniques than nonclumped ones.

Random distribution follows the "normal" or bell-shaped curve on which standard statistical methods are based. This type of dis-

* Dispersion as used in statistics refers to distribution of items around the mean, and more broadly to the pattern of distribution of items within a population. Dispersion should not be confused with dispersal, which refers to exchanges between populations or areas (see page 328).

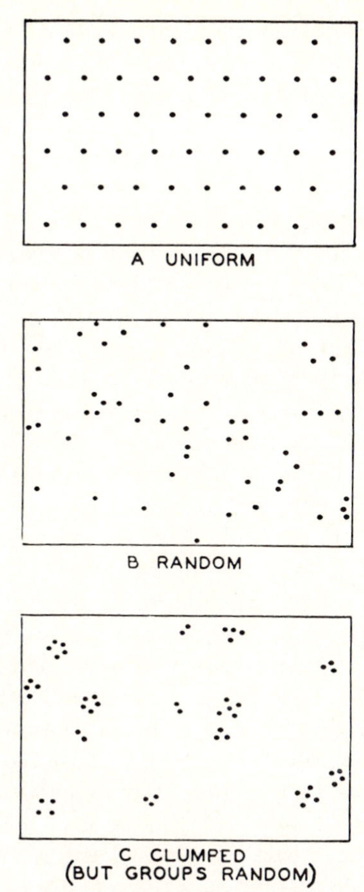

Figure 6–19 Three basic patterns of the distribution of individuals, pairs, or other units in a population.

tribution is to be expected in nature when many small factors are acting together on the population. (Recall discussion in Section 5 of a possible random basis for cyclic oscillations.) When a few major factors are dominating, as is the usual case (recall principle of limiting factors), and when there is such a strong tendency for plants and animals to aggregate for (or because of) reproductive and other purposes, there is little reason to expect a completely random distribution in nature. However, nonrandom or "contagious" distributions of organisms are sometimes found to be made up of intermingled random distributions of groups containing various numbers of individuals (as in Figure 6–19C), or the groups could turn out to be uniformly distributed (or at least more regular than random). In other words, to take an extreme case, it would be much better to determine the number of ant colonies (using the colony as the population unit) by a sample method, then determine the number of individuals per colony, than it would be to try to measure the number of individuals directly by random samples.

Several methods have been suggested that may be used to determine the type of spacing and degree of clumping between individuals in a population (where it is not self-evident), but there is much that must still be done in solving this important problem. Two methods may be mentioned as examples. One method is to compare the actual frequency of occurrence of different-sized groups obtained in a series of samples with a Poisson series that gives the frequency with which groups of 0, 1, 2, 3, 4 . . . n individuals will be encountered together if the distribution is random. Thus, if the occurrence of small-sized groups (including blanks) and large-sized groups is more frequent and the occurrence of middle-sized groups less frequent than expected, the distribution is clumped. The opposite is found in uniform distribution. Statistical tests can be used to determine whether the observed deviation from the Poisson curve is significant, but this general method has the disadvantage that sample size may influence the results. An example of the use of the Poisson method to test for random distribution in spiders is shown in Table 6–5. In all but three of 11 quadrats, spiders were randomly distributed. The nonrandom distributions occurred in quadrats in which the vegetation was least uniform.

A general property of random distributions is that the variance (V) equals the mean (m); variance greater than the mean indicates a clumped distribution, and variance less than mean, a uniform (regular) pattern. Thus, in a random distribution,

$$V/m = 1; \text{ standard error} = 2/n - 1$$

If, on the application of standard significance tests, the variance/mean ratio is found to be significantly greater than 1, the distribution is clumped; if it is significantly less, the distribution is regular; and if it is not different from 1, the distribution is random. This approach is also illustrated in Table 6–5.

Another method, suggested by Dice (1952), involves actually measuring the distance between individuals in some standardized way. When the square root of the distance is plotted against frequency, the shape of the resulting frequency polygon indicates the distribution pattern. A symmetrical polygon (a normal bell-shaped curve, in other words) indicates random distribution; a polygon skewed to the right, a uniform distribution; and one skewed to the

Table 6–5 **Random and Nonrandom Spatial Distribution of Spiders and Clams**

A. Number and Distribution of Wolf Spiders (Lycosidae) on 0.1 Hectare Quadrats in an Old-Field Habitat*

Species	Quadrat	Number per Quadrat	Chi Square from Poisson Distribution
Lycosa timuqua	1	31	8.90†
	2	19	9.58†
	3	15	5.51
	4	16	0.09
	5	45	0.78
	6	134	1.14
L. carolinensis	2	16	0.09
	5	23	4.04
	6	15	0.05
L. rabida	3	70	17.30†
	4	16	0.09

B. Mean, Variance, and Spatial Distribution of Two Species of Small Clams on an Intertidal Mudflat in Connecticut‡

Species and Age	Mean	Variance	Variance-Mean Ratio§
Mulinia lateralis			
All ages	0.27	0.26	Random
Gemma gemma			
All ages	5.75	11.83	Clumped
1st year	4.43	7.72	Clumped
2nd year	1.41	1.66	Random

* After Kuenzler, 1958.

† Significant at 5 percent level, i.e., nonrandom; in all other quadrats the distribution was random.

‡ After Jackson, 1968.

§ Where not significantly different (5 percent confidence level) from 1, random distribution indicated; significantly greater than 1, clumped (aggregated) distribution indicated.

left, a clumped distribution (individuals coming closer together than expected). A numerical measure of the degree of "skewness" may be computed. This method, of course, would be most applicable to plants or stationary animals, but it could be used to determine spacing between animal colonies or domiciles (fox dens, rodent burrows, bird nests, and so forth).

Flour beetle larvae were usually distributed randomly throughout their very uniform environment, since observed distribution corresponded with the Poisson distribution (Park, 1934). Lone parasites or predators, such as the species of spiders in Table 6–5, sometimes show a random distribution (and they often engage in random searching behavior for their hosts or prey). Jackson (1968) has reported that individuals of the clam *Mulinia lateralis* are randomly distributed on an intertidal mudflat, as were second-year individuals of *Gemma gemma*, but not first-year individuals, nor the total population of *Gemma*. They were clumped because of the ovoviviparous reproduction (young retained in the body of the female during larval development). Results of this study are shown in Table 6–5B. Note that variance is close to the mean in the randomly distributed population but significantly larger in the clumped population. The mudflat environment is very homogeneous, and interspecific competition is not severe—two facts favoring random dispersion.

Forest trees that have reached sufficient height to form a part of the forest crown may show a uniform distribution because competition for sunlight is so great that the trees tend to be spaced at intervals "more regular than random." A cornfield, orchard, or pine plantation, of course, would be a better example. Desert shrubs often are very regularly spaced, almost as if planted in rows, apparently because of the intense competition (which may include antibiotics) in the low-moisture environment. A similar more regular than random pattern may occur in territorial animals (Section 9).

Examples of clumped distributions and patchy environments are to be found everywhere. Of several forest floor invertebrates studied by Cole (1946, 1946a), only spiders showed a random distribution. Cole reported on another study in which only four of 44 plants showed a random distribution. All the rest showed varying degrees of clumping.

For more of the statistics of spatial distribution of units within a population, see Skellum (1952), Goodall (1970), and Pielou (1975).

8. Population Structure: Aggregation, Allee's Principle, and Refuging

Statement

As noted in the previous section, varying degrees of clumping are characteristic of the internal structure of most populations at one time or another. Such clumping is the result of individuals aggregating (1) in response to local habitat differences, (2) in response to daily and seasonal weather changes, (3) because of reproductive processes, or (4) because of social attractions (in higher animals). Aggregation may increase competition between individuals for nutrients, food, or space, but this is often more than counterbalanced by increased survival of the group because of its ability to defend itself or find resources or to modify microclimate or microhabitat. The degree of aggregation, as well as the overall density, that results in optimum population growth and survival varies with species and conditions; therefore, undercrowding (or lack of aggregation), as well as overcrowding, may be limiting. This is Allee's principle.

A special type of aggregation has been called "refuging," in which large, socially organized groups of animals establish themselves in a favorable central place from which they disperse and return regularly to satisfy their needs for food or other energy. Some of the most successfully adaptive animals on earth, including starlings and humans, employ this strategy.

Explanation and Examples

In plants, aggregation may occur in response to the first three factors listed in the Statement. In higher animals, spectacular aggregations may be the result of all four factors, but especially social behavior—illustrated, for example, by the herds of reindeer or caribou in the very uneven arctic habitat or the great migratory flocks of birds or herds of antelope on the East African savanna that move from one grazing area to another, thus avoiding overgrazing any one part of the range.

In plants in general and probably in some of the lower animal groups, it is a well-defined ecological principle that aggregation is inversely related to the mobility of disseminules (seeds, spores, and so on), as was brought out in Weaver and Clements' (1929) pioneer plant ecology textbook. In old-fields cedars, persimmons, and other plants with nonmobile seeds are nearly always clumped near a parent or along fences and other places where birds or other animals have deposited the seeds in groups. On the other hand, ragweeds and grasses, and even pine trees, which have light seeds widely distributed by the wind, are by comparison much more evenly distributed over the old-fields.

Group survival value is an important characteristic that may result from aggregation. A group of plants may be able to withstand the action of wind better than isolated individuals or be able to reduce water loss more effectively. With green plants, however, the deleterious effects of competition for light and nutrients generally soon overbalance the advantages of the group. The most marked group survival values are to be found in animals. Allee (1931, 1938, 1951) conducted many experiments in this field and summarized the extensive writings on the subject. He found, for example, that groups of fish could withstand a given dose of poison introduced into the water much better than could isolated individuals. Isolated individuals were more resistant to poison when placed in water formerly occupied by a group of fish than when placed in water not so "biologically conditioned"; in the former case, mucus and other secretions aided in counteracting the poisons, thus revealing something of the mechanism of group action in this case. Bees provide another example of group survival value; a hive or cluster of bees can generate and retain enough heat in the mass for survival of all the individuals at temperatures low enough to kill all the bees if each were isolated. Colonial birds often fail to reproduce successfully when the colony size becomes small (Darling, 1938). Allee points out that these types of primitive cooperation (protocooperation) found even in very primitive phyla are the beginning of social organization, which shows varying degrees of development in the animal kingdom, culminating in the group behavior of human beings. Allee's principle is diagrammed in Figure 6–20.

Actual social aggregations as seen in the social insects and vertebrates (as contrasted with passive aggregation in response to some common environmental factor) have a definite organization involving social hierarchies and individual specializations. A social hierarchy

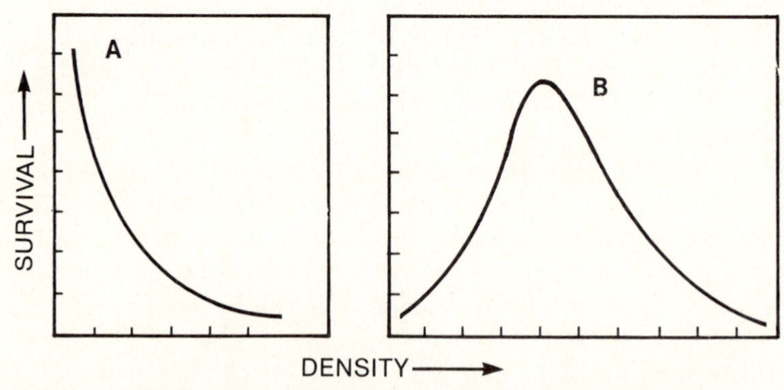

Figure 6–20 Illustration of Allee's principle. In some populations, growth and survival is greatest when population size is small (A), whereas in others, intraspecific protocooperation results in an intermediate-sized population being the most favorable (B). In the latter instance, "undercrowding" is as detrimental as "overcrowding." (After Allee et al., 1949.)

may be a "peck-order" (so-called because the phenomenon was first described in chickens) with clearcut dominance-subordinance between individuals often in linear order (like a military chain of command from general to private), or it may be a more complicated pattern of leadership, dominance, and cooperation as occurs in well-knit groups of birds and insects that behave almost as a single unit. In such units there may be a definite leader, but as often as not no one individual actually leads; members of the group follow the individual that acts forthrightly as if it "knew what it was about." These kinds of social organizations benefit the population as a whole by preventing overgrowth.

Among higher animals, a very successful aggregation strategy has been called "refuging" by W. J. Hamilton III (Hamilton and Watt, 1970). Large numbers of individuals resort to a favorable central place or core, for example, a starling roost or large breeding colony of sea birds. From there, they forage within a large perimeter or life-support area, often daily. Aggregation at a central place is advantageous in ensuring a net energy gain by individuals when good central places are scarce. Disadvantages are stresses of excrement pollution and excessive trampling of vegetation or substrate at the central place. It is obvious that modern human societies are organized along these lines. Exploitation of fossil fuels has extended the dispersal of foraging areas to the far reaches of the earth, so cities and other central places have few energy and fuel constraints on the size of the refuging population. But trampling and pollution become increasingly limiting as population density increases.

The remarkable organizations of social insects are unique in their specialized roles. The most highly developed insect societies are found among the termites (Isoptera) and ants and bees (Hymenoptera). A division of labor is accomplished in the most specialized species by three castes: reproducers (queens and kings), workers, and soldiers. Each is morphologically specialized to perform the functions of reproduction, food gathering, and protection. As will be discussed in the next chapter, this kind of adaptation leads to group selection not only within a species but also in groups of closely linked species.

Allee's principle is very relevant to the human condition. Aggregation into cities and urban districts (the "refuging" strategy discussed previously) is obviously beneficial, but only up to a point (discussed in Chapter 2, Section 7, and in Chapter 3, Section 8, in connection with the law of diminishing returns). A plot of benefit (y-axis) against city size (x-axis) would theoretically have the same hump-backed shape as curve B in Figure 6–20, as does the "subsidy-stress" curve of Figure 3–5. Thus, cities as well as bee or termite colonies can get too big for their own good. Optimum size of the aggregation of social insects is determined by the trial and error of natural selection. Because the optimum size of cities cannot as yet be objectively determined, the cities must depopulate when their costs exceed their benefits. By ecological principles, it is a mistake to maintain or subsidize a city that has grown too large for its life support. In terms of social welfare, who is to say?

9. Population Structure: Isolation and Territoriality

Statement

Forces isolating or spacing individuals, pairs, or small groups in a population are perhaps not as widespread as those favoring aggregation, but these forces are nevertheless very important for enhancing fitness and possibly also regulating population. Isolation usually is the result of (1) competition between individuals for resources in short supply or (2) direct antagonism involving behavioral responses in higher animals and chemical isolating mechanisms (antibiotics and allelopathics) in plants, microorganisms, and lower animals. In both cases, a random or a uniform distribution may result, as outlined in Section 7, because close neighbors are eliminated or driven away. Individuals, pairs, or family groups of vertebrates and the higher invertebrates commonly restrict their activities to a definite area, called the **home range.** If this area is actively defended so that there is little or no overlap of space used by the antagonistic individuals, pairs, and so on, it is called a **territory.** Territoriality seems to be most pronounced in vertebrates and certain arthropods that have complicated reproductive behavior patterns involving nest building, egg laying, and care and protection of young.

Explanation and Examples

Just as aggregation may increase competition but has other advantages, so spacing of individuals in a population may reduce the competition for the necessities of life or provide the "privacy" necessary for complex reproductive cycles (as in birds), but perhaps at the expense of the advantages of cooperative group action. Presumably, the pattern that survives through evolution in a particular case depends on which one gives the greatest long-time advantage. In any event, both patterns are frequent in nature; in fact, some species populations alternate from one to the other. Robins, for example, isolate into territories during the breeding season and aggregate into flocks in the winter and thus obtain advantages from both arrangements. Again, different ages and sexes may show opposite patterns at the same time (adults isolate, young aggregate, for example).

The role of intraspecific competition and "chemical warfare" in bringing about spacing in forest trees and desert shrubs has already been noted in Section 6. Isolating mechanisms of these sorts are widespread among microorganisms and higher plants. We have not yet considered in detail the spacing that results from active antagonism in higher animals. Many animals isolate themselves and restrict their major activities to definite areas or home ranges, which may vary from a few square inches to the many square miles of the puma's range. Since home ranges often overlap, only partial spacing is achieved; territoriality achieves the ultimate in spacing (compare turtles and birds in Figure 6–21).

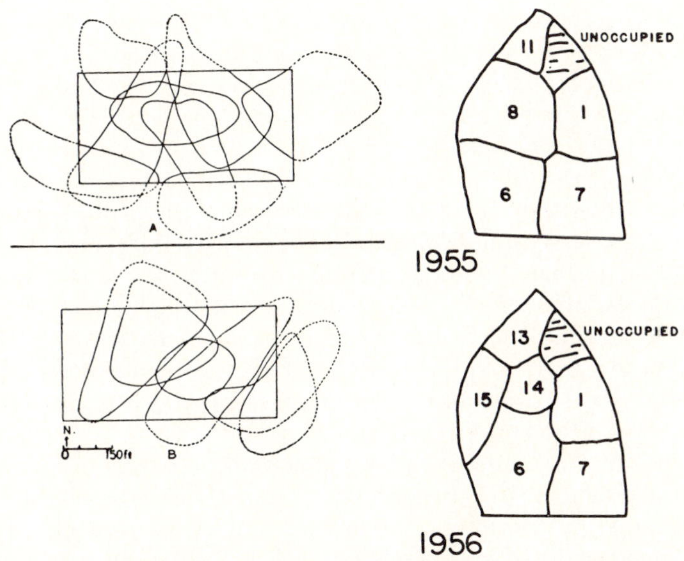

Figure 6–21 Home ranges and territories (i.e., defended home ranges) in various animals. *Left*. Home ranges of 15 box turtles (*Terrapene carolina*) occupying parts of a 5-acre plot; A, males; B, females. In this species there is no defense of the home ranges, which overlap. (From Stickel, 1950.) *Right*. Territories of song thrushes (*Turdus philomelos*) in Oxford Botanic Garden, England, in two successive years. Numbers identify banded (ringed) males of territorial pair; note that three individuals (1, 6, and 7) maintained the same territories both years, while two other individuals, holding territories in 1955, failed to return and were replaced by three new individuals. Generally, successful territorial holders keep the same area year after year as long as they are alive. (Redrawn from Lack, 1966.)

The term "territory," as defined in this section, was first introduced by Elliot Howard in his book *Territory in Bird Life*, published in 1920. Since then, most of the literature on the subject has dealt with birds. However, "isolation by antagonistic behavior" is widespread among other vertebrates and some arthropods, especially among species in which a parent or parents guard nests and young. Territories of birds have been classified by Nice (1941) into several basic types: (A) entire mating, feeding, and breeding area defended; (B) mating and nesting but not feeding area defended; (C) mating area only defended; (D) nest only defended; and (E) nonbreeding areas defended. In Type A, the defended area may be quite large, larger than needed for food supply of the pair and its young. For example, the tiny gnatcatcher (which weighs about 7 grams) establishes a territory averaging 4.6 acres but obtains all the food it needs in a much smaller area around the nest (Root, 1969). In most territorial behavior, actual fighting over boundaries is held to a minimum. Owners advertise their land or location in space by song or displays, and potential intruders generally avoid entering an established domain. Many birds and also fish and reptiles have conspicuous head, body, or appendage markings that can be displayed to intimidate intruders. In most migratory songbirds, males arrive on nesting grounds before the females and devote their time to establishing and advertising territories with loud songs. People always seem disappointed when they learn that the first robin in spring is singing to announce ownership of land and not to please the female.

The fact that the area defended by birds is often larger at the beginning of the nesting cycle than later when the demand for food is greatest, and the fact that many territorial species of birds, fish, and reptiles do not defend the feeding area at all, support the idea that reproductive isolation and control has greater survival value than the isolation of a food supply as such.

Territoriality certainly affects genetic fitness (probability of leaving descendants), since individuals (of territorial species) that cannot secure suitable territories do not breed. Whether territoriality functions to prevent overpopulation and has evolved for this reason, as so strongly argued by Wynne-Edwards (1962), is debatable. Jerram Brown (1969) summarizes the arguments against the population-limitation hypothesis, including the idea that the energy cost of defending an area larger than needed would not produce a selective advantage. Verner (1977), on the other hand, argues that it may be adaptive to occupy a space larger than dictated by immediate needs, since adequate resources for reproductive needs would be ensured should a drought or other harsh condition reduce food availability in the future. An experimental study by Riechert (1981) of a territorial species of desert spider (*Agelenopsis operta*) provides evidence for this view. Riechert found that territorial size was fixed (only so many spiders could occupy the experimental area), and territory size was adjusted to lows in availability of prey in times of greatest stringency. Accordingly, density would not increase beyond an upper limit set by numbers of available favorable territory sites no matter how much food was available in favorable times. Individuals unable to establish territories lost weight and eventually died, as shown in Figure 6–22. Territorial holders occupied the best sites and were most successful in producing young, especially under harsh conditions (unfavorable weather and scarce food). In this case, the potential of territoriality to limit population and to select the most fit individuals seemed to be realized. Part of the problem with the control hypothesis involves the question of group selection, a topic to be considered in Chapter 8.

Some other functions that have been suggested for territoriality include avoidance of predation or disease through spacing of individuals, favorable allocation and preservation of resources, and procurement of mates.

The extent to which humans are territorial by virtue of inherent behavior and the extent to which they can learn land-use control and

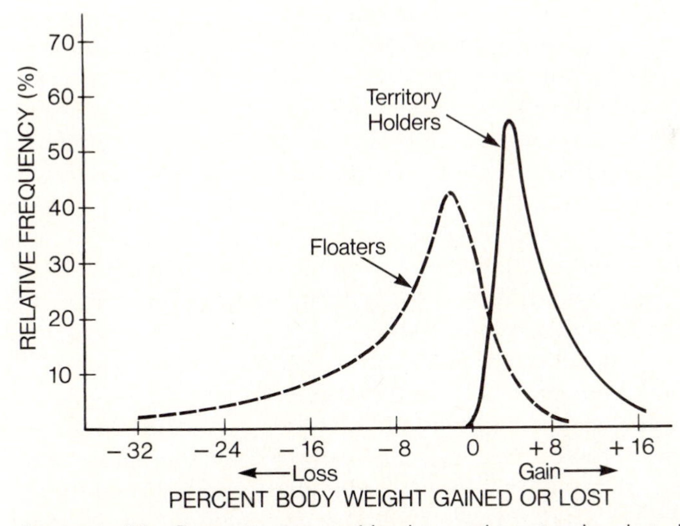

Figure 6–22 Fitness in terms of body weight gained or lost daily of territory-holding spiders as compared with individuals unable to establish and hold territories (floaters). In unfavorable seasons, floaters are at an even greater disadvantage; few survive to produce offspring. (Redrawn from Riechert, 1981.)

planning as safeguards against overpopulation are intriguing questions. Certainly, there are some territorial aspects of human behavior, such as the private-property imperative and laws and customs that make every person's home his or her castle, to be defended against intruders, with guns if necessary. In the widely read book *The Territorial Imperative* (1967), Robert Ardrey argues optimistically that humans are inherently territorial and will eventually resist crowding and thereby avoid the doomsday of overpopulation. Unfortunately, humankind adapts all too well to crowding, as seen in such places as Hong Kong (one of the most densely populated spots on earth).

10. Energy Partitioning and Optimization: r- and K-Selection

Statement

Paralleling the partitioning of energy between P (production) and R (respiration or maintenance) and the concept of net energy for an ecosystem as a whole (discussed in Chapter 3), individual organisms and their populations can grow or reproduce only if they can acquire more energy than is needed for maintenance. Maintenance energy consists of the resting or basal rate of metabolism (which can be measured in the laboratory and varies with size and kind of organism and temperature) plus a multiple of this to cover minimum activity for survival under field conditions. Such **existence energy** must be estimated by time-energy observations in the field, since it varies widely according to whether a species is sedentary or active. Additional or net energy required for reproduction, and therefore for survival of future generations, entails energy devoted to reproductive structures, mating activities, production of offspring (seeds, eggs, young), parental care, and so on. Through natural selection, organisms achieve as favorable a benefit-cost ratio of net energy/time as is possible. For autotrophs, this efficiency involves usable light (convertible to food) minus the energy required to maintain energy-capturing structures (leaves, for example) as a function of the time that light energy is available. What is critical for animals is the ratio of utilizable energy in food minus the energy cost of searching for and feeding on food items, and time required for searching and feeding. Optimization can be achieved in two basic ways: (1) minimizing time (by efficient searching or conversion, for example), or (2) maximizing net energy (by selecting large food items or easily convertible energy sources, for example), or both. Most optimization models indicate that the lower the absolute abundance of food (or other energy sources), the larger the habitat area foraged and the greater the range of food items that should be taken to optimize benefit-cost. However, extrinsic factors such as competition or cooperation with other species can alter this trend.

The ratio of reproductive energy to maintenance energy varies not only according to the size of organisms and life history patterns but also with population density and carrying capacity. In uncrowded environments, selection pressure favors species with a high reproductive potential (high ratio of reproductive to maintenance effort). In contrast, crowded conditions favor organisms with lower growth potential but better capabilities for utilizing and competing for scarce resources (greater energy investment in the maintenance and survival of the individual). These two modes are known as r-selection and K-selection, respectively (and species exhibiting them as r- and K-strategists), based on the r and K constants in growth equations (described in Section 4).

Explanation

As expressed by Cody (1966), partitioning or allocation of energy among various activities of an organism reflects balances between advantages and costs of each activity in producing a change in r_{max}, the intrinsic (genetically determined) rate of increase, to enhance future survival or fitness. The first consideration, of course, is survival and maintenance of the individual (the respiratory component) with additional, or net, energy allocated to growth and reproduction (the production component). Large organisms (like large cities) must allocate a larger portion of their metabolized energy input to maintenance than small organisms, which do not have so much structure to maintain. Natural selection, that uncompromising master forcing function, requires that all organisms find an optimum balance between energy spent on future survival and energy spent on survival in the present.

Figure 6–23 shows four hypothetical allocations of net energy between three major activities: (1) energy to cope with competition from other species striving for the resources, (2) energy expended to avoid being eaten (or grazed) by a predator, and (3) energy to produce offspring. Where competition and predation have a low impact, a large part of energy flow may then go to reproduction and production of offspring, as shown in A. Alternatively, competition avoidance or antipredator activities may take most of the available energy as shown in B and C. In D, all three demands receive approximately equal allocations. A, B, C, and D can represent four different species

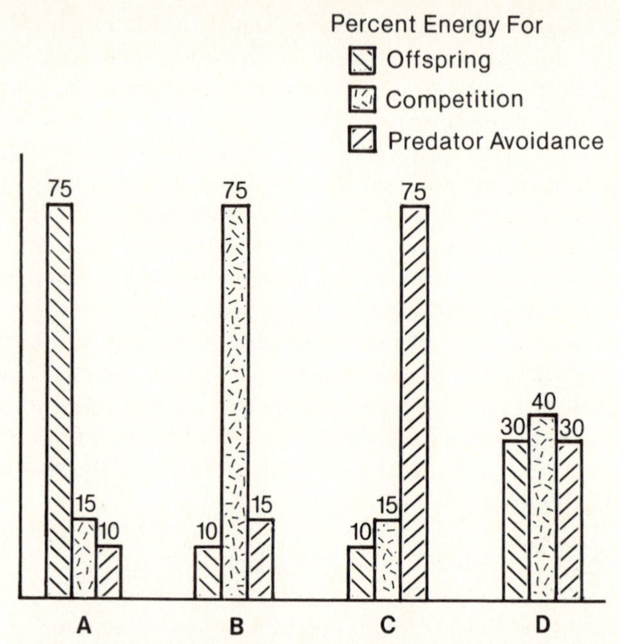

Figure 6–23 Hypothetical allocations of energy to three major activities necessary for survival in four contrasting situations where relative importance of each activity varies. (Modified from Cody, 1966.)

or four different communities where selection pressure produces the illustrated pattern in many species, or, as will be seen in Chapter 8, A represents a common situation in pioneer or colonizing stages of succession where *r*-selection predominates, while B through D are likely patterns in more mature stages where *K*-selection may predominate.

Schoener (1971), Cody (1974), and Pyke et al. (1977), in reviewing how energy partitioning and optimization can be analyzed, suggest that the problem is analogous to cost-benefit analysis in economics, with benefit being increased fitness and costs being energy and time required to ensure future reproductive output. A predator, for example, is under selective pressure to increase the ratio between net energy (utilizable energy minus energy cost of obtaining prey) and time required to search, pursue, and eat the prey. Increasing the energy available for reproduction can, in theory, be accomplished by (1) selecting larger or more nutritious prey or prey easier to catch or (2) reducing search and pursuit time and effort.

Several approaches have been suggested for determining optimum partitioning for a given species, population, or habitat situation. One approach involves observing how ecologically equivalent species maintain fitness through morphological, metabolic, and behavioral adaptations in similar habitats but in different geographical areas. If there are extensive similarities or convergences in fitness measurements, one can infer that selection has reached optimal solutions despite difference in history, time scale, and genetic origin. Cody's (1974) comparison of grassland birds in Kansas and Chile, as shown in Table 7–4, is an example.

Another more specific approach involves graphic "strategic analysis" (Levins, 1968) as illustrated by the graphs in Figure 6–24. A and B are foraging-strategy models for a hypothetical species faced with the problem of how many of six potential food items to use (A) or how many isolated feeding areas or "patches" to forage in (B). If only one food item among many available is sought, greater search effort per item is required, compared with the search effort for feeding on all six items, as shown by ΔS curve in Figure 6–24A. More pursuit effort is required (as harder to catch or smaller prey are sought), as shown by the ΔP curve. The optimum cost-benefit comes when the decreasing and increasing trend curves cross, at four prey items in the hypothetical case of Figure 6–24A. Interactions with other species or other environmental factors can shift the optimum in either direction. Competition with other species can force this hypothetical animal to

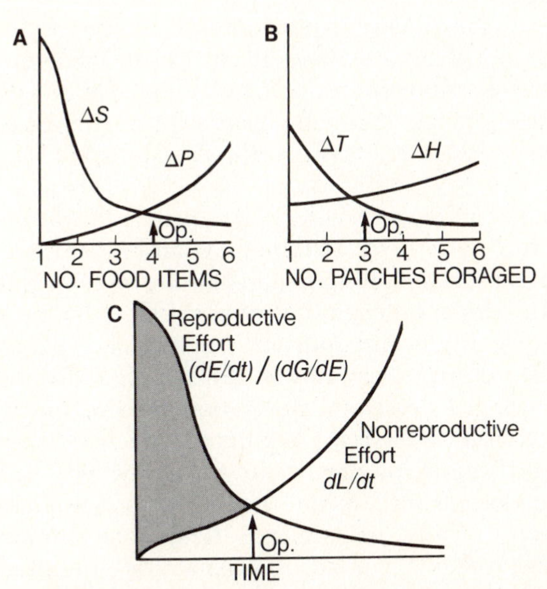

Figure 6–24 Optimization benefit-cost models. See text for explanation.

become a "specialist" and feed on only one item if it has a competitive advantage. Or such selection could be advantageous when food is abundant. Then again, conditions could dictate that becoming a "generalist" is a better strategy. An example of shifting food selection will be cited next.

In Figure 6–24B, hunting effort (ΔH) increases as more feeding areas are explored, but this is balanced by decrease in traveling time per unit food caught (ΔT). Again, the optimum is a compromise between opposing trends—three patches foraged in the situation graphed.

Figure 6–24C is a more general model in which reproductive output per unit energy decreases, and nonreproductive output increases (both monotonically), with time spent in procurement of energy (feeding). The shaded area represents the region where reproductive output is maximized, and optimum feeding time, again, is where the curves cross, providing a favorable balance between the two necessary allocations of energy.

As noted in the Statement, species with a high biotic potential (r) tend to be selected for in uncrowded or uncertain environments subject to periodic stresses (such as storms, droughts, and so on). Species with partition energy in favor of maintenance and enhanced competitive ability do better under K (saturation) densities or stable physical factors (low probability of severe disturbances). To put it another way, species exhibiting the J-shaped population growth form are good pioneers that can quickly exploit unused or recently accumulated resources, and they are resilient to perturbations. Slower-growing species and populations are better adapted to mature communities and are more resistant but less resilient to perturbations (recall discussion of resistant versus resilient stability in Chapter 2, Section 6).

A general model for r- and K-selection as proposed by MacArthur (1972) is shown in Figure 6–25. Although X_1 and X_2 in the diagram were designated as two competing genetic alleles, they can also represent species. In region A (to the left of point C), where density is low and food (or sunlight and nutrients, in the case of plants) is abundant, the faster-growing species or allele X_2 (dX/Xdt being the specific growth rate, as described in Section 2) wins out: there is r-selection. In region B (to the right of point C) species X_1 is growing faster than X_2 and thus wins out: there is K-selection. MacArthur notes that K-selection prevails in the relatively nonseasonal tropics, while r-selection prevails in seasonal temperate environments where population growth is marked by exponential growth followed by catastrophic declines.

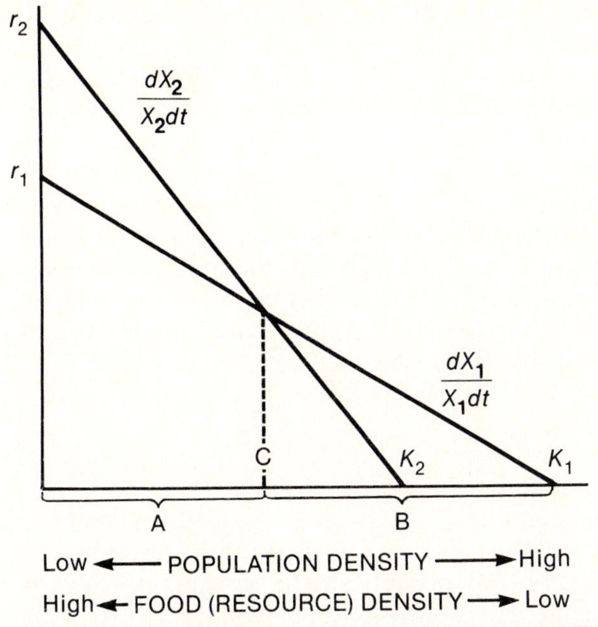

Figure 6–25 MacArthur's (1972) model of *r*- and *K*-selection. The rates of increase of two alleles (or species), X_1 and X_2, are portrayed as functions of population density and resource density. When density exceeds C and resource availability is low, X_1 increases faster than X_2, but X_2 increases faster and will "win out" under natural selection when density is less than C. Accordingly, A represents region of *r*-selection, and B, region of *K*-selection.

Clutch size (number of eggs or young per reproductive period) in birds seems not only to reflect mortality and survivorship but also to mirror *r*- and *K*-selection (see Lack, 1966, 1968; Cody, 1966; Hutchinson, 1978, Chapter 3). Opportunistic birds have a larger clutch size than do equilibrium species, as do temperate birds compared with tropical ones.

The *r*- and *K*-strategist designations can be faulted as an oversimplified classification, since many populations have variable or intermediate modes. However, Pianka (1970) finds an "apparent bimodality in relatively *r*- and relatively *K*-selected organisms in nature" related to body sizes and generation times. He argues that "an either/or strategy is usually superior to some kind of compromise." Schaffer (1976) has begun refining the concept by assessing the advantages and disadvantages of reproduction at different stages and times in the life cycle.

In his book *Evolution in Changing Environments*, Levins (1968) concludes that environmental uncertainty limits specialization in the evolution of species. Under unstable conditions, for purposes of selection it is better to be a generalist as well as have a high r_{max}. Also, under such conditions, communities can be only very loosely organized. Specialization and organization can increase to higher levels only if unpredictability of the environment is low (according to Levins). To what extent can groups of populations and communities, by their concerted action, reduce environmental uncertainty and thereby open the way for organization to proceed to a higher level, as sometimes (but not always) happens with human societies? That question remains to be answered.

Examples

Aspects of energy partitioning as related to trophic levels, size of organism, and endothermy-ectothermy, were discussed in Chapter 3, Section 4. (See especially the universal energy flow model of Figure 3–12.) As a partial review, Table 6–6 compares the allocation of assimilated energy (A in Figure 3–12) between production (P) (growth and reproduction) and respiration (R) (maintenance) in four species representing predators and herbivores, and vertebrates and ar-

Table 6–6 Allocation of Assimilated Energy Between Production (Growth and Reproduction) and Respiration (Maintenance)

	Percent Assimilated Energy to	
	Respiration	Production
Endothermic vertebrates		
Marsh wrens* (insectivore)	99	1
Cotton rat† (herbivore)	87	13
Exothermic arthropods		
Wolf spiders‡ (predator)	75	25
Pea aphid§ (herbivore)	42	58

* Kale, 1965.
† Randolph et al., 1977.
‡ Humphreys, 1978.
§ Randolph et al., 1975.

thropods. In general, predators (marsh wrens and hunting spiders) allocate more of their assimilated energy to maintenance (foraging for food, defending territories, and so on) than do herbivores (cotton rats and caterpillars). Large endotherms (warm-blooded vertebrates) likewise allocate a greater percent of A to R than do small exotherms (arthropods). McNeill and Lawton (1970) report that aquatic species have a higher respiratory cost than terrestrial species, but whether this is a basic difference and why it should be so are open questions.

Comparison of energy partitioning in hunting and web-building spiders is interesting. Since the web has a high protein content, silk formation comes at a high cost in energy, but many spiders recycle the silk by eating it as they rebuild the web, thereby cutting the cost. Peakall and Witt (1976) estimate that silk production in an orb-weaver spider, which recycles its web, requires about one-fourth work calories needed to build the web and keep it in repair. Total energy cost of the web is about one-half basal energy consumption, which is less than energy expended in hunting by some non-web builders. This is a possible lesson for humans: the species that builds expensive, labor-saving devices can reduce energy costs by recycling the materials.

The theory that predators optimize energy cost-benefit by varying the selection of the size of prey according to overall prey abundance has been tested and verified experimentally by Werner and Hall (1974). These investigators presented bluegill sunfish with different combinations of sizes and numbers of cladoceran prey and recorded which size of prey was selected. When the absolute food abundance was low, prey of all sizes were eaten as encountered. When abundance of prey was increased, the smaller-size classes were ignored, and the fish concentrated on the largest-size cladocera. The fish thus switched from feeding "generalists" to "specialists" as food abundance increased (and vice versa when it declined).

As an illustration of r- and K-selection, ragweed (*Ambrosia*), which grows in old-fields and other recently disturbed places, and *Dentaria laciniata*, a herbaceous plant that lives in the relatively stable forest floor, are compared in terms of seed production and peak reproductive effort (Table 6–7). The ragweed produces about 50 times as many seeds as *Dentaria* and allocates a much larger percent of its net energy to reproduction.

Goldenrods provide an example of a range of reproductive strategies between the extremes of r- and K-selection. In Figure 6–26, reproductive effort is plotted against biomass accumulation for six populations (representing four species) of goldenrods of the genus *Solidago*. Population 1, a species that grows in dry open fields or

Table 6–7 Contrasting Reproductive Strategies in Two Herbaceous Plants, and the Field and Forest Communities in Which They Live: An Example of r- and K-Selection*

Community	Average No. Seeds per Individual	Peak Reproductive Effort†
One-year field		
Ambrosia artemisifolia (ragweed)	1190	0.30
Mean of dominants	—	0.21
Rest of community	—	0.24
Forest		
Dentaria laciniata (toothwort or pepper-root)	24	0.01
Mean of dominants	—	0.05
Rest of community	—	0.05

* Data from Newell and Tramer, 1978.

† Dry weight reproductive structures/total dry weight.

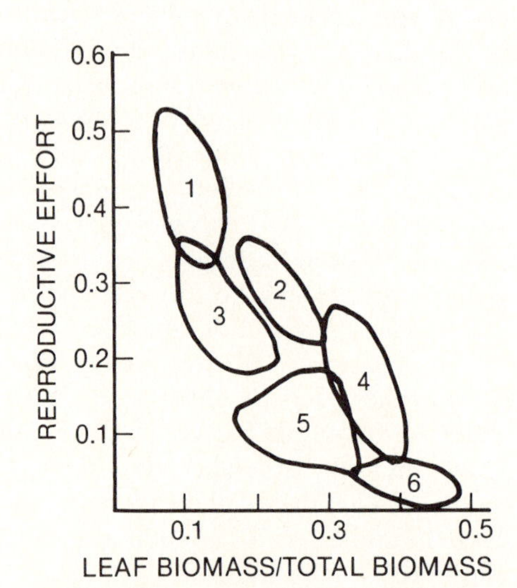

Figure 6–26 Reproductive effort (ratio of dry weight of reproductive tissues to total dry weight of aboveground tissue) plotted against ratio of leaf weight to total weight in six populations of four species of goldenrods (*Solidago*). Population 1 is a species that occurs in dry open fields or disturbed soil, while population 6 occurs in moist hardwood forests; the other populations occur in habitats intermediate in moisture and stability. (After Abrahamson and Gadgil, 1973.)

disturbed sites, maintains a low leaf biomass and allocates about 45 percent of net production to reproductive tissues. In contrast, population 6, which occurs in moist hardwood forests, puts more of its energy into leaves, with only 5 percent allocated to reproduction. The other populations occur in habitats intermediate in moisture and stability and have corresponding intermediate allocations.

Solbrig (1971) reports that *r*- and *K*-strategists can be found within the same species. The common dandelion (*Taraxacum officinale*) has several strains or variations that differ in the mix of genotypes controlling allocation of energy. One strain grows primarily in disturbed areas and produces more but smaller seeds that ripen earlier in the season, compared with another strain found in a less disturbed habitat that allocates more energy to leaves and stems and produces fewer seeds that ripen late. The latter strain shades out the more fecund variety when the two are grown together in good soil. Thus, strain 1 is a better colonizer of new ground and qualifies as an *r*-strategist; strain 2 is a better competitor, a *K*-strategist.

While uncertain or disturbed environments do favor *r*-selection, the *K*-strategist is by no means ruled out. In fire-adapted communities such as the California chaparral (Chapter 5, page 282) "resprout" plant species that allocate large energy reserves to underground parts are just as well adapted to survive periodic fires as are plants that put their future in seeds (see Carpenter and Recher, 1979).

For other examples of energy partitioning from the plant kingdom, see Harper (1977) and Gadgil and Solbrig (1972).

11. Integration: Life History Traits and Tactics

Statement

Selection pressure resulting from the impact of physical environments and biotic interactions shape patterns of life history so that each species evolves an adaptive combination of the population traits already considered in previous sections of this chapter. While each species' life history is unique, several basic life history tactics can be recognized, and the combination of traits characteristic of organisms living in specified circumstances can, to some extent, be predicted.

Explanation and Examples

Stearns (1976) lists four life history traits that are keys to survival tactics: (1) brood size (number of seeds, eggs, young, or other progeny); (2) size of young (at birth, hatching, or germination); (3) age distribution of reproductive effort; and (4) interaction of reproductive effort with adult mortality (especially ratio of juvenile to adult mortality). The following predictive theories have been presented by Gadgil and Bossert (1970), Stearns (1976), and others.

1. Where adult mortality exceeds juvenile mortality, the species should reproduce only once in a lifetime, and conversely, where juvenile mortality is higher, the organism should reproduce several times.
2. Brood size should maximize the number of young surviving to maturity averaged over the lifetime of the parent. Thus, a ground-nesting bird may require a clutch size of 20 eggs to ensure replacement, whereas a bird nesting in a cavity or other protected place will have a much smaller clutch size.
3. In expanding populations (the growth segment of the population growth curve), selection should minimize age at maturity (organisms will breed at an early age); in stable populations (at carrying capacity or *K* level), maturation should be delayed. This principle seems to hold for human populations; in fast-growing countries, reproduction begins at an early age, whereas in stable countries, people postpone having children to a later age.
4. Where there is risk of predation, or scarcity of resources, or both, size at birth should be large; conversely, size of young should decrease with increasing availability of resources and decreasing predation or competition pressure.
5. For growing populations, in general, not only is age of maturity minimized and reproduction concentrated early in life but also brood size should be increased and a large portion of energy flow partitioned to reproduction—a combination of traits that will be recognized as an *r*-selection tactic as described in Section 10. For stable populations, one expects the reverse of the combination of traits, or *K*-selection.
6. When resources are not strongly limiting, breeding begins at an early age.
7. Complex life histories with larval stages enable a species to exploit more than one habitat and niche.

Comparison of the floras of extreme deserts and moist tropical forests provides a good example of how a basic life history trait predominates throughout an ecosystem type. Annual plants predominate in extreme deserts where survival of a perennial plant would be very low because of long drought periods. Conversely, perennial life histories are favored in the tropical rain forest where intense competition and seed predation greatly reduce survival of seedlings. This case can be considered an example of the first predictive theory listed above.

The relationship between size of young at birth and predatory pressure is illustrated by a comparison of antelope and bears: antelopes produce larger young (in proportion to adult body size) than bears. Young antelope can keep up with the herd within a few hours after birth, whereas young bears are well protected from predation by their mothers.

Interest among population ecologists in life history strategies goes back to a pioneer paper by LaMont Cole (1954) entitled "The Population Consequences of Life History Phenomena," which is recommended reading, as is the reexamination of Cole's theories by Charnov and Schaffer (1973).

7
Populations in Communities

1. Types of Interaction Between Two Species

Statement

Theoretically, populations of two species may interact in basic ways that correspond to combinations of 0, +, and −, as follows: 00, −−, ++, +0, −0, and +−. Three of these combinations (++, −−, and +−) are commonly subdivided, resulting in nine important interactions that have been demonstrated [see Burkholder's adaptation (1952) of Haskel's (1949) classification scheme]. These are as follows (see Table 7–1): (1) **neutralism,** in which neither population is affected by association with the other; (2) **competition, mutual inhibition type,** in which both populations actively inhibit each other; (3) **competition, resource use type,** in which each population adversely affects the other indirectly in the struggle for resources in short supply; (4) **amensalism,** in which one population is inhibited and the other not affected; (5) **parasitism** and (6) **predation,** in which one population adversely affects the other by direct attack but nevertheless depends on the other; (7) **commensalism,** in which one popu-

Table 7–1 Analysis of Two-species Population Interactions

Type of Interaction*	Species† 1	Species† 2	General Nature of Interaction
1. Neutralism	0	0	Neither population affects the other
2. Competition: direct interference type	—	—	Direct inhibition of each species by the other
3. Competition: resource use type	—	—	Indirect inhibition when common resource is in short supply
4. Amensalism	—	0	Population 1 inhibited, 2 not affected
5. Parasitism	+	—	Population 1, the parasite, generally smaller than 2, the host
6. Predation (including herbivory)	+	—	Population 1, the predator, generally larger than 2, the prey
7. Commensalism	+	0	Population 1, the commensal, benefits, while 2, the host, is not affected
8. Protocooperation	+	+	Interaction favorable to both but not obligatory
9. Mutualism	+	+	Interaction favorable to both and obligatory

* Types 2 through 4 can be classed as "negative interactions," Types 7 through 9 as "positive interaction," and 5 and 6 as both.

† 0 indicates no significant interaction; + indicates growth, survival, or other population attribute benefited (positive term added to growth equation); − indicates population growth or other attribute inhibited (negative term added to growth equation).

lation is benefited, but the other is not affected; (8) **protocooperation,** in which both populations benefit by the association but relations are not obligatory; and (9) **mutualism,** in which growth and survival of both populations is benefited, and neither can survive under natural conditions without the other.

Three principles based on these categories are especially worthy of emphasis:

1. Negative interactions tend to predominate in pioneer communities or in disturbed conditions where *r*-selection counteracts high mortality.
2. In the evolution and development of ecosystems, negative interactions tend to be minimized in favor of positive symbiosis that enhances the survival of the interacting species.
3. Recent or new associations are more likely to develop severe negative co-actions than are older associations.

Explanation

One population often affects the growth or death rate of another population. Thus, the members of one population may eat members of the other population, compete for foods, excrete harmful wastes, or otherwise interfere with the other population. Likewise, populations may help one another, the interaction being either one-way or reciprocal. Interactions of these sorts fall into several definite categories as listed in the preceding Statement and shown in Table 7–1.

All these population interactions are likely to occur in the average community. For a given species pair, the type of interaction may change under different conditions or during successive stages in their life histories. Thus, two species might exhibit parasitism at one time, exhibit commensalism at another, and be completely neutral at still another time. Simplified communities and laboratory experiments allow ecologists to single out and study quantitatively the various interactions. Also, deductive mathematical models derived from such studies permit ecologists to analyze factors not ordinarily separable from the others.

Growth equation "models," as described in the preceding chapter, make definitions more precise, clarify thinking, and allow a determination of how factors operate in complex natural situations. If the growth of one population can be described by an equation, the influence of another population may be expressed by a term that modifies the growth of the first population. Various terms can be substituted according to the type of interaction. For example, in competition, the growth rate of each population is equal to the unlimited rate minus its own self-crowding effects (which increase as its population increases) minus the detrimental effects of the other species, N_2 (which also increase as the numbers of both species, N and N_2, increase), or

$$\frac{dN}{dt} = rN - \left(\frac{r}{K}N^2\right) - CN_2N$$

Growth rate	=	Unlimited rate	−	Self-crowding effects	−	Detrimental effects of the other species

This equation will be recognized as the logistic equation given on page 322, except for the addition of the last term, "minus detrimental effects of the other species." There are several possible results of this kind of interaction. If C is small for both species so that the interspecific depressing effects are less than intraspecific (self-limiting) effects, the growth rate and perhaps the final density of both species will be depressed slightly; but both species will probably be able to live together because the depressing interspecific effects will be less important than the competition within the species. Also, if the species exhibit exponential growth (with self-limiting factor absent from the equation), interspecific competition might provide the leveling-off function missing from the species' own growth form. However, if C is large, the species exerting the largest effect will eliminate its competitor or force it into another habitat. Thus, theoretically, species having similar requirements cannot live together because strong competition will likely develop, causing one of them to be eliminated. Our models point up some of the possibilities; how these possibilities actually work out will be discussed in the next section.

When both species of interacting populations have beneficial effects on each other instead of detrimental ones, a positive term is added to the growth equations. In such cases, both populations grow and prosper, reaching equilibrium levels that are mutually beneficial. If beneficial effects of the other population (the positive term in the equation) are necessary for growth and survival of both populations, the relation is known as mutualism. If, on the other hand, the beneficial effects only increase the size or growth rate of the population but are not necessary for growth or survival, the relationship comes under the heading of cooperation or protocooperation. (Since this cooperation is not necessarily the result of conscious or "intelligent" reasoning, the latter term is preferable.) In both protocooperation and mutualism, the outcome is similar: the growth of either population is less or zero without the presence of the other population. When an equilibrium level is reached, the two populations exist together stably, usually in a definite proportion.

Consideration of population interactions as shown in Table 7–1, or in terms of the growth equations, avoids the confusion that often results when terms and definitions alone are considered. Thus, the term "symbiosis" is sometimes used in the same sense as mutualism; sometimes, the term is used to cover commensalism and parasitism as well. Since symbiosis literally means "living together," the word is used in this book in its broad sense without regard to the exact nature of the relationship. The term "parasite" and the science of parasitology are generally considered to deal with any small organism that lives on or in another organism regardless of whether its effect is negative, positive, or neutral. Various nouns have been proposed for the same type of interaction, adding to the confusion. When relations are diagrammed, however, there is little doubt about the type of interaction being considered; the word or "label" then becomes secondary to the mechanism and its result.

Note that the word "harmful" was not used in describing negative interactions. Competition and predation decrease the growth rate of affected populations, but this does not necessarily mean that the interaction is harmful either to long-term survival or by evolutionary considerations. In fact, negative interactions can increase the rate of natural selection, resulting in new adaptations. Predators and parasites often benefit populations lacking self-regulation, because they may prevent overpopulation that might result in self-destruction.

2. Interspecific Competition and Coexistence

Statement

Competition in the broadest sense refers to the interaction of two organisms striving for the same thing. Interspecific competition is any interaction that adversely affects the growth and survival of two or more species populations. It can take two forms, as shown in Table 7–1. The tendency for competition to bring about an ecological separation of closely related or otherwise similar species is known as the **competitive exclusion principle.** Simultaneously, competition triggers many selective adaptations that enhance the coexistence of a diversity of organisms in a given area or community.

Explanation

Ecologists, geneticists, and evolutionists have written much about interspecific competition. Generally, the word "competition" is used in situations in which negative influences are due to a shortage of resources used by both species. The more direct reciprocal interferences, such as mutual predation or the secretion of harmful substances, can be placed in another category (Table 7–1), even though there is no generally accepted special term for these types of interactions. However, **allelopathy** (a term first introduced in Chapter 2, Section 4) is now generally used for the secretion of chemical messengers (also called allelochemic substances by Whittaker, 1970), which provide a competitive advantage for one species against another.

The competitive interaction often involves space, food or nutrients, light, waste materials, susceptibility to carnivores, disease, and so forth, and many other types of mutual interactions. The results of competition are of the greatest interest and have been much studied as one of the mechanisms of natural selection. Interspecific competition can result in equilibrium adjustments by two species or, if severe, in one species population replacing another or forcing the other to occupy another space or to use another food, whatever is the basis of competitive action. Closely related organisms having similar habits or morphologies often do not occur in the same places. If they do occur in the same places, they frequently use different resources or are active at different times. The explanation for the ecological separation of closely related (or otherwise similar) species has come to be known as **Gause's principle**, after the Russian biologist who in 1932 first observed such separation in experimental cultures (see Figure 7–2), or the **competitive exclusion principle** as designated by Harden in 1940.

Some of the most widely debated theoretical aspects of competition theory revolve around what have become known as the Lotka-Volterra equations (so called because the equations were proposed as models by Lotka and Volterra in separate publications in 1925–1926). They are a pair of differential equations similar to the one outlined in the preceding section. Such equations are useful for modeling predator-prey, parasite-host, competition, or other two-species interactions. In terms of competition within a limited space where each population has a definite K or equilibrium level, the simultaneous growth equations can be written in the following forms by using the logistic equation as a basis:

$$\frac{dN_1}{dt} = r_1N_1 \frac{K_1 - N_1 - \alpha N_2}{K_1}$$

$$dN_2 = r_2N_2 \frac{K_2 - N_2 - \beta N_1}{K_2}$$

where N_1 and N_2 are the numbers of species 1 and 2, respectively, α is the competition coefficient indicating the inhibitory effect of species 2 on 1, and β is the corresponding competition coefficient signifying the inhibition of 2 by 1. Competition equations of the more general form have been proposed by Wiegert (1974). In the absence of refuges or other modifying conditions, the species with the greatest inhibitory effect on the other will eliminate it from the space, unless coefficients are very small in relation to the ratios of saturation densities (K_1/K_2 and K_2/K_1).

To understand competition one must consider not only conditions and population attributes that may lead to competitive exclusion but also situations under which similar species coexist, since large numbers of species do share common resources in the open systems

Table 7–2 The Case for Competitive Exclusion in Populations of Flour Beetles (*Tribolium*)*

Climate	Temperature (°C)	Relative Humidity (%)	Results of Interspecific Competition (%)†	
			Tribolium castaneum Wins	*Tribolium confusum* Wins
Hot-wet	34	70	100	0
Hot-dry	34	30	10	90
Warm-wet	29	70	86	14
Warm-dry	29	30	13	87
Cool-wet	24	70	31	69
Cool-dry	24	30	0	100

* Data from Park, 1954.

† Twenty to 30 replicate experiments for each of the six conditions. Each species can survive at any of the climates when alone in the culture, but only one species survives when both are present in the culture. The percentages indicate the proportion of replicates in which each species remained after elimination of the other species.

of nature. Table 7–2 and Figure 7–1 present what might be called the **Tribolium-Trifolium** model, which includes an experimental demonstration of exclusion in paired species of beetles (*Tribolium*) and of coexistence in two species of clover (*Trifolium*).

One of the most thorough long-term experimental studies of interspecific competition was carried out in the laboratory of the late Dr. Thomas Park of the University of Chicago. Park, his students, and associates worked with flour beetles, especially those belonging to the genus *Tribolium*. These small beetles can complete their entire life history in a very simple and homogeneous habitat, namely, a jar of flour or wheat bran. The medium in this case is both food and habitat for larvae and adults. If fresh medium is added at regular intervals, a population of beetles can be maintained for a long time. In energy flow terminology, this experimental setup is a stabilized heterotrophic ecosystem in which imports of food energy balance respiratory losses.

The investigators found that when two different species of *Tribolium* were placed in this homogeneous little universe, one species invariably was eliminated sooner or later, while the other continued to thrive. One species always "wins," or to put it another way, two species of *Tribolium* cannot survive in this particular "one-habitat" microcosm. The relative number of individuals of each

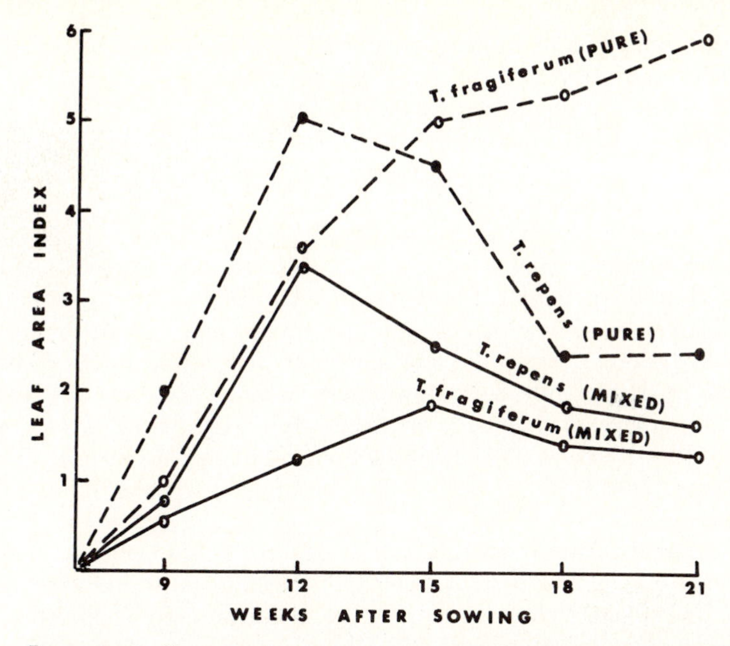

Figure 7–1 The case for coexistence in populations of clover (*Trifolium*). The graph shows population growth of two species of clover in pure (i.e., growing alone) and in mixed stands. Note that the two species in pure stands have a different growth form, reaching maturity at different times. Because of this and other differences, the two species are able to coexist in mixed stands but at reduced density even though they interfere with one another. Leaf area index, which is used as an index of biomass density, is the ratio of leaf surface area to soil surface area (cm^2 per cm^2). (Redrawn from Harper and Clatworthy, 1963.)

species originally placed in the culture (the stocking rate) does not affect the eventual outcome, but the "climate" imposed on the ecosystem does have a great effect on which species of the pair wins out. As shown in Table 7–2, one species (*T. castaneum*) always wins under conditions of high temperature and humidity, while the other (*T. confusum*) always wins under cool-dry conditions, even though either species can live indefinitely at any of the six climates, provided it is alone in the culture. Under intermediate conditions, each species has a certain probability of surviving (for example, the probability is 0.86 that *T. castaneum* would win under warm-wet conditions). Population attributes, as measured in one-species cultures, help explain some of the outcome of the competitive action. For example, the species with highest rate of increase (r) under the conditions of existence in question was usually found to win if the species difference in

r was rather large. If growth rates differed only moderately, the one with the highest rate did not always win. The presence of a virus in one population could easily tip the balance. Feener (1981) describes a case in which a parasitic fly alters the competitive balance between two species of ants. Also, genetic strains within the population differ greatly in competitive ability.

From the *Tribolium* model it is easy to construct conditions that could result in coexistence instead of exclusion. If the cultures were alternately placed in hot-wet and cool-dry conditions (to simulate seasonal weather changes), the advantage one species would have over the other might not continue long enough for either to become extinct. If the culture system were "open," and individuals of the dominant species were to emigrate (or be removed as by a predator) rapidly, the competitive interaction might be so reduced that both species could coexist. Many other circumstances would favor coexistence.

Some of the most interesting experiments in plant competition were reported by J. L. Harper and associates at the University College of North Wales (see Harper, 1961; Harper and Clatworthy, 1963; Clatworthy and Harper, 1962). The results of one of these studies, as shown in Figure 7–1, illustrate how a difference in growth form allows two species of clover to coexist in the same environment (same light, temperature, soil, and so forth). Of the two species, *Trifolium repens* grows faster and reaches a peak in leaf density sooner. However, *T. fragiferum* has longer petioles and higher leaves and can overtop the faster-growing species, especially after *T. repens* has passed its peak, and thus avoids being shaded out. In mixed stands, therefore, each species inhibits the other, but both can complete their life cycle and produce seed, even though each coexists at a reduced density (however, the combined density in mixed stands of two species was approximately equal to the density in pure stands). In this case, the two species, although competing strongly for light, could coexist because the morphology and the timing of growth maxima differed. Harper (1961) concludes that two species of plants can persist together if the populations are independently controlled by one or more of the following mechanisms: (1) different nutritional requirements (legume and nonlegume, for example), (2) different causes of mortality (differential sensitivity to grazing, for example), (3) sensitivity to different toxins, and (4) sensitivity to the same controlling factor (light, water, and so forth) at different times (for example, the clover just described).

Park (1954), Brian (1956), and Crombie (1947), all working with flour beetles, were among the first to distinguish between indirect or

exploitation competition and direct or **interference competition.** On the basis of studies on tadpoles, Steinwasher (1978) concludes that interference mechanisms replace exploitative ones when food becomes scarce. Hutchinson (1978) suggests that interference competition appears more frequently as we move up the phylogenetic "tree" of animal life from simple filter feeding protozoans and cladocerans, which usually compete in gathering food, to vertebrates with their elaborate behavior patterns of aggression, territoriality, and so on. He concludes (as did Slobodkin, 1964, on the basis of competition experiments with *Hydra*) that these two types of competition overlap, but that it is useful to distinguish between the two processes on theoretical grounds because competition equations constructed on the basis of exploitation graph out with rectilinear isoclines, whereas those based on interference have curvilinear isoclines. Thus, coexistence may be more likely in the latter case. For more on competition, see Chapter 4 in Huchinson (1978). For a review of the extensive experimental and genetic work on *Drosophila* populations, see Ayala (1972) and Richmond et al. (1975).

A general pattern emerging from all the literature on competition is that competition is most severe and competition exclusion most likely to occur in systems where immigration and emigration are absent or reduced, such as in laboratory cultures or on islands or other natural situations with substantial barriers to inputs and outputs. In the usual open systems of nature, the probability of coexistence is higher.

Examples

The results of one of Gause's original experiments are illustrated in Figure 7–2. This is a classic example of competitive exclusion. *Paramecium caudatum* and *Paramecium aurelia*, two closely related ciliate protozoans, when in separate cultures exhibited typical sigmoid population growth and maintained a constant population level in culture medium that was maintained constant with a fixed density of food items (bacteria that did not themselves multiply in the media and thus could be added at frequent intervals to keep food density constant). When both protozoans were placed in the same culture, however, *P. aurelia* alone survived after 16 days. Neither organism attacked the other or secreted harmful substances; *P. aurelia* populations simply had a more rapid growth rate (higher intrinsic rate of increase) and thus "out-competed" *P. caudatum* for the limited amount of food under the existing conditions (a clear case of exploitation competition). By contrast, both *Paramecium caudatum* and

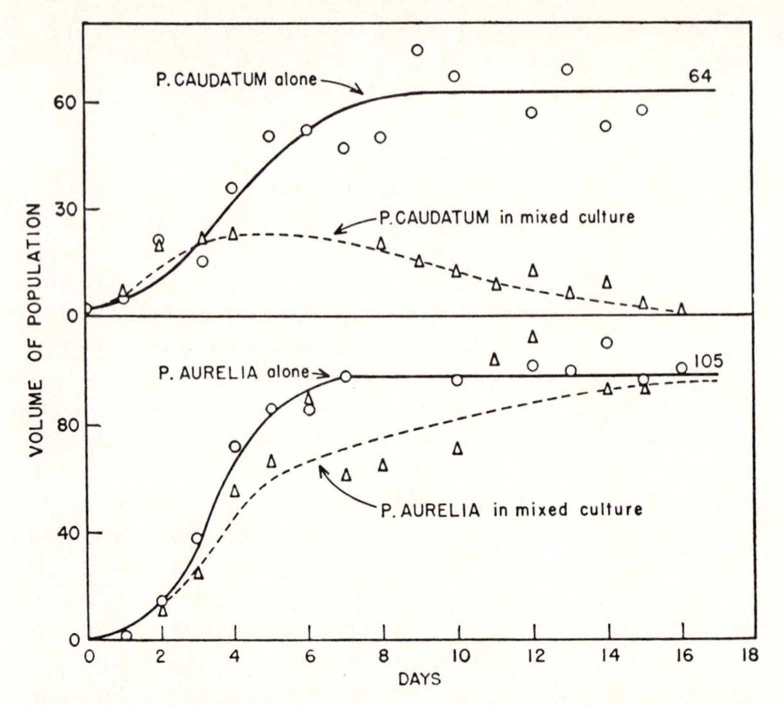

Figure 7–2 Competition between two closely related species of protozoa that have similar niches. When separate, *Paramecium caudatum* and *Paramecium aurelia* exhibit normal sigmoid growth in controlled cultures with constant food supply; when together, *P. caudatum* is eliminated. (From Allee et al., 1949, after Gause.)

Paramecium bursaria were able to survive and reach a stable equilibrium in the same culture medium, because although they were competing for the same food, *P. bursaria* occupied a different part of the culture where it could feed on bacteria without competing with *P. caudatum*. Thus, the habitat proved to be sufficiently different for the two species to coexist, even though their food was identical.

An example of the direct interference type of competition is described by Crombie (1947). He found that *Tribolium* exterminates *Oryzaephilus* (another genus of flour beetle) when both live together in flour, because *Tribolium* is more active in destroying immature stages of the other species. However, if glass tubes are placed in the flour into which immature stages of the smaller *Oryzaephilus* may escape, both populations survive. Thus, when a refuge from direct interference (in this case, predation) is provided, competition is reduced sufficiently for the support of two species.

So much for laboratory examples. It is readily conceded that crowding may be greater in laboratory experiments and, hence, competition exaggerated. Interspecific competition in plants in the field has been much studied and is generally believed to be an important factor in bringing about a succession of species (as will be described in Chapter 8). Keever (1955) reported that a species of tall weed occupying first-year fallow fields in almost pure stands was gradually replaced in these fields by another species previously unknown in the region. The two species, although belonging to different genera, have very similar life histories (time of flowering, seeding) and life forms and were thus brought into intense competition. The follow-up studies of these fallow fields have shown that the new invading species has not eliminated the original one; both coexist but in different ratios depending on soil and the degree and timing of disturbance.

We have already noted that competition between individuals of the same species is one of the most important density-dependent factors in nature, and the same can be said of interspecific competition. Competition appears to be extremely important in determining the distribution of closely related species, although evidence for this is often circumstantial. Gause's rule is difficult to test in nature by the accepted scientific procedure of falsification, since many other variables may affect the chance for either coexistence or exclusion. Nevertheless, closely related species or species that have very similar requirements usually occupy different geographical areas or different habitats in the same area or otherwise avoid competition by differing in daily or seasonal activity or in food. Natural selection seems to operate over the long term to eliminate or avoid prolonged "head-to-head" confrontation between species with similar life histories.

Morphological differences that enhance ecological separation may arise by an evolutionary process termed "character displacement" (Chapter 8, Section 4). For example, in middle Europe, six species of titmice (small birds of the genus *Parus*) coexist, segregated partly by habitat and partly by feeding areas and size of prey, which is reflected in small differences in length and width of the bill. In North America, more than two species of titmice are rarely found in the same locality, even though seven species are present on the continent as a whole. Lack (1969) suggests that "the American species of tits are at an earlier stage in their evolution than the European, and their differences in beak, body size, and feeding behavior are adaptations of their respective habitats, and are not yet adaptations for permitting coexistence in the same habitat."

The general theory of competition's role in habitat selection is

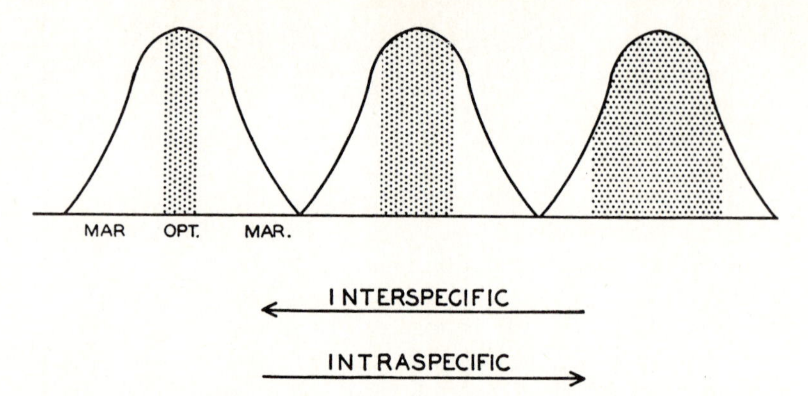

Figure 7–3 The effect of competition on the habitat distribution of birds. When intraspecific competition dominates, the species spreads out and occupies less favorable (marginal) areas; where interspecific competition is intense, the species tends to be restricted to a narrower range comprising the optimum conditions. (Modified from Svardson, 1949.)

summarized in Figure 7–3. The curves represent the range of habitat that can be tolerated by the species, with optimum and marginal conditions indicated. Where there is competition with other closely related or ecologically similar species, the range of habitat conditions that the species occupies generally becomes restricted to the optimum (i.e., to the most favorable conditions under which the species has an advantage in some manner over its competitors). Where interspecific competition is less severe, intraspecific competition generally brings about a wider choice of habitats.

Islands are good places to observe the tendency for wider selection of habitats to occur when potential competitors fail to colonize. For example, meadow voles (*Microtus*) often occupy forest habitats on islands where their forest competitor, the red-backed vole (*Clethrionomys*), is absent (see Cameron, 1964). Crowell (1962) found that the cardinal was more abundant and occupied a more marginal habitat in Bermuda, where many of its mainland competitors are absent.

A good example of ecological separation by feeding areas is the case of two similar fish-eating birds of Britain, the cormorant (*Phalacrocorax carbo*) and the shag (*P. aristotelis*) studied by Lack (1945). These two species commonly feed in the same waters and nest on the same cliffs, yet close study showed that actual nest sites were different and, as shown in Figure 7–4, the food was basically different. The shag feeds on free-swimming fish and eels in the upper waters, whereas the cormorant is more of a bottom feeder, taking flatfish (flounders) and bottom invertebrates (shrimp and others).

Just because closely related species are sharply separated in nature does not, of course, mean that competition is actually operating continuously to keep them separated; the two species may have evolved different requirements or preferences that effectively keep them out of competition.

For example, in Europe, one species of *Rhododendron*, *R. hirsutum*, is found on calcareous soils, while another species, *R. ferrugineum*, is found on acid soils. The requirements of the two species are such that neither can live at all in the opposite type of soil, so there is never any actual competition between them (see Braun-Blanquet, 1932). Teal (1958) made an experimental study of habitat selection of species of fiddler crabs (*Uca*), which are usually separated in their occurrence in salt marshes. One species, *U. pugilator*, is found on open sandy flats; another, *U. pugnax*, is found on muddy substrates covered with marsh grass. Teal found that one species would tend not to invade the habitat of the other even in the absence of the other, because each species would dig burrows only in its preferred substrate. The absence of active competition, of course, does not mean that competition in the past is to be ruled out as a factor in bringing about the isolating behavior.

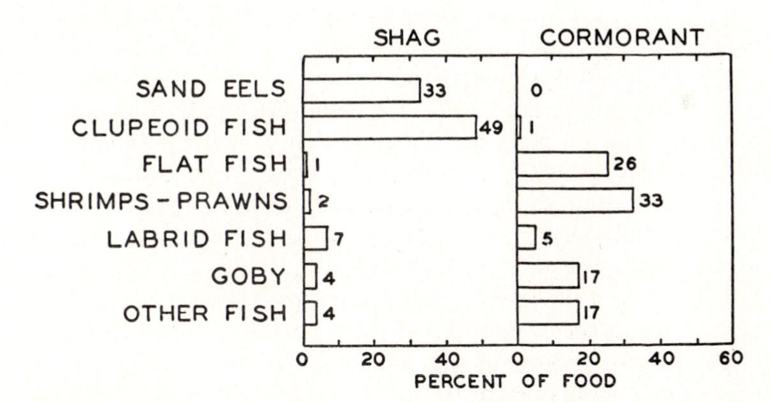

Figure 7–4 Food habits of two closely related species of aquatic birds, the cormorant (*Phalacrocorax carbo*) and the shag (*P. aristotelis*), which are found together during the breeding season. Food habits indicate that although the habitat is similar, the food is different; therefore, the niche of the two species is different and they are not actually in direct competition for food resources. (Data from Lack, 1945.)

Three tentative models proposed by Philip (1955) may serve as a basis for future observation, analysis, and experimentation: (1) imperfect competition, where interspecific competition is a limiting factor, but not to the extent of complete elimination of one of the competitors from the interaction arena; (2) perfect competition, as in the unmodified Gause or Lotka-Volterra model, in which one species is gradually eliminated in competition for common resources as crowding occurs; and (3) hyperperfect competition, in which the depressing effects are great and immediately effective, as in the production of antibiotics. A striking example of direct interference, or "exaggerated" competition, in plants will be given in the next section.

3. Predation, Herbivory, Parasitism, and Allelopathy (Antibiosis)

Statement

Predation and parasitism are familiar examples of interactions between two populations that result in negative effects on the growth and survival of one population and a positive or beneficial effect on the other. When the predator is a primary consumer (usually an animal), and the prey or "host" is a primary producer (plant), the interaction is termed **herbivory.** When one population produces a substance harmful to a competing population, the term **allelopathy** or **antibiosis** is commonly used for the interaction. Accordingly, there are a variety of "+−" relationships.

The negative effects tend to be quantitatively small when the interacting populations have had a common evolutionary history in a relatively stable ecosystem. In other words, natural selection tends to lead to a reduction in detrimental effects or to the elimination of the interaction altogether, since continued severe depression of a prey or host population by the predator or parasite population can only lead to the extinction of one or both populations. Consequently, severe impact is most frequently observed when the interaction is of recent origin (when two populations first become associated), or when large-scale or sudden changes (perhaps temporary) have occurred in the ecosystem (as might be produced by humans).

Explanation and Examples

It is difficult to approach the subject of parasitism and predation (along with the other negative interactions) objectively. We all have a natural aversion to parasitic organisms, whether bacteria or tapeworms. Likewise, although humans are the greatest predators the world has known and the greatest perpetrators of epidemics in nature, we tend to condemn all other predators without bothering to ascertain whether these predators are really detrimental to our interests. The idea that "the only good hawk is a dead hawk" is a common public attitude, but, as we shall see, it is a most uncritical generalization.

The best way to be objective is to consider predation, parasitism, herbivory, and allelopathy from the population rather than from the individual standpoint. Predators, parasites, and grazers certainly kill and injure the individuals that they feed on or secrete toxic chemicals on, and they depress, in some measure at least, the growth rate of target populations or reduce the total population size. But does this mean that populations would be better off without consumers or inhibitors? From the long-term view, are predators the sole beneficiaries of the association? As pointed out in the discussion of population regulation (Chapter 6, Section 5; see especially Figure 6–20), predators and parasites help keep herbivorous insects at a low density so they will not destroy their own food supply and habitat (even though they may be ineffective when the host population erupts or "escapes" from density-dependent control). And in Chapter 3, Section 4, we discussed how animal herbivores and plants evolve almost a ++, or mutualistic relationship. Valerio (1975) describes a case of mutual "predation" in a spider host-parasite relationship; egg parasites are eaten as they emerge from spider eggs by young spiders that have escaped being parasitized. For more on the positive aspects of parasitism, see Smith (1968).

Deer populations are often cited as examples of populations that tend to erupt when predator pressure is reduced. The Kaibab deer herd, as originally described by Leopold (1943) on the basis of estimates by Rasmussen (1941), allegedly increased from 4000 (on 700,000 acres on the north side of the Grand Canyon in Arizona) in 1907 to 100,000 in 1924, coincident with a predator-removal campaign organized by the government. Caughley (1970) has reexamined this case and concludes that though there is no question that deer did increase, overgraze, and then decline, there is doubt about the extent of the overpopulation, and there is no real evidence that it was due solely to removal of predators. Cattle and fire may also have played a part. Caughley believes that eruptions of ungulate populations are more likely to result from changes in habitat or food that enable the population to "escape" from the usual mortality control.

One thing is clear: the most violent eruptions occur when a species is introduced into a new area where there are both unexploited resources *and* a lack of negative interactions. The population explosion of rabbits introduced into Australia is, of course, a well-known example among the thousands of cases of severe oscillations that result when species with high biotic potential are introduced into new areas. An interesting sequel to the attempts to control the irruption of rabbits by introducing a disease organism has provided evidence for group selection in a parasite-host system (Chapter 8, Section 6).

The most important generalization of all is that negative interactions become less negative with time if the ecosystem is sufficiently stable and spatially diverse to allow reciprocal adaptations. Parasite-host or predator-prey populations introduced into experimental microecosystems usually oscillate violently, with a certain probability of extinction. The Lotka-Volterra equation models of predator-prey interaction produce a perpetual undamped oscillation *unless second-order terms are added that induce self-limitations capable of damping the oscillation.* Pimentel and Stone (1968) have shown experimentally that such second-order terms can take the form of reciprocal adaptations. As shown in Figure 7–5, violent oscillations occur when a host such as the house fly and a parasitic wasp are first placed together in a limited culture system. When individuals selected from cultures that had managed to survive the violent oscillations for two years were then reestablished in new cultures, it was evident that through genetic selection an ecological homeostasis had evolved in which both populations had "powered down," so to speak, and could now coexist in a much more stable equilibrium.

In the real world of humans and nature, time and circumstances may not favor such reciprocal adaptation by new associations. There is always the danger that the negative reaction may be irreversible in that it leads to the extinction of the host. The chestnut blight in America is a case in which the question of adaptation or extinction hangs in the balance.

Originally, the American chestnut tree was an important member of the forests of the Appalachian region of eastern North America, often constituting up to 40 percent of total forest biomass. It had its share of parasites, diseases, and predators. Likewise, the oriental chestnut trees in China—a different but related species—had their share of parasites, including the fungus *Endothia parasitica*, which attacks the bark of the stems. In 1904, the fungus was accidentally introduced into the United States. The American chestnut proved to be unresistant to this new parasite. By 1952, all the large chestnut trees had been killed, their gaunt grey trunks becoming a characteris-

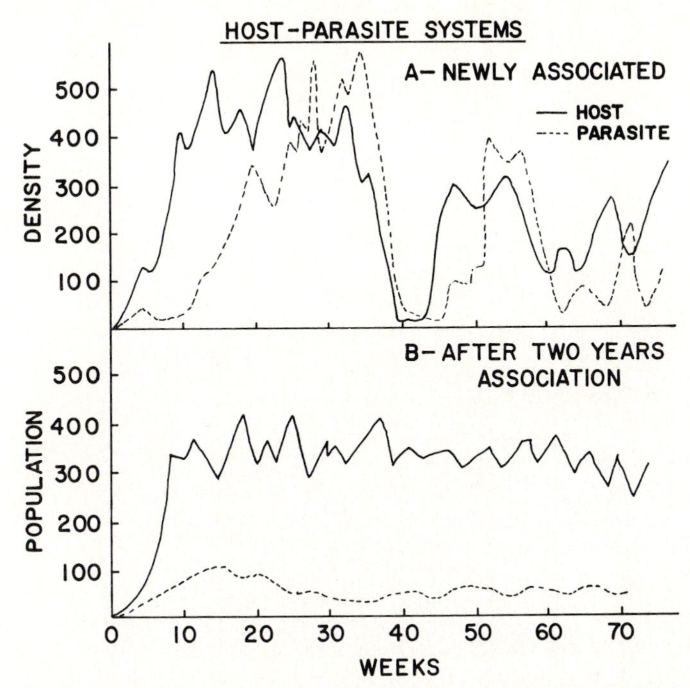

Figure 7–5 Evolution of homeostasis in the host-parasite relationship between house fly (*Musca domestica*) and parasitic wasp (*Nasonia vitropennis*) populations in a laboratory multicell population cage consisting of 30 plastic boxes interconnected with tubes especially designed to slow down parasite dispersal. *A*. Newly associated populations (wild stocks brought together for the first time) oscillated violently, as first the host (fly) and then the parasite (wasp) density increased and "crashed." *B*. Populations derived from colonies in which the two species had been associated for two years coexisted in a more stable equilibrium without "crashes." The adaptive resistance that had evolved in the host was indicated by the fact that the natality of the parasite was greatly reduced (46 progeny per female, compared with 133 in the newly associated system), and the parasite population leveled off at a low density. The experiment demonstrates how genetic feedback can function as both a regulatory and a stabilizing mechanism in population systems. [Redrawn from Pimentel and Stone, 1968; the lower graph (*B*) is a composite of two experimental populations as shown in their Figures 2 and 3. Density figures are number per cell in the 30-cell cage.]

tic feature of Appalachian forests (Figure 7–6). The chestnut tree continues to sprout up from the roots, and such sprouts may produce fruits before they die, but no one can say whether the ultimate outcome will be extinction or adaptation. For all practical purposes, the chestnut tree has been removed, for now at least, as a major influence in the forest. There is some hope that a virus disease that has been

Figure 7–6 Results of the chestnut blight in the southern Appalachian region (Georgia), an example of the extreme effect a parasitic organism (fungus) introduced from the Old World may have on a newly acquired host (American chestnut tree).

discovered can be used to control the blight (see Anagnostakis, 1982). (How the forest responded to removal of its principal dominant is shown in Table 7–7.)

The preceding examples are not just cases hand-picked to prove a point. If the student will do a little reading in the library, he or she can find similar examples showing (1) that where parasites and predators have long been associated with their respective hosts and prey, the effect is moderate, neutral, or even beneficial from the long-term view, and (2) that newly acquired parasites or predators are the most damaging. In fact, a list of the diseases, parasites, and insect pests that cause the greatest loss in agriculture or forestry would include many species that have recently been introduced into a new area, such as the chestnut blight, or that have acquired a new host or prey. The European corn earworm, the gypsy moth, the Japanese beetle, and the Mediterranean fruit fly are just a few introduced insect pests that belong to this category. Much the same principle also applies to severe human diseases; the most feared are the newly acquired.

In summary, what we might call "the principle of the instant pathogen" can be stated as follows: Pathogenicity or pestilence is often induced by (1) sudden or rapid introduction of an organism with a potentially high intrinsic rate of increase into an ecosystem in which adaptive control mechanisms for it are weak or lacking, or (2) abrupt or stressful environmental changes that reduce the energy available for feedback control or otherwise impair the capacity for self-control. The lesson, of course, is to avoid introducing new potential pests and to avoid wherever possible the stressing of ecosystems with poisons that destroy useful as well as pest organisms.

Although predation and parasitism are similar from the ecological standpoint, the extremes in the series, the large predator and the small internal parasite, do have important differences other than size. Parasitic or pathogenic organisms, which usually have a higher biotic potential than do predators, are often more specialized in structure, metabolism, host specificity, and life history. This specialization is necessitated by their special environment and the problem of dispersal from one host to another.

Of special interest are organisms intermediate between predators and parasites, for example, the so-called parasitic insects. These forms often can consume the entire individual prey, as does the predator, yet they have the host specificity and high biotic potential of the parasite. Entomologists have propagated some of these organisms artificially and use them to control insect pests. In general, attempts to make similar use of large unspecialized predators have not been successful. For example, the mongoose introduced in the Caribbean islands to control rats in sugarcane fields has more severely reduced ground-nesting birds than rats. If the predator is small, is specialized in its choice of prey, and has a high biotic potential, control can be effective. A good example, as documented by Huffaker and Kennett (1956), involves a small predator mite that effectively controls a herbivorous mite in California strawberry patches. Herbivorous insects have sometimes been successfully used to control weeds (Huffaker, 1957, 1959).

As predators, human beings are slowly learning how to be what Slobodkin (1962) calls a "prudent predator," one who does not exterminate his prey by overexploitation. The problem of the optimum yield is discussed by Beverton and Holt (1957), Ricker (1958), Menshutkin (1962), Slobodkin (1968), Silliman (1969), Wagner (1969), McCullough (1979), and many others. Theoretically, if sigmoid growth is symmetrical, as in the logistic model, the highest growth rate, dN/dt, occurs when density is $K/2$ (one half the saturation density). This inflection point in the sigmoid curve was referred to as the

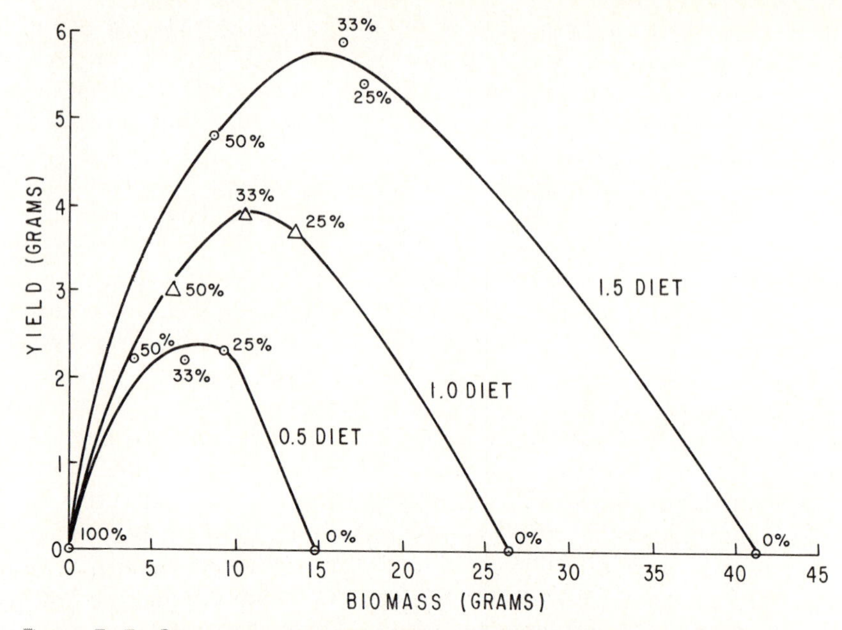

Figure 7–7 Biomass and yield in test populations of the guppy exploited at different rates (shown as percent removal per reproductive period) at three different diet levels. The highest yields were obtained when about one third of the population was harvested per reproductive period and mean biomass was reduced to less than half that of the unexploited population (yield curves skewed to the left). (After Silliman, 1969.)

I-carrying capacity density level in Figure 3–19. However, since the growth form is often skewed to the right (owing to time lags), the peak of the humpbacked or parabolic growth rate curve is often not midway between 0 and *K*, so that maximum sustained yield density is often somewhat less than half the unexploited equilibrium density.

The problem can be attacked experimentally by setting up test populations in microecosystems. In one such experimental model, shown in Figure 7–7, guppies (*Lebistes reticulatus*), small aquarium fish, were used to "mimic" a commercial fish population being exploited by humans. As shown, the maximum sustained yield was obtained when one third of the population was harvested during each reproductive period, which reduced the equilibrium density to slightly less than half the unexploited one. Within the limits of the experiment, these ratios tended to be independent of the carrying capacity of the system, which was varied at three levels by manipulating the food supply.

One-species models often prove to be oversimplifications because they do not account for competing species that may respond to the reduced density of the harvested species by increasing their density and using up food or other resources needed to sustain the exploited species. "Top predators" such as humans (or major grazers such as cows) can easily tip the balance in a competitive equilibrium so that the exploited species is replaced by another species that the predator or grazer may not be prepared to use. In the real world, examples of such shifts are being documented more often as human beings strive to become more "efficient" as fishermen, hunters, and harvesters of plants. This situation poses both a challenge and a danger: *one-species harvest systems, as well as monocultural systems* (such as one-crop agriculture), *are inherently unstable*, because when stressed they are vulnerable to competition, disease, parasitism, predation, or other negative interactions. Some good examples of this general principle are to be found in the fishing industry.

Murphy (1966, 1967) diagnosed the decline of the heavily fished Pacific sardine (*Sardinops caerulea*) and the subsequent increase of the anchovy (*Engraulis mordox*), which has a very similar ecology, as a case of replacement of an overexploited species by a competing species that was not being exploited.

In another example, Smith (1966) describes how a succession of species-specific exploitations combined with introductions and eutrophication have resulted in successive rises and falls in number of commercial fish in Lake Michigan. First there was the lake trout that supported a stable fishery for half a century, but this was virtually eliminated by the combined assault of overharvest, attack by the introduced parasitic lamprey eel, and eutrophication. Then, in rapid succession, lake herring, lake whitefish, chubs, and the exotic alewife exhibited population growth and decline as each in turn was exploited and gave way under the pressure of a competitor, predator, or parasite. Coho salmon were introduced in the 1960s and are thriving on a diet of alewifes, much to the delight of sports fishermen. This bonanza is not likely to last unless harvest and pollution can be better controlled.

The stress of predation or harvest often affects the size of individuals in the exploited populations. Thus, harvesting at the maximum sustained yield-level usually results in reducing the average size of fish, just as maximizing timber yields for volume of wood reduces the size of trees and the quality of wood. As reiterated many times in this volume, one cannot have maximum quality and quantity at the same time. Brooks and Dodson (1965) describe how large species of zooplankton are replaced by smaller species when zooplankton-feeding fish are introduced into lakes that formerly lacked such direct predators. In this case in which the ecosystem is relatively small, both the size and species composition of a whole trophic level may be controlled by one or a few species of predators.

Good examples of allelopathy can be cited from the work of C. H. Muller and associates who studied inhibitors produced by shrubs in the California chaparral vegetation. These investigators have not only examined the chemical nature and physiological action of the inhibitory substances but have also shown that these are important in regulating the composition and dynamics of the community (see Muller, 1966, 1969; Muller et al., 1964, 1968). Figure 7–8 shows how volatile terpenes produced by two species of aromatic shrubs inhibit the growth of herbaceous plants. The volatile toxins (notably cineole and camphor) are produced in the leaves and accumulate in the soil during the dry season to such an extent that when the rainy season comes, germination or subsequent growth of seedlings is inhibited in a wide belt around each shrub group. Other shrubs produce water-soluble antibiotics of a different chemical nature (phenols and alkaloids, for example), which also favor shrub dominance. However, periodic fires, which are an integral part of the chaparral ecosystem, effectively remove the source of the toxins, denaturing those accumulated in the soil and triggering the germination of fire-adapted seeds. Accordingly, fire is followed in the next rainy season by a conspicuous blooming of annuals, which continue to appear each spring until shrubs grow back and the toxins again become effective. Very few herbs are left in the mature chaparral. The interaction of fire and antibiotics thus perpetuates cycle changes in composition that are the adaptive feature of this type of ecosystem.

Whittaker (1970), in a general review of botanical inhibitors, concludes

> Higher plants synthesize substantial quantities of substances repellent or inhibitory to other organisms. Allelopathic effects have a significant influence on the rate and species sequence of plant succession and on species composition of stable communities. Chemical interactions affect species diversity of natural communities in both directions; strong dominance and intense allelopathic effects contribute to low species diversity of some communities, whereas variety of chemical accommodations are part of the basis (as aspects of niche differentiation) of the high species diversity of others.

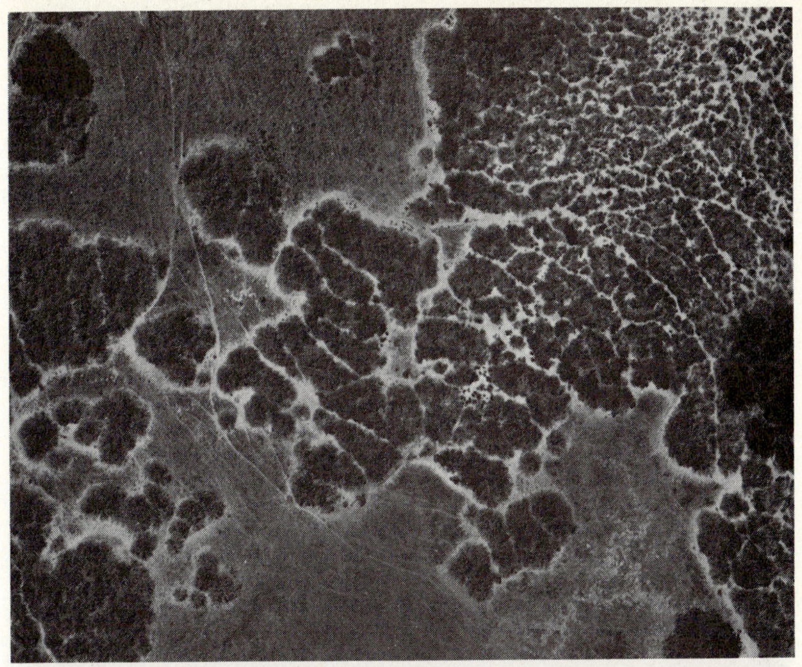

Figure 7–8 *Upper.* Aerial view of aromatic shrubs *Salvia leucophylla* and *Artemisia californica* invading an annual grassland in the Santa Inez Valley of California and exhibiting biochemical inhibition. *Lower.* Closeup showing the zonation effect of volatile toxins produced by *Salvia* shrubs seen to the center-left of A. Between A and B is a zone 2 meters wide bare of all herbs except for a few minute, inhibited seedlings (the root systems of the shrubs that extend under part of this zone are thus free from competition with other species). Between B and C is a zone of inhibited grassland consisting of smaller plants and fewer species than in the uninhibited grassland seen to the right of C. (Photos courtesy of Dr. C. H. Muller, University of California, Santa Barbara.)

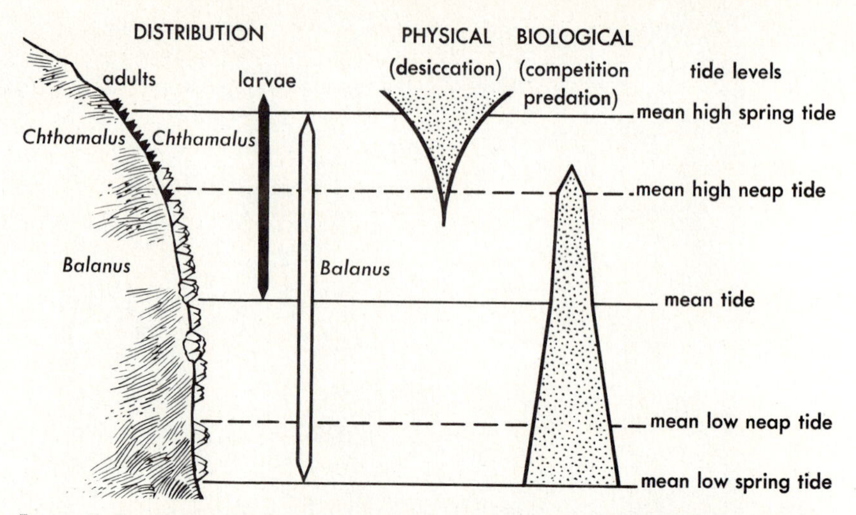

Figure 7–9 Factors that control the distribution of two species of barnacles in an intertidal gradient. The young in each species settle over a wide range but survive to adulthood in a more restricted range. Physical factors such as desiccation control upward limits of *Balanus*, while biological factors such as competition and predation control downward distribution of *Chthamalus*. (After E. P. Odum, 1963; from Connell, 1961.)

Antibiosis, of course, is not restricted to higher plants; numerous examples among microorganisms are known, as illustrated by penicillin, the bacterial inhibitor produced by bread mold and now widely used in medicine. Chemical antibiosis can be considered a form of the direct interference type of competition or, in its milder forms, amensalism. For more on allelopathy, see Rice (1974).

All of these interesting aspects of biological and chemical "warfare" in nature raise questions about the role of negative interactions in the ecosystem as a whole. It often seems that the importance of biotic interactions is a function of the position of the community in physical gradients. In Figure 7–9, a "barnacle model" based on the experimental studies of J. H. Connell is presented. The intertidal zone on a rocky seacoast provides a miniature gradient from a physically stressed to a more biologically controlled environment. In his study in Scotland, Connell (1961) found that the larvae of two species of barnacles settled over a wide range of the intertidal zone but survived as adults in a much more restricted range. The larger species (*Balanus*) was found to be restricted to the lower part of the zone because it could not tolerate long periods of exposure. The smaller species (*Chthamalus*) was excluded from the lower zone by competition with the larger species and by predators that are more active below the high tide mark. Accordingly, the physical stress of desiccation was identified as the main controlling factor in the upper part of the gradient, while interspecific competition and predation were more important controlling factors in the lower zones. This model can be considered to apply to more extensive gradients such as an arctic-to-tropic or a high-to-low-altitude gradient, provided one remembers that all models are, to varying degrees, oversimplifications.

4. Positive Interactions: Commensalism, Cooperation, and Mutualism

Statement

Associations between two populations of species that result in positive effects are exceedingly widespread and probably as important as competition, parasitism, and so forth in determining the nature of populations and communities. Positive interactions may be conveniently considered in an evolutionary series as follows: commensalism—one population benefits; protocooperation—both populations benefit; and mutualism—both populations benefit and become completely dependent on each other.

Explanation

The widespread acceptance of Darwin's idea of "survival of the fittest" as an important means of bringing about natural selection has directed attention to the competitive aspects of nature. As a result, the importance of cooperation between species in nature has been underestimated. Until recently, positive interactions have not been subjected to as much quantitative study as have negative interactions. One might reasonably assume that negative and positive relations between populations eventually tend to balance one another, and that both are equally important in evolution of species and in stabilization of the ecosystem.

Commensalism represents a simple type of positive interaction and perhaps represents the first step toward the development of beneficial relations. It is especially common between sessile plants and animals on the one hand and motile organisms on the other. The ocean is a particularly good place to observe commensalism. Practically every worm burrow, shellfish, or sponge contains various "uninvited guests," organisms that require the shelter of the host but do neither harm nor good in return. Oysters, for example, sometimes have a small delicate crab in the mantle cavity. These crabs are usually commensal, although sometimes they overdo their guest status by partaking of the host's tissues (Christensen and McDermott, 1958). Dales (1957), in his review of marine commensalism, lists 13 species that live as guests in the burrows of large sea worms (*Erechis*) and burrowing shrimp (*Callianassa* and *Upogebia*). This array of commensal fish, clams, polychaete worms, and crabs lives by snatching surplus or rejected food or waste materials from the host. Many commensals are not host-specific, but some apparently are found associated with only one species of host.

It is but a short step to a situation in which both organisms gain by an association or interaction of some kind; this relationship is called protocooperation. The late W. C. Allee (1938, 1951) studied and wrote extensively about this subject. He believed that the beginnings of cooperation between species are to be found throughout nature. He documented many of these instances and demonstrated mutual advantages by experiments. Returning to the sea for an example, crabs and coelenterates often associate with mutual benefit. The coelenterates grow on the backs of the crabs (or are sometimes "planted" there by the crabs), providing camouflage and protection (since coelenterates have stinging cells). In turn, the coelenterates are transported about and obtain particles of food when the crab captures and eats another animal.

In the preceding instance, the crab does not absolutely depend on the coelenterate, nor vice versa. A further step in cooperation results when each population becomes completely dependent on the other. Such cases have been called **mutualism**, or **obligate symbiosis.** Often quite diverse kinds of organisms are associated. In fact, instances of mutualism are most likely to develop between organisms with widely different requirements. (Organisms with similar requirements are more likely to get involved in competition.) The most important examples of mutualism develop between autotrophs and heterotrophs, which is not surprising since these two components of the ecosystem must ultimately achieve some kind of balanced symbiosis. Examples that would be labeled as mutualistic go beyond such general community interdependence to the extent that one particular kind of heterotroph becomes completely dependent on a particular kind of autotroph for food, and the latter becomes dependent on the protection, mineral cycling, or other vital function provided by the heterotroph. The several kinds of partnerships between nitrogen-fixing microorganisms and higher plants have been discussed in Chapter 4. Mutualism is also common between microorganisms that can digest cellulose and other resistant plant residues and animals that do not have the necessary enzyme systems for this.

As previously suggested, mutualism seems to replace parasitism as ecosystems evolve toward maturity, and it seems to be especially important when some aspect of the environment is limiting (infertile soil, for example) to an extent that mutual cooperation has a strong selective advantage. How such mutually beneficial relations get started and become genetically fixed in the world of Darwinian struggle for existence is a controversial subject in population genetics, as will be discussed in Chapter 8.

Examples

Obligate symbiosis between ungulates (such as cattle) and rumen bacteria is a well-studied example (see Hungate, 1963, 1966, 1975). The anaerobic nature of the rumen system is inefficient for bacterial growth (only 10 percent of energy in grass or hay eaten by the cow is assimilated by the bacteria), but the very nature of this inefficiency constitutes the reason that the ruminant can subsist at all on such a substrate as cellulose. The major portion of residual energy of microbial action consists of fatty acids that are converted from cellulose but are not further degraded. These end-products, however, are directly available for assimilation by the ruminant. Accordingly, the partnership is very efficient for the ruminant, since it gets most of the energy in the cellulose, which it could not obtain without the help of the symbionts. In return, of course, the bacteria get a temperature-controlled culture medium.

Mutualism between cellulose-digesting microorganisms and arthropods is quite common and often a major factor in detritus food chains (discussed in Chapter 3, Section 4). The termite–intestinal flagellate partnership is a well-known case, first worked out by Cleveland (1924, 1926). Without the specialized flagellates (a cluster of species of the order Hypermastigina), many species of termites cannot digest the wood they ingest, as shown by the fact that they starve to death when the flagellates are experimentally removed. The symbionts are so well coordinated with their host that they respond to the termite's molting hormones by encysting, thus ensuring transmission and reinfection when the termite molts its gut lining and then ingests it.

In termites, the symbionts live inside the body of the host. However, even more intimate interdependence may develop with the microorganism's partners living outside the body of the animal host, and such associations may actually represent a more advanced stage in the evolution of mutualism (less chance that the relationships might revert to parasitism!). An example is tropical attine ants, which cultivate fungal gardens on the leaves they harvest and store in their nests. The ants fertilize, tend, and harvest the fungal crop in much the same manner as would an efficient human farmer. Martin (1970) has shown that the ant-fungal system short-circuits and speeds up the natural decomposition of leaves. A succession of microorganisms is normally required to decompose leaf litter, with basidiomycete fungi normally appearing during the late stages of decomposition. However, when leaves are "fertilized" by ant excreta in the fungal gardens, these fungi can thrive on fresh leaves as a rapidly growing monoculture that provides food for the ants. A lot of "ant energy," of course, is required to maintain this monoculture, just as much energy is required in intensive crop culture by human beings.

Martin summarizes the ant-fungus mutualism as follows:

> By cultivating as a food crop a cellulose-degrading organism the ants gain access to the vast cellulose reserves of the rain forest for indirect use as a nutrient. What termites accomplish by their endosymbiotic association with cellulose-degrading microorganisms, the attine ants have achieved through their more complex ectosymbiotic association with a cellulose-degrading fungus. In biochemical terms the contribution of the fungus to the ant is the enzymatic apparatus for degrading cellulose. The fecal material of the ant contains proteolytic enzymes which the fungus lacks so that the ants contribute their enzymatic apparatus to degrade protein. The symbiosis can be viewed as a metabolic alliance in which the carbon and nitrogen metabolisms of the two organisms have been integrated.

Coprophagy, or the reingestion of feces, which appears to be characteristic of detritivores, can probably be viewed as a much less elaborate but much more widespread case of mutualism that couples the carbon and nitrogen metabolism of microorganisms and animals, an "external rumen," as it were.

Ants and acacia trees are involved in another striking tropical mutualism symbiosis, as described by Jansen (1966, 1967). The trees house and feed the ants, which nest in special cavities in the branches. In turn, the ants protect the tree from would-be herbivorous insects. When ants are experimentally removed (as by poisoning with an insecticide), the tree is quickly attacked and often killed by defoliating insects.

Mineral cycling as well as food production is enhanced by symbiosis between microorganisms and plants. Prime examples are

mycorrhizae (= fungus-root), comprising the mycelia of fungi, which live in mutualistic association with the living roots of plants (not to be confused with parasitic fungi that kill roots). Like nitrogen-fixing bacteria and legumes, the fungi interact with root tissue to form composite "organs" that enhance the ability of the plant to extract minerals from the soil. In return, of course, the fungi are supplied with some of the plant's photosynthate. So important is the energy flow pathway through mycorrhizae that this route can be listed as a major food chain, as was diagrammed in Figure 3–11A.

Mycorrhizae take several forms, as illustrated in Figure 7–10:

1. Ectotrophic mycorrhizae, mostly basidiomycetes, which form rootlike extensions that grow out from the cortex of the root (Figure 7–10A).
2. Endotrophic mycorrhizae, mostly phycomycetes, which penetrate the cells of the root (Figure 7–10B); these are widespread in tree roots but are difficult to culture and study.
3. Peritrophic (or extramatrical) mycorrhizae, which form mantles or clusters around the roots (Figure 7–10C and D), but mycelia do not penetrate the epidermis of the root; these are thought to be very important in creating a favorable chemical "rhizospheric environment" that transforms insoluble or unavailable minerals to forms that can be taken up by the roots.

Many trees will not grow without mycorrhizae. Forest trees transplanted to prairie soil or introduced into a different region often fail to grow unless inoculated with the fungal symbionts. Pine trees with healthy mycorrhizal associates grow vigorously in soil so poor by conventional agricultural standards that corn or wheat could not survive. The fungi can metabolize "unavailable" phosphorus and other minerals by chelation (see page 42 for explanation of this process) or by other means that are as yet not well understood. When labeled minerals (radioactive tracer phosphorus, for example) are added to the soil, as much as 90 percent may be quickly taken up by the mycorrhizal mass, then slowly released to the plant.

It is fortunate that the pine tree–mycorrhizal system does so well on the millions of acres of the southern United States where topsoil was devastated by the row-crop monoculture and absentee owner system that persisted for so long; otherwise, many of these eroded acres would be deserts today. The remarkable performance of mycorrhizal pines at Copper Hill was previously described (Chapter 2, Section 4).

The role of mycorrhizae in direct mineral recycle, its importance in the tropics, and the need for crops with such built-in recycle sys-

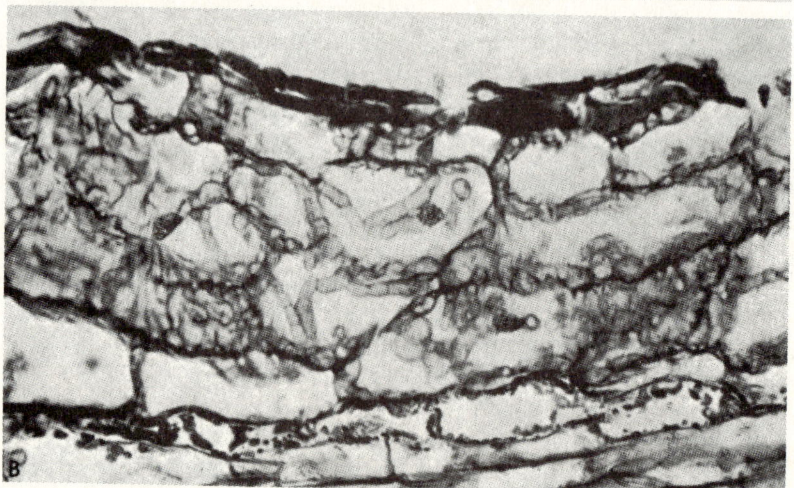

Figure 7–10 Types of mycorrhizae. *A*. Three-year-old white pine (*Pinus strobus*) seedlings devoid of mycorrhizae (left) and with a prolific development of ectotrophic mycorrhizae (right). *B*. Endotrophic mycorrhizae showing fungal mycelia inside cells of the root. *C*. Peritrophic (extramatrical) mycorrhizae forming clusters or masses around the roots of a spruce (*Picea pungens*) seedling. *D*. Peritrophic mycorrhizae forming a sheath or dense fungal mantle around roots imbedded in organic litter of a tropical forest. (Photos courtesy of Professor S. A. Wilde, University of Wisconsin.)

Illustration continued on opposite page

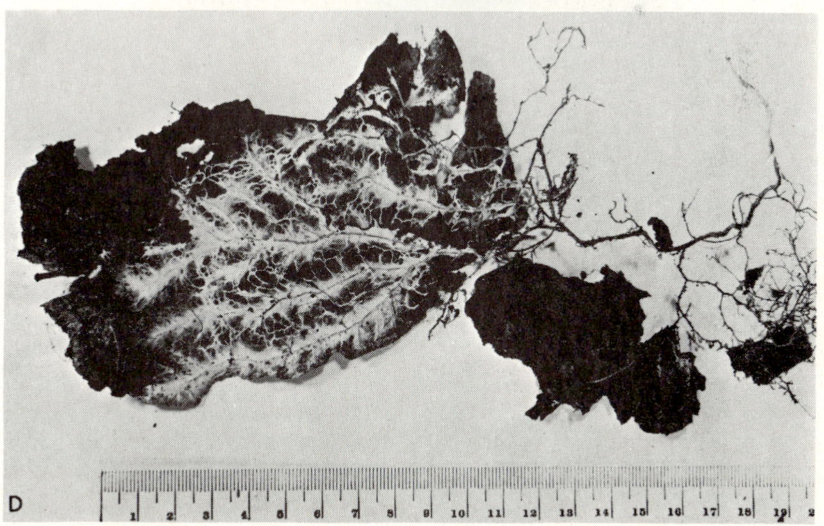

Figure 7–10 (Continued)

tems were all emphasized in Chapter 4, Sections 7 and 8. For additional information on microorganism-root mutualisms, see Harley (1959), Rovira (1965), Wilde (1968), Marks and Kozlowski (1973), Sanders et al. (1975), and Marx and Ruehle (1979).

Lichens are an association of specific fungi and algae so intimate in terms of functional interdependence and so integrated morphologically that a sort of third kind of organism resembling neither of its components is formed. Lichens are usually classified as single "species" even though they are composed of two unrelated species. Although the components can often be cultured separately in the laboratory, the integrated unit is difficult to culture even though it can exist in nature under harsh conditions. In lichens one sees evidence within the group of an evolution from parasitism to mutualism. In some of the more primitive lichens, for example, the fungi actually penetrate the algal cells, as shown in Figure 7–11A, and are thus essentially parasites of the algae. In the more advanced species, the fungal mycelia or hyphae do not break into the algal cells, but the two live in close harmony (Figure 7–11B and C).

For general reviews of symbiotic associations, see the volumes edited by Nutman and Masse (1963), Henry (1966), and Cheng (1971). The relationship between human beings and cultivated plants and domesticated animals, which might be considered a special form of mutualism, is discussed in Chapter 8. In Chapter 2, coral-algal associations were discussed as an emergent property that enhances the recycling of nutrients and the productivity of a whole ecosystem.

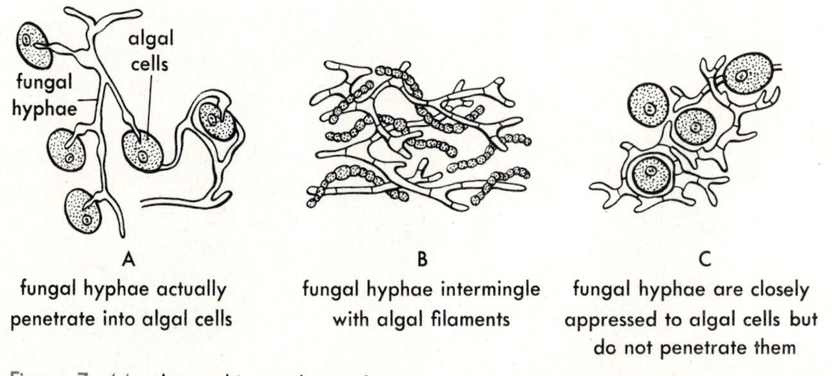

Figure 7–11 A trend in evolution from parasitism to mutualism in the lichens. In some primitive lichens the fungi actually penetrate the algal cells, as in *A*. In the more advanced species the two organisms live in greater harmony for mutual benefit, as in *B* and *C*. (After E. P. Odum, 1963.)

The "lichen model" in Figure 7–11 is perhaps a symbolic one for humans. Until recently, human beings have generally acted as parasites on their autotrophic environment, taking what they need with little concern about the welfare of the planet. Great cities grow and become parasites on the countryside, which must somehow supply food, water, and air and degrade huge quantities of wastes. Human beings must evolve to the mutualism stage in their relationship with nature. If humans do not learn to live mutualistically with nature, then, like the "unwise" or "unadapted" parasite, they may exploit their host to the point of destroying themselves.

5. Concepts of Habitat, Ecological Niche, and Guild

Statement

The **habitat** of an organism is the place where it lives, or the place where one would go to find it. The **ecological niche**, however, includes not only the physical space occupied by an organism but also its functional role in the community (for example, its trophic position) and its position in environmental gradients of temperature, moisture, pH, soil, and other conditions of existence. These three aspects of the ecological niche can be conveniently designated as the **spatial** or **habitat niche**, the **trophic niche**, and the **multidimensional** or **hypervolume niche.** Consequently, the ecological niche of an organism not only depends on where it lives but also includes the sum total of its environmental requirements. The concept of niche is most useful, and quantitatively most applicable, in terms of **differences** between species (or the same species at two or more locations or times) in one of a few major (i.e., operationally significant) features. The dimensions most often quantified are **niche width** and **niche overlap** with neighbors. Groups of species with comparable roles and niche dimensions within a community are called **guilds.** Species that occupy the same niche in different geographical regions are called **ecological equivalents.**

Explanation and Examples

The term "habitat" is used widely, not only in ecology but elsewhere. Thus, the habitat of the water "backswimmer" (*Notonecta*) and the "water boatman" (*Corixa*), as pictured in Figure 7–12, is the shallow,

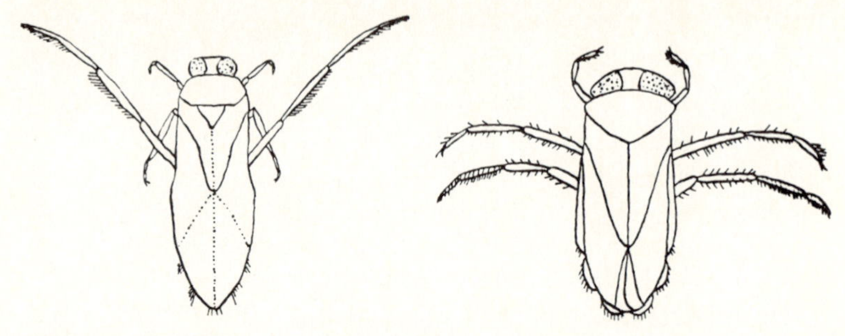

Figure 7–12 *Notonecta* (left) and *Corixa* (right), two aquatic bugs (Hemiptera) that may live in the same habitat but occupy different trophic niches because of differences in food habits.

vegetation-choked areas (littoral zone) of ponds and lakes; one would go there to collect these particular water bugs. However, the two species occupy very different **trophic niches**, since the backswimmer is an active predator, while the water boatman feeds largely on decaying vegetation. Another example of niche separation based on food habits is shown in Figure 7–4. The ecological literature is replete with similar examples of coexisting species that use different energy sources.

Habitat may also refer to the place occupied by an entire community. For example, the habitat of the sand sage grassland community is the series of ridges of sandy soil occurring along the north sides of rivers in the southern Great Plains region of the United States. Habitat in this case consists mostly of physical or abiotic complexes, whereas habitat for the water bugs mentioned previously includes living and nonliving objects. Thus, the habitat of an organism or group of organisms (population) includes other organisms and the abiotic environment. To describe the habitat of the community, one would include only the latter. Recognizing these two possible uses of the term "habitat" helps avoid confusion.

Ecological niche is a more recent concept and is not so generally understood outside the field of ecology. Broad, albeit useful, terms, such as niche, are difficult to define and quantify; the best approach is to consider the component concepts historically. Joseph Grinnell (1917, 1928) used the word "niche" "to stand for the concept of the ultimate distributional unit, within which each species is held by its structural and instinctive limitations. . . . no two species in the same general territory can occupy for long identically the same ecological niche." (Incidentally, the latter statement predates Gause's experimental demonstration of the competition exclusion principle; see

Figure 7–2). Thus, Grinnell thought of the niche mostly in terms of the microhabitat, or what is now called the **spatial niche.** Charles Elton (1927 and later publications) in England was one of the first to begin using the term "niche" in the sense of the "functional status of an organism in its community." Because of Elton's very great influence on ecological thinking, it has become generally accepted that niche is by no means a synonym for habitat. Since Elton emphasized the importance of energy relations, his version of the concept might be considered the trophic niche.

In 1957, G. E. Hutchinson suggested that the niche could be visualized as a multidimensional space or hypervolume within which the environment permits an individual or species to survive indefinitely. Hutchinson's niche, which can be designated the **multidimensional** or **hypervolume niche**, can be measured and mathematically manipulated. For example, the two-dimensional climographs shown in Figure 5–15, which picture the climatic niche of a species of bird and a fruit fly, could be expanded as a series of coordinates to include other environmental dimensions. Hutchinson (1965) also distinguished between the **fundamental niche**—the maximum "abstractly inhabited hypervolume" when the species is not constrained by competition with others—and the **realized niche**—a smaller hypervolume occupied under biotic constraints. The two concepts are illustrated in two dimensions in Figure 7–13A and C.

Vandermeer (1972) has extended Hutchinson's concept of realized niche somewhat to include the likelihood that any given species will have a different realized niche when 0, 1, 2, 3 . . . *n* species are present and interacting. When no other potential competitor is present, the niche is optimal (or fundamental in the Hutchinsonian sense), but in the presence of other similar species, there would be not just one realized niche but a series of partial niches depending on the number of potential competitors. Thus, Vandermeer recommends use of the term **partial niche** in place of "realized niche."

Perhaps a simple analogy from everyday human affairs will help to clarify these overlapping and sometimes confusing ecological uses of the term "niche." To become acquainted with some person in the human community, one would need to know, first of all, his address, that is, where he could be found. "Address" then would represent habitat. To really know the person, however, one would want to know something about occupation, interests, associates, and the person's role in community life. One's profession would be analogous to niche. So in the study of organisms, learning the habitat is just the beginning. To determine the organism's status in the natural com-

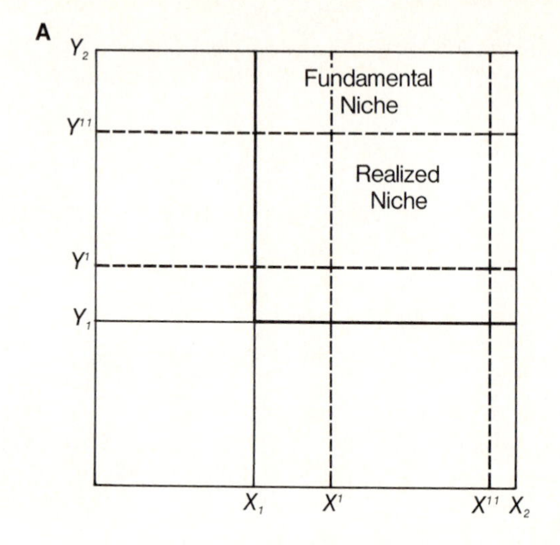

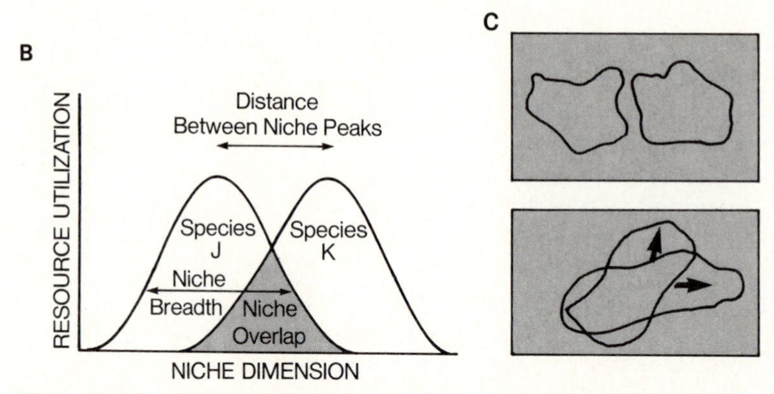

Figure 7–13 Schematic representations of the niche concept. *A* and *C*. The hypervolume concept. In the upper diagram in *C*, two species occupy nonoverlapping niches, while in the lower diagram, niches overlap so much that severe competition results in divergence, as indicated by the arrows. *B*. Activity curves for two species along a single resource dimension illustrate the concepts of niche width and niche overlap.

munity, one would need to know something of its activities, especially its nutrition; energy sources and their partitioning; the relevant population attribute, such as intrinsic rate increase, fitness, and so on; and finally the organism's effect on other organisms with which it comes into contact and the extent to which it modifies or can modify important operations in the ecosystem.

MacArthur (1968) has noted that the ecological term "niche" and the genetic term "phenotype" are parallel concepts in that they both involve a very large number of attributes, both include some or all of the same measurements, and both are most useful in determining differences between individuals and species. Thus, niches of similar species associated together in the same habitat can be precisely compared when only a few operationally significant measurements are involved. MacArthur compares the niches of four species of American warblers (Parulidae) that all breed in the same macrohabitat, a spruce forest, and all feed on insects but forage and nest in different parts of the spruce tree. He constructed a mathematical model, which consisted of a set of competition equations in a matrix from which competition coefficients were calculated for interaction between each species and each of the other three. Two species proved especially competitive, so that if either were absent, the other might be expected to move into the vacated niche space. The general tendency for niches to narrow with interspecific competition was illustrated in Figure 7–3.

In recent ecological literature, the term **guild** is often used for groups or clusters of species, such as MacArthur's warblers, that have a similar or comparable role in the community (see Root, 1967, who first suggested this usage). Wasps parasitizing a herbivore population, nectar-feeding insects, snails living in the forest floor litter, and vines climbing into the canopy of a tropical forest are other examples of guilds. The guild is a convenient unit for studies of interactions between species, but it can also be treated as a functional unit in community analysis, thus making it unnecessary to consider each and every species as a separate entity.

Microhabitat separation of spatial niches in guilds is quite common—almost the rule. Two examples are shown in Figure 7–14. Niche separation in feeding areas is rather sharp for several of the seven species of woodland slugs, but there is considerable overlap in the position of five species of amphipods along a salinity gradient.

Measurements of morphological features of larger plants and animals can often be used as indices in comparison of niches. Van Valen (1965), for example, found that the length and breadth of a bird's bill (the bill, of course, reflects the type of food eaten) are an index of "niche width"; the coefficient of variation in bill width was found to be greater in island populations of six species of birds than in mainland populations, corresponding with the wider niche (wider variety of habitat occupied and food eaten) on islands where competing species are fewer. Within the same species, competition is often greatly reduced when different stages in the organism's life history occupy different niches; for example, the tadpole functions as a herbivore and the adult frog as a carnivore in the same pond. Niche

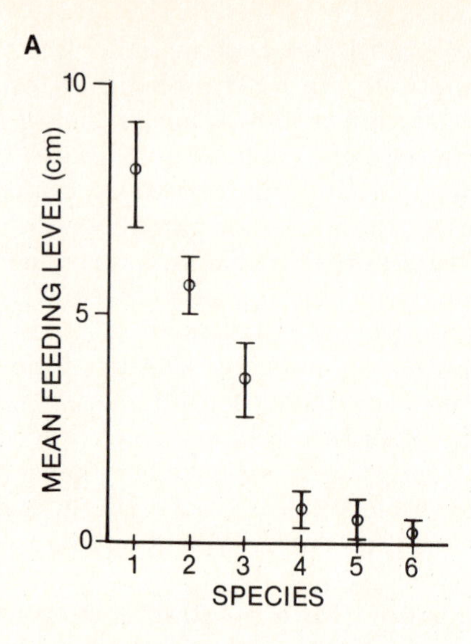

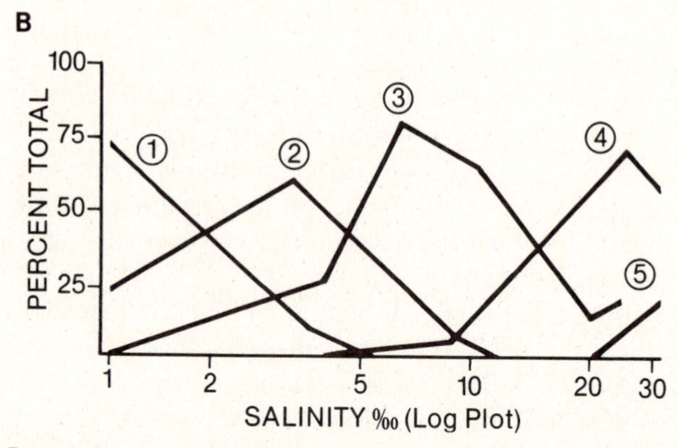

Figure 7–14 *A*. Niche separation in use of vertical space by woodland slugs. (After Jennings and Barkham, 1979.) *B*. Habitat separation in amphipods along a salinity gradient; 1 to 5 are species. (After Fenchel and Kalding, 1979.)

segregation may even occur between the sexes. In woodpeckers of the genus *Picoides* (formerly *Dendrocopus*) male and female differ in size of bill and in foraging behavior (Ligon, 1968). In hawks as well as in many insects, the sexes differ markedly in size and, therefore, in the dimensions of their food niche.

Both nutrient and toxic chemicals introduced into natural ecosystems can be expected to alter niche relations of species most severely affected by the perturbation. In an experimental study of the effect of applying N-P-K commercial fertilizer to old-field vegetation, Bakelaar and Odum (1978) have reported that niche width and niche overlap were not altered for most species as a result of fertilization, but niche width was significantly increased for the major dominant goldenrod (a species of *Solidago*), which increased its coverage at the expense of the codominant plant (a species of *Aster*).

The extent to which species-abundance patterns within trophic levels, taxonomic groups (birds, insects, and so on), and whole communities provide clues to the nature of niche relationships is considered in the next section.

Ecological equivalent species that occupy the same or similar niches in different geographical regions tend to be closely related taxonomically in contiguous regions, but are often not related in noncontiguous regions. Species composition of communities differs widely in different floral and faunal regions, but similar ecosystems develop wherever physical conditions are similar, regardless of geographical location. The equivalent functional niches are occupied by whatever biological groups happen to make up the flora and fauna of the region. Thus, a grassland type of ecosystem develops wherever there is a grassland climate, but the species of grass and grazers may be quite different, especially where regions are widely separated by barriers. The large kangaroos of the Australian grassland are the ecological equivalents of the bison and antelope of the North American grassland (both now largely replaced by domesticated grazers).

Tables 7–3 and 7–4 detail two examples of ecological equivalence in the seashore and the grassland habitat. The periwinkles that replace one another up and down the contiguous coasts of North and South America all belong to the same genus, but the birds in the Kansas and Chilean grasslands belong to different families.

Table 7–3 Ecological Equivalents in Three Major Niches of Four Coastal Zones of North and Central America

Niche	Tropical	Upper West Coast	Gulf Coast	Upper East Coast
Grazer on intertidal rocks (periwinkles)	*Littorina ziczac*	*L. danaxis* *L. scutelata*	*L. irrorata*	*L. littorea*
Benthic carnivore	Spiny lobster (*Panulirus*)	King crab (*Paralithodes*)	Stone crab (*Menippe*)	Lobster (*Homarus*)
Plankton-feeding fish	Anchovy	Pacific herring, sardine	Menhaden, threadfin	Atlantic herring, alewife

Table 7–4 Ecological Equivalent Grassland Birds in a Kansas Field and a Chilean Field*

Ecological Equivalent Pairs of Species	Body Size (mm)	Bill Length (mm)	Ratio of Bill Depth to Length
Eastern meadowlark (*Sturnella magna*) Kansas	236	32.1	0.36
Red-breasted meadowlark (*Pezites militaris*) Chile	264	33.3	0.40
Grasshopper sparrow (*Ammatramus savannarum*) Kansas	118	6.5	0.60
Yellow grass finch (*Sicalis luteula*) Chile	125	7.1	0.73
Horned lark (*Eromorphila alpestus*) Kansas	157	11.2	0.50
Chilean pipit (*Anthus correnderas*) Chile	153	13.0	0.42

* In each field, the three species differ in feeding niches, as shown by differences in body size and bill dimensions, but each pair of equivalents is very closely matched morphologically, indicating very similar niches. The meadowlarks are closely related taxonomically, but the second pair are related only at the family level, and the third pair belong to different families. (After Cody, 1974.)

6. Species Diversity, Pattern Diversity, and Genetic Diversity in Communities

Statement

Of the total number of species in a trophic component, or in a community as a whole, a relatively small percent are often abundant or **dominant** (represented by large numbers of individuals, a large biomass, productivity, or other indication of "importance"), and a large percent are rare (have small "importance" values). Sometimes, though, there are no dominants but many species of intermediate abundance. The concept of **species diversity** has two components: (1)

richness, also called species density, based on the total number of species present, and (2) **evenness**, based on relative abundance (or other measure of "importance") of species and the degree of its dominance or lack thereof. Species diversity tends to increase with the size of area and from high latitudes to the equator. Diversity tends to be reduced in stressed biotic communities, but it may also be reduced by competition in old communities in stable physical environments. Two other kinds of diversity are important: (1) **pattern** diversity, which results from zonation, stratification, periodicity, patchiness, food webs, and other arrangements of component population and microhabitats; and (2) **genetic** diversity, the maintenance of genotypic heterozygosity, polymorphism, and other genetic variability that is an adaptive necessity of natural populations. Many ecologists are becoming concerned that reduction in species and genetic diversity resulting from human activities is jeopardizing future adaptability in both natural ecosystems and agroecosystems.

Explanation

The pattern of a few common or dominant species having large numbers of individuals associated with many rare species having few individuals is characteristic of community structure in northern latitudes and the seasonal tropics (wet-dry seasons), but in the wet tropics with unchanging seasons, one usually finds many species that have low relative abundance. The general trend of an increase in numbers of species from north to south is illustrated in Figure 7–15. A second general trend or natural law is that numbers of species increase with size of area, and probably also with evolutionary time that has been available for colonization, niche specialization, and speciation (see Preston, 1960; Sanders, 1968).

Two broad approaches are used to analyze species diversity in different situations: (1) **dominance-diversity relative abundance curves** and (2) **diversity indices**, which are ratios or other mathematical expressions of species-importance relationships. These approaches will be considered in detail, but it is important first to recognize the two basic components of species diversity, because they may respond differently to geographical, developmental, or physical factors. One major component is **species richness**, or **variety** or **species density.** It is simply the total number of species, usually expressed for the purposes of comparison as a species/area ratio or species/numbers of individuals ratio. Often, but by no means always, species-numbers relationships are logarithmic (straight line on a semilog plot), and number of species is a power function of area

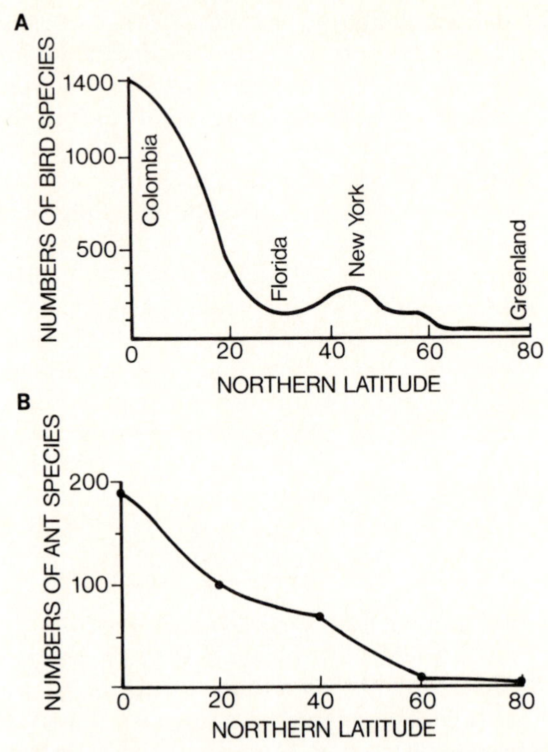

Figure 7-15 Latitudinal gradient in numbers of breeding land birds (*A*) and ants (*B*). (Redrawn from Fischer, 1960.)

(straight line on a double-log plot) as illustrated in Figure 7–16A and B. Both habitat diversity and area per se are involved in the increasing number of species with increasing area trend, as shown in Figure 7–16B. Species-area relationships have been used to determine optimal sample size and to predict number of species in areas larger than that sampled. Connor and McCoy (1979) report that several equations (including power function and double-log transformation) provide a good fit for species-area curves, but no statistical model has any biological significance.

The second major component of diversity is **evenness** or **equitability** in the apportionment of individuals among the species. For example, two systems, each containing ten species and 100 individuals, have the same *S/N* richness index, but these could differ widely in evenness, depending on the apportionment of the 100 individuals among the ten species—for example, 91-1-1-1-1-1-1-1-1-1 at one extreme (minimum evenness and maximum dominance) and ten indi-

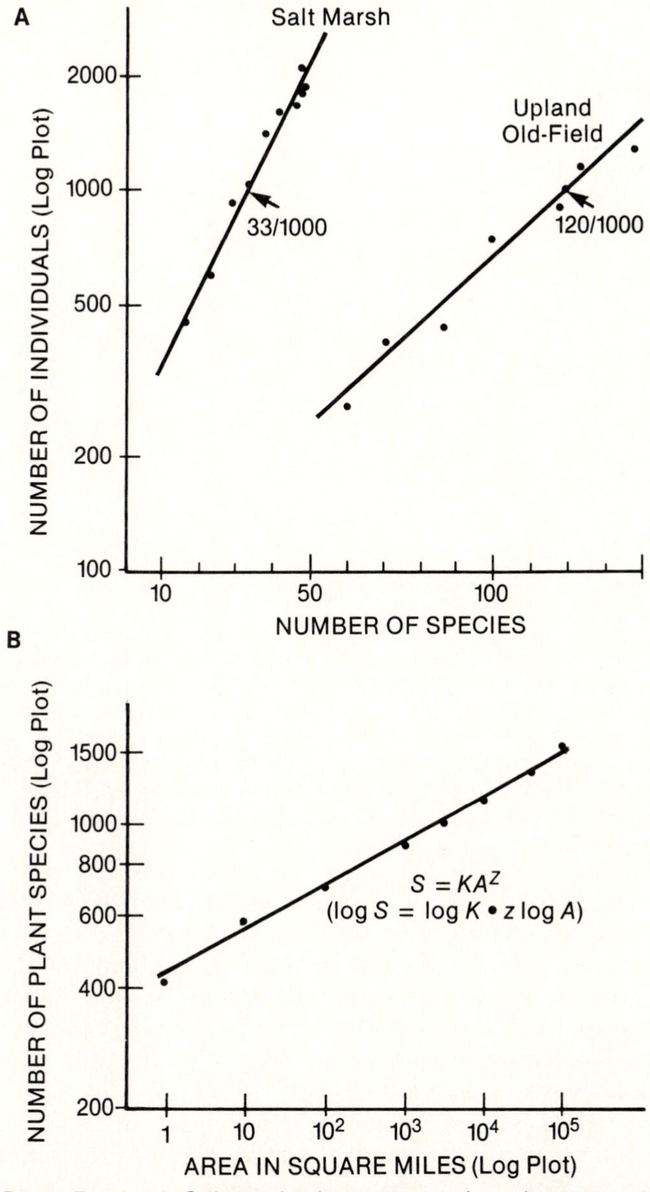

Figure 7–16 *A*. Relationship between number of species and number of individuals in a Georgia salt marsh and old-field. Where the relation is logarithmic, the number of species/1000 individuals can be used as a diversity index. *B*. Species-area relationship in plants of Great Britain. The slopes of species-area regression lines fall between 0.1 and 0.35, with low values on continents and high values on islands. (See Connor and McCoy, 1979.)

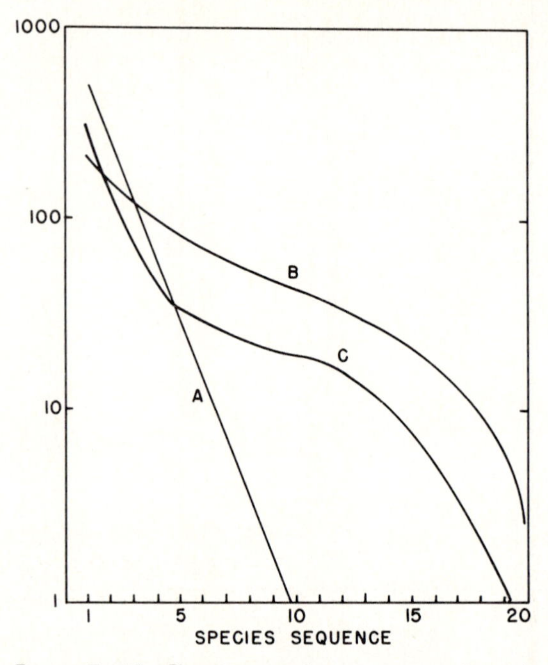

Figure 7–17 Dominance-diversity curves for a hypothetical sample of 1000 individuals in 20 species from a community. Number of individuals in the species (ordinate) are plotted against species number in sequence from the most abundant to the least abundant (abscissa). Curve A—geometric series with niche preemption. Curve B—nonoverlapping, random niche boundaries. Curve C—intermediate sigmoid pattern, with multiple niche dimensions and overlap, which generate a lognormal distribution of species importances. (Redrawn from Whittaker, 1965.)

viduals per species (perfect evenness and no dominance) at the other extreme. Evenness tends to be high and constant in bird populations (probably because of territorial behavior), so differences in various communities and geographical areas are largely due to differences in richness (Tramer, 1969). In contrast, plants and phytoplankton tend to average lower in evenness and to exhibit considerable variation in both components (Austin and Tomoff, 1978).

One of the best ways to picture both components of diversity is to plot the numbers of individuals (biomass, productivity, or other appropriate measure of importance) for each species on semilog graph paper with species arranged on the x-axis in sequence from most to least abundant. A trend line connecting or fitted to the points is called a **dominance-diversity curve** after Whittaker (1965, 1972), who first suggested this designation. "Species importance curves" (Pianka, 1978) would also be a good label for this graphic approach. Figure 7–17 shows three basic patterns for a hypothetical sample of 1000 individuals in 20 species. If the most abundant species is twice as numerous as the next most abundant one, which in turn has twice the density of the third, and so on, one gets a straight line, as shown in curve A of the figure. From this one might assume that the first species occupies half the available niche space, the second species, half the remaining space (25 percent of total), and so on. In other words, each species preempts niche space with no overlaps. If, however, the niche space is divided into random, contiguous, nonoverlapping segments, an entirely different curve is obtained, as in curve B of the figure. This pattern represents the "broken-stick" model of MacArthur (1957). These two possibilities represent what seem to be extremes; most natural distributions appear as some kind of intermediate sigmoid curve (curve C), indicating a more complex pattern of niche differentiation and overlap. A lognormal distribution of species importances is suggested by this pattern (curve C). See Preston (1948), Whittaker (1972), and Figure 7–21B. Groups exhibiting intense interspecific competition and territorial behavior, such as forest birds, tend to conform to the nonoverlapping, random niche pattern of curve B; the simple geometric series (curve A) is found in some plant communities in harsh environments. As emphasized in Section 2, most species in the open systems of nature coexist under conditions of partial rather than direct competition, and many adaptations promote niche differentiation without competition exclusion from a habitat. Accordingly, curve C is the model to be expected, especially in relatively undisturbed communities.

In summary, graphing species importances as in Figures 7–17 and 7–18 not only accurately depicts the richness and the relative abundance of species diversity but also explains how niche space is partitioned. For a given number of species, the higher and flatter the curve, the higher the overall diversity; thus in Figure 7–17, diversity of B > diversity of C > diversity of A. The steeper the curves, the lower the overall diversity and the greater the dominance by one or a few species. Stress, whether natural (severe weather, for example) or anthropogenic (pollution, for example), tends to steepen the curves, so the dominance-diversity curve can be used to assess the effect of perturbations on species structure.

The second approach uses diversity indices. The four most widely used indices are shown in Table 7–5, along with some variants. A is an index of species richness and B, an index of evenness, while B and C combine both components and, therefore, are general indices of diversity. The Simpson index shows "concentration" of dominance, since the higher the value, the greater the dominance by one or a few species. One minus Simpson (or reciprocal) becomes an index of diversity comparable to the others listed.

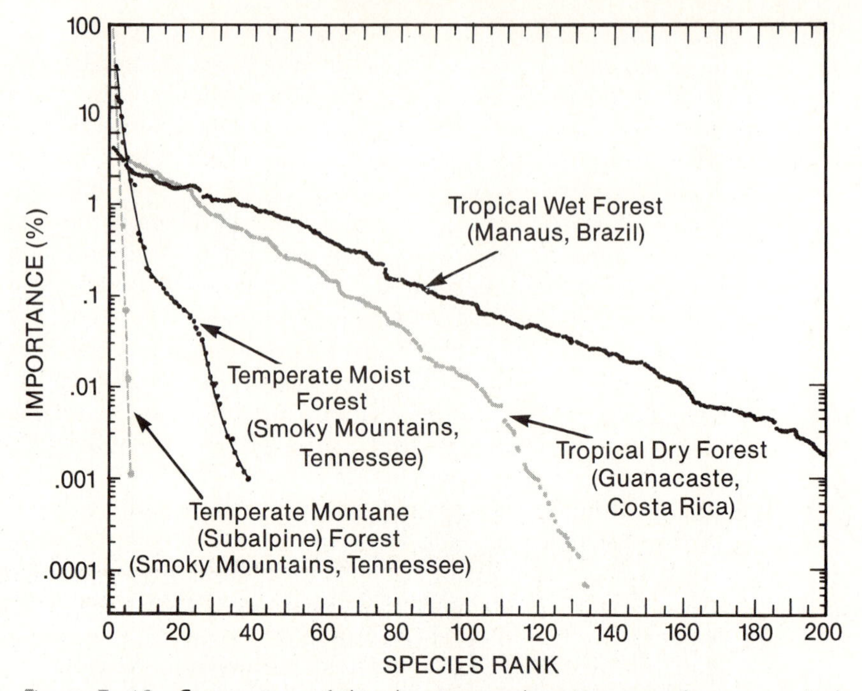

Figure 7–18 Comparison of the dominance-diversity curves for two tropical forests and two temperate forests. Importance values for the temperate forests are based on annual net production; importance values for the dry forest in Costa Rica are from basal area (cross-sectional area of all stems of a given species); importance values in the Amazonian forest are based on above-ground biomass. (After Hubbell, 1979.)

Of the two general indices, the Simpson gives greater weight to common species (since squaring small n_i/N ratios results in very small numbers). The Shannon index gives greater weight to rare species. Since the Shannon index is derived from information theory and represents a type of formulation widely used in assessing complexity and information content of all kinds of systems, it is one of the best to use in making comparisons if one is not interested in separating the two components of diversity. And once $\bar{H}$ is calculated, evenness can be separated out quickly by dividing by the log of number of species (D, Table 7–5). The Shannon is also reasonably independent of sample size and is normally distributed, provided N are integers (Bowman et al., 1970; Hutcheson, 1970), so routine statistical methods can be used to test for significance of differences between means. Lyons (1981) has shown that biomass or productivity (noninteger quantities), which are often more ecologically appropriate, can be used in such statistical tests if numbers of individuals are also known.

Table 7–5 Some Useful Diversity Indices

A. Species Richness Index (d)*

$$d = \frac{S-1}{\log N} \text{ (also: } S/N \text{ and } N \text{ per 1000 individuals)}$$

where

S = number of species
N = number of individuals

B. Simpson's Index (c)†

$$c = \sum (n_i/N)^2 \text{ (or } \sum n_i \left(\frac{n_i(n_i-1)}{N(N-1)}\right) = \text{dominance index form}$$

and

$$1 - \sum (n_i/N)^2 \text{ and } \frac{1}{\sum (n_i/N)^2} = \text{diversity index forms}$$

where

n_i = importance value for each species (number, biomass, and so on)
N = total of importance values

C. Shannon Index ($\overline{H}$)‡

$$\overline{H} = -\sum (n_i/N \log (n_i/N) \text{ or } -\sum P_i \log P_i$$

where

n_i = importance value for each species
N = total of importance values
P_i = importance probability for each species = n_i/N

D. Pielou's Evenness Index (e)§

$$e = \frac{\overline{H}}{\log S}$$

where

$\overline{H}$ = Shannon index
S = number of species

* See Simpson, 1949.

† See Margalef, 1958 (log index); Menhinick, 1964 (square root index); H. T. Odum, Cantlon, and Kornicker, 1960 (S/1000 individuals index).

‡ See Shannon and Weaver, 1949; Margalef, 1968.

§ See Pielou, 1966. For another type of equitability index, see Lloyd and Ghelardi, 1964.

Note: In d, $\overline{H}$, and e, natural logarithms ($\log_e$) are usually employed, but $\log_2$ is sometimes used to calculate $\overline{H}$ to obtain "bits" per individual.

Since stable ecosystems such as rain forests or coral reefs have high species diversity, it has been tempting to conclude that diversity enhances stability. As Margalef, in 1968, expressed it, "The ecologist sees in any measure of diversity an expression of the possibilities of constructing feedback systems." However, recent analyses and critical reviews (beginning with the 1969 symposium "Diversity and Stability in Ecological Systems," edited by Woodwell and Smith, and including papers in the Proceedings of the First International Congress of Ecology, edited by Van Doblen and Lowe-McConnell, 1975) have suggested that the relationship between species diversity and stability is complex, and a positive relationship may sometimes be secondary and not causal in that stable ecosystems promote high diversity but not necessarily the other way around. Huston (1979) concludes that what he calls "non-equilibrium" ecosystems, that is, systems that are periodically perturbed, tend to have a higher diversity than "equilibrium" ecosystems where dominance and competition exclusion are more intense. On the other hand, McNaughton (1978) concludes from his studies of old-fields and East African grasslands that species diversity does mediate functional stability in the community at the primary producer (vegetation) level. Like all desirable entities, there perhaps can be too much of a good thing; a large number of species fighting and competing with one another can be destabilizing, like "too many cooks in the kitchen," as a familiar expression goes (see May, 1973). Nowhere in nature does one find the maximum theoretical diversity: many species are all equally important; some species are always rarer than others. For situations of high diversity, the average seems to be around 80 percent of maximum evenness (Odum, 1975). Species diversity tends to increase during ecological succession, but this trend does not necessarily continue into the older or mature stages, as will be documented in the next chapter (Sections 1 and 2).

A major problem with studies of species diversity is that, so far, they deal only with parts of communities, usually a taxonomic segment (for example, birds or insects) or at most a single trophic level. Estimating the diversity of whole communities requires that all the different sizes and niche roles be "weighted" in some manner by some common denominator such as energy. For some possibilities, see Odum, 1982.

Stability is likely related more closely to functional than to structural ("standing" crop) diversity (discussed previously in Chapter 2, Section 6).

Since diversity within a habitat or community type is not to be confused with diversity of a landscape or region containing a mixture of habitats, Whittaker (1960) has suggested the following terms: (1) alpha diversity, for within-habitat or within-community diversity; (2) beta diversity, for between-habitat diversity; and (3) gamma diversity, for diversity of a large regional area, biome, continent, island, and so on.

In trophic levels, well-studied zones (such as benthic aquatic populations), and other *parts* of communities, species diversity is very much influenced by the functional relationships between the trophic levels. For example, the amount of grazing or predation greatly affects the diversity of the grazed or prey populations. Moderate "predation" often reduces the density of dominants, thus providing less competitive species with a better chance to use space and resources. Harper (1969) reports that the diversity of herbaceous plants on the English chalk downs declined when grazing rabbits were fenced out. Severe grazing, on the other hand, acts as a stress and reduces the number of species to an unpalatable few. Paine (1966) found that species diversity of sessile organisms in the rocky intertidal habitat (where space is generally more limiting than food) was higher in both temperate and tropical regions where first- and second-order predators were active. Experimental removal of predators in such situations reduced species diversity of all sessile organisms whether directly preyed upon or not. Paine concluded that "local species diversity is directly related to the efficiency with which predators prevent monopolization of major environmental requisites by one species." This conclusion does not necessarily hold for habitats in which competition for space is less severe. Though human activities tend to reduce diversity and encourage monocultures, they do often increase diversity of habitat in the general landscape (openings created in the forest, trees planted in the prairie, new species introduced, and so on). The diversity of small song birds and plants is much greater in older established residential districts than in many natural areas (noted in Chapter 2, Section 7).

The species may not always be the best ecological unit for measurements of diversity, since life history stages or life forms within the species often occupy different habitats and niches and thereby contribute to variety in the ecosystem. A caterpillar and a butterfly of the same species or a frog and its tadpole stage are more diverse in their roles in the community than are two species of caterpillars or adult frogs. As noted by Harper (1977), "the life cycle stages of a plant contribute a diversity to an ecosystem as great as that of many species." Accordingly, one does not need to be an expert taxonomist (or require the help of one) to estimate diversity. In a collection of insects obtained with a sweep net, for example, one need only separate individuals that show distinctive morphological and size differences without needing to know species names or worrying about whether the different "kinds" belong to the same species or not. In other words, the n_i in the diversity formulas can represent distinct morphological types that will probably represent different ecological kinds (since, as already noted, size and morphological differences indicate niche differences).

Consider genetic diversity that also is hidden when communities are described only in terms of species present. As explained by Vida (1978), the classical theory has been that members of natural populations should be homozygous for the "fittest" allele, as shown by the following scheme in which all but one of six allele pairs are homozygous (the same):

$$\frac{a_1b_1c_1d_1e_1f_1}{a_1b_1c_1d_1e_2f_1}$$

The alternative theory is that most loci are heterozygous, and much polymorphism is maintained by various forms of balancing selection, as illustrated by the following scheme in which all but one of the six loci are heterozygous (different) and there is a much greater variety of alleles:

$$\frac{a_1b_5c_2d_1e_1f_5}{a_3b_2c_6d_1e_2f_4}$$

The latter or "balanced polymorphism" theory seems the more appropriate, as determined by modern biochemical methods that can detect hidden genetic variability, and it also appears that polymorphism is maintained by natural selection. In the absence of such genetic variability, a species will not be able to adapt to new situations and, therefore, will become extinct in a changing environment.

Variety of species, life history stages, and genetic types are by no means the only elements involved in community diversity. The structure that results from the distribution of organisms in, and their interaction with, their environment has been called **pattern** by Hutchinson (1953). Many different kinds of arrangements in the standing crop of organisms contribute to what might be called **pattern diversity.*** For example:

1. Stratification patterns (vertical layering, as in vegetation and soil profiles).

* Pielou (1966a) has used the term "pattern diversity" in a much more restricted sense, namely, to refer to the degree of segregation of individuals of one population from those of another.

2. Zonation patterns (horizontal segregation, as in mountains or in the intertidal zone).
3. Activity patterns (periodicity).
4. Food-web patterns (network organization in food chains).
5. Reproductive patterns (parent-offspring associations, plant clones, and so on).
6. Social patterns (flocks and herds).
7. Coactive patterns (resulting from competition, antibiosis, mutualism, and so on).
8. Stochastic patterns (resulting from random forces).

Diversity is also enhanced by "edge effects"—junctions between or patches of contrasting types of vegetation or physical habitats, to be discussed in the next section.

Examples

Dominance-diversity curves for trees in forests ranging from the wet tropics to the subalpine mountains, as shown in Figure 7–18, illustrate very well the arctic-to-tropic gradient in species diversity as well as the several types of curves (geometric, lognormal) shown in the generalized model of Figure 7–17.

The use of diversity curves and indices to assess the impact of sewage effluent on stream benthos (organisms living on the bottom) is illustrated in Figure 7–19. Dominance-diversity curves and Shannon indices for bottom fauna in three parallel small rivers that flow out of the Atlanta metropolitan district are shown in Figure 7–19A. One stream flows out of a densely populated suburban area and receives a lot of sewage effluent, the second is further away from the city and receives only moderate amounts, and the third flows through a rural area. Both curves and the indices show how diversity decreases with increasing load of domestic wastes. In a longitudinal study of an Oklahoma stream receiving inadequately treated municipal wastes, Wilhm (1967) found that benthic diversity was depressed for more than 60 miles downstream, as shown in Figure 7–19B. From these and many other studies it is clear that benthic diversity is a good tool for monitoring water pollution.

Before diversity indices are used to compare one situation with another, the effect of sample size should be determined. One way to do this is shown in Figure 7–20A, in which indices are calculated for cumulative 0.1 m^2 samples of arthropods in a uniform grain field. The Shannon index leveled off at four to five samples (even though only a small portion of the species present had been sampled), but the vari-

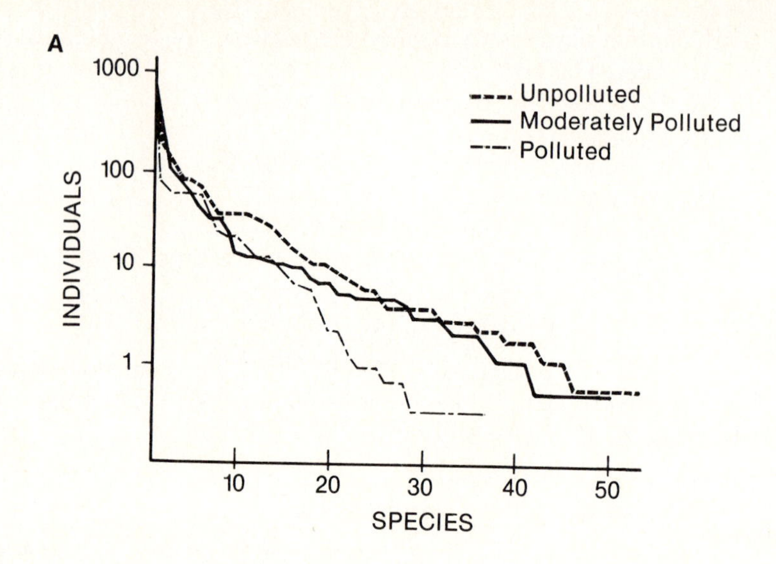

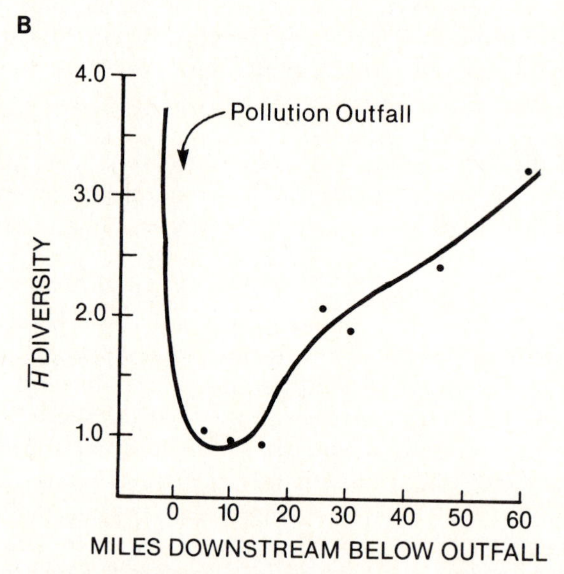

Figure 7–19 *A*. Dominance-diversity profiles for three parallel streams in the same watershed that differ in their degree of pollution from urban domestic wastes. The Shannon diversity indices for the streams are as follows: unpolluted, 3.31; moderately polluted, 2.80; polluted, 2.45. (After E. P. Odum and Cooley, 1980.) *B*. Changes in the Shannon index of diversity ($\overline{H}$) of the benthos (organisms living on the bottom) downstream from a pollution outfall (mixed domestic and industrial sewage from a small city), illustrating the marked effect of chronic pollution of a stream by inadequately treated wastes. (Redrawn from Wilhm, 1967.)

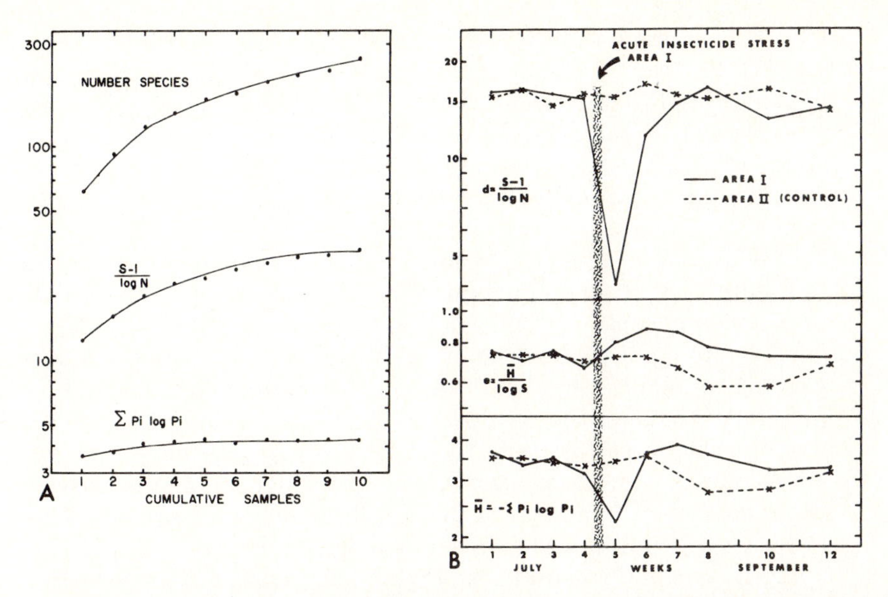

Figure 7–20 *A*. The effect of increasing sample size on two diversity indices plotted on semilog graph paper together with the number of species cumulated in successive 0.1 m^2 samples of the arthropods in the vegetation of a millet (*Panicum*) field in Georgia. *B*. The effect of a single application of the insecticide Sevin (an organophosphate insecticide that remains toxic for only about ten days) on the arthropod population on a 1-acre plot of the millet field. Two components of diversity (d and e) and a general index of total diversity ($\bar{H}$) are based on ten 0.1 m^2 samples taken from the treated area and a control area at weekly or biweekly intervals from early July through September. The semilog plots facilitate a direct comparison of relative deviations resulting from the acute insecticide stress. (After Barrett, 1969.)

ety index ($S - 1/\log N$) did not begin to level off until nine to ten samples had been pooled.

Figure 7–20B shows the effect of an acute insecticide stress on diversity of arthropods in the millet field. Although the variety index was greatly reduced by the treatment, evenness increased and remained elevated for most of the growing season. In this example, the diversity of the pretreatment population of arthropods was low, with strong dominance by a few species in each trophic level. When the insecticide killed many of the dominant species, a greater evenness in the abundance of the surviving populations resulted. The Shannon index expresses the interaction of these two components of diversity and, therefore, shows an intermediate response. Although the insecticide used in this experiment remained toxic for only ten days, and the acute depression lasted only about two weeks, overshoots and oscillations in the diversity ratios were evident for many weeks. This study illustrates several points of interest: (1) it is desirable to separate species richness and relative abundance, (2) a generalized or moderate perturbation may increase rather than decrease diversity when there is strong dominance, and (3) recovery is rapid when small areas are perturbed, because replacements come in quickly from surrounding areas. Applying insecticides to large areas is quite another matter.

Two additional graphic treatments of species diversity are shown in Figure 7–21. In the upper graph of the diversity of major invertebrates, components or "infauna" of sediments in different marine habitats are compared by the "rarification" method, which involves plotting cumulative number of species against cumulative number of individuals counted and evaluating differences in diversity from the shape of the curve (Sanders, 1968). Surprisingly, the deep sea sediments proved to house a diversity of animals higher than the adjacent continental shelf, even though the density (number per m^2) was quite low. To explain this finding, Sanders proposed a "time-stability" hypothesis: competitive interactions over long evolutionary time in a physically stable environment lead to many species with narrow niches. However, as Abele and Walters (1979) point out, high diversity in the deep sea could just as well be explained by this habitat's very large area (the species-area hypothesis).

When the number of species is plotted against the number of individuals by geometric interval (that is, 1–2, 2–4, 4–8, 8–16, and so on), a truncated normal curve is often obtained, as shown in the lower graph of Figure 7–21. This is the lognormal distribution mentioned earlier in this section. Again, pollution or other stress tends to lower and flatten the curve (reduce the height of the mode in this case).

Table 7–6 presents a comparison of density and diversity of arthropod populations in a grain field and a natural herbaceous community that replaced the field one year later. The values shown are means of ten samples taken over the growing season. After only one year under "nature's management," the following changes had occurred:

1. The number of herbivorous (phytophagous) insects was greatly reduced, as was total arthropod density.
2. The variety component of diversity and the index of total diversity were significantly increased in each guild, as well as for the total community of arthropods.
3. Evenness also increased.
4. The number, diversity, and percent composition of predators and parasites were greatly increased; predators and parasites made up only 17 percent of the population density in the grain field,

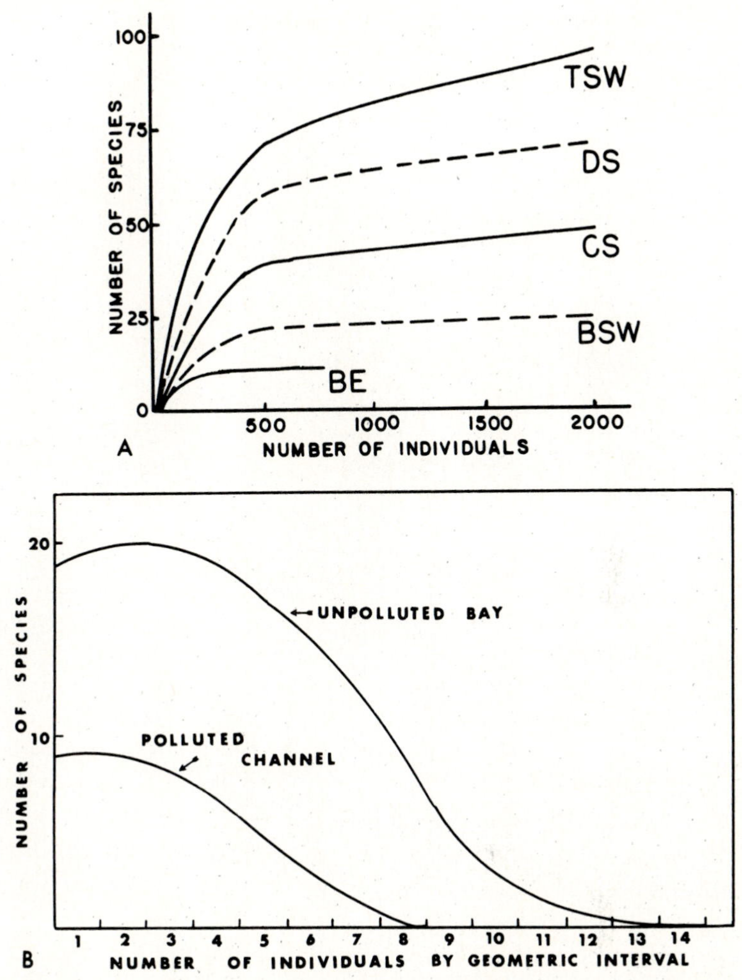

Figure 7–21 The use of species abundance curves to compare diversity in different habitats. *A*. The "rarefaction" method, involving the construction of curves based on using the percentage composition of species in single samples of marine sediments large enough to contain 500 to 3000 individuals of polychaetes and bivalves to determine the hypothetical number of species in successively smaller samples. The habitats in sequence from high to low species diversity are as follows: tropical shallow water (TSW), deep sea (DS), continental shelf (CS), boreal shallow water (BSW), and boreal estuary (BE). (Redrawn from Sanders, 1968.) *B*. The species structure of the diatom component of two estuarine communities in Texas as shown by truncated normal curves obtained by plotting the number of species in successive geometric intervals of abundance; that is, 1 to 2 individuals compose the first interval, 2 to 4 the second, 4 to 8 the third, 8 to 16 the fourth, and so on. In the polluted channel (the Houston ship channel) the number of species in all abundance classes was sharply reduced. (Redrawn from Patrick, 1967.)

Table 7–6 Density and Diversity of Arthropod Populations in an Unharvested Millet Crop Compared with the Natural Successional Community that Replaced It One Year Later*

Indices	Populations	Cultivated Millet Field	Natural Successional Community
Density no./m^2	Herbivores	482	156‡
	Predators	82	117
	Parasites	24	51‡
	Total arthropods	624	355‡
Variety index (d)	Herbivores	7.2	10.6‡
	Predators	3.9	11.4‡
	Parasites	6.3	12.4‡
	Total arthropods	15.6	30.9‡
Evenness	Herbivores	0.65	0.79‡
	Predators	0.77	0.80
	Parasites	0.89	0.90
	Total arthropods	0.68	0.84‡
Total diversity index ($\bar{H}$)†	Herbivores	2.58	3.28‡
	Predators	2.37	3.32‡
	Parasites	2.91	3.69‡
	Total arthropods	3.26	4.49‡

* The stand of millet (*Panicum*) was the control plot in the experiment graphed in Figure 7–20. Fertilizer was applied at time of planting in the prescribed agricultural manner, but no insecticide or other chemical treatment was applied, and the crop was not harvested. All figures are means of weekly samples taken during growing season, July through September.

† For formulas of diversity indices, see Table 7–5.

‡ Differences between the two communities significant at 99 percent level.

compared with 47 percent in the natural field (where they actually outnumbered the herbivores).

This comparison gives some clues as to why artificial communities often require chemical or other control of herbivorous insects, whereas natural areas should not require such control if only human beings would give nature a chance to develop its own self-protection.

What happened to diversity of tree species when the chestnut blight removed the principal dominant from the southern Appalachian forest is shown in Table 7–7. The chestnut tree, which constituted 30 to 40 percent of biomass of the original stand, has been replaced by several, not just one, species of oak, and several subdominant or pioneer species (such as tulip poplar) increased in response to the opening up of the canopy. These changes combined to reduce dominance and increase general diversity. In 1970, 25 years after the

Table 7–7 Forest Tree Diversity Before and After the Chestnut Blight Removed the Chief Dominant in Southern Appalachian Forests*

		Simpson Index of	
	Basal Area	Dominance	Diversity
1934; before blight	25	.25	.75
1953; after blight	22	.13	.87
1970; after blight	26	.17	.83

* Based on data of Nelson (1955) and Day and Monk (1974).

chestnut tree was removed from the canopy, the total basal area, but not diversity, had returned to a preblight level.

Forests and lakes are good places to observe stratification and its contribution to diversity. An interesting example of a stratified population in lakes is illustrated by the depth distribution of three species of game fish in impoundments in midsummer by the Tennessee Valley Authority (Figure 7–22). As in many deep lakes in temperate regions, a distinct physical stratification develops during the summer, with a layer of warm, oxygen-rich, circulating water lying over a deeper, colder, noncirculating layer that often becomes depleted of oxygen. As indicated in the figure, the large-mouth bass is the most tolerant of high temperatures (as is also indicated by its occurrence in nature farther south than the other two species) and is found near the surface. The other two species aggregate in deeper waters, the sauger selecting the deepest (and, therefore, the coldest) water that still contains an adequate supply of oxygen. By determining the depth distribution of oxygen and temperature, one can predict where the fish will be found in greatest numbers. Diagrams such as Figure 7–22 have, in fact, been published in local newspapers to aid the fisherman in deciding how deep he should fish to catch the desired species. As every fisherman realizes, simply knowing where the fish are does not guarantee success, but it might help.

In the ocean, schools of fish are often remarkably stratified, so much so that they create sharply delineated sound scattering layers or false bottom echoes on ships' sonar.

The Concern About Loss of Species and Genetic Diversity

Reduction in species and genetic diversity in historical times has produced short-term benefits in agriculture and forestry, as evidenced by the propagation of specialized, high-yielding varieties

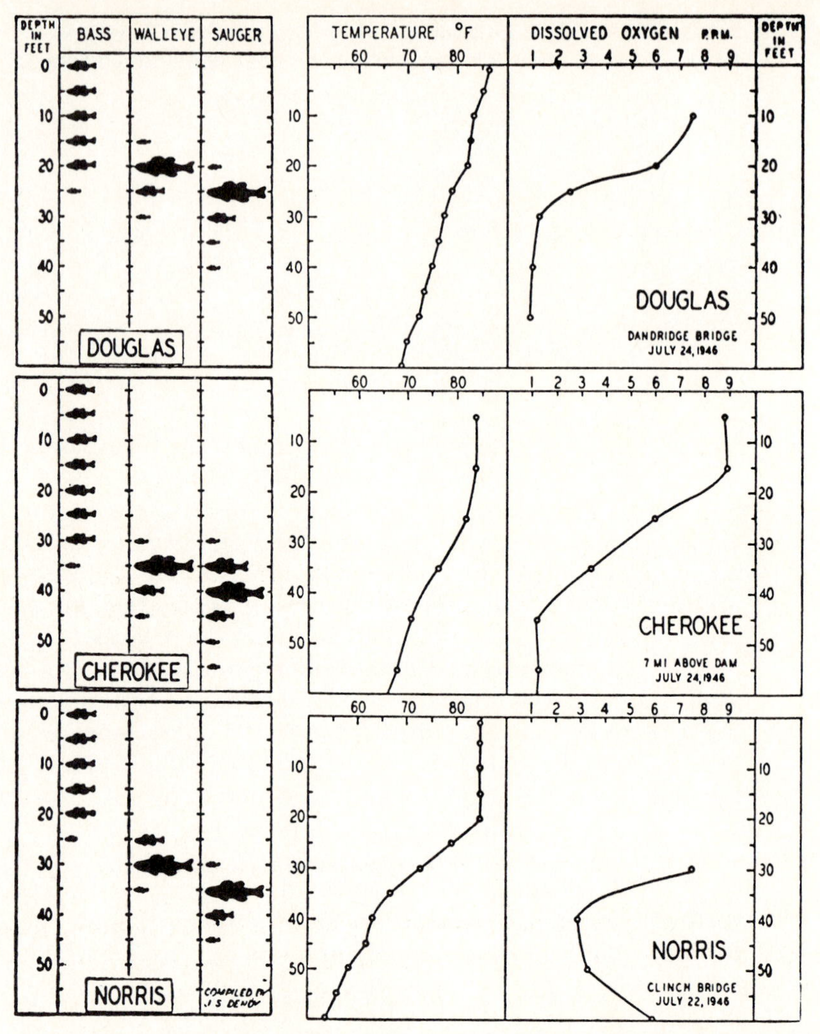

Figure 7–22 Depth distribution of three species of game fish in three Tennessee Valley Authority (TVA) impoundments in midsummer. Oxygen and temperature conditions that determine the level at which different species aggregate are shown on the right. (After Dendy, 1945.)

over large areas of the world's crop and forest land. Depending too much on these varieties, however, invites future catastrophe should climates change, or should the energy and chemical subsidies needed to maintain these varieties become scarce, or should new diseases and pests attack a vulnerable variety. That agriculturalists are deeply concerned about loss of crop diversity is seen in recent efforts to

tions at the expense of these other vegetation types to increase the yield of paper pulp and timber. However, conversion would be expensive economically (questionable cost-benefit) and would drastically reduce beta diversity (thus jeopardizing future maintenance and stability).

Figure 7–23 shows how landscapes can be planned to preserve diversity and yet accommodate urban and industrial development. The new town of Columbia, Maryland, is a good example of successful planning that was developed within the private, free-market sector with the blessings of the state and federal governments, but with a minimum of financial assistance from either. More about this example in Chapter 8, Section 7.

Many ecological concepts relating to diversity are controversial and need more study, but everyone agrees that diversity is necessary for the future survival of humans and nature.

In Chapter 2, we argued strongly that natural areas must be preserved for their essential role in life support. We can now add a second compelling reason—to preserve and safeguard the diversity required for future adaptation and survival.

7. Populations and Communities in Geographical Gradients; Ecotones and Concept of Edge Effect

Statement

Description of how populations and communities are arranged within a given geographical region or area of landscape has featured two contrasting approaches: (1) the zonal approach, in which discrete communities are recognized, classified, and listed in a sort of checklist of community types, and (2) the gradient analysis approach, which involves the arrangement of populations along a uni- or multidimensional environmental gradient or axis with recognition of community based on frequency distributions, similarity coefficients, or other statistical comparisons. The term **ordination** is frequently used to designate the ordering of species populations and communities along gradients, and the term **continuum** to designate the gradient containing the ordered populations or communities. In general, the steeper the environmental gradient, the more distinct or discontinuous are communities, not only because abrupt changes are more probable in the physical environment but also because boundaries are sharpened by competition and coevolutionary processes between interacting and interdependent species.

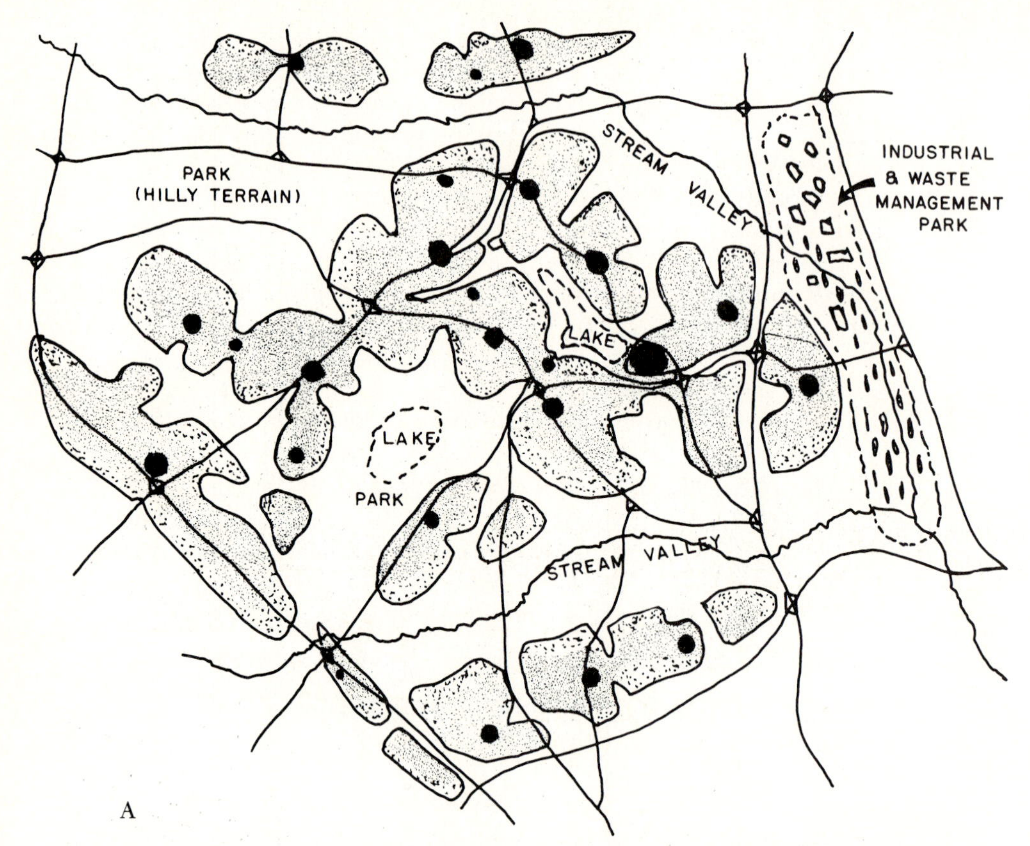

Figure 7–23 Designs for maintaining landscape diversity (beta and gamma levels) and environmental quality. *A*. Planned urban development for 110,000 people on 45,000 acres which preserves environmental quality and natural beauty and provides ample room for recreation and pollution abatement. A large amount of "open space" (unshaded areas in the diagram) is preserved (1) by cluster development of residential housing (two densities indicated) around town and village centers (black circles), separated by wide green belts or parks, and (2) by leaving stream valleys, lakes, and scenic areas unencumbered. Industrial plants are sited in a very large waste management park (right margin of the diagram) with room for waste treatment plants, oxidation ponds, land fills, and other means of degrading, recovering, containing, or recycling wastes and water. (Redrawn from Wallace, McHarg, Roberts and Todd Associates bulletin "Plan for the Valleys," page 59, with addition of the waste management park.) *B*. Schematic design for a waste management park for a power plant or (with appropriate change) a chemical or other large industrial plant. Toxic wastes are held in solid form in a burial ground (W), and degradable wastes and heated water are assimilated by a series of holding ponds and natural or agricultural areas. The chief inputs and outputs for this environmental system include (see numbered marginal arrows) 1, input of sunlight and rainfall; 2, export of nuclear wastes to burial grounds; 3, electric power to cities; 4, input of nuclear and other fuels; 5, output of food, fibers, clean air; 6, downstream flow of clean water for agriculture, industry, and cities; 7, public and professional use for recreation, education, and environmental research.

Illustration continued on opposite page

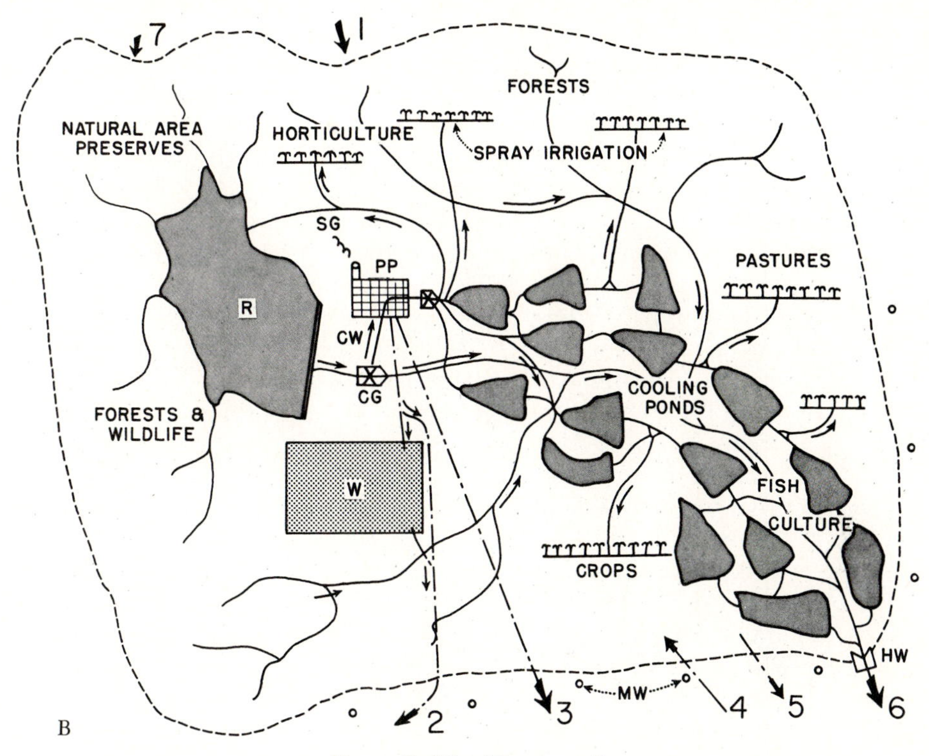

Figure 7–23 (Continued)

A sharp transition between two or more diverse communities, for example, between forest and grassland or between a soft-bottom and a hard-bottom marine substrate, is known as an **ecotone.** This junction zone or tension belt may have considerable linear extent but is narrower than the adjoining community areas themselves. The ecotonal community commonly contains many of the organisms of each of the overlapping communities and, in addition, organisms characteristic of and often restricted to ecotone. Often, both the number of species and the population density of some of the species are greater in the ecotone than in the communities flanking it. The tendency for increased variety and density at community junctions is known as the **edge effect.**

Explanation and Examples

Deciding where to draw boundaries is not too much of a problem in ecosystem analysis (Chapter 2, Section 2), as long as input and output environments are considered part of the system, no matter how the system is delimited (whether by natural features or by arbitrary lines drawn for convenience). However, when one delineates biotic communities by their component species populations, there is a problem. During the past half century, plant ecologists have debated whether land plant communities are to be thought of as discrete units with definite boundaries, as suggested by Clements (1905, 1916), Braun-Blanquet (1932, 1951), and Daubenmire (1966), or whether populations respond independently to environmental gradients to such an extent that communities overlap in a continuum, so that recognition of discrete units is arbitrary, as viewed by Gleason (1926), Curtis and McIntosh (1951), Whittaker (1951), Goodall (1963), and others. Whittaker (1967) illustrates these contrasting viewpoints with the following example. If at the peak of autumn coloration in the Great Smoky Mountain National Park one were to select a vantage point along the highway to obtain a view of the altitudinal gradient from valley floor to ridge top, one would observe five zones of color: (1) a multihued cove forest, (2) a dark green hemlock forest, (3) a dark red oak forest, (4) a reddish-brown oak-heath vegetation, and (5) a light green pine forest on the ridges. These five zones could be viewed as discrete community types, or all five could be considered part of a single continuum to be subjected to a gradient analysis that would emphasize the distribution and response of individual species populations to changing environmental conditions in the gradient. This situation is illustrated in Figure 7–24, which shows the frequency distribution (as hypothetical bell-shaped curves) of 15 species of dominant trees (a through o) that overlap along the gradient, and which presents the somewhat arbitrary designation of five community types (A, B, C_1, C_2, and D) based on the peaks of one or more dominants. Much can be said for considering the whole slope as one major community, since all the forests are linked together by exchanges of nutrients, energy, and animals as a watershed ecosystem, which, as emphasized in Chapter 2, is the smallest ecosystem unit amenable to functional studies and overall human management. On the other hand, recognizing the zones as separate communities is useful to the forester or land manager, for example, since each type of community differs in timber growth rate, timber quality, recreational value, vulnerability to fire and disease, and other aspects.

Concepts and approaches are often a function of geography; thus, ecologists working in areas of gentle gradients and uniform soils or substrates favor the continuum concept and various ordination techniques (statistical means of ordering species populations and communities along gradients), whereas ecologists working in areas of steep gradients or topographic discontinuities prefer the zonal con-

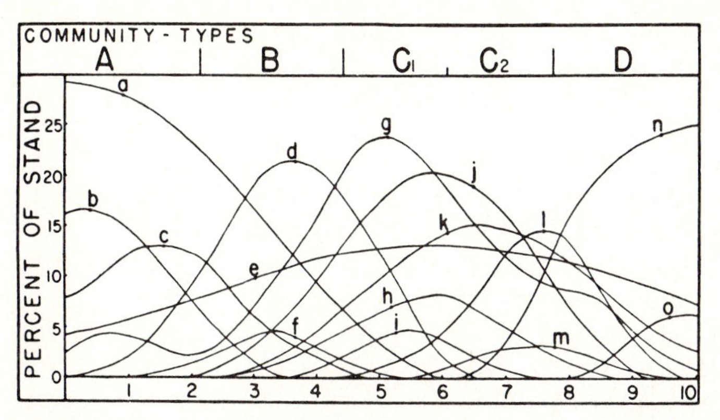

Figure 7–24 Distribution of populations of dominant trees along a hypothetical gradient, 0 to 10, illustrating the arrangement of component populations within a "continuum" type of community. Each species shows a "bell-shaped" distribution with a peak of relative abundance (percent of stand) at a different point along the gradient; some species show a wider range of tolerance (and usually a lesser degree of dominance) than other species. Within the large community, subcommunities may be delimited (as indicated by A to D above the graph) on the basis of combinations of two or more dominants, indicators, or other features. Such divisions will be somewhat arbitrary but useful for description and comparison. The curves have been patterned after data from several studies of tree distribution along an altitude gradient. (After Whittaker, 1954.)

cept. Ordination techniques often require ecologists to compare the similarity (or dissimilarity) of successive samples taken along an environmental gradient by using an index of the following general form:

$$\text{Similarity index } (S) = \frac{2C}{A + B}$$

where A = number of species in sample A, B = number of species in sample B, and C = number of species common to both samples. For a review of ordination and other gradient analyses, see Whittaker (1967) and McIntosh (1967).

Beals (1969) directly compared vegetational changes along a steep and a gentle altitudinal gradient in Ethiopia; two aspects of this study are shown in Figure 7–25. There was more discontinuity in the steep gradient as shown by the several sharp peaks in the plot of dissimilarity indices ($1 - S$ as shown above) calculated for pairs of adjacent samples (upper diagram). Furthermore, there was a distinct trend toward more sudden appearance and disappearance of species along the steep gradient than along the gentle one. As shown in the lower diagram, the frequency distribution curves for dominant species tended to be normal and bell-shaped (resembling the

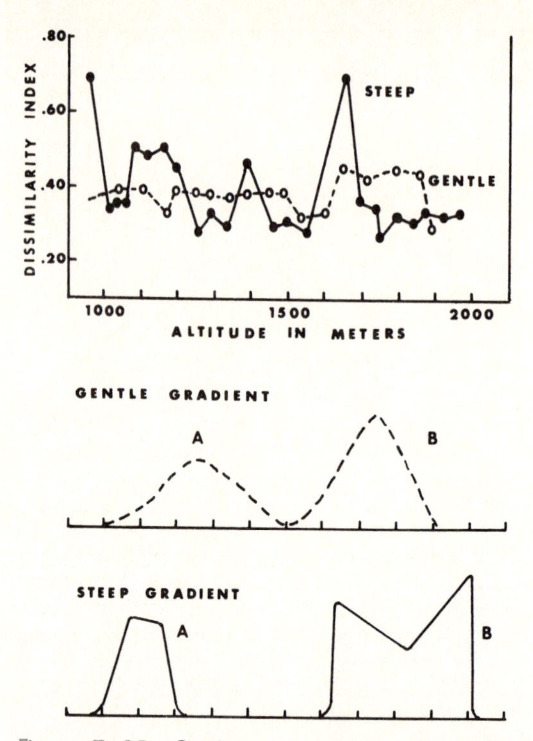

Figure 7–25 Gradient analysis of vegetative changes along a steep slope (solid lines) as compared with a gentle slope (broken lines), both having the same altitudinal range, in Ethiopia. *Upper.* A plot of dissimilarity indices for adjacent segments along the gradients. *Lower.* Frequency distribution of two species, *Acacia senegal* (A) and *Carissa edulis* (B). Species and community groups are more sharply delimited along the steep gradient. (Redrawn from Beals, 1969.)

hypothetical ones in Figure 7–24) along the gentle gradient, but they were steep-sided and truncated in the steep gradient, indicating sharper delimitation of populations. Beals concluded that "along a steep gradient the vegetation itself can impose disjunctions on an extrinsically continuous environmental gradient, whereas along a gentle gradient it may not do so." Coactions between populations can contribute to separating one community from another, as, for example, (1) competitive exclusion, (2) symbiosis between groups of species that depend on one another, and (3) coevolution of groups of species (to be considered in detail in the next chapter). Also, such factors as fire and antibiotic production can create sharp boundaries. Buell (1956) described a situation at Itasca Park, Minnesota, where, within the general maple-basswood forest, islands of spruce-fir forest maintain rather sharp boundaries unassociated with changes in topography. Marine benthic communities similarly show rather sharp zonation in steep gradients, as does vegetation on a mountainside.

Where abrupt changes occur along a gradient or where two distinctly different habitats or communities border one another, the resulting ecotone or transition zone often supports a community with characteristics besides those of the adjoining communities because many species require, as part of their habitat or life history, two or more adjacent communities that differ greatly in structure. For example, the American robin requires trees for nesting and open grassy areas for feeding. Since well-developed ecotonal communities may contain organisms characteristic of each of the overlapping communities plus species living only in the ecotone region, the variety and density of life are greater in the ecotone. This condition is what is meant by **edge effect.**

In a pioneer study demonstrating edge effect, Beecher (1942) found that the population density of birds increased as the number of feet of edge per unit area of community increased. From general observation, most people have noticed that the density of song birds is greater on estates, campuses, residential districts, and similar settings, which have mixed habitat and consequently much "edge," than on large, unbroken tracts of forest or grassland.

Ecotones may have characteristic species not found in the communities forming the ecotones. For example, in a study of bird populations along a community developmental gradient (see Table 8–2), study areas were selected to minimize the influence of junctions with other communities. Thirty species of birds were found to have a density of at least five pairs per 100 acres in some one of these stages. However, about 20 additional species were known to be common breeding birds of upland communities of the region as a whole; seven of these were found in small numbers, whereas 13 species were not even recorded on the uniform study areas. Among those not recorded were such common species as robin, bluebird, mockingbird, indigo bunting, chipping sparrow, and orchard oriole. Many of these species require trees for nest sites or observation posts, yet feed largely on the ground in grass or other open areas; therefore, their habitat requirements are met in ecotones between forest and grass or shrub communities, but not in areas of either alone. Thus, in this case, 40 percent (20 of 50) of the common species known to breed in the region may be considered primarily or entirely ecotonal.

David Patton (1975) has proposed a kind of diversity index for comparing the ratio of edge to area in different situations. Since the lowest ratio of edge to area is a circle, this ratio is given an index value of 1, so an edge index (*EI*) can be calculated as follows:

$$EI = \frac{TP}{2 \cdot A\pi}$$

where TP = total perimeter around area plus any linear edge within the area, A = area, and π = 3.14. A large square of one kind of vegetation accordingly would have an index of 1.13. If a square contained four different vegetation types of equal size, the additional internal edge would increase the index to 1.69. If half of two of the four vegetation types were clearcut to produce additional edge, the edge index would rise to 1.97. Taylor (1977) discusses two other edge indices.

An important general type of ecotone for human beings is the **forest edge.** A forest edge may be defined as an ecotone between forest and grass or shrub communities. Wherever humans settle, they tend to maintain forest-edge communities near their habitations. Thus, if human beings settle in the forest, they reduce the forest to scattered small areas interspersed with grasslands, cropland, and other more open habitats. Hawkins (1940) shows with maps how this change occurred in Wisconsin during the century following the appearance of the first settlers in 1838. If humans settle on the plains, they plant and water trees, creating a similar pattern. The preferred habitat of *Homo sapiens* can be said to be forest edge, since the species likes the shelter of trees and shrubs but gets its food from grassland and cropland. Some of the original organisms of the forest and plains can survive in the manmade forest edge, whereas those organisms especially adapted to the forest edge, notably many species of weeds, birds, insects, and mammals, often increase in numbers and expand their ranges because humans have created vast new forest-edge habitats.

By and large, game species such as deer, rabbits, grouse, pheasants, and so on can be classified as edge species, so a good part of game management involves creating edge by planting food or cover patches, patch clearcutting, patch burning (see Figure 5–25), and so on. Aldo Leopold, who is generally credited with introducing the concept of edge effect, wrote in his pioneering game management text (1933) that "wildlife is a phenomenon of edges." More recently, Hansson (1979) has commented on the importance of landscape heterogeneity in survival of northern warm-blooded animals that are active all year. Agricultural and other disturbed areas offer more food in winter than do mature, undisturbed forests, which, however, offer more in spring and summer.

An increase in density at ecotones is by no means a universal phenomenon. Many organisms, in fact, may show the reverse. Thus, the density of trees is obviously less in a forest-edge ecotone than in the forest. Breaking up the vast stretches of tropical rain forest will almost certainly reduce species diversity and cause the extinction of many species adapted to large areas of similar habitat. Ecotones appear to assume their greatest importance where humans have greatly modified natural communities and domesticated the landscape for many centuries, thus allowing evolutional time for adaptation. In Europe, for example, where most of the forest has been reduced to forest edge, thrushes and other forest birds live in cities and suburbs to a greater extent than do related species in North America. But, of course, many other European species do not adapt and have become rare or extinct.

As with most positive or beneficial phenomena, the humpbacked subsidy-stress performance curve (see Figure 3–5) is relevant to edge-diversity relations. Excessive edge (many small blocks of habitat) causes diminishing returns in diversity. Although increasing edge often increases diversity, decreasing size of habitat area decreases diversity (the diversity-area trend). Theoretically, maximum beta species diversity occurs when habitat "blocks" are large, or fairly large, and total edge in the region is also large (Thomas et al., 1979). These countertrends need to be considered in forest and wildlife management and landscape design in general. Island biogeographical theory (to be discussed in the next chapter) can help somewhat in determining just how large the "blocks" should be for preservation of particular species as well as for preservation of overall diversity.

8. Paleoecology: Community Structure in Past Ages

Statement

Since we know from fossil and other evidence that organisms were different in past ages and have evolved to their present status, it naturally follows that the structure of communities and the nature of environments must have been different also. Knowledge of past communities and climates contributes greatly to our understanding of present communities. This is the subject of paleoecology, a borderline field between ecology and paleontology that has been defined by Stanley Cain (1944) as "the study of past biota on a basis of ecological concepts and methods insofar as they can be applied," or, more broadly, as the study "of the interactions of earth, atmosphere, and biosphere in the past." The basic assumptions of paleoecology are (1) that the operation of ecological principles has been essentially the same throughout various geological periods and (2) that ecology of fossils may be inferred from what is known about equivalent or related species now living.

Explanation

Since Charles Darwin proposed the theory of evolution, reconstruction of life in the past through the study of the fossil record has been an absorbing scientific pursuit. The evolutionary history of many species, genera, and higher taxonomic groups has now been pieced together. For example, the story of the skeletal evolution of the horse from a four-toed animal the size of a fox to its present status is pictured in most elementary biology textbooks. But what about the associates of the horse in its developmental stages? What did it eat, and what was its habitat and density? What were its predators and competitors? What was the climate like at the time? How did these ecological factors contribute to the natural selection that must have shaped the structural evolution? Some of these questions, of course, may never be answered. However, given quantitative information on fossils associated together at the same time and place, scientists should be able to determine something of the nature of communities and of their dominants in the past. Likewise, such evidence, together with that of a purely geological nature, may help determine climatic and other physical conditions existing at the time. The development of radioactive dating and other new geological tools has greatly increased our ability to establish the precise time when a given group of fossils lived.

Until recently, little attention was paid to the questions listed in the preceding paragraph. Paleontologists were busy describing their finds and interpreting them in the light of evolution at the taxonomic level. As such information accumulated and became more quantitative, however, it was only natural that interest in the evolution of the group should develop, and thus paleoecology was born.

In summary, then, the paleoecologist attempts to determine from the fossil record how organisms were associated in the past, how they interacted with existing physical conditions, and how communities have changed in time. The basic assumptions of paleontology are that "natural laws" were the same in the past as they are today, and that organisms with structures similar to those organisms living today had similar behavior patterns and ecological characteristics. Thus, if the fossil evidence indicated that a spruce forest occurred 10,000 years

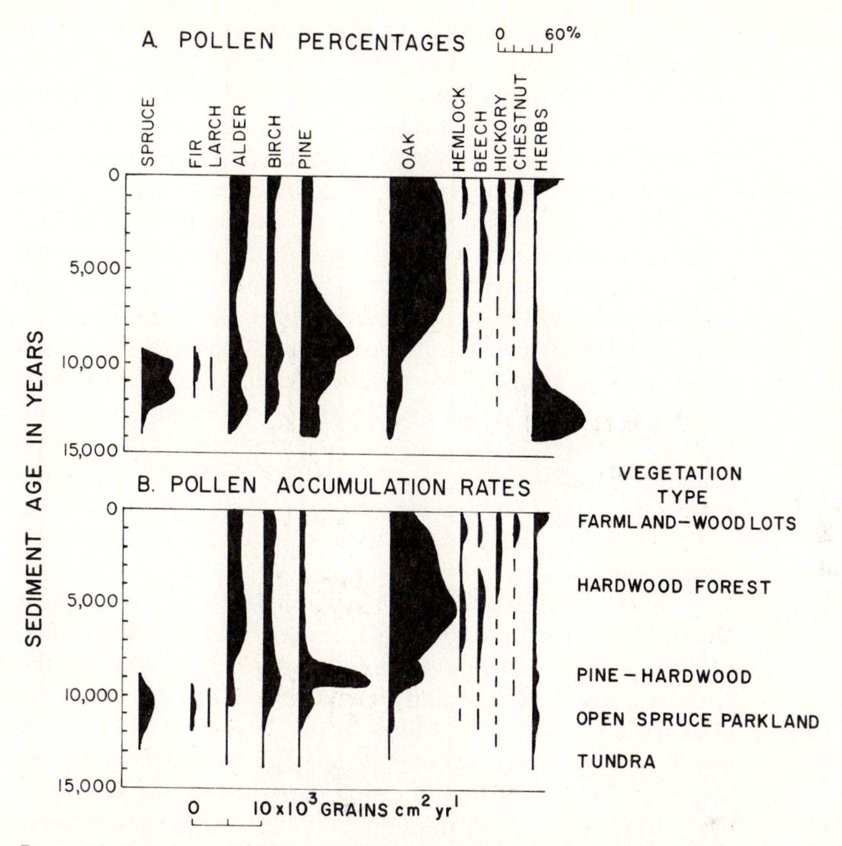

Figure 7–26 Fossil pollen profiles from dated layers in lake sediment cores from southern New England. In *A* the number of pollen grains of each species group is plotted as a percentage of the total number in the sample, while in *B* the estimated rate of pollen deposition for each plant group is plotted. The "rate" profile gives a better indication of the quantitative nature of the post-Pleistocene vegetation than does the percentage profile. (Redrawn from Davis, 1969.)

ago where an oak-hickory forest is now, one has every reason to think that the climate was colder 10,000 years ago, since modern species of spruce are adapted to colder climates than are oaks and hickories.

Examples

Fossil pollen provides excellent material for the reconstruction of terrestrial communities that have existed since the Pleistocene. Figure 7–26 shows how the nature of postglacial communities and climates can be reconstructed by determining the dominant trees. As the glacier retreats, it often leaves scooped-out places, which become lakes. Pollen from plants growing around the lake sinks to the bottom and becomes fossilized in the bottom mud. Such a lake may fill up and become a bog. By taking a core sample from the bog or lake bottom, a chronological record is obtained from which the percentage of various kinds of pollen can be determined. Thus, in Figure 7–26, the "oldest" pollen sample comprises chiefly spruce, fir, larch, birch, and pine, indicating a cold climate. A change to oak, hemlock, and beech indicates a warm moist period several thousand years later, whereas oak and hickory suggest a warm, dry period still later and a return to slightly cooler and wetter conditions in the most recent part of the profile. Finally, the pollen "calendar" clearly reflects recent human effects. For example, clearing a forest is accompanied by an increase in herbaceous pollen. According to Davis (1969), pollen profiles in Europe even show the effects of the Black Plague, when agriculture declined, resulting in a decrease of herbaceous pollen in sediment layers dated at the same time as the widespread death of human beings.

Figure 7–26 also illustrates how improved quantification can change interpretations of the fossil record. When pollen abundance is plotted as a percentage of the total amount in the sample (the conventional approach, at least until recently), one concludes that New England was covered with a dense spruce forest 10,000 to 12,000 years ago. However, when carbon dating made it possible to determine the *rate of pollen deposition in dated layers* and rates plotted as in Figure 7–26B, it became evident that trees of all kinds were scarce 10,000 years ago and that the vegetation existing then was actually an open spruce parkland probably not unlike the present one along the southern edge of the tundra. This is a good example of what statisticians often warn of: beware of percentage analyses; they may be misleading!

In the ocean, the shells and bones of animals often provide the best record. Shell deposits are especially good for assessing diversity in the past. Valentine (1968) has stressed the importance of distinguishing between "in-community" and "geographical" or "in-gradient" diversity. Thus, in past ages when there was no ice at the poles, there were many more species in northern sea bottoms than there are now. However, in the whole pole-to-equator gradient, there are twice as many species of benthic mollusks now—when the poles are ice-covered—presumably because the sharper gradient increases the variety of habitats and niches.

Looking at core samples from lake bottoms is one way to read the recent history of human disturbance of the watershed. A study of fossilized diatoms, midges, and zooplankton and of chemical composition of dated sections of cores from Linsley Pond, a small lake in Connecticut, has led Brugam (1978) to identify three stages of human impact: (1) early farming in the late 1700s and 1800s had little effect on the lake; (2) intensive agriculture after about 1915 resulted in a flow of agricultural chemicals into the lakes and an increase in eutrophic species of diatoms and midges; (3) suburbanization from the 1960s to the present resulted in still more nutrient enrichment (hypereutrophication) and in soil erosion that brought large amounts of minerals and metals (Fe, Cu) into the sediments. These inputs produced major changes in the composition of the biota, especially the zooplankton.

9. From Populations to Communities to Ecosystems

The two approaches to studying, understanding, and, where necessary, managing ecosystems (discussed in Chapter 2, Section 3) are the holological and the merological. In the former approach, one first delimits the area or system of interest in some convenient way as a sort of "black box." Then, the energy and other inputs and outputs are examined (see Figure 2–2), and major functional processes within the system are assessed (see Figure 2–1). Following the parsimonious principle (frugal effort), one then examines operationally significant populations and factors as determined by observing, by modeling, or by perturbing the ecosystem itself. In this general approach, one goes into the details of population components within the "box" only as far as necessary to understand or manage the system as a whole.

There are so many exciting things to be learned about individual and interacting populations that it becomes a challenge to go from these levels of organization to the ecosystem level. It is obviously not practical to study every population in detail. Also, as abundantly illustrated in this chapter, populations may behave very differently when functioning in communities than when isolated in the laboratory or by enclosures in the field. Once studied, how does one reassemble the components into communities and ecosystems and consider new holistic properties that may emerge as parts functioning together in the intact ecosystem? Cody (1974) suggests two approaches: (1) study experimentally induced environmental variations to determine the operational significance of population patterns as observed in the unperturbed system and (2) study variations that occur naturally in the environment to reveal significant patterns. Foin and Jain (1977) have suggested a kind of "road map" for integrating population biology and ecosystem science as follows: (1) make a descriptive analysis, but not in depth, of those community-level properties considered relevant to goals of the study (diversity, species composition, dominance, biomass, productivity), (2) generate a hypothesis about community structure and function for those properties considered important, (3) select for detailed study those populations involved in the community processes as identified in Steps 1 and 2, and (4) construct a mathematical model from the population data to see how well one can mimic the ecosystem in question. Whatever the procedures, value judgments have to be made along the way (for example, deciding which populations are important), and these judgments should be based on firsthand field experience.

The relationship between the parts and the whole may well depend on the level of complexity. At one extreme, ecosystems subjected to severe physical limitations (as on the arctic tundra or in a hot spring) have relatively few biotic components. Such "low-numbers systems" can be studied and understood by focusing on the parts; the whole is probably very close to the sum of the parts, with few, if any, emergent properties. In contrast, "large-numbers systems" (such as the biosphere) have a great many components that act synergistically to produce emergent properties; the whole is definitely not just a sum of the parts. Studying all the parts separately is out of the question, so one must focus on the properties of the whole. Most ecosystems as delimited in practice (a lake or a forest, for example) are "middle-numbers systems" that can best be studied by a multilevel or black box approach, as described in the first paragraph of this section. For more on levels-of-complexity theory, see Allen and Starr (1982).

How to deal with parts and wholes has long confounded philosophers and bedeviled society. Scientists in all disciplines are split on the matter of atomism versus holism. The difficulty in dealing simultaneously with the part and the whole is perhaps best reflected in the conflict between the individual "good" and the public "good." Numerous economic and political approaches designed to deal with this conflict have been suggested or tried, but as yet with little success. In the United States, elected governments over the years have shifted back and forth from strong attention to the individual (the conservative stance) to emphasis on public well-being (the liberal stance), so parts (individual) and whole (public) get attention, but not at the same time. Perhaps the study of how natural ecosystems develop, to be considered in Chapter 8, may help resolve the problem.

During the 1970s a number of symposia on population and community ecology were published, including a memorial volume dedicated to Robert MacArthur (edited by Cody and Diamond, 1975) and one dedicated to David Lack (edited by Stonehouse and Perrins, 1977). Others referenced in the Bibliography are edited by Goulden (1977), Bernardi (1979), Soule and Wilcox (1980), and Tounsend and Calow (1981). Since symposium authors generally present comprehensive and often detailed reviews of their special subjects, these volumes are primarily useful for advanced study. For more general reference, Hutchinson (1978) and Begon and Mortimer (1981) are especially recommended.

8
Development and Evolution in the Ecosystem

1. The Strategy of Ecosystem Development

Statement

Ecosystem development, or what is more often known as **ecological succession**, involves changes in species structure and community processes with time. When not interrupted by outside forces, succession is reasonably directional and, therefore, predictable. It results from modification of the physical environment by the community and from competition-coexistence interactions at the population level; that is, succession is community-controlled even though the physical environment determines the pattern and the rate of change and often limits how far development can go. If successional changes are largely determined by internal co-actions, the process is known as **autogenic** (= self-generated) **succession**. If outside forces in the input environment (for example, storms and fire) regularly effect or control change, there is **allogenic** (= externally generated) **succession**. When new territory is opened up or becomes available for colonization (after a volcanic laval flow, or in an abandoned crop field or new water impoundment), autogenic succession usually begins with an unbalanced community metabolism where gross production (P) is either greater than, or less than, community respiration (R) and proceeds toward a more balanced condition, $P = R$. The ratio of biomass to production (B/P) increases during succession until a stabilized ecosystem is achieved in which maximum biomass (or high information content) and symbiotic function between organisms are maintained per unit of available energy flow.

The whole sequence of communities that replace one another in a given area is called the **sere**; the relatively transitory communities are variously called **seral stages** or **developmental stages** or **pioneer stages.** The terminal stabilized system is the **climax,** which persists, in theory, until affected by major disturbances. Succession beginning with $P > R$ is **autotrophic succession,** which contrasts with **heterotrophic succession** that begins with $P < R$. Succession on a previously unoccupied substrate (laval flow, for example) is called **primary succession**, whereas that starting on a site previously occupied by a community (forest clearcut or abandoned crop field, for example) is known as **secondary succession.**

Even though, as already emphasized, ecosystems are not "super organisms," their development has many parallels in the developmental biology of organisms, as well as in the development of human society.

Explanation and Examples

The descriptive studies of succession on sand dunes, grasslands, forests, marine shores, or other sites, and more recent functional considerations, have led to a partial understanding of the developmental process and generated a number of theories about its cause. H. T. Odum and Pinkerton (1955), building on Lotka's (1925) "law of maximum energy in biological systems," were the first to point out that succession involves a fundamental shift in energy flows with increasing energy relegated to maintenance (respiration) as the standing crop of biomass and organic matter accumulates. Margalef (1963, 1968) has more recently documented this bioenergetic basis for succession and has extended the concept. The role that population interactions play in shaping the course of species replacement, a characteristic feature of ecological succession, has been much discussed during the past decade (see Connell and Slayter, 1977, and McIntosh, 1980, for reviews).

Changes that may be expected to occur in major structural and

Table 8–1 Trends to Be Expected During the Course of Autogenic Succession

Energetics
1. Biomass (B) and organic detritus increase
2. Gross production (P) increases in primary; little change in secondary
3. Net production decreases
4. Respiration (R) increases
5. P/R ratio moves toward unity (balance)
6. B/P ratio increases (conversely, P/B decreases)

Nutrient Cycling
7. Element cycles increasingly closed
8. Turnover time and storage of essential elements increase
9. Cycling ratio (recycle/thruput) increases
10. Nutrient retention and conservation increases*

Species and Community Structure
11. Species composition changes (relay floristics and faunistics)
12. Diversity-richness component increases
13. Diversity-evenness component increases
14. r-Strategists largely replaced by K-strategists
15. Life cycles increase in length and complexity
16. Size of organism and/or propagule (seed, offspring, and so on) increases
17. Mutualistic symbiosis increases*

Stability
18. Resistance increases*
19. Resilience decreases*

Overall Strategy
20. Increasing efficiency of energy and nutrient utilization*

* Trend based on theoretical considerations, yet to be validated in the field.

functional characteristics of autogenic development are listed in Table 8–1. Twenty attributes of ecological systems are grouped for convenience of discussion under five headings. Trends contrast the situation in early and in late development. The degree of absolute change, the rate of change, and the time required to reach a steady state may vary not only with different climatic and physiographic situations but also with different attributes of the ecosystem in the same physical environment. When good data are available, rate-of-change curves are usually convex, with changes occurring most rapidly at the beginning, but bimodal or cyclic patterns may also occur.

The trends listed in Table 8–1 represent those observed to occur when internal, within-community (autogenic) processes predominate. The effect of external (allogenic) disturbances may reverse or otherwise alter these developmental trends, as will be discussed later.

Bioenergetics of Ecosystem Development The first six attributes in Table 8–1 relate to the bioenergetics of the ecosystem. In the early stages of autotrophic succession in an inorganic environment, the rate of primary production or total (gross) photosynthesis (P) exceeds the rate of community respiration (R), so that the P/R ratio is typically greater than 1. The P/R ratio is less than 1 in the special case of an organic environment (sewage pond, for example) so succession in such cases is termed "heterotrophic" because bacteria and other heterotrophs are the first to colonize. In both cases, however, the theory is that P/R approaches 1 as succession occurs. In other words, the energy fixed tends to be balanced by the energy cost of maintenance (that is, total community respiration) in the mature or climax ecosystem. The P/R ratio, therefore, is a functional index of the relative maturity of the system. The P/R relationships of several familiar ecosystems are graphically displayed in Figure 8–1A; the direction of autotrophic and heterotrophic succession is also indicated.

As long as P exceeds R, organic matter and biomass (B) will accumulate in the system, with the result that the ratio B/P, B/R, or B/E (where $E = P + R$) will increase (or conversely, the P/B ratio will decrease). Recall that these ratios were discussed in Chapter 3 in terms of thermodynamic order functions (page 90). Theoretically, then, the amount of standing-crop biomass supported by the available energy flow (E) increases to a maximum in the mature or climax stages. As a consequence, the net community production, or yield, in an annual cycle is large in the early stages and small or zero in mature stages.

Comparison of Succession in a Laboratory Microcosm and a Forest One can readily observe bioenergetic changes by initiating succession in experimental laboratory microecosystems of the type derived from natural systems as described in Chapter 2 (see Figure 2–17, III). In Figure 8–2, the general pattern of a 100-day autotrophic succession in a microcosm based on data of Cooke (1967) is compared with a hypothetical model of a 100-year forest succession presented by Kira and Shidei (1967).

During the first 40 to 60 days in a typical flask microcosm experiment, daytime net production (P) exceeds nighttime respiration (R), so that biomass (B) accumulates in the system. After an early "bloom" at about 30 days, both rates decline and become approximately equal at 60 to 80 days. The B/P ratio, in terms of grams of carbon supported per gram of daily carbon production, increases from less than 20 to more than 100 as the steady state is reached. Not only are autotrophic and heterotrophic metabolism balanced in the climax

A

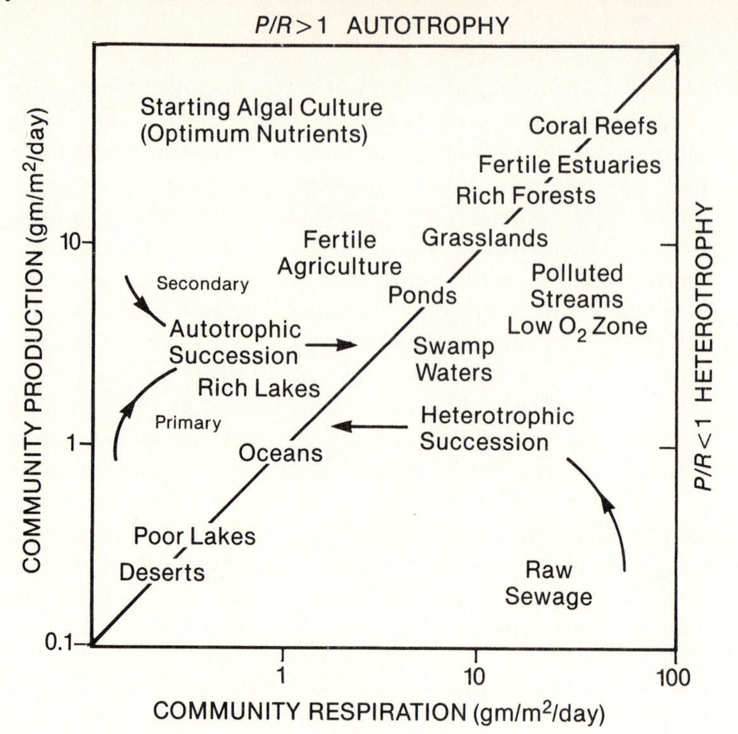

B

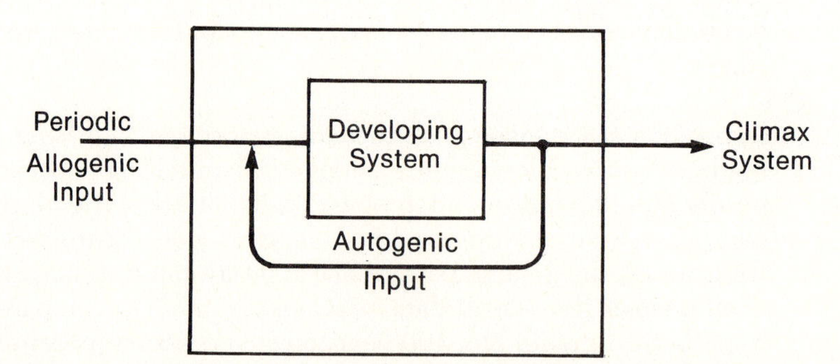

Figure 8–1 *A*. Position of various community types in a classification based on community metabolism. Gross production (P) exceeds community respiration (R) on the left side of the diagonal line (P/R greater than 1 = autotrophy), while the reverse situation holds on the right (P/R less than 1 = heterotrophy). The latter communities import organic matter or live on previous storage or accumulation. The direction of autotrophic and heterotrophic succession is shown by the arrows. Over a year's average, communities along the diagonal line tend to consume about what they make and can be considered metabolic climaxes. (Redrawn from H. T. Odum, 1956.) *B*. A general systems model of ecosystem development.

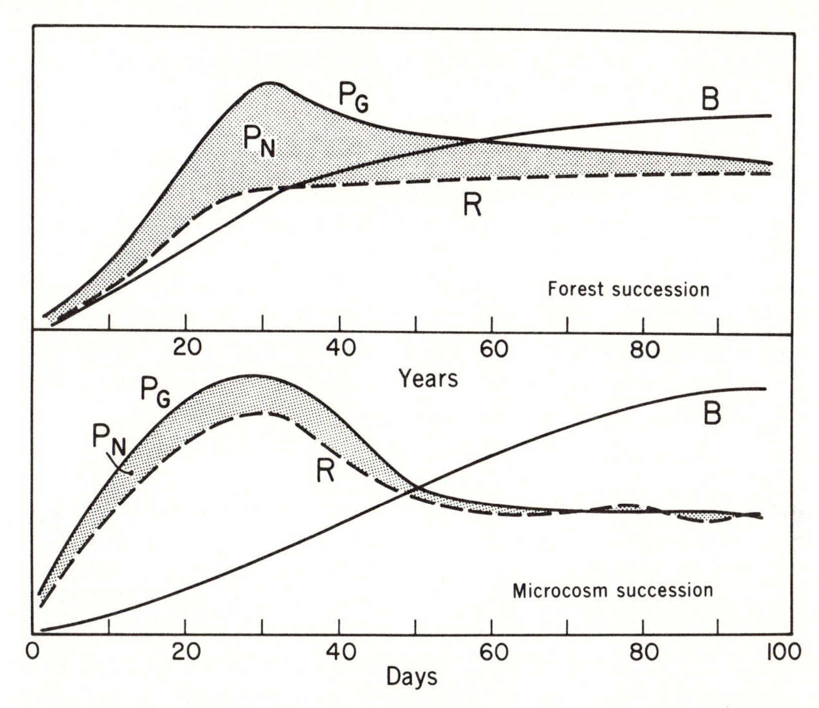

Figure 8–2 Comparison of the energetics of ecosystem development in a forest (redrawn from Kira and Shidei, 1967) and a laboratory microcosm (redrawn from Cooke, 1967). P_G, gross production; P_N, net production; R, total community respiration; B, total biomass.

but also a large organic structure is supported by small daily production and respiratory rates. The relative abundance of species also changes, so that different kinds of bacteria, algae, protozoa, and small crustaceans dominate at the end than at the beginning of the 100-day succession (Gorden et al., 1969).

In "flow-thru" laboratory microcosms (that is, culture medium flows slowly through the microcosm container, simulating an input and output environment of water and nutrients), succession is slower, with more time (2 to 3×) required for biomass to accumulate and for P/R to approximate 1 (Hendrix et al., 1982).

Direct projection from small laboratory microcosms to open nature is not possible because the former are limited to small organisms with simple life histories and of necessity have a reduced species and chemical diversity. Nevertheless, the same basic trends seen in the microcosm are characteristic of succession on land (Figure 8–2, upper diagram) and in large bodies of water. Seasonal succession also often follows the same pattern—an early seasonal bloom characterized by rapid growth of a few dominant species, followed later in the season by development of high *B/P* ratios, increased diversity, and a relatively steady, although temporary, state in terms of *P* and *R* (Margalef, 1963). Open systems may not experience a decline at maturity in total or gross productivity, as the space-limited microcosms do, but the general pattern of bioenergetic change in the latter seems to mimic nature quite well.

It is also interesting to note that peak net primary production (P_N), which represents maximum "yield" possibility, comes at 30 days in the microcosm and 30 years in the forest. Short rotation forestry is based on harvest at the peak of P_N, which on many sites comes between 20 and 40 years.

Allogenic versus Autogenic Influences The discussion so far has concerned the changes brought about by biological processes within the ecosystem in question. Imported materials or energy, geological forces, storms, human disturbances, and so on can and do alter, arrest, or reverse the trends shown in Table 8–1. For example, eutrophication of a lake, whether natural or cultural, results when nutrients and soil enter the lake from outside, that is, from the watershed. This is equivalent to adding nutrients to the laboratory microecosystem or fertilizing a field; the system is "set back," in successional terms, to younger "bloom" states. Allogenic succession of this type is, in many aspects, the reverse of autogenic succession. When the effect of allogenic processes consistently exceeds that of autogenic ones, as in the case of many ponds and small lakes, the ecosystem not only cannot stabilize but also may become "extinct" by filling up with organic matter and sediments and becoming a bog or a terrestrial community. Such is the ultimate fate of man-made lakes subjected to accelerated man-made erosion.

Studies on lake sediments by Mackereth (1965), Cowgill and Hutchinson (1964), and Harrison (1962), as well as theoretical considerations, have indicated that lakes can and do progress to a more oligotrophic (less enriched) condition when the nutrient input from the watershed slows or ceases. Thus, there is hope that the troublesome cultural eutrophication, which reduces water quality and shortens the life of the water body, can be reversed if the inflow of nutrients from the watershed can be greatly reduced. An example is the "recovery" of Lake Washington in Seattle, described by Edmondson (1968, 1979). For 20 years, treated, nutrient-rich sewage was discharged into the lake, which became increasingly turbid and full of nuisance algal blooms. As a result of public outcry, sewage effluent was diverted from the lake, which quickly returned to a more oligotrophic condition (clearer water and no blooms).

The interaction of external and internal forces can be summarized in a general systems model (Figure 8–1B) of the form first introduced in Figure 1–4. Autogenic forces are depicted as internal input or feedback, which, in theory, tends to drive the system toward some sort of equilibrium state. Allogenic forces are depicted as periodic external input disturbances, which set back or otherwise alter the state trajectory.

Where ecosystem development takes a long time to run its course, as in a forest development starting from bare ground, periodic disturbances will affect the successional process, especially in the variable environments of the temperate zone. Oliver and Stephens (1977) have reported on a study of the vegetative history of a small area of the Harvard Forest (Massachusetts). Fourteen natural and human-caused disturbances of varying magnitudes occurred periodically (at irregular intervals) between 1803 and 1952. There was evidence of two hurricanes and a fire before 1803. Small disturbances did not bring in new species of trees but often allowed species already in the understory, such as black birch, red maple, and hemlock, to emerge to the canopy. Large-scale disturbances (such as a hurricane or large fire) created openings into which early successional species (white birch or pin cherry, for example) invaded or into which a new age class developed from seeds or seedlings already present on or in the forest floor (northern red oak was a species that often filled such openings and grew to canopy dominance after several decades). Replacement and succession in forest clearings has been called "gap phase succession" by Bray (1956). Oliver and Stephens concluded from their study that the present composition of the forest was more the result of allogenic influences than of autogenic development. In a subsequent review paper, Oliver (1981) concluded that severity and frequency of disturbance are the major factors determining forest structure and species composition in many areas of North America.

If disturbances are rhythmic (come at more or less regular intervals), either because of a cyclic input environment or because of community development itself, the ecosystem undergoes what can logically be called **cyclic succession** (Watt, 1947). The fire-chaparral vegetation cycle described previously (Chapter 5, Section 4) is a good example of a self-generated cyclic succession, since the accumulation of undecomposed litter builds up fuel for the periodic fires in the dry season.

Figure 8–3 Wave-generated succession in a balsam fir forest.

Another example is the "wave-generated" succession in balsam fir forests at high altitudes in the northeastern United States (Sprugel and Bormann, 1981). As trees reach their maximum height and density in the thin soils, they become vulnerable to strong winds that uproot and kill old trees, thereby starting a secondary succession. As shown in Figure 8–3, a series of bands of young, mature, and dead trees (the latter appearing as light-colored bands in the photo) cover the mountainside. Because of the continuous cyclic succession, the bands move as "waves" across the landscape in the general direction of prevailing winds. At any one time, all stages of succession are present, providing a variety of habitats for animals and smaller plants. The whole mountainside constitutes a steady state or "cyclic climax" in equilibrium with the surrounding environment.

The natural pattern of alternating bands of young and mature stands suggests that strip or patch clearcutting could prove to be a good commercial harvest procedure for large forested areas, since natural regeneration would be facilitated (thus avoiding expensive tree replanting) and the soil and animal populations would be little disturbed compared with the disturbance in a massive clearcut of the whole forest. Furthermore, mixtures of different successional stages provide a lot of edge that benefits many forms of wildlife.

Still another example of cyclic succession is the cycle of spruce and budworms (described in Chapter 6, Section 5). In this case the periodic disturbance is not a physical force but a herbivore that defoliates and kills older growth, thus bringing on a succession of young growth.

The adjective **perturbation-dependent** is frequently used to designate ecosystems especially adapted to recurrent disturbances by virtue of a make-up of quick-recovery processes and species (see Vogl, 1980, for a review). In predicting and managing recovery after a disturbance, such as strip mining, one must know in detail the succession pattern and recovery potential of the particular ecosystem in question so that reclamation efforts will help and not hinder the natural recovery processes (see McIntosh, 1980).

The theoretical trajectories of ecological succession in benign and uncertain physical environments are compared in Figure 8–4. As the biotic community develops, there is a rapid increase in organizational complexity. Increase in information content is indicated by an

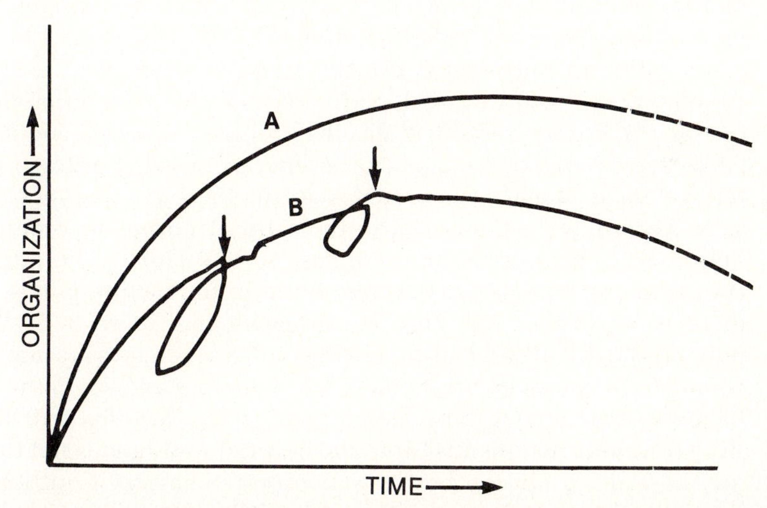

Figure 8–4 The trajectory of ecological succession in a benign environment with low probability of catastrophic perturbations (curve A) and a stressful environment subject to periodic perturbations that disrupt and set back the developmental process (curve B). In situation B, two hypothetical set-back loops are shown following disturbances indicated by the arrows; the biotic community becomes adaptively perturbation-dependent by maintaining a lower level of organization that would be achieved in the absence of stress input. (After Odum, 1981.)

increase in biomass and the B/P ratio, along with increased diversity and efficiency of various functional processes relating to energy flow and nutrient cycling. If the physical environment is benign, with a low probability of catastrophic disturbance, then a high level of structure is achieved and maintained as a steady state, or climax, for a long time, as shown in curve A, Figure 8–4. In contrast, a trajectory leading to a lower level of organization is likely in a stressful environment subject to periodic disturbances, which disrupt and set back the successional process, as shown in curve B of the figure. In situation B, two set-back loops are shown following two perturbations (indicated by the arrows). Presumably, this lower level is more adapted to perturbation and more resilient. A tentative hypothesis follows that older stages of succession, in general, are more resistant to nominal or short-term stress (such as a one-year drought) than are younger stages, but the latter are more resilient (recover more quickly) to catastrophic stress such as a large storm or fire (see items 18 and 19 in Table 8–1).

Nutrient Cycling Important trends in successional development involve an increase in turnover time, greater storage of materials, and increased cycling, all of which contribute to the closing or "tightening" of the biogeochemical cycling of major nutrients, such as nitrogen, phosphorus, and calcium (Table 8–1, items 7 to 10). The extent to which conservation of nutrients is a major trend or strategy in ecosystem development is controversial, partly because there are different ways to index it. Figure 8–5 illustrates the problem. Vitousek and Reiners (1975) have noted that biogenic nutrients will likely be stored within the system as biomass accumulates during the early stages of succession. According to their theory, the ratio of output to input drops below 1 as nutrients go into biomass accumulation. The ratio then rises again to 1 as output balances input in the mature climax when there is no further net growth, as shown in Figure 8–5. But, as pointed out by Woodmansee (1978), nutrients may continue to accumulate in soil even after plants are no longer adding to the living biomass. Also, as noted by Henderson (1975), Westman (1975), and Finn (unpublished manuscript), the output/input ratio is not the only way or perhaps not the best way to assess behavior of nutrients. As shown in Figure 8–5, the cycling index (CI = ratio of recycled input to output; see Chapter 4, Section 7) increases steadily as the system matures; accordingly, nutrients are retained for a longer period and reused, thereby reducing the input requirement, even though input and output are balanced. Also, the ratio of amount stored (S) to amount lost (O) is likely to be low in the early stages and increase in later stages. In summary, there are theoretical reasons and some ob-

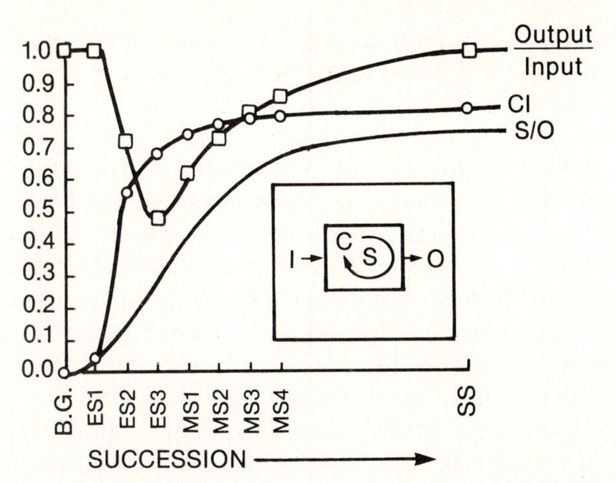

Figure 8–5 Hypothetical trends of output/input ratio, cycling index (CI), and storage/output (S/O) ratio of nutrients during succession. I = input; O = output; S = storage; C = cycling; B.G. = bare ground; ES = early stages; MS = middle stages; SS = steady state. (Based on a diagram by J. T. Finn.)

servational evidence that storage and recycling of nutrients increase during ecosystem development, so that the requirement for input nutrients per unit of biomass supported is reduced. No such conservation would be expected for nonessential or toxic elements.

In a series of papers, Rice and Pancholy (1971, 1972, 1974) presented indirect evidence for the allelochemical inhibition of nitrification in mature or climax stages, which suggests that there may be a shift in nitrogen source from nitrate to ammonia during succession. Pioneer plants use nitrate primarily, whereas later stages, particularly forest stages, use ammonia as a nitrogen source. Peter McRoy (personal communication) reports that mature stages of seagrass succession recycle nitrogen by way of ammonia through anaerobic microbes. All of this is intriguing theory, since a shift from nitrate to ammonia reduces the amount of energy necessary to recycle nitrogen (see discussion of nitrogen cycle in Chapter 4) and thereby increases the efficiency of energy use. However, Robertson and Vitousek (1981) could not find experimental evidence of the nitrate-to-ammonia shift, so the question remains open.

Also unresolved is the question of whether nitrogen fixation, mycorrhizal symbiosis, and other mutualisms that enhance efficiency of nutrient cycling increase or decrease during the course of succession (item 17 of Table 8–1). These mutualisms are widespread and appear in all stages of succession, but few attempts have been made to quantify their activity in different seral stages. Perhaps nutrient-conserving mutualisms respond more to "demand" (nutrient scarcity) than to ecosystem development as such.

Replacement of Species A more or less continuous replacement of species with time is characteristic of most successional series (seres). The changing species composition of vegetation has been called "relay floristics" by Egler (1954), and, of course, there is also "relay faunistics" since animal species also replace one another in the sere.

If development begins on an area previously unoccupied by a community (such as a newly exposed rock or sand surface or a lava flow), the primary succession that ensues may be slow to begin and may require a long time to reach steady-state maturity. The classic example of primary ecological succession occurs on the Indiana dunes at the south end of Lake Michigan, which was once much larger than its present size. In retreating to its present boundaries, the lake left successively younger and younger sand dunes. Because of the sand substrate, succession is slow, and a series of communities of various ages are available for observation: pioneer stages at the lake shore and increasingly older seral stages as one proceeds away from the shore. In this "natural laboratory of succession" H. C. Cowles (1899) made his pioneer studies of plants and V. E. Shelford (1913) his studies of animal succession. Both studies showed that species of both plants and animals changed with the increasing age of dunes; species present at the beginning were completely replaced by other, quite different species in the older communities. Olson (1958) has restudied ecosystem development on these dunes and has given us updated information on rates and processes. Because of encroachment of heavy industry, conservationists are hard pressed in their efforts to preserve the dune series, but fortunately, some parts of the "Indiana Dunes" are now in state and national parks. The public should support such preservation efforts, because these areas not only have a priceless natural beauty that can be enjoyed by urban dwellers but also constitute a natural teaching laboratory in which the visual display of ecological succession is dramatic.

The pioneer colonists on the dunes are beach grasses (*Ammophila, Agropyron, Calamovilfa*); willow, sand cherry, and cottonwood trees; and animals such as long-legged tiger beetles that flit along the sand, burrowing spiders, and grasshoppers. The pioneer community is followed by open dry forests of jack pine, then black oak, and finally, on the oldest dunes, moist forests of oak-hickory or beech-maple. Although the community began on a very dry and sterile sort of habitat, development eventually results in a closed canopy forest, moist and cool in contrast with the bare dune. The deep humus-rich soil, with earthworms and snails, contrasts with the dry sand on which it developed. Thus, the original relatively inhospitable pile of sand is eventually transformed completely by the action of a succession of communities.

Succession on dunes in the early stages is often arrested when the wind piles up the sand over the plants and the dune begins to move, entirely covering the vegetation in its path. Here is an example of the arresting or reversing effect of allogenic perturbations discussed earlier in this section. Eventually, however, as the dune moves inland it becomes stabilized, and pioneer grasses and trees again become established. Using modern methods of carbon dating, Olson (1958) estimates that about 1000 years is required to reach a forest climax on the Lake Michigan dunes, about five times longer than required for mature forest development on a more hospitable site, as seen in the next example.

An example of secondary succession is illustrated in Table 8–2, which shows the sequence of plant communities and bird populations that develop on abandoned upland agricultural fields on the piedmont region of the southeastern United States. Pioneer colonists are *r*-strategist annual plants such as crabgrass (*Digitaria*), horseweed (*Erigeron*), and ragweed (*Ambrosia*) that spend a large part of their energy on dispersal and reproduction (see Table 6–7). After two or three years, perennial forbs (asters and goldenrods), grasses (especially broomsedge, *Andropogon*), and shrubs such as blackberry (*Rubus*) move in. If there is a good seed source nearby, pines invade and soon form a closed canopy shading out the early pioneers. Several species of fast-growing deciduous trees, such as sweetgum and tulip tree, often come in with the pines. Since all these species are long-lived, the pine stage (with scattered broad-leaved trees) persists for a long time, but gradually an understory of shade-tolerant oaks and hickories develops. Since pines cannot reproduce under their own shade, the oaks and hickories rise to canopy dominance as the pines die (from disease, old age, and storms).

As shown in the body of Table 8–2, bird populations change with each major seral stage; the most pronounced change occurs as the life form of the plant dominants changes (herb to shrub to pine to hardwood). Selection of habitat by birds is more targeted to vegetative life-form than to species of plant. No species of plant or bird can thrive from one end of the sere to the other; species have their maxima at different points in the time gradient.

Animals are not just passive agents in community change. Birds disperse seeds necessary for establishment of shrub and hardwood stages, and herbivores, parasites, and predators often control the se-

Table 8–2 Secondary Succession on the Piedmont Region of the Southeastern United States*

AGE IN YEARS		1 2	3–20	25–100	150 +
COMMUNITY TYPE	Bare Field	Grassland	Grass-Shrub	Pine Forest	Oak-Hickory Forest Climax

Crabgrass Horseweed Aster Broomsedge Shrubs Pine Hardwood Understory Oak Hickory

Plant Dominants	Forbs	Grass	Grass-Shrub		Pine Forest				Oak–Hickory Climax
Age in Years of Study Area	1–2	2–3	15	20	25	35	60	100	150–200
Bird Species (having a density of 5 or more in a given stage)									
Grasshopper sparrow	10	30	25						
Meadowlark	5	10	15	2					
Field sparrow			35	48	25	8	3		
Yellowthroat			15	18					
Yellow-breasted chat			5	16					
Cardinal			5	4	9	10	14	20	23
Towhee			5	8	13	10	15	15	
Bachman's sparrow				8	6	4			
Prairie warbler				6	6				
White-eyed vireo				8		4	5		
Pine warbler					16	34	43	55	
Summer tanager					6	13	13	15	10
Carolina wren						4	5	20	10
Carolina chickadee						2	5	5	5
Blue-gray gnatcatcher						2	13		13
Brown-headed nuthatch							2	5	
Wood pewee							10	1	3
Hummingbird							9	10	10
Tufted titmouse							6	10	15
Yellow-throated vireo							3	5	7
Hooded warbler							3	30	11
Red-eyed vireo							3	10	43
Hairy woodpecker							1	3	5
Downy woodpecker							1	2	5
Crested flycatcher							1	10	6
Wood thrush							1	5	23
Yellow-billed cuckoo								1	9
Black and white warbler									8
Kentucky warbler									5
Acadian flycatcher									5
Totals: (including rare species not listed above)	15	40	110	136	87	93	158	239	228

* After Johnston and E. P. Odum (1956). Figures are occupied territories or estimated pairs per 100 acres. The principal plant dominants of the upland sere which follows abandonment of cropland are shown in pictorial fashion along the upper edge of the table. The sere has been described in great detail by Oosting (1942), and some of the plant interactions have been studied by Keever (1950). The succession of common breeding birds is shown in the body of the table.

quence of species. And, of course, in shallow-water marine habitats, large animals rather than plants often provide the structural matrix. Glemarec (1979) describes a secondary succession of benthic animals off France's Brittany coast. After storms caused a redistribution of sediments and disruptions of bottom fauna, a period of relative calm followed. During this period in the absence of outside interference, a more or less directional and predictable sequence of populations established dominance. First were bivalve suspension feeders, then bivalve deposit feeders, and finally the benthos became dominated by polychaete worm detritus feeders, thus confirming the theory that uninterrupted succession converts an inorganic environment to a more organic one.

Secondary plant succession is as striking in grassland regions as in forests, even though only herbaceous plants are involved. In 1917, Shantz described succession on the abandoned wagon roads used by pioneers crossing the grasslands of central and western United States, and virtually the same sequence has been described many times since. Although the species vary geographically, the same pattern holds everywhere. This pattern involves four successive stages: (1) annual weed stage (2 to 5 years), (2) short-lived grass stage (3 to 10 years), (3) early perennial grass stage (10 to 20 years), and (4) climax grass stage (reached in 20 to 40 years). Thus, starting from bare or plowed ground, 20 to 40 years is required for nature to "build" a climax grassland, the time depending on the limiting effect of moisture, grazing, and other factors. A series of dry years or overgrazing causes the succession to go backward toward the annual weed stage; how far back depends on the severity of the effect.

Succession is equally apparent in aquatic as well as terrestrial habitats. However, as already emphasized, the community development process in shallow-water ecosystems (ponds and small lakes) is usually complicated by strong inputs of materials and energy that may speed up, arrest, or reverse the normal trend of community development that would occur in the absence of such strong allogenic influences. The complex interaction of autogenic and allogenic succession is illustrated by rapid changes in artificial ponds and impounded lakes. When a reservoir is created by flooding rich soil or an area with a large amount of organic matter (as when a forested area is flooded), the first stage in development is a highly productive "bloom" stage characterized by rapid decomposition, high microbial activity, abundant nutrients, low oxygen on the bottom, but often rapid and vigorous growth of fish. Fishermen are very happy with this stage. However, when the stored nutrients are dispersed and the accumulated food used up, the reservoir stabilizes at a lower rate of

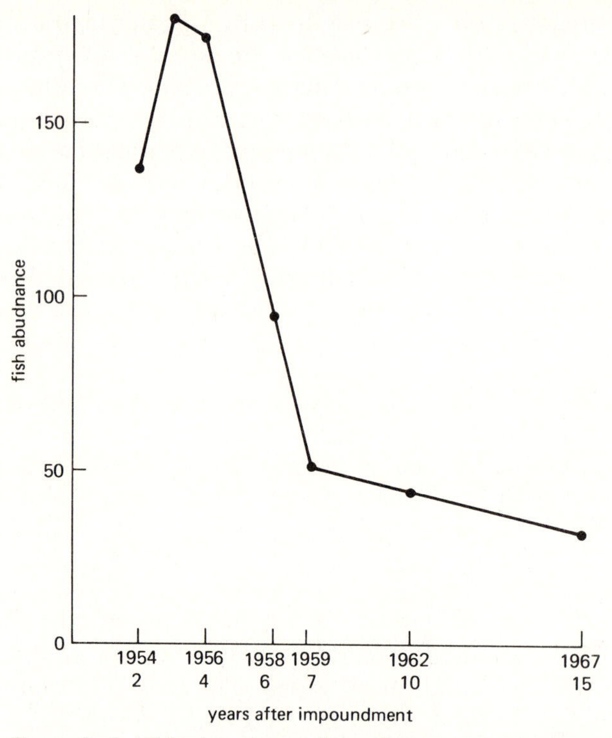

Figure 8–6 Fish abundance (based on mean of two sampling methods) in a new mainstream reservoir on the upper Missouri River from second to fifteenth year after completion of the dam and full impoundment of water in Lake Francis Case, South Dakota. (Data from Gasaway, 1970.)

productivity, greater benthic oxygen, and lower fish yields. Fishermen become unhappy with this stage (see Figure 8–6). If the watershed is well protected by mature vegetation, or if the soils of the watershed are infertile, the stabilized stage may last for some time—a "climax" of sorts. However, erosion and various human-accelerated nutrient inputs usually produce a continuing series of "transient states" until the basin fills up. Impoundments in impoverished watersheds or primary sterile sites will, of course, have a reverse pattern of low productivity at the start. Failure to recognize the basic nature of ecological succession and the relationships between the watershed and the impoundment has resulted in many failures and disappointments in human attempts to maintain such artificial ecosystems. Shallow-water systems can be "pulse stabilized" at high productivity levels by high energy water level fluctuations (to be described in Section 3 of this chapter).

Because the oceans are, generally speaking, in a steady state, and because they have been chemically and biologically stabilized for centuries, oceanographers have not been concerned with ecological succession. However, with pollution threatening to disturb equilibria in the sea, the interaction of autogenic and allogenic processes will undoubtedly receive greater attention from marine scientists. Successional changes are evident in coastal waters, as already noted in the example of development of benthic communities after severe storms have disrupted the sea bottom. Margalef (1967) summarized his observation of the changes that occur in a successional gradient in the coastal water column as follows:

1. Average size of cell and relative abundance of mobile forms among the phytoplankton increase.
2. Productivity, or rate of multiplication, slows down.
3. Chemical composition of the phytoplankton, as exemplified by the plant pigments, changes.
4. The composition of the zooplankton shifts from passive filter feeders to more active and selective hunters in response to a shift from numerous small suspended food particles to scarcer food concentrated in bigger units and dispersed in a more organized (stratified) environment.
5. In the later stages of succession, total energy transfer may be lower, but its efficiency seems to be improved.

Jassby and Goldman (1974) have used the rate of plankton species change to plot seasonal succession in lakes. Species replacement is rapid in the spring but occurs at a much slower rate in summer. When nutrients were introduced in June, the rate of change in species composition was elevated to the spring level for about a month.

The succession of organisms on artificial substrates in aquatic environments has received a great deal of attention because of the practical importance of fouling of ship bottoms and piers by barnacles and other sessile marine organisms. Also, small replicated substrates such as glass slides or squares of plastic, wood, or other material are widely used to assess the effect of pollutants on biota in both fresh and salt water (see Patrick, 1954; Cairns and Dickson, 1971). Such substrates are a kind of microcosm on which one would expect ecological succession to occur, but as with any restricted or simplified model, one must be cautious about projecting hypotheses to larger, less space-limited, open systems that possess many kinds of substrates. In general, the first species to colonize are those that have abundant propagules available in the water when and where the surfaces are put out or become available for colonization. Sometimes, the pioneers change the physical or chemical nature of the substrate in a way that may facilitate the invasion of other species, but just as often the pioneers resist encroachment by other species and endure until replaced by a better competitor. As already noted in discussing intertidal communities on rocky coasts, negative interaction (competition, predation) plays a greater role than positive interaction (coexistence, mutualism) in determining replacement of species in confined or space-limited habitats.

Heterotrophic Succession Woodruff's (1912) early experiments with hay infusions can be cited as an example of heterotrophic succession. When a culture medium made by boiling hay is allowed to stand, a thriving culture of bacteria develops. If some pond water (containing seed stock of various protozoa) is then added, a definite succession of protozoan populations with successive dominants occurs, as shown in Figure 8–7. In this situation, energy is maximal at the beginning and then declines. Unless new medium is added, or an autotrophic re-

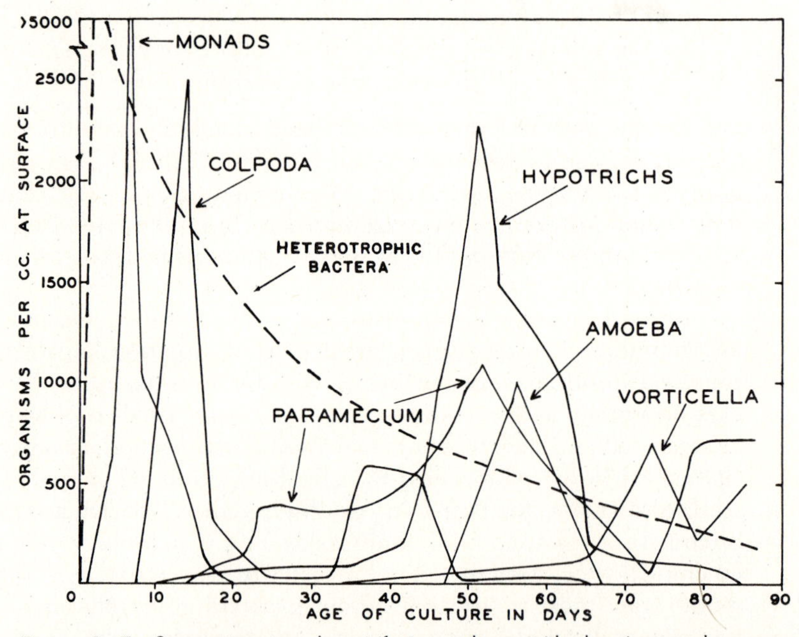

Figure 8–7 Succession in a hay-infusion culture with dominance by successive species. This is a laboratory example of heterotrophic succession. (After Woodruff, 1912, with hypothetical curve for heterotrophic bacteria added.)

gime takes over, the system eventually runs down, and all the organisms die or go into resting stages (spores, cysts, and so on). This is quite different from the autotrophic succession, in which energy flow is maintained indefinitely. The hay-infusion microcosm is a good model for the kind of succession that occurs in decaying logs, animal carcasses, fecal pellets, and the secondary stages of sewage treatment. It might also be considered a model for the "downhill" succession that must be associated with a society dependent on fossil fuels. In all these examples, there is a series of transient stages in a declining energy gradient with no possibility of a steady state being achieved.

Heterotrophic and autotrophic successions can be combined in a laboratory microecosystem model if samples from a derived system are added to media enriched with organic matter. Succession in such a system has been described, and energy flow measured, by Gorden and colleagues (1969). The system first becomes "cloudy" as heterotrophic bacteria "bloom," and then it turns bright green as nutrients and growth substances (especially the vitamin thiamine) required by algae are released by the activities of the bacteria. This succession, of course, is a good model of cultural eutrophication resulting from organic pollution such as incompletely treated municipal sewage.

Selection Pressure: Quantity versus Quality Stages of colonization of islands, as described by MacArthur and Wilson (1967), provide direct parallels with stages in ecological succession on continents. In the early, uncrowded stages of island colonization, as in the early stages of succession, *r*-selection predominates, so that species with high rates of reproduction and growth are more likely to colonize. In contrast, selection pressure favors *K*-strategist species with lower growth potential but better capabilities for competitive survival under the equilibrium density of late stages of both island colonization and succession (item 14 in Table 8–1).

Genetic changes involving the whole biota may be presumed to accompany the successional change from quantity production to quality production, as indicated by the tendency for size of organism to increase (item 16 in Table 8–1). For plants, change in size appears to be an adaptation to the shift of nutrients from inorganic to organic. In a mineral- and nutrient-rich environment, small size is of selective advantage, especially to autotrophs, because of the greater surface-to-volume ratio. As the ecosystem develops, however, inorganic nutrients tend to become more and more tied up in the biomass (that is, to become intrabiotic), so that the selective advantage shifts to larger organisms (either larger individuals of the same species or larger species, or both), which have greater storage capacities and more complex life histories. They are thus adapted to exploiting seasonal or periodic releases of nutrients or other resources.

Diversity Trends Although both components of diversity (items 12 and 13 in Table 8–1) almost always increase in the early stages of ecosystem development, peak diversity seems to come somewhere in the middle of the sere in some cases and again near the end in other situations. Not all trophic or taxonomic groups exhibit the same trend of diversity change with successional time. Nicholson and Monk (1974) determined richness and evenness of plant species for four lifeforms—herbs, vines, shrubs, and trees—in major seral stages in the Georgia piedmont old-field succession already briefly described (Table 8–2). Richness increased rapidly in each stratum after its establishment, then at a decreasing rate throughout the remainder of succession. Evenness, on the other hand, increased to near maximum level immediately and changed very little thereafter. Dominance-diversity curves for another old-field succession (southern Illinois) are shown in Figure 8–8. Diversity of plant species generally increased with succession, reaching a maximum during the early forest stages. The distribution curves of species are geometric (approaching a straight line in the semilog plot) during the first few years of the secondary succession, and then gradually change to lognormal as more species are added. The process results in a high degree of evenness (for a discussion of these distributions, see Chapter 7, Section 6). The sequence of curves shown in Figure 8–8 closely resembles the arctic-to-tropic sequence shown in Figure 7–18.

One of the most extensive studies of diversity trends in long-term succession is that of Auclair and Goff (1971), who measured species diversity of trees in forests of different ages in the western Great Lakes region. On moist sites diversity tended to be highest in middle-aged forests and lower in the oldest forests. In contrast, on dry sites diversity continued to increase with the age of the forest.

Whether or not species diversity continues to increase during succession or peaks at some intermediate stage may well depend on whether the increase in potential niches resulting from increased biomass, stratification, and other consequences of biological organization exceeds the countereffects of increasing size of organisms (item 16 in Table 8–1) and competition exclusion by well-adapted, long-lived dominants, which would tend to reduce the variety of species. No one has yet been able to catalogue all the species in any sizable area, much less follow *total* species diversity in a successional series. Studies on diversity and succession have so far dealt with segments of

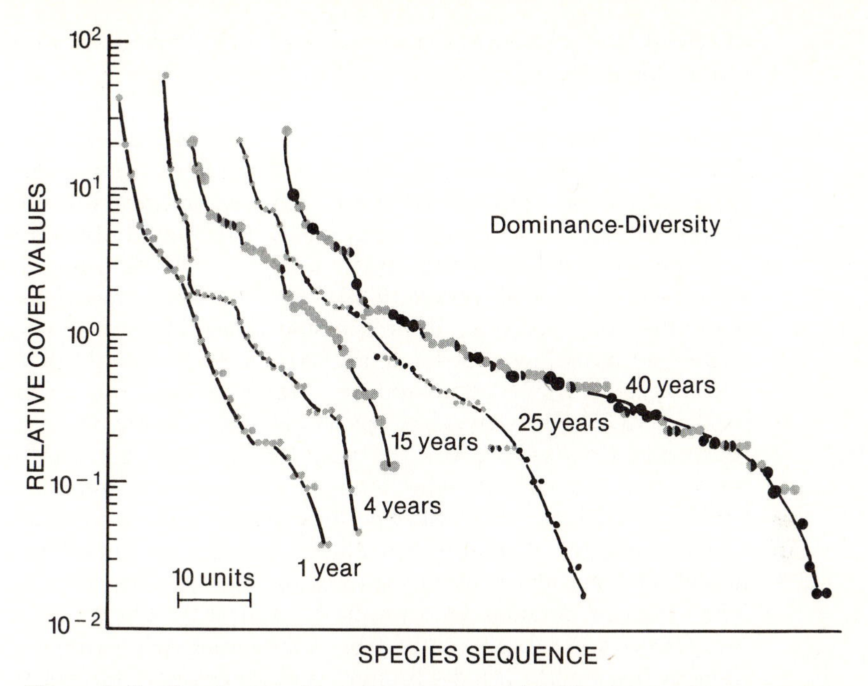

Figure 8–8 Dominance-diversity curves for old-fields of five different ages of abandonment in southern Illinois. Symbols are black for trees, grey for herbs, and a combination of black and grey for shrubs. (After Bazzaz, 1975.)

the community (trees, birds, and so on). One would expect that the pattern of change in species composition will vary widely according to the group under consideration and the geographical situation, which determines what species are available for colonization.

As discussed in the section on stability (Chapter 2, Section 6), the consensus among ecologists is that changes in species diversity are more an indirect consequence of increasing organic development and complexity than a direct causal factor in succession. The level of diversity achieved may well depend on energetics, since maintenance of high diversity has an energy cost and can be destabilizing (the "too much of a good thing" syndrome again).

Although little studied, aspects of organic diversity other than species variety and relative abundance would logically show increasing trends during the course of autogenic ecosystem development. Jeffries (1979), for example, reports that as marine communities mature and become more complex, so do the fatty-acid compositions of the plankton and benthos. This may be an example of an increase in biochemical diversity.

Some Theoretical Considerations At the beginning of this section it was stated that ecosystem development resulted from (1) modification of the physical environment by the community acting as a whole and (2) the interaction of competition and coexistence between component populations. Although one could logically assume that both ecosystem-level and population-level processes contribute to the many-faceted successional progressions described in this section, some ecologists have chosen to argue either/or but not both. Connell and Slayter (1977) compared two theories: (1) the "facilitation model," in which early seral species change the conditions of existence and thereby prepare the way for later invaders, and (2) the "inhibition model," in which the first species resist invasion and remain until they are replaced because of competition, predation, and disturbance. Connell and Slayter strongly favored the second model, at least for secondary succession. Proponents of population-level theories of causation essentially argue that if observed successional trends can be explained by interactions at the species level, there is no need to invoke higher-level processes. Conversely, other theorists argue that species succession is only a part of the process and that self-organizing development is a property of whole ecosystems, and hence there is no need to look into interaction of component populations to explain basic trends.

The idea that ecological succession is a holistic phenomenon goes back to Frederick E. Clements and his 1916 monograph entitled "Plant Succession" (subsequently reprinted in 1928 under the title "Plant Succession and Indicators"). Although his notions that a community repeats in its development the sequence of stages of development of an individual organism and that all communities in a given climatic area develop toward a single climax (the "monoclimax" idea; see next section) are deemphasized or modified today, Clements' main thesis—that ecological succession is a developmental process and not just a succession of species each acting alone—remains one of the most important unifying theories in ecology. Margalef (1963, 1963a, 1968) and E. P. Odum (1969) have reworked and extended Clements' basic theory to include functional attributes such as community metabolism. Basic to the holistic approach is the question of self-organization as a property of nonequilibrium thermodynamic systems (Prigogine's basic theory; see page 90). Following is a sample of recent statements by those who find at least a theoretical basis for viewing ecological succession as a self-organizing strategy:

> The general process [succession] is construed as one in which the system accrues diversity and specialization until the level of environmental unpredictability makes further organization counter-productive. (Valiela, 1971)
>
> It is a hard fact that the diverse ecological systems successfully compete with and displace the simplified systems when environmental conditions are stable. Unless special energies are directed to prevent succession, the complex system with its specializations performs more work functions for a total effort and displaces systems whose energies are going into storage instead of useful work for competitive survival. (H. T. Odum, 1971)
>
> The trade-off between power energy conversions does seem to apply to the relative efficiencies of successive species in a sere of secondary terrestrial plant succession. The application of the power trade-off to species in a sere predicts the common general properties of growth and reproduction in succession. (Christopher Smith, 1976)
>
> A theoretical feedback dynamics model of the mutual causalities that generate secondary successional behavior [is in accordance with] Odum's tabular model of ecological succession. (Gutierrez and Fey, 1975)
>
> Although populations change as succession proceeds, productivity is a property of the system measureable through time. Succession is not a sequence of different systems, but a single system which exchanges transient species and populations through time. (O'Neill and Reichle, 1980)
>
> Self-organizing, dissipative systems are hypothesized to develop over time so as to optimize their ascendancy [complexity of biomass and network flows]. These assumptions appear to be supported by trends in ecosystem development. (Ulanowicz, 1980)

The contrary concept that ecological succession does not have an organizational strategy, but results from the interactions of individuals and species as they struggle to occupy space goes back to H. A. Gleason's papers, especially the 1926 paper entitled "The Individualistic Concept of Plant Succession." Gleason's writings, recently reviewed by McIntosh (1975), have provided a point of departure for the development of new population-level theories of succession that consider new insights into evolutionary biology and the importance of consumer as well as producer influences. Reviews by Drury and Nisbet (1973) and Horn (1974, 1975) explore theories of succession that are based on properties of organisms rather than emergent properties of the ecosystem. The basic premise, as expressed by Pickett (1976), is that evolutionary strategy (Darwinian selection, competitive exclusion, and so on) and characteristics of the life cycle determine the position of species in successional gradients that are constantly changing, depending on disturbances and physical gradients. Since the holistic Clementian theory can be viewed as an evolutionary theory of communities, population, and ecosystem, ecologists may not be so far apart as a reading of their respective position papers might indicate. This position, in general, is the one taken by the late Robert Whittaker (1957, 1975; Whittaker and Woodwell, 1971). Also, in a recent review, Glasser (1982) notes that though the early colonization phase is often stochastic (chance establishment of opportunistic organisms), later stages are much more organizational and directional.

Sooner or later, theories get tested in the practical world of applied science, for example, in forest management. Foresters, by and large, find that forest succession is directional and predictable. To assess future timber potential, they often develop models that combine natural successional trends with disturbance and management scenarios that modify natural development. For example, on the Georgia piedmont, the natural forest succession is from pines to hardwood. Because pines are now more valuable commercially than hardwoods, efforts are made to arrest this succession so that the pine stages can be retained and regenerated, especially in areas under commercial timber management. Johnson and Sharpe (1976) report that, despite silvicultural efforts to maintain pine, hardwoods (both the early sweetgum–tulip tree stage and the later oak-hickory stage) increased in area between 1961 and 1972, and they project, on the basis of a 30-year model, that hardwood stages will continue to increase in area coverage, although at a slower rate than would be the case if only natural succession were involved. Urbanization and suppression of fire, both of which favor hardwoods over pines, are important factors in the model projection. Johnson and Sharpe conclude that although composition of the piedmont forest is strongly influenced by human management, projected future composition will follow trends of natural succession.

2. Concept of the Climax

Statement

The final or stable community in a developmental series (sere) is the climax community. In theory, the climax community is self-perpetuating because it is in equilibrium within itself and with the physical habitat. In contrast to a developmental or other transient community, the annual production plus import is balanced by the annual community consumption plus export (see Figures 8–1 and 8–2). For a given region, it is convenient, although quite arbitrary, to recognize (1) a single **regional** or **climatic climax**, which is in equilibrium with the general climate, and (2) a varying number of **local** or **edaphic climaxes**, which are modified steady states in equilibrium with special local conditions of the substrate. Succession ends in an edaphic climax where topography, soil, water, and regular disturbance such as fire are such that development of the ecosystem does not proceed to the theoretical end-point.

Explanation and Examples

The assumption that autogenic development eventually produces a stable community is generally accepted as being based on sound observation and theory. However, there have been two schools of interpretation. According to the monoclimax idea (which goes back to F. E. Clements, as noted in the previous section), any region has one theoretical climax toward which all communities are developing, however slowly. According to the more realistic polyclimax idea, it is unlikely that all communities in a given climatic region will end up the same when conditions of physical habitat are by no means uniform. Nor can all habitats be molded to a common level by the community within a reasonable length of time, as measured by the life-span of a person (or of a few multiples thereof!). A good compromise between these viewpoints is to recognize a single theoretical climatic climax and a variable number of edaphic climaxes, depending on the variation in the substrate.

The polyclimax concept can best be illustrated by a specific example. Topographical situations in southern Ontario and the stable biotic communities associated with the various physical situations are shown in Figure 8–9. On level or moderately rolling areas where the soil is well drained but moist, a maple-beech community (sugar maple and beech trees being the dominant plants) is found to be the

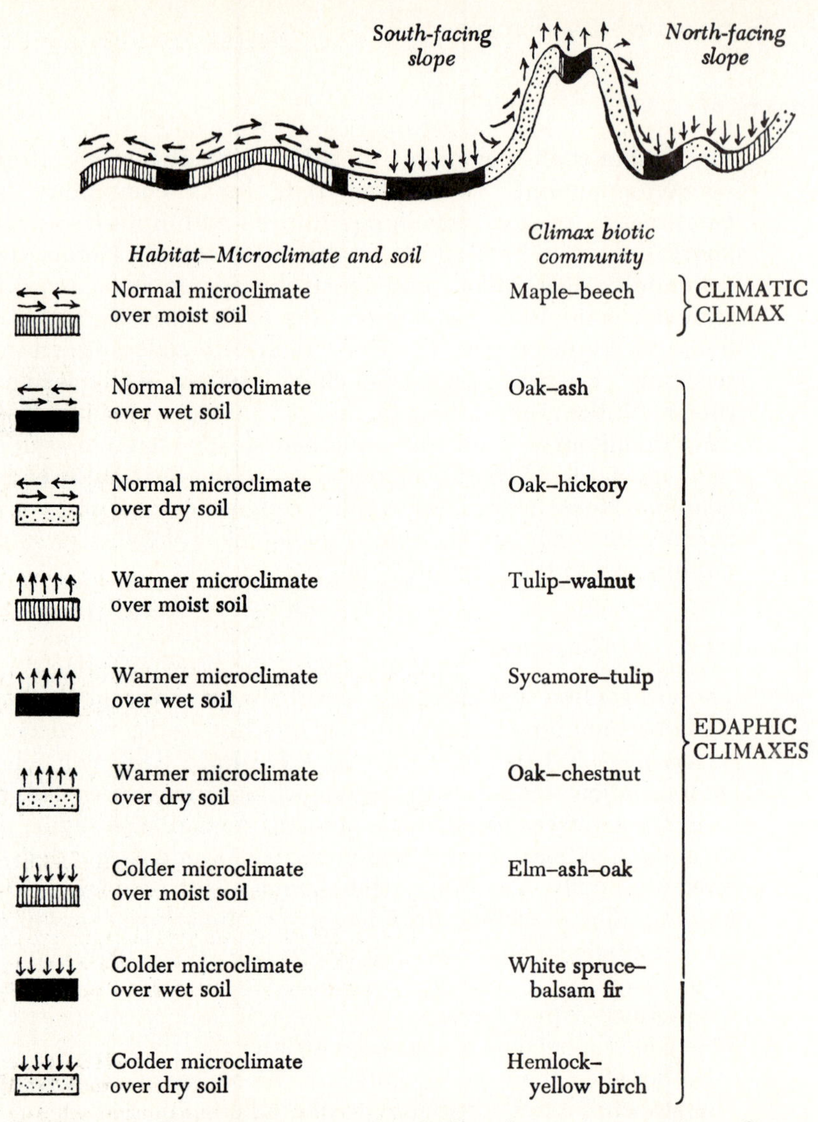

Figure 8–9 Climatic climax and edaphic climaxes in southern Ontario. (Simplified from Hills, 1952.)

terminal stage in succession. Since this type of community is found again and again in the region wherever land configuration and drainage are moderate, the maple-beech community can be arbitrarily designated as the climatic climax of the region. Where the soil remains wetter or drier than normal (despite the action of communities), a somewhat different end-community occurs, as indicated. Still greater deviations from the climatic climax occur on steep south-facing slopes where the microclimate is warmer, or on north slopes and in deep ravines where the microclimate is colder. These latter climaxes often resemble climatic climaxes found farther south and north, respectively. Accordingly, if you live in eastern North America or in the mountainous parts of the west and wish to see what a climax forest would be like further north, find an undisturbed north-facing slope or ravine. Similarly, a south-facing slope will likely exhibit the type of forest to be found further south.

Theoretically, a forest community on dry soil would, if given indefinite time, gradually increase the organic content of the soil and raise its moisture-holding properties, and thus eventually give way to a more moist forest such as the maple-beech community. Actually, whether this would occur or not is unknown, since little evidence of such change has been seen, and since records of undisturbed areas have not been kept for the many human generations that probably would be required. The question is academic, anyway, because long before any autogenic change could occur, some climatic, geological, or anthropogenic force would intervene. The alternative to recognizing a series of climaxes and seres associated with physiographic situations in the case of a landscape mosaic like that shown in Figure 8–9 would be some form of gradient analysis (as discussed in Chapter 7, Section 7). Ecological succession is essentially a gradient in time that interacts with spatial topographical and climatic gradients.

Since autogenic ecological succession results from the changes in the environment brought about by the organisms themselves, the more extreme the physical substrate, the more difficult modification of the environment becomes and the more likely that community development will stop short of the theoretical regional climax. Regions vary considerably in the percentage of area that can support climatic climax communities. On the deep soils of the central plains of the United States, early settlers found a large proportion of the land covered with a climax grassland. In contrast, on the sandy, geologically young lower coastal plain of the southeast United States, the climatic climax (a broad-leaved evergreen forest) was originally as rare as it is today. Most of the coastal plain is occupied by edaphic climax pine or wetland communities or their seral stages. In contrast, the oceans, which occupy geological ancient basins, can be considered to be in a climax state insofar as community development is concerned. However, seasonal succession and succession following disturbance occur, especially in inshore waters, as already mentioned.

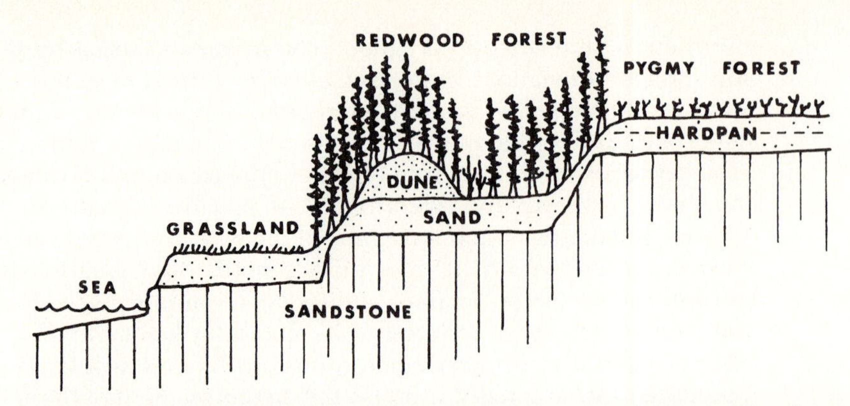

Figure 8–10 Edaphic climaxes on the coast of northern California (Mendocino region). Although the parent material (C horizon) is the same (beach deposits and sandstone), forests of tall redwood and dwarf conifers grow side by side on adjacent marine terraces. The stunted nature of the pygmy forest is due to an iron-cemented B-horizon hardpan located about 18 inches below the surface. The soil above the impervious hardpan is extremely acid (pH 2.8 to 3.9) and low in Ca, Mg, K, P, and other nutrients. At least one of the dominant pygmy pines is an ecotype adapted especially to this extreme soil condition. (After Jenny, Arkley, and Schultz, 1969.)

A dramatic example of a contrast between regional and edaphic climax is shown in Figure 8–10. In a certain area on the coast of northern California, giant redwood forests occur side by side with pygmy forests of tiny, stunted trees. As shown in the figure, the same sandstone substrate underlies both forests, but the pygmy forest occurs where an impervious hardpan close to the surface greatly restricts root development as well as movement of water and nutrients. The vegetation that reaches an equilibrium, or climax, condition with this special situation is almost totally different in species composition and structure from that of adjacent areas that lack the hardpan.

Where cycle succession, as described in the previous section, prevails, there is no climax in the sense of a self-perpetuating steady state, unless one wishes to consider the end-point in each cycle as a short-lived climax or, as some have suggested, a "catastrophic climax," one that is "programmed" for destruction.

Usually, species composition has been used as a criterion to determine whether a given community is climax. However, this criterion alone is often not a good one, because species composition can change appreciably in response to seasons and short-term fluctuations of weather, even though the ecosystem as a whole remains stable. As already indicated, the *P*/*R* ratio or other functional criteria may provide a better index. Various statistical ratios—for example, turnover times (longer in climax than in developmental stages), coefficients of variation, indices of similarity, and other indicators of stability—should also be useful.

Human beings, of course, greatly affect the progress of succession and the achievement of climaxes. When a stable community that is not the climatic or edaphic climax for a given site is maintained by people or their domestic animals, it may conveniently be designated a **disclimax** (= **disturbance climax**) or **anthropogenic subclimax** (= human-generated). For example, overgrazing by stock may produce a desert community of creosote bushes, mesquite, and cactus where the local climate actually would allow a grassland to maintain itself. The desert community would be the disclimax, and the grassland, the climatic climax. In this case, the desert community is evidence of poor management by humans, whereas the same desert community in a region with a true desert climate would be a natural condition. An interesting combination of edaphic and disturbance climaxes occupies extensive areas in the California grassland region where introduced annual species have almost entirely replaced native prairie grasses (see McNaughton, 1968, for an account of this disclimax).

Agricultural ecosystems that have been stable for a long time can certainly be regarded as climaxes (or disclimaxes), since on an annual average, imports plus production balance respiration plus exports (harvest), and the landscape remains the same from year to year. Agriculture in the Low Countries (Holland and Belgium) and age-old rice cultures in the Orient are examples of long-term anthropogenic steady states. Unfortunately, many crop systems, especially as currently managed in the tropics and on irrigated deserts, are by no means stable, since they are subject to erosion, leaching, accumulation of salt, and irruptions of pests. Maintaining high productivity in such systems requires an increasing human subsidy, and too much subsidy becomes a stress (Chapter 3, Section 3, Figure 3–5).

3. Evolution of the Biosphere

Statement

As with short-term development, described in Section 1 of this chapter, the long-term evolution of the biosphere is shaped by (1) allogenic (outside) forces such as geological and climatic changes and (2) autogenic (inside) processes resulting from activities of the organisms of the ecosystem. The first ecosystems 3 billion years ago were populated by tiny anaerobic heterotrophs that lived on organic matter synthesized by abiotic processes. Then came the origin and population explosion of algal autotrophs, which are believed to have played a dominant role in converting a reduced atmosphere into an oxygenic one. Since then, organisms have evolved through long geological ages into increasingly complex and diverse systems that (1) have achieved control of the atmosphere and (2) are populated by larger and more highly organized multicellular species. Evolutionary change is believed to occur principally through *natural selection at or below the species level*, but natural selection above this level is also important, especially (1) **coevolution**, that is, the reciprocal selection between interdependent autotrophs and heterotrophs, and (2) **group** or **community selection**, which leads to the maintenance of traits favorable to the group even when disadvantageous to the genetic carriers within the group.

Explanation

The broad pattern of the evolution of organisms and the oxygenic atmosphere, two factors that make the biosphere unique among the planets of our solar system, are pictured in Figure 8–11A and B. Scientists generally believe that when life began on earth more than 3 billion years ago, the atmosphere contained nitrogen, ammonia, hydrogen, carbon monoxide, methane, and water vapor, but no free oxygen (see Berkner and Marshall, 1964, 1965; Drake, 1968; Tappen, 1968; and Calvin, 1969). The atmosphere also contained chlorine, hydrogen sulfide, and other gases that would be poisonous to much of present life. The composition of the atmosphere in those early days was largely determined by the gases from volcanos, which were much more active then than now. Because of the lack of oxygen, no ozone layer (O_2 acted on by short-wave radiation produces O_3, which in turn absorbs ultraviolet radiation) shielded out the sun's deadly ultraviolet radiation, which therefore penetrated to the surface of the land and water. Such radiation would kill any exposed life, but, strange to say, this radiation is thought to have created a chemical evolution leading to complex organic molecules such as amino acids, which became the building blocks for primitive life. The very small amount of nonbiological oxygen produced by ultraviolet dissociations of water vapor may have provided enough ozone to form a slight shield against ultraviolet radiation. Yet as long as the atmospheric oxygen and ozone remained scarce, life could develop only under the protective cover of water. The first living organisms, then, were aquatic yeast-like anaerobes that obtained the energy necessary for respi-

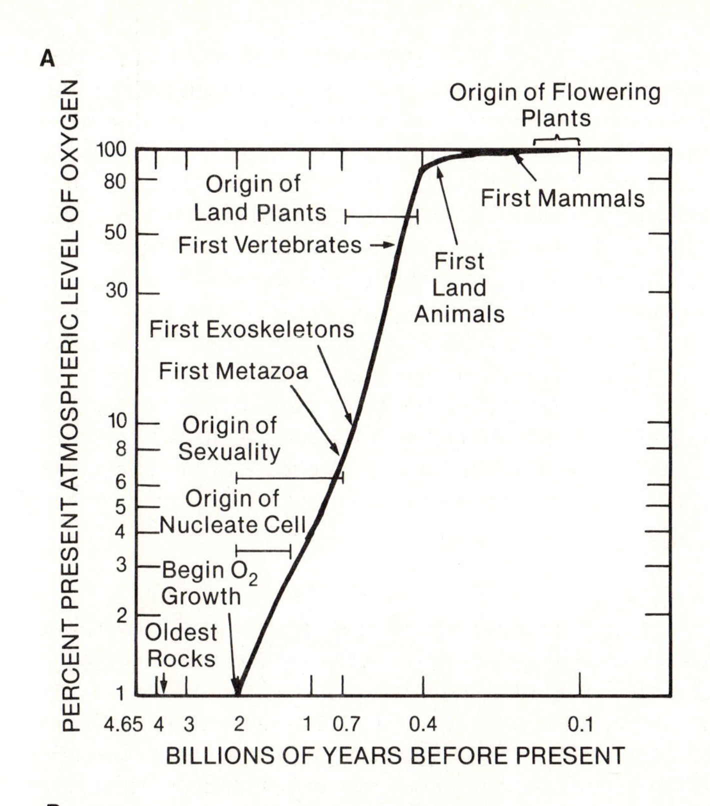

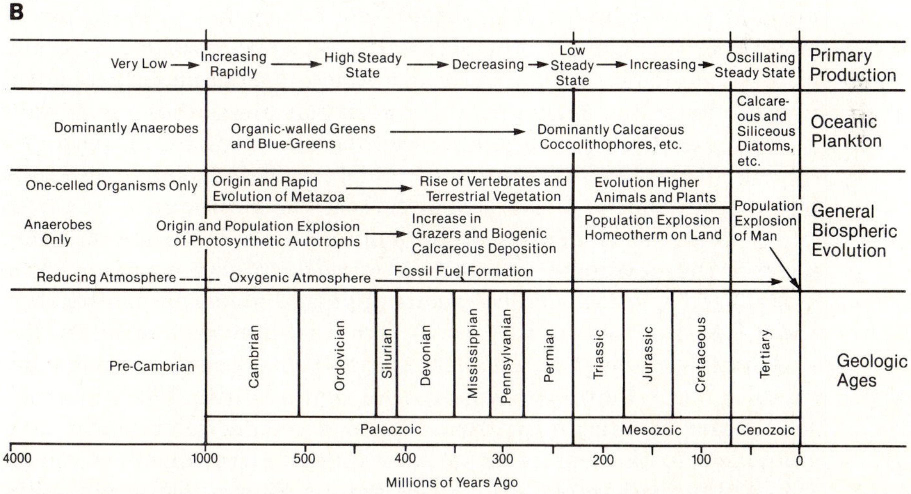

Figure 8–11 The evolution of the biosphere and its oxygenic atmosphere. *A*. Apparent timing of events in biological evolution compared with hypothetical levels of oxygen. (After Cloud, 1978.) *B*. Geological ages and origin of major biotic components as revealed by fossil record.

ration by the process of fermentation. Because fermentation is so much less efficient than oxidative respiration (see page 37), the primitive life could not evolve beyond the prokaryote (nonnucleate) single-cell stage. Such primordial life also had a very limited food supply, since it would depend on the slow sinking of organic materials synthesized by radiation in the upper water layers where the hungry microbes could not venture! Thus, for millions of years, life must have existed in a very limited and precarious condition. Berkner and Marshall (1966) describe the situation:

> This model of primitive ecology calls for pool depths sufficient to absorb the deadly ultraviolet, but not so deep as to cut off too much of the visible light. Life could have originated on the bottom of pools or shallow, protected seas fed, perhaps, by hot springs rich in nutrient chemicals.

The origin of photosynthesis is shrouded in mystery. Perhaps selection pressure exerted by scarcity of organic food played a part. The gradual buildup of photosynthetically produced oxygen and its diffusion into the atmosphere about 2 billion years ago (Figure 8–11B) brought about tremendous changes in the earth's geochemistry and made possible the rapid expansion of life and the development of the eukaryote (nucleated) cell, which led to evolution of larger and more complex living systems. Many minerals, such as iron, were precipitated from water and formed characteristic geological formations. As the oxygen in the atmosphere increased, the layer of ozone formed in the upper atmosphere thickened sufficiently to screen out the DNA-disrupting ultraviolet radiation. Life could then move more freely to the surface of the sea. Then followed what Cloud (1978) calls the "greening of the lands." Aerobic respiration made possible the development of complex multicellular organisms. It is thought that the first nucleate cells appeared when oxygen reached about 3 to 4 percent of its present level (or about 0.6 percent of the atmosphere, compared with the present 20 percent), a time now dated at least 1 billion years ago. Margulis (1981, 1982) makes a strong case for the theory that the eukaryote cell originated as a mutualistic coming-together of once independent microbes, analogous to the modern evolution of lichens as shown in Figure 7–11.

The first multicellular organisms (metazoa) appeared when oxygen content reached about 8 percent some 700 million years ago (Figure 8–11B). The term "Precambrian" is used to cover that vast period of time when only the small, prokaryote single-celled life existed. During the Cambrian there was an evolutionary explosion of new life, such as sponges, corals, worms, shellfish, seaweed, and the ancestors of seed plants and the vertebrates. Thus, the fact that the tiny green plants of the sea were able to produce an excess of oxygen over the respiration needs of all organisms allowed the whole earth to be populated in a comparatively short time, geologically speaking. In the following periods of the Paleozoic era, occupation of all the biosphere was completed. The developing green mantle of terrestrial vegetation provided more oxygen and food for the subsequent evolution of large creatures such as dinosaurs, mammals, and then humans. At the same time, calcareous and then siliceous forms were added to the organic-walled phytoplankton of the oceans (Figure 8–11A).

When oxygen use finally caught up with oxygen production sometime in the mid-Paleozoic (about 400 million years ago), the concentration in the atmosphere reached its present level of about 20 percent. From the ecological viewpoint, then, evolution of the biosphere seems to be very much like a heterotrophic succession followed by an autotrophic regime, such as one might set up in a laboratory microcosm starting with culture medium enriched with organic matter. During the late Paleozoic there appears to have been a decline of O_2 and an increase of CO_2 accompanied by climatic changes. The increase of CO_2 may have triggered the vast "autotrophic bloom" that created the fossil fuels on which human industrial civilization now depends. After a gradual return to a high O_2–low CO_2 atmosphere, the O_2/CO_2 balance remained in what might be called an oscillating steady state. Anthropogonic CO_2 and dust pollution may be making this precarious balance still more "unsteady" (discussed in Chapter 2 and again in Chapter 4).

Incidentally, the story of the atmosphere, briefly described here, should be told to every citizen and school child because it dramatizes the absolute dependence of human beings on other organisms in the environment. According to the Gaia Hypothesis (Chapter 2, Section 4), this control, especially by microorganisms, developed very early in the history of the biosphere. A contrary hypothesis is that early life was merely able to adapt to conditions resulting from radiation-induced physical-chemical changes. In other words, the question is: Was early evolution more autogenic than allogenic, or vice versa? Also much debated is whether evolution of life occurs gradually or is strongly pulsed (short period of rapid change alternating with long periods with little or no change), as is suggested by the fossil record. This question will be considered in the next section.

The constant shifting of the continents through time in a process known as "continental drift" has an important bearing on the evolution of life. For a further discussion of this process, see Wilson (1972).

For a very readable account of the evolution of the biosphere, try the paperback by Cloud (1978); for a fascinating account of the evolution of life, see Margulis' little book entitled *Early Life* (1982).

4. Natural Selection: Allopatric and Sympatric Speciation: Microevolution versus Macroevolution

Statement

The species is a natural biological unit tied together by the sharing of a common gene pool. Evolution involves changing gene frequencies resulting from (1) selection pressure from the environment and interacting species, (2) recurrent mutations, and (3) genetic drift (stochastic or "chance" changes in gene structure). Speciation, the formation of new species and the development of species diversity, occurs when gene flow within the common pool is interrupted by an isolating mechanism. When isolation occurs through geographical separation of populations descended from a common ancestor, **allopatric** (= different fatherland) **speciation** may result. When isolation occurs through ecological or genetic means within the same area, **sympatric** (= joint fatherland) **speciation** is a possibility. At present, it is uncertain to what extent speciation is a slow, gradual process (microevolution) or a matter of periodic, rapid changes (macroevolution).

Explanation

Biological evolution, involving Darwinian natural selection and genetic mutation at the species level, is widely accepted as fact, not theory, by scientists, but there is much debate about the mechanisms. Especially uncertain is the relative role played by the three mechanisms listed in the Statement (selection, mutation, chance) and the role played by selection at the higher levels of biological organization (coevolution and group selection). The latter aspects will be considered in Sections 5 and 6.

Ever since Darwin, biologists have generally adhered to the theory that evolutionary change is a slow, gradual process involving many small mutations and continuous natural selection of those that provide competitive advantage at the individual level. However, gaps in the fossil record and failure to find transitional forms ("missing links") has led many paleontologists to believe in what Gould and Eldredge (1972, 1977) have called "punctuated equilibria." According to this theory, species remain unchanged in a sort of evolutionary equilibrium for long periods. Then, once in a while, the equilibrium is "punctuated" when a small population splits off from the parent species and rapidly evolves into a new species without there being transitional forms in the fossil record. The new species may be sufficiently different to coexist rather than replace the parent, or both may become extinct. The punctuated evolutionary theory does not emphasize competition at the individual level as the driving force, but as yet there is no explanation of what might cause a population to suddenly split off to form a new, genetically isolated unit. For more on macroevolution versus microevolution, see essays by Gould (1977) and Rensberger (1981, 1982) and the books by Stanley (1979, 1981).

Species that occur in different geographical regions or are separated by a spatial barrier are said to be allopatric; those occurring in the same area are said to be sympatric. Allopatric speciation has been generally assumed to be the primary mechanism by which species arise. According to this conventional view, two segments of a freely interbreeding population become separated spatially (as on an island or separated by a mountain range). In time, sufficient genetic differences accumulate in isolation so that the segments will no longer interchange genes (interbreed) when they come together again, and thereby coexist as distinct species in different niches. Sometimes these differences are further accentuated by **character displacement.** When two closely related species have overlapping ranges, they tend to diverge in one or more morphological, physiological, or behavioral characters in the area of overlap and to converge (to remain or become similar to each other) in parts of their range where each species occurs alone. See Brown and Wilson (1956) for further explanation and examples of the phenomenon of character displacement.

Evidence is mounting that strict geographical separation is not necessary for speciation and that sympatric speciation may be more widespread and important than previously believed. Populations can become genetically isolated within the same geographical area as a result of behavioral and reproductive patterns such as colonization, restricted dispersal of propagules, asexual reproduction, selection, predation, and so on. In time, sufficient genetic differences accumulate in the local population segments to prevent interbreeding.

Examples

A classic example of allopatric speciation (resulting from geographical isolation) with subsequent character displacement is the well-documented case of Galapagos finches, first described by Darwin, who visited the Galapagos Islands during his famous voyage of the Beagle. From a common ancestor a whole group of species evolved in isolation on the different islands and adaptively radiated so that a variety of potential niches was eventually exploited on reinvasion. Species now present include slender-billed insect eaters and thick-billed seed-eaters, ground- and tree-feeders, large- and small-bodied finches, and even a woodpecker-like finch, which although hardly able to compete with a real woodpecker, survives in the absence of invasion by woodpecker stock (see Lack, 1947). Figure 8–12 shows how beak size varies in one of the Galapagos finches according to whether the species is alone on an island or coexists with two other closely related species on another larger island. In the latter situation (B in Figure 8–12), the beak is "displaced" to an increased depth so that it does not overlap with the beak size of its two competitors. As a result, competition for food is reduced since each of the three species is adapted to feed on different-sized seeds.

British salt marshes provide an example of sympatric speciation resulting from hybridization and polyploidy. When the American salt marsh grass *Spartina alterniflora* was introduced into the British Isles, it crossed with the native species, *S. maritima,* to produce a new polyploid species, *S. townsendii,* which has now invaded formerly bare tidal mud flats not occupied by native species.

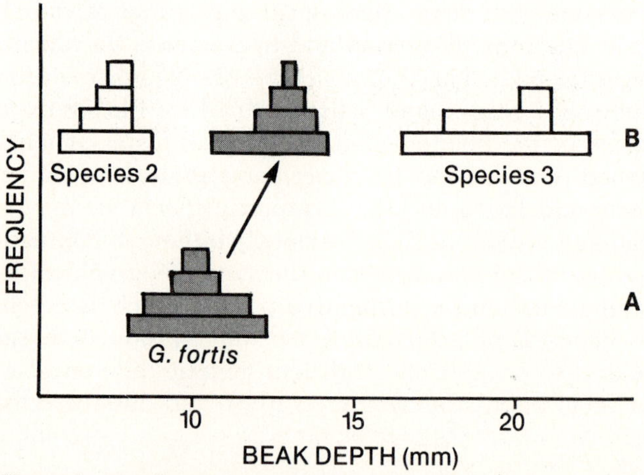

Figure 8–12 Beak size of *Geospiza fortis* (one of the Darwin's finches) when alone on a Galapagos island (*A*) and when with a competing *Geospiza* species on other islands (*B*). Beak size is increased when competitors are present (an example of character displacement). (After Lack, 1947.)

Another example of rapid natural selection resulting from human interference is what has come to be known as "industrial melanism." Dark-pigmented tree-trunk moths have evolved in industrial areas of England where the bark of trees has become greatly darkened by industrial pollution, which kills the lichens that give normal bark a light appearance. Kettlewell (1956) demonstrated experimentally that dark moths survive better in dark (polluted) woods and light moths in natural woods, presumably because predation by birds is selective for the individuals not protectively colored. This example leads us naturally onto the subject of direct or purposeful selection by humans.

Artificial Selection: Domestication Selection carried out by people to adapt plants and animals to their needs is known as artificial selection. Domestication* of plants and animals involves more than modifying the genetics of a species, because reciprocal adaptations between the domesticated species and the domesticator are required. Accordingly, domestication leads to a special form of mutualism. Because humans are egotistical, we fall into the trap of thinking that domesticating another organism through artificial selection means merely "bending" nature to suit our purposes. Actually, of course, domestication produces changes (ecological and social, if not genetic) in people as well as in the domesticated organisms. Thus, we are just as dependent on the corn plant as the corn is dependent on us. A society that depends on corn develops a very different culture than one that depends on herding cattle. It is a real question as to who becomes the slave of whom! The same question can be posed about the relationships of humans with machines. Perhaps the relationship between a person and his tractor is not too different from that of a person and his plowhorse, except that the tractor requires a higher quality and a greater energy input (fuel) and produces more poisonous wastes!

Artificial selection in crops, a major basis for the green revolution, is a good example of interdependence between the domesticated species and the human domesticator. Increased yield is obtained by selecting for an increased ratio of grain (or other edible parts) to supporting tissues (leaves, roots, stems, and so forth). Above a certain point, increased yield must come at a sacrifice of the plant's adaptive, self-sustaining capacity. Therefore, highly bred strains require massive subsidies of energy, fertilizers, and pesticides, which bring

* The term "cultivation" is preferred by many for artificial selection of plants. Here we use the term "domestication" in the general sense for both plants and animals.

about profound changes in the social, economic, and political structure of human society. Many poor countries are finding these socioeconomic and resource requirements to be the greatest obstacle in using high-yielding varieties to increase food supply.

Throughout history, some of the worst environmental problems have been caused by domesticated plants and animals that "escape" back into nature (become feral) and become major pests. A feral organism differs from its wild ancestor in that it has experienced a period of artificial selection during which some new traits may have been acquired and some of the original "wild" traits lost. On returning to the wild, the feral species again comes under natural selection that favors traits necessary for survival on its own. For example, spotted or light-colored coats and large body size of domestic pigs are selected against when they revert to the wild, so the feral pig becomes slender and dark. The combination of artificial and natural selection seems to produce plants and animals that thrive in habitats that have been partially altered or disturbed. Feral pigs and goats, for example, can wreak havoc on the vegetation of islands or areas from which large predators have been removed.

Island Biogeography Islands provide natural laboratories for studying evolution, and as such they have fascinated biologists and ecologists ever since Darwin visited the Galapagos Islands. The interplay of isolation, natural selection, and speciation has attracted attention recently, especially after MacArthur and Wilson published their theory of island biogeography (1963, 1967). Simply stated, the theory holds that the number of species on an island is determined by the equilibrium between immigration of new species and extinction of those already present. Since species increase and decrease in an approximate logarithmic manner, and since rates of immigration and extinction depend on the size of the island and the distance from a mainland species reservoir, a general equilibrium model can be diagrammed, as in Figure 8–13. Four equilibrium points are shown, representing (1) a small, distant island predicted to have a few species; (2) a large, close-in island that should support many species; (3) a large, distant island; and (4) a small, nearby island intermediate in terms of species richness. The equilibrium theory has been tested by Simberloff and Wilson (1969), who removed all arthropods (by insecticide treatment) from a small mangrove island in the Florida Keys and observed recolonization. For more on island biogeography, see the review by Simberloff (1974).

Wilson and Willis (1975), Diamond and May (1981), and others have suggested that the theory of island biogeography provides a

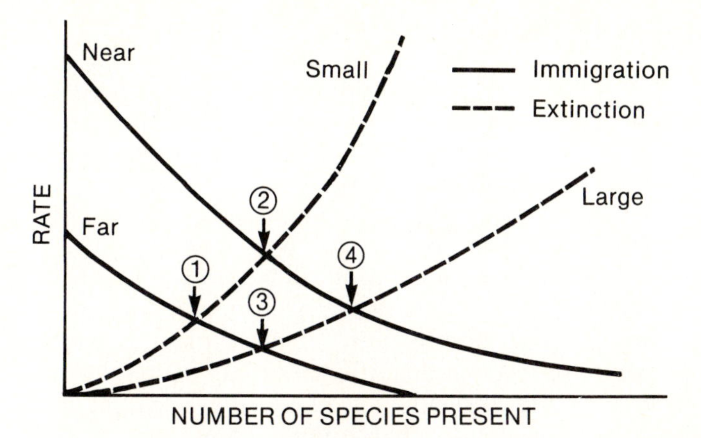

Figure 8–13 Theory of island biogeography. The number of species on an island is determined by equilibrium between rate of immigration and rate of extinction. Four equilibrium points are shown, representing different combinations of large and small islands near and far from continental shores. (After MacArthur and Wilson, 1963.)

basis for the design of natural reserves set up to preserve natural diversity or endangered species, or both. Accordingly, a large reserve is preferable to a group of smaller reserves with the same area. If one must settle for small parks, they should be close together or connected with corridors to facilitate immigration. A circular shape that maximizes area-to-perimeter ratio is preferable to an elongated reserve. Since reserves or parks set up within a continental area are rarely as isolated as oceanic or coastal islands, it is uncertain whether the MacArthur-Wilson model applies. Simberloff and Abele (1976) and Kushlan (1979) listed several other problems in applying island theory to the design of nature preserves.

5. Coevolution

Statement

Coevolution is a type of community evolution (i.e., evolutionary interactions among organisms in which exchange of genetic information among the kinds is minimal or absent). It involves reciprocal selective interaction between two major groups of organisms with a close ecological relationship, such as plants and herbivores, large organisms and their microorganism symbionts, or parasites and their hosts.

Explanation

Using their studies of butterflies and plants as a basis, Ehrlich and Raven (1965) were among the first to outline the theory of coevolution as it is now widely accepted by students of evolution. Their hypothesis may be stated as follows: Plants, through occasional mutations or recombinations, produce chemical compounds not directly related to basic metabolic pathways (or perhaps as waste byproducts generated in these pathways) that are not inimical to normal growth and development. Some of these compounds, by chance, reduce the palatability of the plants to herbivores. Such a plant, protected from phytophagous insects, would in a sense have entered a new adaptive zone. Evolutionary radiation of the plants might follow, and eventually what began as a chance mutation or recombination might characterize an entire family or group of related families. Phytophagous insects, however, can evolve in response to physiological obstacles, as shown by widespread development of immune strains. Indeed, response to secondary plant substances and the evolution of resistance to insecticides seem to be intimately connected (Gordon, 1961). If a mutant or recombinant appeared in a population of insects that enabled individuals to feed on the previously protected plant, selection would carry the line into a new adaptive zone, allowing it to diversify in the absence of competition with other herbivores. Thus, the diversity of plants not only may tend to augment the diversity of phytophagous animals, but the converse may also be true. In other words, the plant and the herbivore evolve together in the sense that the evolution of each depends on the evolution of the other.

Pimentel (1968) has used the expression "genetic feedback" for this kind of evolution, which leads to population and community homeostasis within the ecosystem.

Examples

Numerous examples of coevolution are discussed and analyzed in the symposium volume edited by Gilbert and Raven (1980). One example from this volume will illustrate how plants and animals have evolved to benefit one another.

Bumblebees, widely distributed species in the genus *Bombus*, are very important pollinators of both wild plants and important cultivated crops such as alfalfa, clover, beans, and blueberries. Heinrich (1979, 1980) has assessed the interactions of flowers and bumblebees on the basis of energetics. He measured nectar production in terms of sugar available per flower and counted the number of visits by bees and the rate of removal of nectar in relation to time of day and temperature. Bees, unlike butterflies, have a high metabolic rate and must visit flowers rapidly to "make an energy profit." To attract these necessary pollinators and ensure their survival, many species of flowers have evolved the "strategy" of either blooming synchronously or growing in colonies.

That reciprocal natural selection is not limited to two-species interactions is shown by Colwell (1973), who describes how ten diverse species—four flowering plants, three hummingbirds, one coerebid bird, and two mites—have coevolved to produce a fascinating tropical subcommunity. Coevolution can also involve more than one step in the food chain. Brower and colleagues (1968), for example, have studied the monarch butterfly (*Danaus plexippus*), which is well known for its general unpalatability to vertebrate predators. They found that this insect can sequester the highly toxic glycosides present in the milkweed plants on which they feed, thereby providing a highly effective defense against bird predators (not only for the caterpillar but also for the adult butterfly). Thus, this insect has evolved the ability to feed on a plant that is unpalatable to other insects, and it also "uses" the plant poison for its own protection against predators.

6. Evolution of Cooperation and Complexity: Group Selection

Statement

To account for the incredible diversity and complexity of the biosphere, scientists have postulated that natural selection operates beyond the species level and beyond coevolution. **Group selection**, accordingly, is defined as natural selection between groups or organisms not necessarily closely linked by mutualistic associations. Group selection theoretically leads to the maintenance of traits favorable to populations and communities that may be selectively disadvantageous to genetic carriers within populations. Conversely, group selection may eliminate, or keep at low frequencies, traits unfavorable to the survival of the species but selectively favorable within populations. Group selection involves positive benefits that an organism may exert on the community organization required for that organism's continued survival.

Explanation and Examples

The "struggle for existence" and "survival of the fittest" (to use Darwin's words) are not just a matter of "dog eat dog." In many cases, survival is based on cooperation rather than competition. How cooperation and elaborate mutualistic relationships get started and become genetically fixed has been difficult to explain in evolutionary theory, because when individuals first interact it is nearly always advantageous for each individual to act in its own interest rather than to cooperate. Axelrod and Hamilton (1981) have analyzed the evolution of cooperation and devised a model based on the "prisoner's dilemma game" (see Rapoport and Chammah, 1970) and on the theory of reciprocation as an extension of the conventional competition-based survival-of-the-fittest genetic theory. In the prisoner's dilemma game, two "players" decide whether to cooperate or not on the basis of immediate benefits. On first encounter, a selfish decision not to cooperate yields the highest reward for each individual regardless of what the other individual does. However, if both choose not to cooperate, they both do worse than if both had cooperated. If individuals continue to interact (i.e., the "game" continues), the probability is that cooperation may be selected on a trial basis and its advantages recognized. Deductions from the model and the results of a computer tournament show how cooperation based on such reciprocity can get started in an asocial environment and then develop and persist once fully established. Ability to recognize individuals in the higher animals enhances opportunities for the development of reciprocity, while constant close contact between numerous individuals of microorganisms or microorganisms and plants enhances the possibilities for interaction with mutual benefit, such as has evolved in nitrogen-fixing bacteria and legumes.

Axelrod and Hamilton also suggest that altruism—sacrifice of fitness by one individual for the benefit of another—in related individuals (parent and offspring, for example) can be the start of an evolution toward cooperation (even in unrelated species). Once genes favoring reciprocity have become established by kin-selection, cooperation can spread into circumstances of less and less relatedness.

David Sloan Wilson (1975, 1977, 1980) states the case for group selection as follows (see page 97 in his 1980 book):

> Populations routinely evolve to stimulate or discourage other populations upon which their fitness depends. As such, over evolutionary time an organism's fitness is largely a reflection of its own effect on the community and the reaction of the community to that organism's presence. If this reaction is sufficiently strong, only organisms with a positive effect on their community persist.

Wilson argues that selection between "structured demes" (closely knit genetic segments of a population) facilitates group selection. He also draws an analogy between the paradox of individual versus community fitness in biological communities and private good versus public good in human communities.

Since predator-prey and parasite-host interaction tend to become less negative in time (Chapter 7, Section 3), Gilpin (1975) has proposed group selection in the development of a "prudence" trait that leads predators and parasites not to overexploit their prey or hosts. The history of the myxomatosis virus introduced to control European rabbits (actually, "hares") in Australia is discussed by Levin and Pimentel (1981) as an example of selection for reduced virulence. When first introduced, the parasite killed the rabbit within a few days. Subsequently, the virulent strain was replaced by a less virulent one that took two to three times as long to kill the host; hence, mosquitoes that transmit the virus had a longer time to feed on infected rabbits. Because the avirulent strain did not destroy their food resource (rabbit) as rapidly as the virulent strain, more and more avirulent-type parasites were produced and were available for transmission to new hosts. Thus, interdemic selection favored the avirulent over the virulent; otherwise, both parasite and host would eventually become extinct.

Though few doubt that group selection occurs, its importance in evolutionary history remains controversial. Saunders and Ho (1976), among others, argue that the complexity that has developed in the natural world cannot possibly be explained solely by selection at the individual and species level; hence, higher level selection has to play a major role. In contrast, G. C. Williams (1966) and Ernst (1982), among others, argue that everything can be explained within the framework of traditional Darwinian natural selection; consequently, according to this view, it is unlikely that group selection is a major evolutionary mechanism.

For more on group selection, see E. O. Wilson (1973), Maynard Smith (1976), and Alexander and Borgia (1978).

7. Relevance of Ecosystem Development and Biosphere Evolution Theory to Human Ecology

Statement

The principles of ecosystem development bear importantly on the relationships between human beings and nature, because the developmental trend in natural systems that involves increasing structure

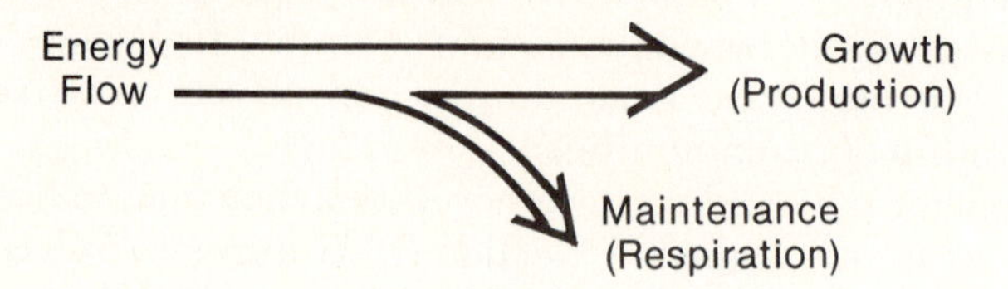

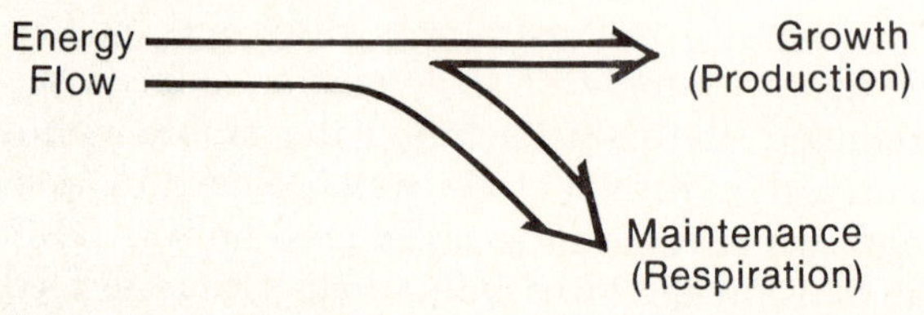

Figure 8–14 The contrast in energy partitioning between developing and mature systems.

and complexity per unit of energy flow (a maximum protection strategy, as it were) contrasts with the human goal of maximum production (trying to obtain the highest possible yield). Recognizing the ecological basis for this conflict between humans and nature is a first step in establishing rational policies for managing the environment.

Explanation

Figure 8–14 (as well as Figure 8–2) depicts a basic conflict between the strategies of humans and of nature. The energy partitioning exhibited in early development, such as the 30-day microcosm or the 30-year forest, illustrates how political and economic leaders think nature should be directed. For example, the goal of agriculture or intensive forestry, as now generally practiced, is to achieve high rates of production of readily harvestable products with little standing crop left to accumulate on the landscape—in other words, a high *P*/*B* efficiency. Nature's strategy, on the other hand, as seen in the outcome of the successional process, is directed toward the reverse efficiency—a high *B*/*P* ratio. Human beings have generally been preoccupied with obtaining as much production from the landscape as possible by developing and maintaining early successional types of ecosystems, often monocultures. But, of course, we do not live by food and fiber alone; we also need a balanced CO_2/O_2 atmosphere, the climatic buffer provided by oceans and masses of vegetation, and clean (that is, unproductive) water for cultural and industrial uses. Many essential life-cycle resources, not to mention recreational and esthetic needs, are best provided by the less productive landscapes. In other words, the landscape is not just a supply depot but is also the *oikos*—the home—in which we must live. Until recently we have taken for granted the gas-exchange, water-purification, nutrient-cycling, and other protective functions of self-maintaining ecosystems, that is, until our numbers and our environmental manipulations became great enough to affect regional and global balances. The most pleasant and certainly the safest landscape to live in is one containing a variety of crops, forests, lakes, streams, roadsides, marshes, seashores, and "waste places"—in other words, a mixture of communities of different ecological ages. As individuals, we more or less instinctively surround our houses with protective, nonedible cover (trees, shrubs, grass) at the same time that we strive to coax extra bushels from our cornfields. The cornfield is a "good thing," of course, but most people would not want to live in the middle of one. It would be suicidal to cover the whole land area of the biosphere with crops, since we would not have the nonedible life-support buffer that is vital for biospheric stability and for esthetic pleasure. It also would invite disaster from epidemic disease.

Since it is impossible to maximize for conflicting uses in the same system, two possible solutions to the dilemma suggest themselves. We can continually compromise between quantity of yield and quality of living space, or we can deliberately compartmentalize the landscape to maintain both highly productive and predominantly protective types as separate units subject to different management strategies (strategies ranging, for example, from intensive cropping to wilderness management). If ecosystem development theory is valid and applicable to planning, then the multiple-use strategy, about which we hear so much, will work only through one or both of these approaches, because in most cases the projected multiple uses conflict with one another. For example, dams on large rivers are often touted as providing a whole range of benefits such as power generation, flood control, water supply, fish production, recreation, and so on. But these uses actually conflict, because for flood control to be achieved, the water must be "drawn down" before the flood season, an action that reduces power generation and interferes with recreation. Accordingly, one can maximize for a single use (or perhaps for several closely coupled uses) while reducing other uses, or one can settle for some of all, i.e., compromise. It is appropriate, then, to examine some examples of the compromise and the compartmental strategies.

Pulse Stability A more or less regular but acute physical perturbation imposed from without can maintain an ecosystem at some intermediate point in the development sequence, resulting in, so to speak, a compromise between youth and maturity. What might be termed "fluctuating water-level ecosystems" are good examples. Estuaries, and intertidal zones in general, are maintained in an early, relatively fertile stage by the tides, which provide the energy for rapid cycling of nutrients. Likewise, freshwater marshes, such as the Florida Everglades, are held at an early successional stage by the seasonal fluctuations in water levels. The dry-season drawdown speeds up aerobic decomposition of accumulated organic matter, releasing nutrients that, on reflooding, support a wet-season bloom in productivity. The life histories of many organisms are intimately coupled to this periodicity, for example, the timing of breeding in the wood stork (see Figure 6–2). Stabilizing water levels in the Everglades by means of dikes, locks, and impoundments destroys rather than preserves the Everglades as we know them just as surely as complete drainage would. Without periodic drawdowns and fires, the shallow basins would fill up with organic matter, and succession would proceed from the present pond-and-prairie condition toward a scrub or swamp forest.

It is strange that humans do not readily recognize the importance of recurrent changes in water level in a natural situation such as the Everglades, even though similar pulses are the basis for some of our most enduring systems of food culture. Alternate filling and draining of ponds has been a standard procedure in fish culture for centuries in Europe and the Orient. The flooding, draining, and soil-aeration procedure in rice culture is another example. The rice paddy is thus the cultivated analogue of the natural marsh or the intertidal ecosystem.

Fire is another physical factor whose periodicity has been of vital importance over the centuries. As described in Chapter 4, whole biotas, such as those of the African grasslands and the California chaparral, have become adapted to periodic fires, producing what ecologists often call "fire climaxes." For centuries, people have used fire deliberately to maintain such climaxes or to set back succession to some desired point. The fire-controlled forest yields less wood than a tree farm does (that is, young trees, all of about the same age, planted in rows and harvested on a short rotation), but it provides a greater protective cover for the landscape, wood of higher quality, and a home for game birds (quail, wild turkey, and so on) that could not survive in a tree farm. The fire climax, then, is an example of a compromise between production and simplicity on the one hand and protection and diversity on the other.

Pulse stability works only if a complete community (including not only plants but also animals and microorganisms) is adapted to the particular intensity and frequency of the perturbation. Adaptation (operation of the selection process) requires times measurable on the evolutionary scale. Most physical stresses introduced by human beings are too sudden, too violent, or too arrhythmic for adaptation to occur, so severe oscillation rather than stability results. In many cases, at least, modification of naturally adapted ecosystems for cultural purposes would seem preferable to complete redesign.

Prospects for a Detritus Agriculture Heterotrophic use of primary production in mature ecosystems involves largely a delayed consumption of detritus. There is no reason why humans cannot make greater use of the detritus food chain to obtain food or other products while retaining the protective functions of the ecosystem. Again, this choice would represent a compromise, since the short-term yield could not be as great as the yield obtained by direct exploitation of the grazing food chain. A detritus agriculture, however, would have some compensating advantages. Present agricultural strategy is based on selection for rapid growth and edibility in food plants, which, of course, makes the plants vulnerable to attack by insects and disease. Consequently, the more we select for succulence and growth, the more effort we must invest in the chemical control of pests. This effort, in turn, increases the likelihood of our poisoning useful organisms, not to mention ourselves. Why not also practice the reverse strategy—that is, select essentially unpalatable plants, or ones that produce their own systemic insecticides while they are growing, and then, if necessary, convert the net production into edible products by microbial and chemical enrichment in food factories? We could then devote our biochemical genius to the enrichment process instead of fouling our living space with chemical poisons. The production of silage by fermentation of low-grade fodder is an example of such a procedure already in widespread use. The cultivation of detritus-eating fishes in the Orient is another example. Organic farming and limited-till agriculture (discussed in Chapter 2, Section 7) are also steps in this direction.

Compartment Models for Land Use In thinking about how principles of ecosystem development relate to the landscape as a whole, consider the compartment models as shown in Figure 8–15. The upper diagram depicts three types of environments that constitute the life-support systems for the fourth compartment, the urban-industrial heterotrophic systems. The human productive "environment" comprises early successional or growth-type ecosystems such as crop-

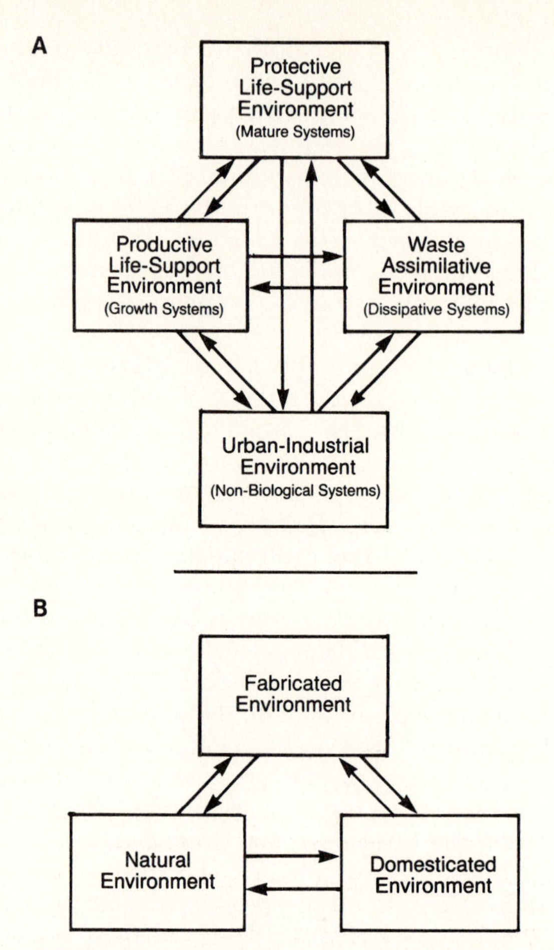

Figure 8–15 Compartment models for environmental-use planning. *A*. Partitioned according to ecosystem theory. *B*. As viewed by environmental architects and designers. See text for explanation.

lands, pastures, tree plantations, and intensively managed forests that provide food and fiber. Mature ecosystems such as old-growth forests, climax grasslands, and oceans are more protective than productive. They stabilize substrates, buffer air and water cycles, and moderate extremes in temperature and other physical factors, while at the same time often providing products. The third category of natural or seminatural ecosystems that bear the brunt of assimilating the vast wastes produced by the urban-industrial and agricultural systems consists of waterways (inland and coastal), wetlands, and other strongly impacted environments. Ecosystems in this admittedly arbitrary category are, in the developmental sense, mostly in intermediate, eutrophicated, or arrested stages of succession. All of these components interact continually in terms of input and output (as shown by arrows in the figure).

Partitioning the landscape into three environmental components, the "natural," "domesticated," and "fabricated," as is traditional with landscape architects (Figure 8–15B), provides another convenient way to consider needs and interrelations between these necessary parts of our household.

Though the urbanized or fabricated environment is "parasitic" on the life-support environment (natural and domesticated) for basic biological necessities, it does create and export other, mostly nonbiotic, resources such as fertilizers, money, processed energy, and goods that both benefit and put stress on the life-support environment. Much more can be done to increase the subsidy and reduce the stress of outputs from high energy and densely populated "hot spots," and this has to be a major goal for humanity from now on. But no known feasible technology can substitute, on a global scale, for the basic biotic life-support goods and services provided by natural ecosystems (recall the discussion of self-contained spacecraft, Chapter 2, Section 7).

Before we consider the prospect of land-use planning or "organic development" replacing the present haphazard, mostly short-term–oriented economic and political policies that determine land and water use (see Watt et al., 1977, Chapter 2, for a comparison of haphazard and organic developmental philosophies), let us look at the partitioning or compartmentalization of the environment in the United States as of 1980. Of the land area alone (Figure 8–16A), about 24 percent is in agroecosystems and 6 percent in urban-industrial systems (including rural transportation corridors), leaving about 70 percent in natural ecosystems variously affected by grazing, timber harvesting, recreation, and pollution. The 4 percent in national or state parks, wilderness, and wildlife refuges is the most protected from human disturbance, although some of this is managed so that particular vegetation, wildlife species, or scenic features may be maintained. Public forests and rangelands (grasslands and deserts), which compose about 30 percent of the area of the contiguous 48 states, are supposed to be managed to maintain a steady state of vegetative structure and a sustained yield balance between harvest or grazing and replacement growth. Such a national policy is becoming more difficult to maintain in the face of increasing human population growth and demands.

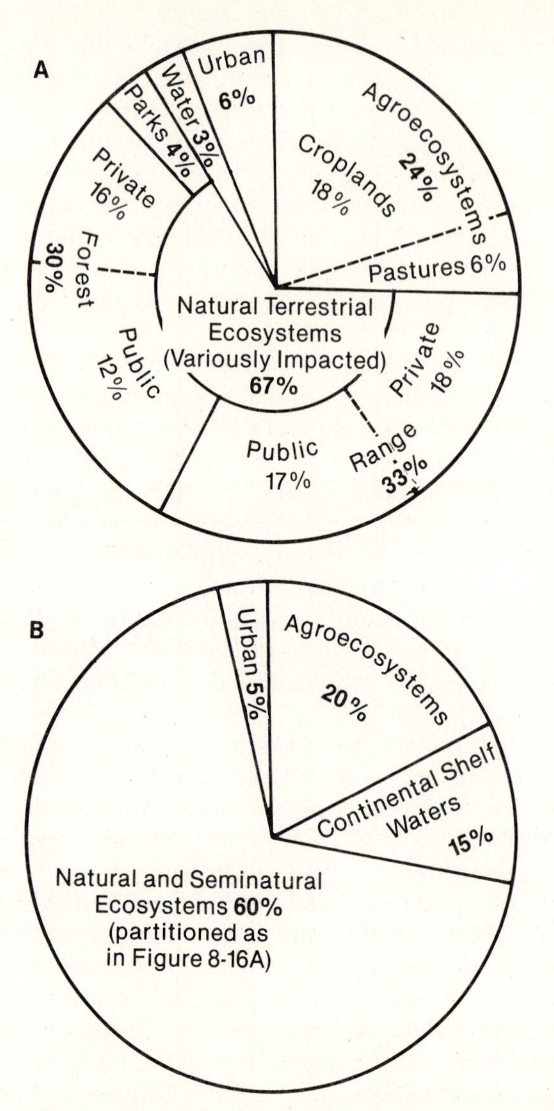

Figure 8–16 Land use in the United States as of 1980. *A*. Continental U.S. land area. *B*. Including continental shelf marine waters.

Since the ocean water over the continental shelf functions as a large waste-assimilative and climatic buffering system and also provides seafood, it is a vital part of the life-support environment of adjacent continental areas. When this area is included, as shown in Figure 8–16B, close to 75 percent of the area of the United States is in natural or seminatural condition (variously affected) and 25 percent in urban-agricultural systems.

Before concluding that a 3 : 1 to 5 : 1 ratio between natural and human-made environment indicates that the life-support environment is sufficient (perhaps more than needed), ponder these three constraints:

1. Since the power level of urban-industrial systems is something like 100 times greater than that of any natural ecosystem (see Table 3–18), a very large area of the natural system is required to dissipate the disorder output of a small area of the urban system, and, as already noted, dissipative capacity is necessary for the development and maintenance of highly organized structure.
2. The life-support capacity of the natural environment can vary over several orders of magnitude according to its productivity (work potential) and the degree of stress already imposed. Thus, a hectare of desert is nowhere as effective as a hectare of fertile coastal marsh, and a lake already heavily polluted will have very little additional life-support capacity.
3. Densely populated and highly industrialized nations, such as Japan and many European countries, depend on a large area outside the country to provide the necessary inflow of energy, materials, food, and general life-support goods and services. These input environmental requirements are what Borgstrom (1969) calls "ghost acres." For food alone, he estimates that for every acre of cropland in Japan, 5 acres of land and ocean outside Japan are required to feed the nation's dense population. The ratio of ghost acres to internal acres is nearly as high in the Low Countries and elsewhere in Europe. In contrast, the United States not only feeds itself but also exports food, thereby providing ghost acres for other countries unable to feed themselves. Accordingly, one cannot compare land-use patterns and ratios in small, densely populated countries with those in large, more sparsely populated countries unless one includes the external support area required for the former. When such areas are included, the ratio of total support area to urban-industrial areas is high in all situations.

All these considerations make it extremely difficult to determine objectively just how much natural environment must be preserved within a given political unit (state or country) to support a given level of human societal development. One approach, as suggested by Odum and Odum (1972), is to set up a model of land management in which the proportion of developed and natural lands can be varied, as shown in Figure 8–17. If all inputs and outputs of the model components are quantified in energy units as a common denominator,

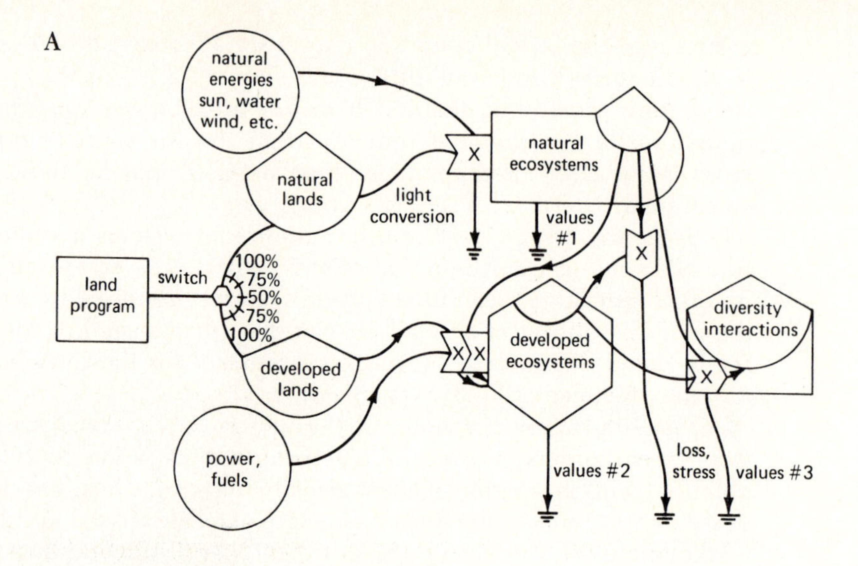

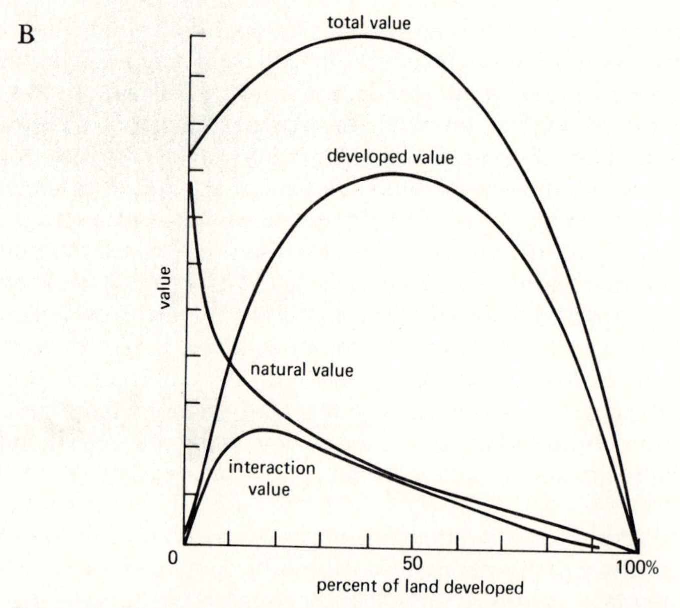

Figure 8–17 *A*. A model for land management in which the proportion of natural and developed lands can be varied in order to determine the optimum balance in terms of the value of the total environment. *B*. Performance curves based on the use of the model for a hypothetical region with extensive urban development. (After Odum and Odum, 1972.)

different options in land-use planning can be simulated with analog or digital computers. For the purposes of the model, "natural lands" are defined as that part of the regional landscape that is self-supporting in that only minimal human management is required for maintenance. In terms of function, "natural environment" is that part of a human life-support system that operates without energetic or economic input from humans. Performance curves in terms of net energy values are shown for one such simulation of a state-sized land area in Figure 8–17. When developed land exceeds 40 percent, the value of developed land and the total value of the whole landscape decline sharply as natural life-support goods and services cannot meet the demands of the intensive agro-urban development. The situation has diminishing returns of scale (discussed in Chapter 3, Section 8; see especially Figure 3–20). In the words of Odum and Odum,

> Even the simplest models clearly demonstrate that high powered systems such as cities require an abundant life-support from nature. If large areas of natural environment are not preserved to provide the needed input from nature then the quality of life in the city declines and the city can no longer compete economically with other cities that have an abundant life-support input. Frequently, it is not energy itself that becomes the limiting factor, but some basic natural resource required to maintain the high rate of energy flow. In many areas water seems now to be that limiting resource. Continued urban, industrial and/or agricultural growth in many parts of the U.S. (and the world, in general) will depend on developing special new water sources such as desalination or pumping from underground or distant sources, all of which are energy-expensive. If such sources have to be developed, the city's energy cost rises until it can no longer compete with cities that do not have to pay this extra cost. It is a sad situation when cities grow beyond their means and can no longer pay for their own maintenance. They borrow money or demand federal grants in order to grow even larger and more demanding of their life-support systems when they ought to be diverting more of their energy to maintaining the quality and efficiency of the environment already developed and to reducing the stress on vital life-supporting environment.

Determining the optimum proportion of different kinds of landscapes is, of course, only the first and most elementary step in comprehensive planning. Spatial patterns are equally important, that is, the location of power plants, industrial parks, agricultural fields, and natural areas to achieve the most favorable interaction. In other words, if power plants, airports, and industrial plants can be surrounded by natural areas or "green belts," the buffering capacity of the latter is more effective than if the same acreage of natural area were located at a distance. Social problems, especially those relating to inequities, economic status, and social justice have overriding importance, especially where population density is high and industrial development intense. The impact of planning options must be determined not only with regard to the whole (that is, per capita), but also with regard to different social, racial, and economic groups that may be affected very differently by a given planning proposal. Thus, a highway projected as a benefit to the regional economy may be a net loss in the long term if it has a serious negative impact on a residential neighborhood. The cost of social disruption must be considered along with projected economic gain.

My students, my colleagues, and I have been asked on several occasions to set up holistic models to assist in making decisions about super highways. One case involved selecting between alternate routes; another case involved a decision not to build a tollway through an Atlanta residential neighborhood; and a third case involved rerouting an interstate highway to reduce the impact on a natural area that had recreational, scenic, and life-support value. In each case, a computerized linear vector model was used in which all perceived values—economic, ecological, social, and others—were scaled to a numerical common denominator and weighed according to the best available expert assessment (see E. P. Odum et al., 1976). In each case, the business community or other special interests opposed such an effort to estimate *total* impact, but when all the "cards were laid on the table," so to speak, all interests agreed to base the decision on the output of the holistic model. As one newspaper editorial expressed it, "This is one way a consensus for the good of the whole can be achieved in a democracy."

Returning to land-use patterns in the United States, as illustrated in Figure 8–16, the one-third "developed" and two-thirds "natural" ratio seems, in light of theoretical considerations just discussed, a favorable situation. One can say that for the country as a whole, Americans are fortunate to have such an abundance of the goods and services of nature. However, the large part of the United States that is arid does not have the support capacity of well-watered regions, as indicated by the severe shortages and increasing cost of water in most western states. Hence, life-support area in the United States is by no means excessive, and Americans need to be concerned with continued preservation and with the quality and capacity of existing natural environments. Also, the life-support capacity of some regions, such as southern Florida, southern California, the industrial coastal Northeast, and the industrial Great Lake region, is obviously being exceeded, as indicated by the increasing spread of acid rain, smog, and other pollution.

Comprehensive Planning from a Perspective of Ecological and Evolutionary Development Three examples will illustrate several innovative approaches to landscape design based on holistic ecological principles.

Ian McHarg's beautifully illustrated book *Design with Nature* (1969) has become a classic and the inspiration for many attempts to design according to the natural features of the landscape. For the first time in a single, easily readable book, a case is made for holistic land-use planning as an alternative to uncontrolled development, which, to paraphrase McHarg, spreads without discrimination and obliterates the landscape with congestion and pollution that irrevocably destroys all that is beautiful and memorable, no matter how well designed is the individual home or subdivision. Table 8–3 contrasts the consequence of unplanned land use with McHarg's planned land use for an urban area projected to grow from a population of 20,000 to 100,000 people. By judicious planning of residential and other development, a third of the area, including hilly terrain and all the stream valleys, is preserved as open space. Also desirable is constructing industrial plants in a large waste-management park that has room for ponds and other seminatural waste treatment systems for efficient and inexpensive tertiary treatment, as shown in Figure 7–23B.

A very successful "new town" that combines the best of commercial development with the kind of open-space design suggested in Table 8–3 is Columbia, Maryland, the brain child of master planner James Rouse. Rouse started out by designing and constructing enclosed shopping malls but soon became enthusiastic about innovation on a larger scale, including inner-city rehabilitation and new towns. In the process, he and his company have become very ardent but practical conservationists.* Columbia, Maryland, comprises 22 square miles (about the size of Manhattan) with 60,000 people and 30,000 jobs within the city. It has five self-sufficient village centers (instead of a single downtown area), each made up of three or four

* For a story of the accomplishments and philosophy of James Rouse, see the August 24, 1981, issue of *Time* (Vol. 118, No. 8, pp. 42–53), which has his picture on the cover.

Table 8–3 Comparison of Unplanned (Uncontrolled) and Planned (Optimum Land-Use) Development of a Rapidly Growing Urban-Suburban Area*

	Year 1970, Population 20,000	Projected Year 2000, Population 110,000	
		Unplanned (Uncontrolled) Development	Optimum Use Plan
Developed area	13,000 acres	38,000 acres	30,000 acres
Residential	7,500	26,000	21,300
Commercial	500	700	630
Industrial	70	300	70
Institutional	2,500	5,500	3,000
Roads	2,500	5,500	5,000
Open space ("undeveloped area")	32,000	7,000†	15,000
Waste disposal parks	0	0	1,000‡
Recreation parks	500	2,000	5,000
Farming and forestry	11,500	0	2,000§
Natural areas	20,000	5,000	7,000‖
Total acres	45,000	45,000	45,000
Percent open space	71	16	33

* Data adapted from a plan for a Maryland urban area prepared by Wallace-McHarg Associates, 2121 Walnut Street, Philadelphia, with addition of new concepts of open space plans by E. P. Odum.

† Projected uncontrolled development for a population density 150,000 would reduce open space to zero!

‡ Land surrounding or adjacent to industrial and municipal sewage plants on which extensive waste-treatment ponds and other pollution abatement facilities can be located for efficient and low cost tertiary treatment of all wastes.

§ Could include not only truck farms but also demonstration farms and forest as educational laboratories for schools and colleges.

‖ Including all steep slopes, ravines, flood plans, marshes, and lakes existing in 1963 together with samples of existing mature forest areas and marginal farmlands restored to natural-area use.

neighborhoods with mixed housing ranging from low to high income. One third of the city is permanent "open space" not to be built on, including all the stream valleys, old-growth woods, and other scenic places. The city is entirely private enterprise; there is only a small government subsidy for low-income housing. The Columbia Association, which manages the community, even employs a staff ecologist who looks after vegetation, streams, lakes, trails, and wildlife, including the barn owls that live in the old barns now converted to stores or recreation facilities. The cost of maintaining the open space is under $10,000 a year because mowing and other expensive maintenance is greatly reduced except on designated playgrounds.*

A third example of innovative thinking about environmental use is contained in a little book called *The California Tomorrow Plan*, edited by Alfred Heller (1972). This report, prepared by a citizens' group, assisted by various professional groups, proposes that the legislature establish a state commission to design a land-use plan for public debate and eventual adoption. Goals would be to identify and preserve the life-support natural environment, to preserve the quality of life in cities, and to regulate as far as feasible the kind and intensity of development so that carrying capacity is not exceeded. A unique idea presented would involve the setting aside of large reserves of undeveloped land; decisions about use of this land would be postponed until sometime in the future, thus providing flexibility for future contingencies. As pointed out by the Heller report, no such plan can work until some kind of coordinated regional control of environmental use replaces current fragmented bureaucratic control by dozens of different city and county commissions. While control of general environmental use is being shifted to a state or regional level, control of housing and other structures should remain under local control to ensure individual freedom for making decisions about local use of private property, operation of schools, housing, civil order, and so on.

Some Considerations for Integrating Ecological and Economic Concerns Given strong public opinion, political difficulties can probably be overcome, but economic considerations remain the chief obstacle to any kind of comprehensive, long-range planning for environmental use. The problem derives from the sharp dichotomy between market and nonmarket values. Regardless of the political systems in different countries, manufactured market goods and services, such as automobiles or electricity, are accorded high economic values, whereas the equally vital goods and services of nature, such as air and water purification and recycling, remain mostly external to the economic system and are accorded little or no monetary value (hence the designation "nonmarket" value). The reason for this dichotomy is illustrated by the evaluation of the total worth of an estuary, as shown in

* Letter from Charles Rhodehamel, Ecologist, Columbia Association, June 24, 1981.

Figure 3–23A. Lester Brown, in a review of "global economic prospects" (1978), comments as follows:

> Economists are unaccustomed to thinking about the role of biological systems in the economy, much less the condition of these systems. The economist's desk may be covered with references containing the latest indicators of the health of the economy but rare indeed is the economist concerned with the health of the earth's principal biological systems. This lack of ecological awareness has contributed to some of the shortcomings in economic analysis and policy formulations.

Brown singles out four biological systems—fisheries, forests, grasslands, and croplands—as the foundation of the global economy and continues his commentary:

> The condition of the economy and of these biological systems cannot be separated. As the global economy expands, pressures on the earth's biological systems are mounting. In large areas of the world, human claims on these systems are reaching an unsustainable level, a point where their productivity is being impaired. When this happens, fisheries collapse, forests disappear, grasslands are converted into barren wastelands and croplands deteriorate along with quality of air, water and other life-support resources.

Brown's scenario for "Building a Sustainable Society" (1981) calls for a systematic, government-backed, global conservation of resources, involving using less with greater efficiency and recycling more. In startling contrast, economist Simon (1981) complains that people like Brown are perpetuating the "myth" of growing scarcity of resources. Simon downgrades the role of biological factors and states that "our supplies of natural resources are not finite in any economic sense" and that "there is no reason why human resourcefulness and enterprise cannot forever continue to respond to impending shortages and existing problems with new expedients that, after an adjustment period, leave us better off than before the problem arose." Very few scholars in any discipline ascribe to this extreme view. Kenneth Boulding (1982) has written an incisive review of the Brown and Simon books that is well worth reading.

Most economists recognize that there is a market failure when it comes to allocating many natural resources. As defined in the economic literature, market failure occurs when society holds some values more or less desirable than market prices indicate. Bator (1958) defines market failure in terms of "allocation theory" as the "failure of a more or less idealized system of price-market institutions to sustain 'desirable' activities and to stop 'undesirable' activities" (in reference, of course, to society as a whole). In the traditional view of the economist, political intervention in the market place is required to protect human value and to allocate scarce resources or those for which there is no substitute (land and water, for example). Three examples from the past are illustrative: (1) legal abolition of "sweat shop" industries, which were profitable to manufacturers but socially undesirable, (2) zoning that restricts land use in the public interest, and (3) establishment of a National Park system by which land is either removed from or never allowed to enter the marketplace. Economists also emphasize that humans do not just consume or utilize resources but also create resources, so they are not entirely limited by existing natural resources (see Straffa, 1973). Although this is undoubtedly true for some natural goods and services, the basic life-support resources cannot be created artificially.

In many ways, the National Environmental Protection Act (NEPA), passed by the U.S. Congress, is a first attempt to provide a nationwide legal basis for extending value systems to include the natural environment. The Act requires that "impact statements" be prepared for all large, proposed human alterations. Slowly, if somewhat painfully, this stopgap approach must lead to improved procedures for total assessment that includes environmental and social cost-benefits along with the conventional economic ones.

A classification of values, as they relate to the problem of market failure, is outlined in Table 8–4. Nonmarket values are divided into two categories: attributable and nonattributable. Attributable nonmarket values can, in the opinion of many economists, be assigned monetary values within the conventional language of market economics. For example, the replacement value of a natural environment can be estimated in terms of what it would cost to provide an artificial substitute for the free goods and services (waste treatment, for example) provided by a natural ecosystem (see Westman, 1977, for a review of possible ways to "internalize" nature's services into anthropogenic economics). In contrast, intangible or nonattributable values cannot be dealt with in conventional economic cost-accounting or cost-benefit analysis.

A. C. Pigou, in 1920, was among the first economists to challenge

Table 8–4 **Comparison of Market and Nonmarket Values***

Market Values Referring chiefly to manufactured goods and services—the output of factories, farm products, and commercial services—which in a free enterprise market are allocated through uninhibited competition and supply and demand. In theory, market price reflects society's evaluation of an item and leads to efficient allocation of resources. In practice, this is not always the case, so some governmental regulation is deemed necessary. The "market model" goes back to Adam Smith's book *The Wealth of Nations*, published in 1776.
Nonmarket Values Referring especially to the goods and services of nature, sometimes called "free" or "common" or "public" goods and services. In general, these "unpriced" values are external to market economics. *Attributable and assignable nonmarket values* The value of a river for waste assimilation. The concepts and language of market economics can be applied and monetary values assigned. *Intangible, nonimputable* or *nonattributable values* Life-support values of natural ecosystems such as forests, grasslands, rivers, lakes, and oceans that operate, buffer, and stabilize atmospheric, mineral, and hydrological cycles are in this category, along with intrinsic values of species, indigenous human culture, natural beauty, and a vast array of esthetic values prized by the human beings through the ages. Nonattributable categories are of individual and public rather than (and very often in conflict with) private market value.

* After Farnworth et al., 1981.

the laissez-faire market as an efficient allocator of resources and thus to pinpoint the market failure that occurs when businesses pursue their self-interest at the expense of the public interest. Pigou was ahead of his time in his concern for urban blight, and he wrote that only the state can "set rules of liability and enforce them to attack uncompensated damage of air and water pollution." He suggested taxes and subsidies as means to equalize private and social costs. (For a discussion of Pigounian analysis of externalities and public goods, see Breit, 1982.)

Since action by the "state" becomes increasingly difficult or is too little, too late in the face of increasing market pressure on land use, and since economists, ecologists, and political scientists tend to deal with only a part of the problem (and are inclined to blame each other for failures), there is perhaps some merit in the idea of establishing a new discipline that would deal with the whole of the market-nonmarket problem. **Bioeconomics** is a term that has been

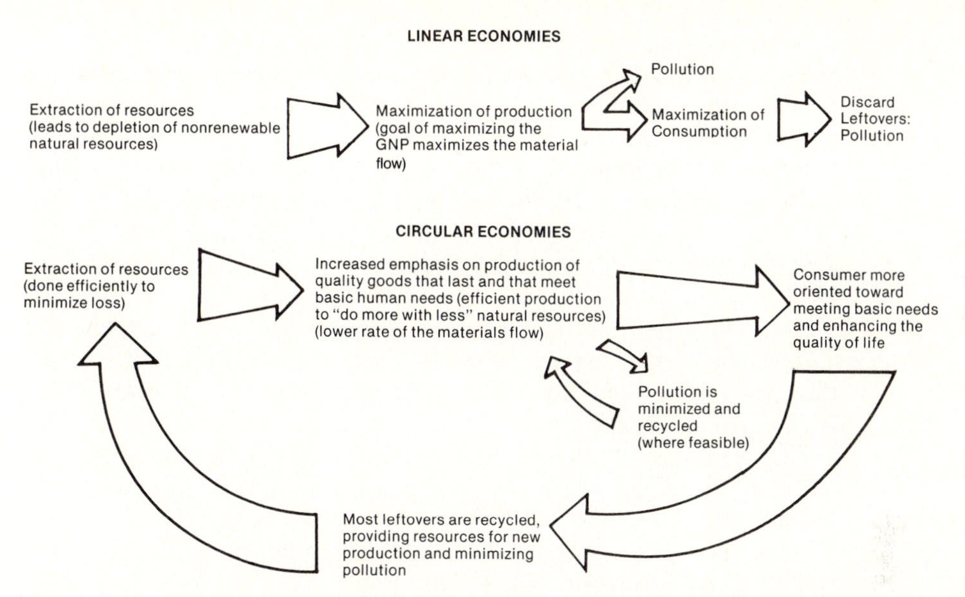

Figure 8–18 Resource use could be tremendously improved if current linear economics could be converted to a circular economics. (After T. E. Jones, 1977.)

suggested for a discipline that would consider the role of biological systems as well as humanmade abiotic systems in supporting the overall economy (Georgescu-Roegen, 1977; Clark, 1981). A better term is **holoeconomics**, which goes back to the efforts of a group of economists to expand the field more than half a century ago (see Grunchy, 1947, and the Epilogue following this chapter).

One promising basis for such a new discipline involves using "embodied" energy as a common denominator (discussed in Chapter 3, Section 10, where the fundamentals of energy-money relationships are described). Another promising approach would be to convert current linear economics to a circular economics, as shown in Figure 8–18. In this manner, the economic system would correspond to the general systems model that has internal feedback as diagrammed in Figure 1–4. Bernstein, an anthropologist, notes that in many isolated cultures that must survive on local resources, actions that would be detrimental to the environment in the future are perceived and avoided (Bernstein, 1981). Local feedback in decision-making is lost when isolated cultures are incorporated into large and complex industrial societies. Bernstein comments that "economics must develop a coherent theory of decision-making behavior that is applicable at all levels of group organization. This will necessitate *defining self-interest in terms of survival rather than consumption."* Such a shift would bring economic behavior under something akin to natural selection, which has worked so well to ensure the perpetuation of life on earth over the eons.

For additional ideas and discussion on economics and environment, the following articles and books are recommended:

On economic growth: Galbraith, 1958; Mishan, 1967, 1970; Krutilla, 1967; Wagner, 1970; Barkley and Seckler, 1972; Ridder, 1972.

On nonmarket values and market failures: Coase, 1960; Barnett and Morse, 1963; Bator, 1958; Pearse, 1968; Alfred Kahn, 1966; Randall, 1972; Lugo and Brinson, 1978; Sinden and Worrell, 1979; Farnworth et al., 1981.

On resource economics: Kneese, Ayres, and d'Arge, 1970; Schurr (ed.), 1972; Fisher, Krutilla, and Cicchetti, 1972; Kneese, 1973, 1979; Page, 1977; Ayres, 1978; Smith and Krutilla, 1979.

On economic indices (GNP, NEW): Samuelson, 1950; Nordhaus and Tobin, 1972; Nordhaus, 1979.

On steady-state economics: Ayres and Kneese, 1971; Daly, 1973.

On energy analysis economics: H. T. Odum, 1973; Hannon, 1973; Gilliland, 1975, 1978; Huetner, 1976; Slesser, 1974; Costanga, 1980; Odell, 1980.

Kenneth Boulding and Nicholas Georgescu-Roegen, the pioneers in the economics-ecology interface, have been cited a number of times in this book. See the Bibliography for a listing of some of their writings.

Epilogue
The Predicament of Humankind: Futuristics

Predicting the future is a fascinating game, especially popular in times of crisis. In fact, no one can really predict what will happen next year, much less 25 years or more in the future. There are just too many unknowns, too many new events, technological innovations, and other factors that cannot be foreseen. Nevertheless, it is instructive to consider a range of alternate scenarios that could come to pass. We then may be able to estimate their probability given current conditions, understanding, and knowledge. More important, we might be able to do something now to reduce the probability of an undesirable future.

As we approach the year 2000, about the only certainty is that human beings will continue to increase in numbers, at least for another century, and industrialized societies will be making a major and very painful transition in energy use as the fossil fuels decrease in quantity, decline in quality, and increase in cost. The energy transition has already begun, so we can at least speculate on some of the "ripple effects" that will accompany drastic changes in civilization's energy-input environment. Most futurists believe that we will have to reduce current prodigious waste and become more efficient and conservation-minded in order to do more with less high-quality energy. Most also agree that increasing per capita energy consumption above current levels in industrialized countries would not improve the quality of life but in fact would have the opposite effect (see Nader and Beckerman, 1978). Likewise, most forecasters see a need to harvest more food with less energy input from the world's limited area of good agricultural land, which even now may be declining in quality with heavy use (erosion, salination).

There is no shortage of studies, reports, and popular books that assess the current predicament of humankind. Most paint a rather grim picture of present global problems, but many are optimistic about the future. We can indeed be optimistic *if* (and this is a big if) humanity can confront and do something about the current predicament. Among the most comprehensive reports are those prepared by agencies of the United States government, notably two reports prepared by the Council on Environmental Quality (CEQ), 1980 and 1981, and one prepared by the Department of Agriculture, 1979. Especially recommended reading is the 1980 CEQ report entitled "The Global 2000 Report to the President; Entering the Twenty-first Century." In three volumes, population, income, resources, and environmental impacts are considered in detail. Also included are a chapter comparing this assessment with other global studies, and the "Government's Model," which explores future scenarios.

The way scholars, as well as people in general, view the future ranges from complete confidence in new technology (a "more of the same" philosophy) to a belief that society must completely reorganize, power down, and develop new international and holistic political and economic mechanisms for dealing with a world of limited resources. Herman Kahn is a well-known spokesman for the former view (see Kahn and Wiener, 1967; Kahn et al., 1976). According to the latter view, we need to adopt the strategy used by a coral reef or a tropical forest on leached soil to prosper on limited resources (see pages 5, 6, and 214). In Chapter 8, Section 7, we contrasted these opposing scenarios by reviewing 1981 books by resource scientist Lester Brown and economist Julian Simon. Watt et al. (1977) devoted a whole chapter (Chapter 2) to contrasting what they call the "technoculture belief" and a "contrary position." The contrary position is essentially an ecological paradigm of humankind as a part of global ecosystems subject to natural environmental laws and resource limitations. Schumacher, in his widely read book *Small Is Beautiful* (1973), makes a good case for a larger investment in small-scale technology that is harmonious with human and environmental values (the term "appropriate technology" has come into wide use in this connection). Margalef (1979) points out, however, that though small may often be beautiful, big is powerful! When should we think big, and when should we think small? For example, when or where do we invest in huge central power plants for generation of electricity, or opt instead for rooftop photovoltaic cells, mini-hydros, wind-powered generators, and similar decentralized modes? Cost-benefit analyses are so difficult that we will have to experiment with many options before we can find a good mix for long-term benefit. Common sense tells us that having drawn deeply from the oil well, we will now have to draw from many different wells, figuratively speaking.

Another way to assess the predicament of humankind is to consider the gaps that must be narrowed if humans and the environment, as well as industrial and nonindustrial nations, are to be brought into harmonious balance. Among the gaps already mentioned in this book are the following:

1. The income gap—the rich and the poor, both within nations and between the 30 percent industrialized and 70 percent nonindustrialized peoples (page 115).
2. The food gap—the well-fed and the underfed (page 117).
3. The value gap—market and nonmarket goods and services (page 501).
4. The huge gap in level of education—the literate and the illiterate, the schooled and the unschooled (also the skilled and unskilled)—has not been mentioned previously, but will be discussed in this section.

None of these gaps has been narrowed very much. Neither the efforts of the United Nations (which spends most of its efforts trying to stop brush wars) nor the large capital outlays for foreign aid by the wealthy nations have had much effect. In fact, the income and value gaps have gotten worse during the past several decades. Morehouse and Sigurdson (1977) point out that transfer of industrial technology from the rich to the poor nations too often benefits the small modern sector but not the masses of rural poor. Wealth does not "trickle down" when there are profound cultural, educational and resource differences within a population. One cannot transfer a high-energy technology to a poor country without also providing the high-quality energy input needed to sustain it. Farvar and Milton (1972) document a number of cases of failed technology transfers.

René Dubos, scientist, humanist, and confirmed optimist, wrote eloquently on the theme that the way to close the human "gaps" is to "domesticate" all of the biosphere so that the beauty and usefulness of farms, fields, forests, water bodies, parks, and so on can not only be preserved but also improved and interfaced with human edifices for mutual benefit. Dubos' idea of utopia would be a global ecosystem powered by renewable resources and organized along the lines of the European village that he remembered so fondly from his boyhood in France (Dubos, 1976, 1980).

If civilization is a system, not an organism, then, contrary to what Toynbee says in *A Study of History* (1961), civilizations do not necessarily grow, mature, become senescent, and die as do organisms, even though this process has happened in the past (the rise and fall of the Roman Empire, for example). According to Butzer (1980), civilization becomes unstable when the high cost of maintenance results in bureaucracy that makes excessive demands on the productive sector. Breakdowns then result "from chance concatenations of mutually reinforcing processes, not senility or decadence." Such a view coincides with ecological theory regarding energy flow and complexity (outlined in Chapter 3).

The Club of Rome Reports

In April 1968, a group of 30 individuals from ten countries gathered in Rome at the instigation of Dr. Arillio Peccei, an Italian industrial manager, economist, and man of vision. This group included scientists, educators, economists, humanists, industrialists, and national and international civil servants. They gathered to discuss a subject of staggering scope, the present and future predicament of humankind, and they called their group "The Club of Rome." Since that time, this group has sponsored a series of "Reports to the Club of Rome," which have the general label "The Predicament of Mankind." The first of these reports, *The Limits to Growth* (1972), was prepared by a group of scientists from the Massachusetts Institute of Technology, headed by Dennis and Donella Meadows. The group set up global models based on the pioneer systems-analysis techniques of Jay Forrester (1968, 1971). These models were designed to predict what the future will be like if our present ways of doing things, that is, our present economic and political methods, remain the same. Model outputs showed that in many cases or in connection with many vital resources, cycles of boom and bust would occur. Growth and consumption of resources would continue to increase rapidly along with the population increase and the increase in energy use and so on until a limit was reached; there would then be a strong downturn or a "bust." This report denounced society's obsession with growth, in which at every level—individual, family, corporation, nation—the goal is to get richer and bigger and more powerful without considering the ultimate cost of exponential growth.

Although *The Limits to Growth* was intended to show simply what *could* happen if we did not change our ways, many people, including most political leaders and a large segment of the public, treated the report as if it was predicting doomsday for civilization. Accordingly, there was a storm of criticism. Many reviewers pointed out that the models did not take into consideration new technology, discovery of new resources, replacement of used-up resources with a new resource, and so on. Most people seemed to feel that humanity would be too smart to fall for any boom and bust cycle and that we would stop or change our style before reaching that point.

In spite of the criticism and general denial by most politicians that anything like a boom and bust could possibly happen, the book did have a tremendous impact. It served as a warning that we should look more closely at where mankind is going. *The Limits to Growth* can be found in the paperback section of most good bookstores, along with the older classic warnings to humankind such as Vogt (1948), Osborn (1948), Leopold (1949), and Carson (1962).

This first report was followed by a series of additional reports that attempted not only to describe more details about present situations and possible future scenarios but also to suggest actions that should be taken so that the doomsday scenario could be avoided. The Meadows also published a second book (1973) with the latter objective.

The second report to the Club of Rome is entitled *Mankind at the Turning Point*. It was prepared and edited by Mihajlo Mesarovic, Professor and Director of the Systems Research Center at Case Western Reserve University, and Eduard Pestel, Professor of Engineering at Hanover University in Germany. Responding to the criticisms of *The Limits to Growth,* this report divides the earth into ten interdependent regions. The rationale was that it should be more practical and more efficient to deal with global problems on a regional basis because there are considerable differences in the predicament of humankind in different parts of the world. Again, this report analyzes the nature of the global crisis and the dangers of delay. The authors of the report conclude that the different regions of the world and ultimately the world as a whole must proceed rapidly from a condition of "undifferentiated (haphazard) growth" to a condition they call "organic growth." In other words, the world can no longer afford to continue the haphazard growth pattern, because such a passive course simply leads to disaster.

The second report also emphasizes two steadily widening gaps that appear to be at the real heart of the present crisis. The first gap is between humans and nature, the second gap between rich and poor. These gaps, incidentally, are the same ones that ecologists have been pointing at for many years. Both gaps must be narrowed if world-shattering catastrophes are to be avoided, but they can be narrowed only if some kind of global unity can be achieved so that the interdependence of humans and nature is reaffirmed and earth's finiteness is explicitly recognized by all nations.

The concept of organic and sustainable growth of the world system is not to be construed as a simplistic one-world "monolithic" concept of world development. The report emphasized that the one-world concept, which has been suggested many times by utopians, just will not work in the complex human situation. Indeed, the homogeneous one-world concept is essentially incompatible with a truly global approach to better the predicament of humans. The recommended approach must really start from recognition of the need to deal with and preserve the world's regional diversity. Development must be specific to the region, yet globally oriented rather than based on narrow national interest. Thus, the world model is viewed as a flexible, computer-based planning instrument that contains a multilevel regional model of the world's systems. This model is fundamentally different from any one previously developed in *The Limits to Growth* because it recognizes the diversity that exists in the world, and which is deeply rooted in past cultural traditions that will undoubtedly prevail in the future. Here again, the world is viewed as a system, that is, a collection of mutually interacting and interdependent parts. Culturally and geographically distinct regions are the compartments in such a proposed model that would provide decision-makers in various parts of the world with a comprehensive, global planning tool.

The regional approach suggested by Mesarovic and Pestel is very similar to the sociological theory of regionalism developed several decades earlier by the late Howard W. Odum (Odum, 1936, 1951; Odum and Moore, 1938). Regionalism as an approach to the study of society is based on the recognition of distinct differences in both cultural and natural attributes of different areas that, nevertheless, are interdependent. Regional social study was motivated in part by the desire to upgrade "backward" regions (such as the South in the 1930s) so they could contribute to rather than detract from the economic well-being of the nation. The biologist's concept of a natural linkage and interdependence between humans and the rest of the biota, of course, goes back more than 100 years. See, for example, Thomas Huxley's essay "Evidence as to Man's Place in Nature"

(1863; reprinted in 1959). This theme has been restated many times by animal ecologists who write about "human ecology," for example, Charles C. Adams (1935), F. Fraser Darling (1951), Marston Bates (1952), and W. C. Allee (1931 and 1951). The behavioral linkage between humans and other animals is also the basis for the new and controversial theory of sociobiology developed by Edward O. Wilson (1975, 1980) and others.

The basic theme, then, of the second Club of Rome report is not new, but is put in a broader context: All of our crises, which seem to draw public attention one after another, are really part of one syndrome that signals a need to evolve a gradual change from undifferentiated growth to a multilevel organic growth with sufficient feedback warnings and actions to prevent that boom and bust cycle. The report suggests that we use all these crises as error detectors and begin to act as any good organic system does—cybernetically, with strong negative feedback coming when growth begins to approach the state of diminishing returns (see Chapter 3, Section 8). Hirsch (1978) optimistically believes that social constraints and maintenance costs will provide the feedback that will limit growth before shortage of material goods does.

The third report to the Club of Rome is entitled *The Rio: Reshaping the International Order* (1977). This report, coordinated by Nobel Prize–Winning economist Jan Tinbergen of The Netherlands, pays special attention to the undeveloped nations whose leaders by and large reject any idea of limits of growth. Tinbergen and associates conclude that before an organic world order, as proposed in the second report, can be implemented, all nations and all regions must combine their local goals with certain global goals that will work for the benefit of humanity as a whole. It has to be shown that local and global goals can be interfaced for the long-term benefit. For example, if a particular country requires a certain amount of industrial development to achieve a reasonably good standard of living for citizens, this can be a worthwhile local goal. If such a local goal is attained by recycling scarce resources as much as possible and by treating waste as completely as possible, then the "commons" that are of global concern, such as the air, the oceans, and so on, will be less damaged. The problem, of course, is to convince citizens and national leaders that matching their own goals with those of humankind as a whole is in their interest. Slowing the armaments race is so vitally important because energy expended in this negative direction makes it more difficult for all nations to have the will, the energy, and the money to work with each other for the common goal. When people are preoccupied with conflict, they give little thought to cooperating for mutual benefit. One is reminded of the evolutionary trend from competition to mutualism in natural systems (Chapter 7, Section 4; and Chapter 8, Section 6).

The fourth report is entitled *Goals for Mankind* (1977), compiled by Ervin Laszlo, a professor of philosophy known for his books *The Systems View of the World* and *A Strategy for the Future* (cited in Chapter 2 of this text). This volume addresses the two basic questions that underlie all the efforts and discussions of the earlier reports. (1) "What are, in reality, the goals of humankind?" and (2) "Are we willing to give human development precedence over material growth?" In other words, will humankind use its enormous techno-scientific power with a long view to benefit all of humanity without damage to the planet? To do this, of course, requires us to forgo some immediate desires in favor of stability and decent conditions for future generations.

Goals for Mankind is based on an inventory or atlas of national and regional goals proposed by working groups from various nations and regions. Thus, there are chapters on goals for the United States and Canada, Western Europe, Eastern Europe, Latin America, Africa, the Middle East, and so on. These chapters are followed by chapters on goals as perceived by international groups such as the United Nations, multinational corporations, and the World Council of Churches. Goals are then assessed in terms of global security, energy, food, and resources. Chapter 16 of the report discusses the "current goals gap" and includes a series of charts in the form of bars and columns on which goals are scored from 0 (exclusive focus on short-term national concerns) to 10 (inclusion of long-term global considerations). For the United States, business scored low and government relatively high for long-term conservation, economic, energy, and agricultural policies. "Life style" policies scored low, as would be predicted. In all cases, intellectual and religious or spiritual groups scored highest (more global concern). National averages on the 0 to 10 scale for key countries are as follows: United States, 4.0; Canada, 4.7; Belgium, 3.7; Netherlands, 4.7; Sweden, 4.6; Japan, 5.5; China, 7.0; Algeria, 6.4; Saudi Arabia, 4.2; Brazil, 4.0; Egypt, 4.6; Ghana, 6.5; Pakistan, 4.7; and India, 4.3. Interestingly, people in undeveloped countries are more optimistic about the future than are people in developed countries, but only China scored high on long-term global concern. The final chapters in the book conclude that global goals and ultimate world solidarity are achievable, and a start is being made toward such goals. A religion- and science-led scenario, a government-led scenario, and a business- and science-led scenario are discussed and compared. The report states, however, that the question of whether a move to world solidarity will come in time cannot be answered with certainty.

Subsequent Club of Rome reports published between 1978 and 1980 focus on important components of global dilemmas, for example, waste (Gabor et al., 1978), energy (de Montbrial, 1979), societal organization (Hawrylyshyn, 1980), and wealth and welfare (Giarini, 1980). A key report deals with the education gap and is entitled *No Limits to Learning: Bridging the Human Gap* (Botkin *et al.*, 1980). This report contrasts what the authors call micro- and macro-learning. Most education, they point out, concentrates on individual learning processes that aim to make the individual knowledgeable and rational. This is the micro-learning level. In contrast, there has been little effort worldwide to promote collective or societal learning, the macro-level. In other words, how do we create awareness and understanding of regional and global predicaments among people who have had little or no schooling? How does society as a whole learn to avoid the boom and bust cycle, the entropy trap, and other predicaments now understood by only the highly educated few? Most of all, how can society become aware of organic growth policies, so eloquently documented in the second and third reports, and develop a will to adopt them? Those who hoped that television would become a macro-learning tool have so far been disappointed, but the potential remains. Developing a societal-learning technology is indeed worth our best efforts.

Laszlo (1977a) has assessed the overall impact of the Club of Rome reports:

> Thanks largely to the efforts of the Club of Rome, international awareness of the world *problematique* has rapidly grown. The Club pioneered the way (to continue the medical analogy) from diagnosis (Meadows, Mesarovic and Pestel), to prescription (Tinbergen, Laszlo and the other reports). But not withstanding heroic efforts on the part of Aurelio Peccei, relatively little has been achieved in the area of therapy. To use another metaphor, the Club helped point the way, but did little to generate the will to take it. If it is true that where there is a will there is a way, then the cart has been put before the horse. Of course, it is easier for a group of concerned world citizens to point the way than it is to generate the great upsurge of will needed to set out on it. Nevertheless, there is a real danger that the work so far done will achieve the fine patina that only comes with prolonged disuse. It would be better—and in this the present writer believes that he is joined by the other Club of

Rome authors—for the reports to go down in the heat of constructive controversy than to gather dust as the respected but unfollowed documents of an era.

Global Models and World Futures

Between 1971 and 1981, ten or so large-scale global models have been completed. These models are computerized mathematical simulations of the world's physical and socioeconomic systems, with projections into the future that are the logical consequences of the data and assumptions that went into the model—and it is to be emphasized that each model differs in these respects. These models have been reviewed and compared as a group in a report issued by the Congressional Office of Technology Assessment (OTA) (1982), in a book by Meadows et al. (1983), and in an article by Donella Meadows (1982). The latter is especially recommended as a nontechnical review that also includes an illustrated section entitled "A Child's Guide to the Systems Viewpoint."

Four of the global models, namely, Forrester, Meadows et al., Mesarovic-Pestel, and Global 2000, have been mentioned previously in this Epilogue. Several of the others focus on socioeconomic considerations rather than on resources and population per se, notably the Latin American World Model (LAWM), the British SARUM model, the Japanese FUGI model, and the United Nations World Model (UNWM). In fact, construction of these models was stimulated by criticisms of the earlier limits-of-growth projections, and the models start with the assumption that political, social, and economic inequities have as much to do with our unhappy predicament as do resource limitations and population pressure.

Despite differing assumptions and biases, the models as a set do lead to consensus on some points. The following items are condensed from the conclusions reached in the OTA report and Meadows' 1982 review:

1. Technological progress is expected and is vital, but social, economic, and political changes will also be necessary.
2. Populations and resources cannot grow forever in a finite planet.
3. There is no reliable or complete information on the degree to which the earth's physical environment and life-support system can meet the needs and demands of future growth in the human population (i.e., carrying capacity unknown), but a sharp reduction in growth rate will greatly reduce the probability of overshoots or major breakdowns (ecodisasters).

4. Continuing "business-as-usual" will not lead to a desirable future, but rather will result in further widening of undesirable gaps (rich-poor, for example).
5. Cooperative long-term approaches will be more beneficial for all parties than competitive short-term policies.
6. Because the interdependencies among peoples, nations, and the environment are much greater than commonly imagined, decisions should be made within a holistic context. Actions intended to reach only narrowly defined goals are likely to be counterproductive.
7. The nature of the future global state, whether better or worse than at present, is not predetermined. Much depends on how soon current undesirable trends can be altered. Actions taken soon (within the next couple of decades) are likely to be more effective and less costly than the same actions taken later. This calls for strong leadership and more macroeducation (as discussed in connection with the Club of Rome report, *No Limits to Learning*), since by the time a problem is obvious to everybody, it may be too late.

And so, for the young in age as well as the young in heart, there is much to excite the imagination and many difficult goals to achieve!

The Ecological Assessment

The wisdom of the many contributors to the Club of Rome reports, as well as the output of global models, conforms rather well with basic ecosystem theory, especially two paradigms: (1) A holistic approach is necessary when dealing with complex systems, and (2) cooperation has greater survival value than competition when limits (resources or otherwise) are approached. The learned discourses also conform to the age-old wisdom in common-sense proverbs such as "look before you leap," "don't put all your eggs in one basket," "haste makes waste," "an ounce of prevention is worth a pound of cure," "power corrupts," and many more. For more on common sense and ecology, see E. P. Odum (1977).

Future human population growth is a major uncertainty that affects any predictive model. Figure 3–21 presented three trend curves based on the assumption that births and deaths would balance each other sometime in the next century. Most futurists agree that all our problems will be more manageable the sooner the rate of growth can be slowed, even though doing so on a global basis means changing deep-seated religious and cultural beliefs about birth control, abortion, and other controversial issues. In 1971, the National Academy of Science issued a landmark report entitled "Rapid Population Growth; Consequences and Policy Implications." It concluded that rapid growth has no economic or any other benefits, but tends to create social and environmental problems faster than they can be solved. Rapid growth tends to develop a momentum that leads to overshoots that are difficult to avoid. Many demographers have faith in the *demographic transition*—the theory that human population growth slows as people become more affluent and less dependent on child labor—but there is much controversy (see Teitelbaum, 1975).

The ecologist contends that one of the obstacles to achieving some kind of world order is the overly narrow economic theory and policies that dominate world politics (discussed in Chapter 3, Section 10, and again in Chapter 8, Section 7). Around the turn of the century, a group of scholars calling themselves "holistic economists" formed an active "school" critical of the economic models of that time. Grunchy (1947) devotes a book to the history of this school that included Thorstein Veblen, J. R. Commons, Wesley Mitchell, John M. Clark, Rexford Tugwell, and Gardiner Means, names that are little known by today's students of economics. The holists complained that classical economists paid little attention to the economy as a dynamic functioning whole, but instead were content to center attention on separate parts. As a result, their models tended to be rigid, mechanistic, and very poor predictors of the real world.

Any further development of a holistic economics was submerged by the flood of oil that spawned rapid growth in monetary and material wealth. Classical growth theory served quite well as long as the supply of oil far exceeded demand. Now that the oil boom is abating, it is time to redevelop a holoeconomics, one that will include cultural and environmental values along with monetary ones. Historian John Haag (1981), in a review of economic thought since 1945, sees the emergence of a new breed of economists who are not "intoxicated" by the "materialistic growth-at-any-price ethic." Wassily Leontief, Nobel Laureate in Economics, in a recent commentary (1982) complains about the failure of economists to recognize that the economy is "a self-regulating system of a great many different but interrelated, and therefore interdependent activities." He wonders "how long will researchers in adjoining fields such as demography, sociology, and political science on the one hand and ecology, biology, health science, engineering and other applied physical sciences on the other, abstain from expressing concern about the splendid isolation in which academic economics now finds itself?" Leontief's criticism is much the same as that of the turn-of-the-century school cited in the previous paragraph. The same complaint of overly narrow viewpoints can also be lodged against many academic ecologists. To an increasing extent, however, ecologists are beginning to make a constructive contribution to economics. As a result, insights from ecological theory are being incorporated into current economic writings, and vice versa. It is always a good sign when researchers in different fields begin to use the same analytical techniques. For example, Leontief's input-output modeling approach is being used by ecologists Hannon and Finn, among others, and H. T. Odum's systems ecology approach is attracting the attention of economists.

Many futurists believe that an extension and reinforcement of ethics to include life-support and other nonmarket values is necessary if we are to close the gap between the know-how of science and the willingness of political leaders to make long-range decisions for the good of the whole. The importance of Leopold's "Land Ethic" was stressed in Chapter 1. No one knows how self-interest can be extended to global concerns, but many write about it (Callahan, 1972; Hardin and Boden, 1977; Cahn, 1978; Callicott, 1979; Fritsch, 1980).

When both the "study of the household" (Ecology) and the "management of the household" (Economics) can be merged, and when Ethics can be extended to include the environment as well as human values, then we can indeed be optimistic about the future of humankind.

APPENDIX
Brief Description of Major Natural Ecosystem Types of the Biosphere

1. The Terrestrial Biomes

The **life form** (grass, shrub, deciduous tree, coniferous tree, and so on) of the climatic climax vegetation is the key to delimiting and recognizing terrestrial biomes. Thus, the climax vegetation of the grassland biome is grass, although the species varies in different parts of the biomes and on different continents. Although the climatic climax vegetation is the key to classification, edaphic climaxes and developmental stages that may be dominated by other life forms are an integral part of a biome. For example, grassland communities are developmental stages in a forest biome, and riparian forests are a part of a grassland biome.

Mobile animals, called "permeants" by V. E. Shelford, couple together different vegetative strata and stages. Birds, mammals, reptiles, and many insects move freely between subsystems and between developmental and mature stages of vegetation, and migratory birds move seasonally between biomes on different continents. In many cases, life histories and seasonal behavior are organized so that a given species will occupy several, often quite different, vegetative types. Large mammalian herbivores—deer, antelope, bison, domestic cattle—are a characteristic feature of terrestrial biomes. Many of these herbivores are ruminants, which possess a remarkable nutrient-regenerating microecosystem, the rumen, in which anaerobic microorganisms can break down and enrich the lignocellulose that constitutes a large part of terrestrial plant biomass. Likewise, the detritus food chain, featuring fungi and soil decomposer animals, is a major energy flow pathway, as is the mutualism between plant roots and mycorrhizae, nitrogen fixers, and other microorganisms.

The biomes of the world are shown in Figure A–1, and climographs for six major biomes are compared in Figure A–2.

During the 1970s and continuing into the 1980s, several of the major biomes (grassland, temperate deciduous forest, northern coniferous forest, tundra, and desert) were subjected to interdisciplinary team research as part of the United States' contribution to the International Biological Program (IBP). Results of these intensive, long-term studies are being published in numerous papers and a series of books. For a general review of this program, see Blair (1977) and Loucks (1983). In terms of flora and fauna, biogeographers divide the world into five or six major regions that correspond roughly to the major continents. Australia and South America are the most isolated regions; accordingly, ecologically equivalent species in biomes on these continents can be expected to be taxonomically quite different (see Chapter 7, Section 5, and Table 7–3).

Tundras—Arctic and Alpine (Figure A–3)

Between the forests to the south and the polar ice cap to the north lies a circumpolar band of about 5 million acres of treeless country (Figure A–3). Smaller but ecologically similar regions found above tree limitation on high mountains, even in the tropics, are called alpine tundras. In both North America and Eurasia, the boundary between tundra and forest lies further north in the west where the climate is moderated by warm westerly winds.

Low temperatures and short growing seasons are the major limiting factors to life; precipitation may also be low but is not limiting because of low evaporation rate. All but the upper few inches of ground remains frozen during the summer. The permanently frozen deeper soil is called **permafrost.** The tundra is essentially a wet arctic grassland with vegetation consisting of grasses, sedges, dwarf woody plants, and lichens ("reindeer moss") on drier locations. "Low tundra" (as on the Alaskan coastal plain) is characterized by a thick spongy mat of living and very slowly decaying vegetation, often saturated with water and dotted with ponds. "High tundra," especially where there is considerable relief, is covered by a much scantier growth of lichens and grasses. Although the growing season is short, the long summer photoperiods allow a respectable amount of primary production (up to 5 gm dry matter/day) on favorable sites such as at Point Barrow, Alaska.

Combined aquatic (including fertile arctic oceans) and terrestrial productivity supports not only large numbers of migratory birds and insects during the open season but also permanent residents that remain active throughout the year. Large animals such as musk ox, caribou, reindeer, polar bears, wolves, foxes, and predatory birds, along with lemmings and other small animals that tunnel about in the vegetation mantle, are some of the permanent residents. The large animals are highly migratory,

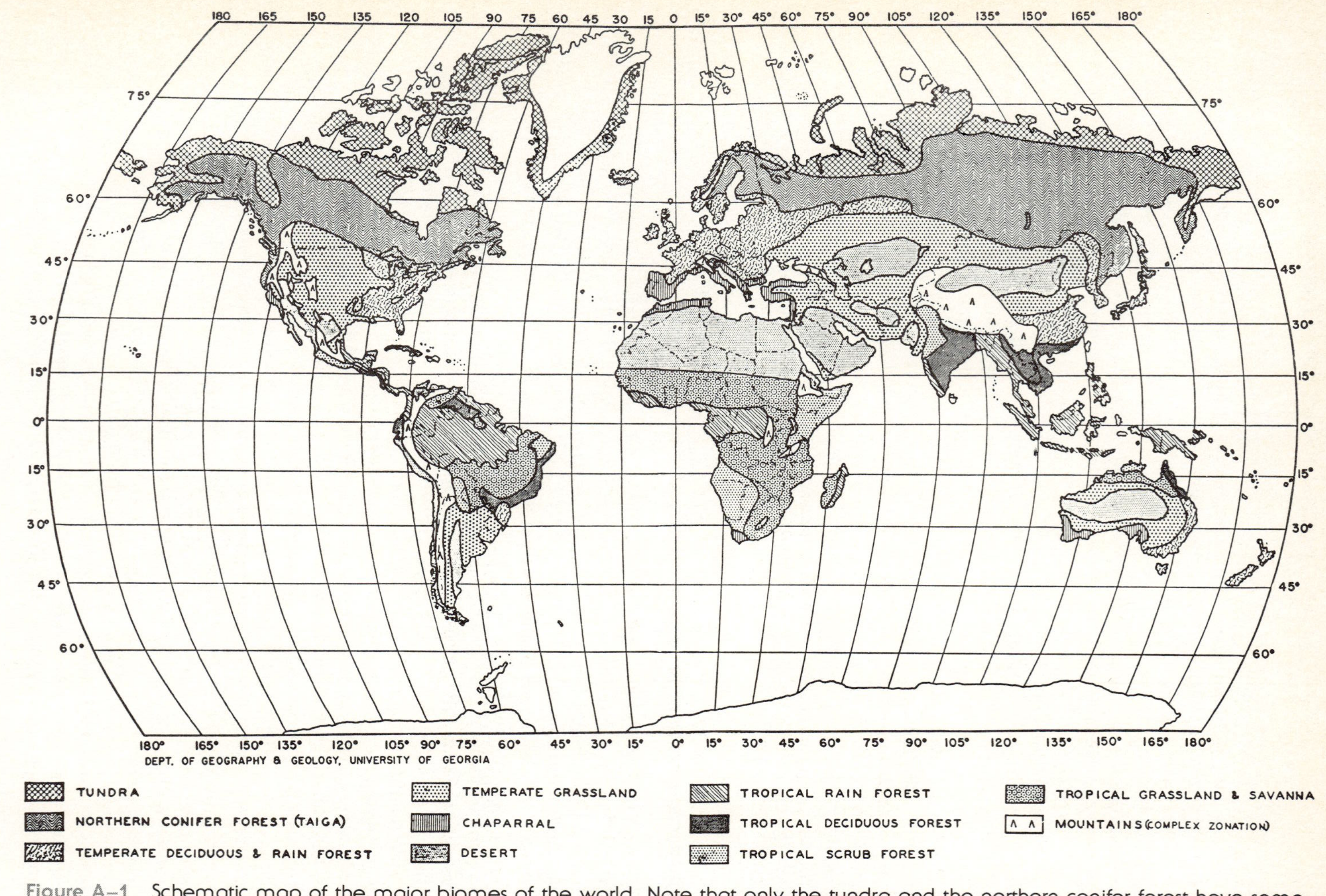

Figure A–1 Schematic map of the major biomes of the world. Note that only the tundra and the northern conifer forest have some continuity throughout the world. Other biomes of the same type (temperate grassland or tropical rain forest, for example) are isolated in different biogeographical regions and, therefore, may be expected to have ecologically equivalent but often taxonomically unrelated species. The pattern of the major biomes is similar to but not identical with that of the primary soil groups as mapped in Figure 5–22. Based on map of original vegetation in Finch, V. C., and G. T. Trewartha, *Physical Elements of Geography*. New York, McGraw-Hill, 1949.)

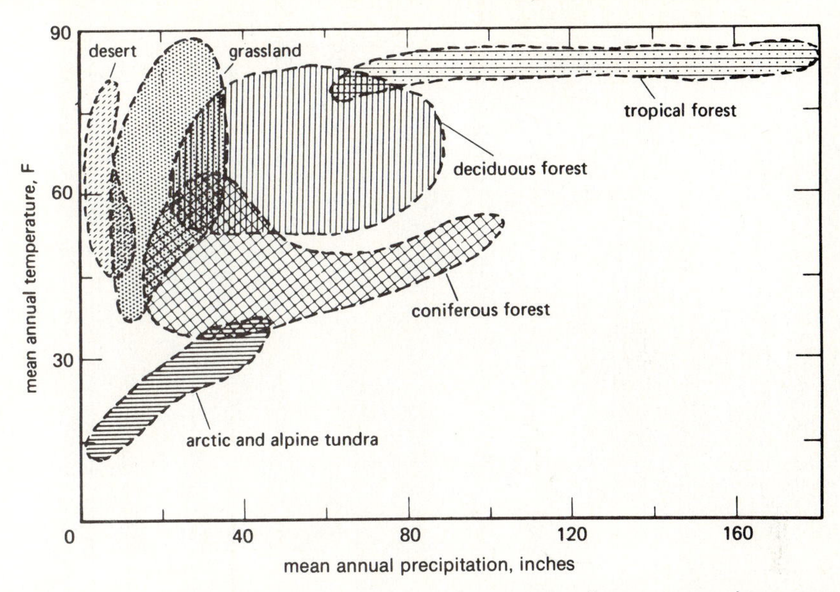

Figure A–2 Distribution of six major biomes in terms of mean annual temperature and mean annual rainfall (in inches). (Courtesy of National Science Foundation.)

whereas many of the smaller animals "cycle" in abundance as described in Chapter 6, Section 5. The special fragility of the tundra needs to be recognized as mineral exploration and other human impacts increase; the thin living mat is easily broken and is slow to recover. The building of the Alaskan pipeline has provided many object lessons.

Northern Coniferous Forest Biomes (Figure A–4)

Stretching as broad belts all the way across both North America and Eurasia are the vast northern evergreen forest regions. Extensions occur in the mountains even in the tropics. The identifying life form is the needle-leaved evergreen tree, especially spruce, fir, and pine (*Picea, Abies, Pinus*). A dense shade thus exists the year around, often resulting in poor development of shrub and herb layers. However, the continuous blanket of chlorophyll results in a fairly high annual production rate despite low temperature during half the year. The coniferous forests are among the great lumber-producing regions of the world. Coniferous needles decay very slowly, and the soil develops a highly characteristic podzolic profile (see Figure 5–18). The soil may contain a fair population of small organisms but few larger ones compared with deciduous forest or grassland soils. Many of the larger herbivorous vertebrates, such as the moose, snowshoe hare, and grouse, depend, at least in part, on broad-leaved developmental communities for their food. The seeds of the conifers provide important food for many animals such as squirrels, siskins, and crossbills.

As in tundra, seasonal periodicity is pronounced and populations tend to oscillate. The snowshoe hare–lynx cycles are classic examples (see Figure 6–15). Coniferous forests are also subject to outbreaks of bark beetles and defoliating insects (sawflies,

Figure A–3 Two views of the tundra in July on the coastal plain near the Arctic Research Laboratory, Point Barrow, Alaska. *Upper.* A broad swale at the head of a stream system about 2 miles from the coast. The arctic grass *Dupontia fischeri* and the sedge *Carex aquatalis* are the dominant plants, rooted in a peaty layer of a half-bog soil. The area is typical of what may be reasonably regarded as "climax" on low tundra sites near the coast. Shown in the picture are sample quadrats and an "enclosure" (fenced area) to keep out lemmings. *Lower.* A site about 10 miles inland, showing characteristic polygonal ground. Ice wedges underlying the troughs contribute to the raised polygons. The white fruits seen in the foreground are cottongrass, *Eriophorum scheuzcheri.* (Photos by the late Royal E. Shanks, E. E. Clebsch, and John Koranda.)

budworms, and so forth), especially where stands have only one or two dominant species; however, as pointed out in Chapter 6, Section 5, such outbreaks are part of the continuous cycle of development to which the coniferous forest ecosystem is adapted.

Coniferous forests of a distinctive type occur along the west coast of North America from central California to Alaska where temperatures are higher, seasonal range is relatively small, and the humidity is very high. Although dominated by the conifer life form, these forests are quite different floristically from the northern coniferous forest. Accordingly, these "temperate rain forests," as they are often known, could well be considered a separate type of biome.

Western hemlock (*Tsuga heterophylla*), western arborvitae (*Thuja plicata*), grand fir (*Abies grandis*), and Douglas fir (*Pseudotsuga*), the latter on drier sites or subclimax on wet sites, are the four most dominant trees in the Puget Sound area, where the forest reaches its greatest development. Southward, the magnificent redwoods (*Sequoia*) are found, and northward the sitka spruce (*Picea sitchensis*) is prominent. Unlike the drier and more northern coniferous forests, the understory vegetation is well developed wherever any light filters through; mosses and other moisture-loving lesser plants are abundant. Epiphytic mosses are the "ecological equivalent" of the epiphytic bromeliads of moist tropical forests. The "standing crop" of producers is indeed impressive, and as can be imagined, the production of lumber per unit area is potentially very great if the harvest-regeneration and nutrient cycles can be maintained. As with all ecosystems in which such a large percentage of nutrients may be tied up in the biomass, overexploitation may reduce future productivity.

Other subdivisions of the northern coniferous forest biome that could be considered separate biomes are the piñon-juniper woodland, or "pigmy conifers," and the ponderosa pine forests that occur at intermediate altitudes between the grassland and spruce zones in Colorado, Utah, New Mexico, and Arizona (see diagram, Figure A–11). These stands are generally open and have a considerable amount of herbaceous growth, especially where fire is part of the environment.

Temperate Deciduous Forests (Figure A–5)

Deciduous forest communities (Figure A–5) occupy areas with abundant, evenly distributed rainfall—30 to 60 inches (75 to 150 cm)—and moderate temperatures which have a distinct seasonal pattern. Temperate deciduous forests originally covered eastern North America, all of Europe, and part of Japan, Australia, and the tip of South America. Deciduous forest biomes are thus more isolated from one another than the tundras and northern coniferous forests, and species composition will, of course, reflect the degree of isolation. Since leaves are off the trees and shrubs for part of the year, the contrast between winter and summer is great. Herb and shrub layers tend to be well developed, as is the soil biota. Many plants produce pulpy fruits and nuts, such as acorns and beechnuts. Animals of the original forest of North America included the Virginia deer, bear, gray and fox squirrels, gray fox, bobcat, and wild turkey. The red-eyed vireo, wood thrush, tufted titmouse, ovenbird, and several woodpeckers are characteristic small birds of the mature stages. Development of temperate deciduous forests involves lengthy ecological successions, as reviewed in

Figure A–4 Three kinds of coniferous forests. *A*. A high-altitude forest of Engelmann spruce and Alpine fir in Colorado. Note the large amount of litter that accumulates because of low temperatures and long seasonal periods of snow cover. *B*. A spruce forest in Idaho with one of its chief developmental stages, aspen, a broad-leaved species whose leaves turn golden yellow in the fall (light-colored stand, left and center of the photo).

Illustration continued on the opposite page

Figure A–4 *C*. A fine example of the moist coniferous forest, often called temperate rain forest, in the Olympic National Forest, Washington. Note large size of the trees, luxuriant ground cover of ferns and other herbaceous plants, and the epiphytic mosses that festoon the limbs of trees. (Photos courtesy of U.S. Forest Service.)

Chapter 8, Section 1 (see Table 8–2). Transient grassland or "old-field" vegetation is characteristic of early seral stages, and pines are often prominent in intermediate stages.

Deciduous forest regions are one of the most important biotic regions of the world because Western civilization has achieved its greatest development in these areas. This biome is therefore greatly modified, and much of it is replaced by cultivated and forest edge communities.

The deciduous forest biome of North America has many important subdivisions, which have different climax forest types. Some of these are

The beech-maple forest of the north central region
The maple-basswood forest of Wisconsin and Minnesota
The oak-hickory forest of western and southern regions
The oak-chestnut forest of the Appalachian Mountains (now chiefly an oak forest with the chestnut wiped out by fungus diseases)
The diverse mixed mesophytic forest of the Appalachian plateau
The pine edaphic forest of the southeastern coastal plain maintained by fire and nutrient-poor sandy soils (evergreen forests have a competitive advantage over deciduous forests on poor soils)

Figure A-5 Climatic, edaphic, and fire climaxes in the eastern United States. *A*. A virgin stand of deciduous forest in western North Carolina—a climatic climax. (Courtesy of U.S. Forest Service.) *B*. An edaphic climax gum (*Nyssa*) swamp forest festooned with epiphytic "Spanish moss," *Tillandsia*, on the flood plain of the Savannah River. (Courtesy of E. I. Du Pont de Nemours & Co.)

Illustration continued on the opposite page

Figure A–5 *C.* A remnant of fire climax virgin longleaf pine forest on Millpond Plantation near Thomasville, Georgia. Use of frequent controlled burning has maintained the open, park-like condition and prevented the invasion of fire-tender hardwood trees. (Photo by Roy Komarek, Tall Timbers Research Station.)

Each of these has distinctive features, but many organisms, especially the larger animals, range through two or more of the subdivisions ("binding species").

Deciduous forests of western Europe have relatively few species (diversity presumed to have been reduced by glaciation). In contrast, the southern Appalachian mixed mesophytic type (not glaciated) and comparable forests in eastern Asia are the richest in species of the world's temperate forests.

Temperate Grasslands (Figure A–6)

Temperate grasslands occur where rainfall is intermediate between that of desert lands and that of forest lands, that is, where annual precipitation is between 10 and 30 inches (25 to 75 cm), depending on temperature, seasonal distribution of the rainfall, and the water-holding capacity of the soil. The previously mentioned studies by the International Biological Program have shown that soil moisture is a key factor, especially because it limits microbial decomposition and recycling of nutrients. Large grassland areas occupy the interior of the North American and Eurasian continents, southern South America (the Argentine pampas), and Australia.

Figure A–6 Natural temperate grassland in central North America with two original mammalian herbivores. *A*. Lightly grazed grassland in the Red Rock Lakes National Wildlife Refuge, Montana, with small herd of pronghorn antelope. *B*. Short-grass grassland, Wainwright National Park, Alberta, Canada, with herd of bison. The animal in the center is wallowing; old "buffalo wallows" often may be detected in the grassland many years after the bison have been extirpated.

In North America the grassland biome is divided into east-west zones: tall grass, mixed grass, short grass, and bunch grass prairies. These zones are determined by the rainfall gradient, which is also a gradient of decreasing primary productivity. Some of the important perennial species classified according to the height of the aboveground parts are as follows:

1. Tall grasses (5 to 8 feet)—big bluestem (*Andropogon gerardi*), switchgrass (*Panicum virgatum*), Indian grass (*Sorghastrum nutans*), and in the bottomlands, the sloughgrass (*Spartina pectinata*).
2. Mid grasses (2 to 4 feet)—little bluestem (*A. scoparius*), needlegrass (*Stipa spartea*), dropseed (*Sporobolus heterolepis*), western wheatgrass (*Agropyron smithii*), June grass (*Koeleria cristata*), Indian rice grass (*Oryzopsis*), and many others.
3. Short grasses (0.5 to 1.5 feet)—buffalo grass (*Buchloe dactyloides*), blue grama (*Bouteloua gracilis*), other gramas (*Bouteloua* sp.), the introduced bluegrass (*Poa*), and cheat grass (*Bromus* sp.).

Roots of most species penetrate deeply (up to 6 feet), and the weight of roots of healthy climax perennials will be several times that of the tops. The growth form of the roots is important. Some of the aforementioned species—for example, big bluestem, buffalo grass, and wheatgrass—have underground rhizomes and are thus sod-formers. Other species, such as little bluestem, June grass, and needlegrass, are bunchgrasses and grow in clumps. These two life forms may be found in all zones, but bunchgrasses predominate in the drier regions where grassland grades into desert.

Forbs (composites, legumes, and so forth) generally constitute only a small part of the producer biomass in climax grasslands but are consistently present. Certain species are of special interest as indicators of stress. Increased grazing or drought, or both, tends to increase the percentage of forbs, which are also prominent in early seral stages. Secondary succession in the grassland biome and the rhythmic changes in vegetation during wet and dry cycles have been described in Chapter 8, Section 1. The automobile traveler through the middle of the United States should bear in mind that the conspicuous annual forbs of the roadsides, such as the Russian thistle tumblewood (*Salsola*) and sunflowers (*Helianthus*), owe their luxuriance to the continual disturbance of the soil by highway maintenance machinery.

Very extensive areas of grassland, especially the tall grass prairies, have now been replaced by grain agriculture or cultivated pastures or have been invaded by woody vegetation. Original or virgin tall grass prairie is hard to find, and where preserved for study (as in the University of Wisconsin Arboretum), it must be burned to preserve its prairie character.

A well-developed grassland community contains species with different temperature adaptations, one group growing in the cool part of the season (spring and fall) and another in the hot part (summer). The grassland as a whole "compensates" for temperature, thus extending the period of primary production. The role of the C_3 and C_4 types of photosynthesis in this adaptation was discussed in Chapter 2, Section 5, Table 2–2.

The grassland community builds an entirely different type of soil compared with that of a forest, even when both start with the same parent mineral material. Since grass plants are short-lived compared with trees, a large amount of organic matter is added to the soil. The first phase of decay is rapid, resulting in little litter but much humus; in other words, humification is rapid but mineralization is slow. Consequently, grassland soils may contain five to ten times as much humus as forest soils, as shown in comparing the profiles in Figure 5–18. The dark grassland soils are among those best suited for growing of corn, wheat, and other grains, which, of course, are species of cultivated grasses.

Fire helps maintain grassland vegetation in competition with woody vegetation in warm or moist regions. Large herbivores are a characteristic feature of grasslands. The "ecological equivalence" of bison, antelope, and kangaroos in grasslands of different biogeographical regions was noted on page 407. The large grazers come in two "life forms": running types, such as those just mentioned, and burrowing types, such as ground squirrels and prairie dogs.

When natural grasslands become pasture, the native grazers are replaced by the domestic kind: cattle, horses, sheep, and goats. Since grasslands are adapted to heavy energy flow along the grazing food chain, such a switch is ecologically sound. However, humans persistently suffer from the "tragedy of the commons" by allowing overgrazing and overplowing. Many grasslands, therefore, are now human-made deserts that are difficult to restore to the grassland state. For example, Morello (1970) reports that intensive cattle grazing on the Argentine pampas reduces the combustible matter to such an extent that fires, which are necessary to maintain grass cover, can no longer burn. As a result, thorny shrubs, formerly kept in check by periodic fires, take over. The only way to restore grazing productivity is to expend fuel energy in mechanical removal and burning of woody vegetation. This is an example of an anthropogenic vegetation change reversible only at great cost.

Tropical Grasslands and Savannas (Figure A–7)

Tropical savannas (grasslands with scattered trees or clumps of trees) are found in warm regions with 40 to 60 inches of rainfall but with one or two prolonged dry

Figure A–7 A view of the tropical savanna of East Africa. Grass, scattered trees with picturesque shapes, dry-season fires, and numerous species of large mammalian herbivores (Thomson's gazelle is the species shown here) are unique features of this biome. (Photography by Donald I. Ker, Ker & Downey Safaris Ltd., Nairobi, East Africa.)

seasons when fires are an important part of the environment. The largest area of this type is in central and east Africa, but sizable tropical savannas or grasslands also occur in South America and Australia. Since both trees and grass must be resistant to drought and fire, the number of species in the vegetation is not large, in sharp contrast to adjacent equatorial forests. Grasses belonging to such genera as *Panicum*, *Pennisetum*, *Andropogon*, and *Imperata* provide the dominant cover, whereas the scattered trees are of entirely different species from those of the rain forest. In Africa, the thorny and picturesque acacias and other leguminous trees and shrubs, the large-trunked baobab trees (*Adansonia*), arborescent euphorbias (ecological equivalent of cacti), and palms dot the landscape. Often, single species of both grass and trees may be dominant over large areas.

In number and variety, the population of hoofed mammals of the African savanna is unexceeded anywhere in the world. Antelope (numerous species, including the wildebeest), zebra, and giraffe graze or browse and are sought by lions and other predators in areas in which the "big game" has not been replaced by humans and their cattle. The mutualism between grazers and grass was discussed in some detail in Chapter 3, page 141. Insects are most abundant during the wet season when most birds nest, whereas reptiles may be more active during the dry season. Thus, seasons are regulated by rainfall rather than by temperature as in the temperate grasslands.

The earliest human fossils have been found in East Africa, but it is not certain whether this region was wetter or drier at the "dawn of man" than it is now.

Chaparral and Sclerophyllous Woodland (Figure A–8)

In mild temperate regions with abundant winter rainfall but dry summers, the vegetation consists of trees or shrubs, or both, with hard, thick evergreen leaves (Figure A–8). Under this heading is included a range of vegetation from coastal chaparral, in which shrubs predominate, to broad-sclerophyll woodland dominated by small to medium-sized evergreen trees. A "picture" of the climate of the winter rain region is shown in Figure 5–14. Chaparral communities are extensive in California and Mexico, along the shores of the Mediterranean Sea, in Chile, and along the southern coast of Australia. Many plant species may serve as dominants, depending on the region and local conditions. All species have mycorrhizae, and some have nitrogen-fixing actinomycete nodules. Both mutualisms enhance survival under harsh conditions. Fire is an important factor that tends to perpetuate shrub dominance at the expense of trees. Chaparral as a "fire type" ecosystem and as an example of a cyclic climax was noted in Chapter 5, page 282; Chapter 7, page 390; and Chapter 8, page 451.

In California, about 5 to 6 million acres of slopes and canyons are covered with chaparral. Chamiso (*Adenostoma*) and manzanita (*Arctostaphylos*) are common shrubs that often form dense thickets, and several evergreen oaks are characteristic, also, as either shrubs or trees. The rainy or growing season generally extends from November to May. Mule deer and many birds inhabit the chaparral during this period, then move north or to higher altitudes during the hot, dry summer. Resident vertebrates are generally small and dull-colored to match the dwarf forest; the small brush rabbits (*Sylvilagus bachmani*), wood rats, chipmunks, lizards, wren-tits, and

Figure A–8 *A*. General view of the chaparral-covered hills of California. *B*. Fighting a small fire in this "fire-type" vegetation. (Photos courtesy of U.S. Forest Service.)

brown towhees are characteristic. The population density of breeding birds and insects is high as the growing season comes to a close, then decreases as the vegetation dries out in late summer. At this time, fires may sweep the slopes with incredible swiftness (and into suburbia in southern California). After a fire, the chaparral shrubs sprout vigorously with the first rains and may gain maximum size in 15 to 20 years.

The sclerophyll woodland of the winter rain areas of the Mediterranean region is locally called "macchie"; similar vegetation in Australia, in which trees and shrubs of the genus *Eucalyptus* are dominant, is called "mallee scrub." It is not surprising that Australian "eucalypts" do well in California, where they have been widely introduced and largely replace the native woody vegetation in urban areas.

Deserts (Figure A–9)

Regions having less than 10 inches (25 cm) of rainfall, or sometimes regions with greater rainfall that is very unevenly distributed, are generally classed as deserts (Figure A–9). Scarcity of rainfall may be due to (1) high subtropical pressure, as in the Sahara and Australian deserts, (2) geographical position in rain shadows (see Figure 5–11), as in the western North American deserts, or (3) high altitude, as in Tibetan, Bolivian, and Gobi deserts. Most deserts receive some rain during the year and have at least a sparse cover of vegetation, unless edaphic conditions of the substrate happen to be especially unfavorable (moving sand dunes, for example). Apparently the only places where little or no rain falls are the central Sahara and northern Chile. A "picture" of the climate of New Mexico's desert is shown in Figure 5–14 (see also Figure A–2).

Walter (1954) has measured net production of a series of deserts and semiarid communities that lie along a rainfall gradient in southwest Africa. The annual production of dry matter was a linear function of rainfall, illustrating the sharpness with which moisture acts as an overall ("master") limiting factor. The annual net primary productivity of deserts is less than 1000 kg dry matter/hectare, averaging about 200 kcal/m^2 (see Figure 3–7 and Table 3–7).

When deserts are irrigated, and water is no longer a limiting factor, the type of soil becomes a prime consideration. Where texture and nutrient content of the soil are favorable, irrigated deserts can be extremely productive because of the large amount of sunlight. However, the cost per pound of food produced may be high because of the high cost of developing and maintaining irrigation systems. Very large volumes of water must flow through the system; otherwise, salts may accumulate in the soil (as a result of rapid evaporation rate) and become limiting. As the irrigated ecosystem "ages," increased water demands may bring on an "inflationary spiral," requiring the building of more aqueducts, higher costs of production, and greater exploitation of the underground or mountain water sources. Old World deserts are full of ruins of old irrigation systems. In many cases, no one knows why they failed and why the "Garden of Eden" became a desert again. At least, these ruins should warn us that the irrigated desert will not continue to bloom indefinitely without due attention to the basic laws of the ecosystem.

Three life-forms of plants are adapted to deserts: (1) the annuals, which avoid drought by growing only when there is adequate moisture (see page 239), (2) the succulents, such as the cacti, which have the moisture-conserving CAM photosynthesis (page 34) and also store water, and (3) the desert shrubs, which have numerous branches ramifying from a short basal trunk bearing small, thick leaves that may be shed during the prolonged dry periods. The desert shrub presents very much the same appearance throughout the world, even though species may belong to diverse taxa (another striking example of ecological equivalence).

Figure A–9 Three types of deserts in western North America. *A*. A low-altitude "hot" desert in southern Arizona dominated by creosote bush (*Larrea*). Note the characteristic growth form of the desert shrub (numerous branches ramifying from ground level) and the rather regular spacing. (Photo by R. R. Humphrey.) *B*. An Arizona desert at a somewhat higher altitude with several kinds of cacti and a greater variety of desert shrubs and small trees. This is one of the sites selected for IBP interdisciplinary study. (Photo by R. H. Chew.)

Illustration continued on the opposite page

Figure A–9 *C.* A "cool" desert in eastern Washington dominated by sagebrush (*Artemisia*). The picture was taken during the spring at a time of peak primary production when annual grasses and forbs carpet the spaces between and around the shrubs. In progress was a radioactive tracer experiment designed to compare uptake of shrubs and annuals. (Hanford Atomic Products Operation photo.)

Desert shrubs have a highly characteristic "spaced" distribution in which individual plants are scattered thinly with large bare areas in between. In some cases, antibiotics help to keep plants spaced apart. In any event, spacing reduces competition for a scarce resource; otherwise, intense competition for water might result in the death or stunting of all the plants.

Although arbitrary, it is convenient to recognize two types of desert on the basis of temperature: hot deserts and cool deserts. In North America, the creosote bush (*Larrea*) is a widespread dominant of the southwestern hot desert, and sagebrush (*Artemisia*) is the chief plant over large areas of the more northern cool deserts of the Great Basin. Bur sage (*Franseria*) is also widespread in southern areas, while at higher altitudes where moisture is a little greater, the giant cactus (*Sahuaro*) and palo verde are conspicuous. Eastward, a considerable amount of grass is mixed with desert shrubs to form a desert grassland type. In the cool deserts, especially on the alkaline soils of the internal drainage regions, saltbushes of the family Chenopodiaceae, such as shadscale (*Atriplex*), hop sage (*Grayia*), winter fat (*Eurotia*), and greasewood (*Sarcobatus*), occupy extensive zones. In fact, "chenopods" are also widely distributed in arid regions in other parts of the world. The succulent life form, including the cacti and the arborescent yuccas and agaves, reaches its greatest development in the Mexican desert; some species of this type extend into the shrub deserts of Arizona and California, but this life form is unimportant in the cool deserts. In all deserts, annual forbs and grasses may make quite a show during brief wet periods. The extensive "bare ground" in deserts is not necessarily free of plants. Mosses, algae, and lichens may be present, and on sands and other finely divided soils they may form a stabilizing crust; the blue-green algae (often associated with lichens) are also important as nitrogen fixers.

Desert animals as well as plants are adapted in various ways to lack of water. Reptiles and some insects are "preadapted" because of relatively impervious integuments and dry excretions (uric acid and guanine). Desert insects are "waterproofed" with substances that remain impermeable at high temperatures. Although evaporation from respiratory surfaces cannot be eliminated, it is reduced to the minimum in insects by the internally invaginated spiracle system. It should be pointed out that the production of metabolic water (from breakdown of carbohydrates), often the only water available, is not in itself an adaptation; it is the conservation of this water that is adaptive, as is, in the case of tenebrionid beetles (a characteristic desert group), the ability to produce more metabolic water at low humidities. Mammals, by contrast, are not very well adapted as a group (because they excrete urea, which involves the loss of much water), yet certain species have developed remarkable secondary adaptations. Among these mammals are rodents of the family Heteromyidae, especially the kangaroo rat (*Dipodomys*) and the pocket mouse (*Perognathus*) of the New World deserts and the jerboa (*Dipus*, family Dipodidae) of Old World deserts. These animals can live indefinitely on dry seeds and do not require drinking water. They remain in burrows during the day, and conserve water by excreting very concentrated urine and by not using water to regulate body temperature. Thus, adaptation to deserts by these rodents is as much behavioral as physiological. Other desert rodents—wood rats (*Neotoma*) for example—cannot live solely on dry food but survive in parts of the desert by eating succulent cacti or other plants that store water. Even the camel must drink, but camels can go for long periods without water because their body tissues can tolerate elevation of body temperature and a degree of dehydration that would be fatal to most animals. (Incidentally, camels do not store water in their humps, as popularly supposed.)

Semi-evergreen, Seasonal Tropical Forests (Figure A–10A)

Tropical seasonal forests, including the monsoon forests of tropical Asia, occur in humid tropical climates with a pronounced dry season during which some or all of the trees lose their leaves (depending on length and severity of the dry season). The key factor is the strong seasonal pulse of a fairly large annual rainfall. Where wet and dry seasons are of approximately equal length, the seasonal appearance is the same as that of a temperate deciduous forest, with "winter" corresponding to the dry season. In a Panamanian seasonal forest, shown in Figure A–10A, the tall emergent trees lose their leaves during the dry season, but palms and other understory trees retain theirs (hence the term "semi-evergreen"). Seasonal tropical forests have a species richness second only to that of the rain forests.

Figure A–10 The tropical forests. *A*. View from above of a lowland seasonal forest in Panama. The tall emergent trees (with white trunks), which lose their leaves in the dry season, project above the general canopy of broad-leaved evergreen hardwoods and palms. *B*. Interior of a montane rain forest in Puerto Rico, showing the abundance of epiphytes that characterize the rain forest in the mountains of the humid tropics. *C*. Close-up view of the base of a buttressed tree in the same Puerto Rican montane forest, showing the aboveground mass of small roots that permeate the litter (rather than the soil as in temperate forests). (*A*, photo by George Child, Institute of Ecology, University of Georgia; *B* and *C*, photos by E. P. Odum.)

Tropical Rain Forests (Figures A–10B and C)

The variety of life perhaps reaches its culmination in the broad-leaved evergreen tropical rain forests that occupy low-altitude zones near the equator. Rainfall exceeds 80 or 90 inches (200 to 225 cm) a year and is distributed over the year, usually with one or more relatively "dry" seasons (5 inches per month or less). Rain forests occur in three main areas: (1) the Amazon and Orinoco basins in South America (the largest continuous mass) and the Central American isthmus; (2) the Congo, Niger, and Zambezi basins of central and western Africa, and Madagascar; and (3) the Indo-Malay-Borneo-New Guinea regions. These differ from each other in the species present (since they occupy different biogeographical regions), but the forest structure and ecology are similar in all three areas. The variation in temperature between winter and summer is less than that between night and day. Seasonal periodicity in breeding and other activities of plants and animals is largely related to variations in rainfall or is regulated by inherent rhythms. For example, some trees of the family Winteraceae apparently show continuous growth, while other species in the same family show discontinuous growth with formation of rings. Rain-forest birds may also require periods of "rest," since reproduction often exhibits periodicity unrelated to season.

The rain forest is highly stratified. Trees generally form three layers: (1) scattered, very tall emergent trees that project above the general level of the (2) canopy layer, which forms a continuous evergreen carpet 80 to 100 feet tall, and (3) an understory stratum that becomes dense only where there is a break in the canopy. The tall trees are shallow rooted and often have swollen bases or "flying buttresses." A profusion of climbing plants, especially woody lianas and epiphytes, often hides the outline of the trees. The "strangler figs" and other arborescent vines are especially noteworthy. The number of species of plants is very large; often, there are more species of trees in a few acres than in the entire flora of Europe.

A much larger proportion of animals lives in the upper layers of the vegetation than in temperate forests, where most life is near the ground level. For example, more than 50 percent of the mammals in British Guiana are arboreal. Besides the arboreal mammals, there is an abundance of chameleons, iguanas, geckos, arboreal snakes, frogs, and birds. Ants and the orthoptera, as well as butterflies and moths, are ecologically important. Symbiosis between animals and ephiphytes is widespread. As is true of the flora, the fauna of the rain forest is incredibly rich in species. For example, in a 6-square-mile area on Barro Colorado, a well-studied bit of rain forest in the Panama Canal Zone, there are 20,000 species of insects, compared with only a few hundred in all of France. Numerous archaic types of both animals and plants survive in the many niches of the unchanging environment. Many scientists believe that the rate of evolutionary change and speciation is especially high in the tropical forest regions, which, therefore, have been a source of a number of species that have invaded more northern communities. The need to preserve large areas of tropical forests as a "gene resource" is a matter of increasing concern in the scientific community (see pages 425 to 429).

Fruit and termites are staple foods for animals in the tropical forest. One reason why birds are often abundant is that so many of them, such as fruit-eating parakeets, toucans, hornbills, contingas, trogons, and birds-of-paradise, are herbivorous. Because the "attics" of the jungle are crowded, many bird nests and insect cocoons are of a hanging type, enabling their inhabitants to escape from army ants and other predators. Although some spectacularly bright birds and insects occupy the more open areas, the majority of rain-forest animals are inconspicuous, and many of them are nocturnal.

In the mountainous areas of the tropics is a variant of the lowland rain forest, the **montane rain forest,** which has some distinctive features. The forest becomes progressively less tall with increasing altitude, and epiphytes make up an increasingly larger proportion of the autotrophic biomass, culminating in the dwarf **cloud forests**. A functional classification of rain forests can be based on **saturation deficit** because this determines transpiration, which in turn determines root biomass and the height of trees. Still another variant of the rain forest occurs along banks and flood plains of rivers and is called the **gallery forest,** or sometimes the **riverine forest**.

Efficient direct nutrient cycling by mutualistic microorganisms is a remarkable property of rain forests that enables them to be as luxuriant on poor soils as on more fertile sites. This feature was discussed in some detail in Chapter 4, Section 7.

When the rain forest is removed, a secondary forest often develops that includes softwood trees, such as *Musanga* (Africa), *Cecropia* (America), and *Macoranga* (Malaysia). The secondary forest looks lush but is quite different from the virgin forest in both ecology and flora. The "climax" is usually very slow to return, especially on sandy or other nutrient-poor sites, since all the nutrients in the original forest were lost with the removal of the biomass and the disruption of microbial recycling networks.

How to develop rain forests for human use continues to be controversial and frustrating to those who look upon these great forests as the last frontier for colonization and as a potential source of wealth. The large size of trees fooled the early European explorers of the Amazon into believing that the soil was rich. There followed many unsuccessful attempts to convert the forest to agricultural and commercial forestry. Despite the failures, developers and governments continue to try to transfer temperate-zone agricultural and forest technology to the region, although it is definitely not appropriate technology. In the late 1960s, billionaire D. K. Ludwig acquired an area of the Brazilian Amazon (Jara) about the size of Connecticut. He floated in a pulp mill and converted mature forests to plantations of exotic species. Jordan (1982) comments on this expensive venture as follows: "Now, less than 15 years after the project began, it is failing. Yet, there is still refusal to believe that the problem is related to soils. Failure is blamed on [economic] mismanagement." A better way to utilize the forest resources, Jordan suggests, is to develop a system of strip cutting in such a way that retention of nutrients by the root mat in cut areas is not greatly disturbed and that seedlings from adjacent uncut areas can become quickly established in cleared areas. Horticulture might be similarly organized. Clearly, humans must design with, not against, natural adaptations in regional ecosystems.

Tropical Scrub or Thornwoods

Where moisture conditions are intermediate between desert and savanna on one hand and seasonal or rain forest on the other, tropical scrub or thorn forests may be found. These cover large areas in central South America, southwest Africa, and parts of southwest Asia. The key climatic factor is the imperfect, irregular distribution of a moderate total rainfall. Thorn forests, which are often referred to as "the bush" in Africa or Australia and the "Caatinga" in Brazil, contain small hardwood trees often grotesquely twisted and thorny; leaves are small and are lost during dry seasons. Thorn trees may occur in dense stands or in a scattered or clumped pattern. In some locations, it is not always certain whether thorn woodlands are natural or the product of generations of use by pastoralists and their cattle.

Zonation in Mountains (Figure A–11)

The distribution of biotic communities in mountainous regions is complicated, as would be expected because of the diversity of physical conditions. Major communities generally appear as irregular bands, often with very narrow ecotones. A small-scale map such as that in Figure A–1 is inadequate to depict this feature. On a given mountain, four or five major biomes with many zonal subdivisions may be

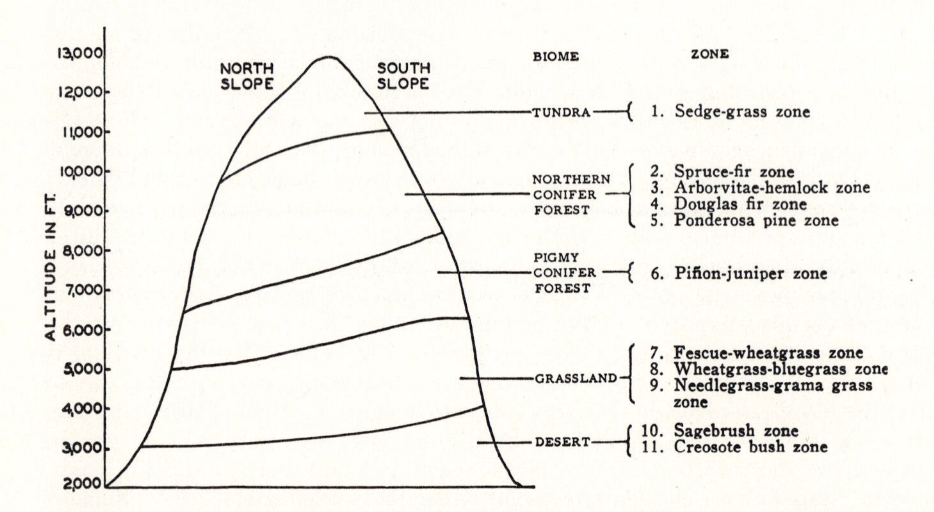

Figure A–11 Zonation in the mountains of western North America. The diagram does not refer to a specific mountain but shows general conditions that might be expected in the central Rockies (so-called intermountain region) of Utah. North, south, east, or west of this region, zonation may be expected to vary somewhat. Thus, the creosote bush and piñon-juniper zone are absent northward and eastward. Eastward, an oak–mountain mahogany zone, and westward, various zones in the chaparral biome, occur between the grassland and the northern conifer forest. (Diagram modified from Woodbury, 1945; zone data from Daubenmire, 1946.)

present, as shown in Figure A–11. Consequently, there is closer contact between biomes and more interchange of biota between different biomes than occurs in nonmountainous regions. On the other hand, similar communities are more isolated in the mountains, since mountain ranges are rarely continuous.

In general, many species that are characteristic of a biome in its broad, nonmountainous phase are also characteristic of the belt-like extensions in the mountains. Because of isolation and topographical differences, many other species and varieties are unique to the mountain communities. As shown in Figure A–11, the altitudinal boundaries of zones are higher on the warmer south slope than on the north side.

2. Freshwater Ecosystems

Freshwater habitats may be conveniently considered in three series as follows:

1. standing-water or lentic (from *lenis,* "calm") ecosystems: lakes and ponds
2. running-water or lotic (from *lotus,* "washed") ecosystems: springs, streams, and rivers
3. wetlands, where water levels fluctuate up and down, often seasonally as well as annually: marshes and swamps

An example of each type of habitat is pictured in Figure A–12.

Ground water, although a large freshwater reservoir and a very important resource for humans (Chapter 5, Section 4), is not generally thought of as an ecosystem since it contains little or no life (sometimes bacteria). Ground water does interconnect with all three major aboveground ecosystems (see Figure 4–10) and is thereby an important part of the input and output environments of lentic, lotic, and wetland ecosystems.

Freshwater habitats occupy a relatively small portion of the earth's surface compared with marine and terrestrial habitats, but their importance to humans is far greater than their area for the following reasons:

1. They are the most convenient and cheapest source of water for domestic and industrial needs (we can and probably will get more water from the sea, but at a great cost in terms of energy required and salt pollution created).

2. The freshwater components are the "bottle-neck" in the hydrologic cycle (see Figure 4–9B).

3. Freshwater ecosystems, along with estuaries, provide the more convenient and cheapest tertiary waste disposal systems. Almost without exception, the world's largest cities are located on large rivers, lakes, or estuaries that serve as "free" sewers. Because this natural resource is so abused, a major effort to reduce this stress must come quickly; otherwise, water will become *the* limiting factor for our species (see page 259).

Water has several unique thermal properties that combine to minimize temperature changes; thus, the range of variation is smaller and changes occur more slowly in water than in air. The most important of these thermal properties are

1. High specific heat; that is, a relatively large amount of heat is involved in changing the temperature of the water. One gram-calorie (gcal) of heat is required to raise 1 millimeter (or 1 gram) of water 1 degree Centigrade (between 15° and 16°). Only ammonia and a few other substances have values higher than 1.
2. High latent heat of fusion. Eighty calories are required to change 1 gram of ice into water with no change in temperature (and vice versa).
3. Highest known latent heat of evaporation. Five hundred thirty-six calories per gram are absorbed during evaporation, which occurs more or less continually from vegetation, water, and ice surfaces. A major portion of the incoming solar radiation is dissipated in the evaporation of water from the ecosystems of the world (see Chapter 3, Section 1). This energy flow moderates climates and makes possible the development of life in all its fantastic diversity.
4. Water has its greatest density at 4°C; it expands and hence becomes lighter both above and below this temperature. This unique property prevents lakes from freezing solid.

Lentic Ecosystems (Lakes and Ponds)

In the geological sense, most basins that now contain standing fresh water are relatively young. The life span of ponds ranges from a few weeks or months for small seasonal ponds to several hundred years for larger ponds. Although a few lakes, such as Lake Baikal in Russia, are ancient, most large lakes date back only as far as the ice ages. Standing-water ecosystems may be expected to change with time at rates more or less inversely proportional to size and depth. Although geographical discontinuity of fresh waters favors speciation, the lack of isolation in time does not. Generally speaking, the species diversity is low in freshwater communities, and many taxa (species, genera, families) are widely distributed within a continental mass and even between adjacent continents. A pond was considered in some detail in Chapter 2, Section 7, as an example of a convenient-sized ecosystem for introducing the study of ecology.

Distinct zonation and stratification are characteristic features of lakes and large ponds. Typically, we can distinguish a **littoral zone** containing rooted vegetation along shore, a **limnetic zone** of open water dominated by plankton, and a deep-water **profundal zone** containing only heterotrophs. In temperate regions, lakes often become thermally stratified during summer and again in winter, owing to differential heating and cooling. The warmer upper part of the lake, or **epilimnion** (from Greek *limnion*, "lake") becomes temporarily isolated from the colder lower water, or **hypolimnion,** by a **thermocline** zone that acts as a barrier to exchange of materials. Consequently, the supply of oxygen in the hypolimnion and nutrients in the epilimnion may run short. During spring and fall, as the entire body of water approaches the same temperature, mixing again occurs. "Blooms" of phytoplankton often follow these seasonal turnovers, since nutrients in the bottom become available in the photic zone. In warm climates, mixing may occur only once a year (winter); in the tropics, mixing is a gradual or irregular process.

Primary production in standing-water ecosystems depends on the chemical nature of the basin and on the nature of imports from streams or land, and it is generally inversely related to depth. Accordingly, the yield of fish per unit of water surface area is greater in shallow than in deep lakes (see Table 3–10), but the latter may have larger individual fish. Lakes are often classified as either oligotrophic (few foods) or eutrophic (good foods) on the basis of productivity. Since a biologically poor lake is preferable to a fertile one, from the standpoint of water quality for domestic use and recreation, we have a paradox. In some parts of the biosphere, humans are increasing its fertility to feed themselves, whereas in other places they are preventing its fertility (by removing nutrients, poisoning plants, and so on) to maintain a pleasant environment. A fertile green pond that can produce many fish is not considered a good recreational swimming pool.

By constructing artificial ponds and lakes (impoundments), humans have changed the landscape in regions that lack natural bodies of water. In the United States, almost every farm now has at least one farm pond, and large impoundments have been constructed on practically every river. Much of this activity is beneficial, but the impoundment idea can be carried too far; covering up fertile land with a body of water that cannot yield much food may not always be the best long-term land use. Standing waters are generally less efficient at oxidizing waste than are running waters. Unless the watershed is well vegetated, erosion may fill up an impoundment in a human generation. More comprehensive cost-benefit studies are needed before new impoundments are constructed.

The heat budget of impoundments may differ greatly from that of natural lakes, depending on the design of the dam. If water is released from the bottom, as would be the case with dams designed for hydroelectric power generation, cold, nutrient-rich but oxygen-poor water is exported downstream while warm water is retained in the lake. The impoundment then becomes a **heat trap and nutrient exporter,** in contrast to natural lakes, which discharge from the surface and therefore function as **nutrient traps and heat exporters.** Accordingly, the type of discharge greatly affects downstream conditions.

Lotic Ecosystems (Streams and Rivers)

Differences between running and standing waters generally revolve around a triad of conditions: (1) current is much more of a major controlling and limiting factor in streams; (2) land-water interchange is relatively more extensive in streams, resulting in a more "open" ecosystem and a "heterotrophic" type of community metabolism where size of stream is small; and (3) oxygen tension is generally high and more uniform in streams, and there is little or no thermal or chemical stratification except in large, slow-moving rivers.

The "river continuum" involving longitudinal changes in the community metabolism, biotic diversity, and particle size from headwater to river mouth is shown in Figure 4–11. Within a given stretch of stream, two zones are generally apparent:

1. A rapids zone has current great enough to keep the bottom clear of silt or other loose material, thus providing a firm substrate. This zone is occupied by specialized organisms that become firmly attached or cling to the substrate (such as black fly and caddis fly larvae) or, in the case of fish, that can swim against the current (trout, darters, and so on).

2. A pool zone has deeper water where velocity of current is reduced so that sand and silt settle, providing a soft bottom favorable for burrowing and swimming animals, rooted plants, and, in large pools, plankton. In fact, communities of pools in large rivers resemble those of ponds.

Rivers in their upper reaches are generally eroding: they cut into the substrate, so a hard bottom predominates. As rivers reach base level in the lower reaches, sediments are deposited and floodplains and delta that are often extremely fertile are built up.

In terms of chemical composition of the water, Livingstone (1963) divides the world's rivers into two types: (1) hard-water or carbonate rivers with 100 or more ppm dissolved inorganic solids and (2) soft-water or chloride rivers with less than 25 ppm dissolved solids.

Water chemistry of carbonate rivers is controlled largely by rock weathering, whereas atmospheric precipitation is the dominant factor in chloride rivers. Humic or black-water streams with high concentrations of dissolved organic material represent still another class of streams found in warm lowlands.

Several excellent studies of food-chain energetics of streams with emphasis on fish have been compiled, for example Allen (1951), Horton (1961), Cummins (1974), and Mann on the River Thames (1965).

Springs hold a position of importance as study areas that is far out of proportion to their size and number. Some of the best whole-system studies have been made on springs, for example, the large limestone springs of Florida (H. T. Odum and colleagues), the small cold-water springs of New England (Teal, Tilly), and the hot springs of Yellowstone (Wiegert, Brock and colleagues).

Freshwater Wetlands (Marshes and Swamps) (Figure A–12A)

A wetland is defined as any area covered by shallow, fresh water for at least part of the annual cycle; accordingly, soils are saturated with water continually or for part of the year. The key factor that determines productivity and species composition of the wetland community is the **hydroperiod,** that is, the periodicity of water-level fluctuations. Water flow as an energy subsidy was discussed in Chapter 3, Section 3, and Chapter 5, Section 4. The relationship between productivity and hydroperiod in swamp forests was diagrammed in Figure 5–16. Freshwater wetlands can thus be classed as "pulse-stabilized fluctuating water level ecosystems" (as are intertidal marine and estuarine ecosystems; Chapter 8, Section 7).

Wetlands tend to be very open systems and can be conveniently classified according to their interconnections with deep water or upland ecosystems, or both, as follows:

1. Riverine wetlands located on lowlying depressions (ox-bows) and floodplains associated with rivers. The bottomland hardwood forests on the floodplains of large rivers are among the most productive of natural ecosystems, as are the freshwater tidal marshes along the lower reaches of large coastal plain rivers.

2. Lacustrine (from *lacus,* "lake") wetlands are associated with lakes, ponds, or dammed river channels. They are periodically flooded when these deeper bodies of water overflow.

3. Palustrine (from *paludis*, "marsh") wetlands include what are variously called marshes, bogs, fens, wet prairies, and temporary ponds that occur in depressions not directly connected with lakes or rivers (although they may be in old river beds or filled-up ponds or lake basins). Such wetlands are widely scattered over the landscape in glaciated regions. They are generally vegetated with various submerged aquatic macrophytes, emergent marsh plants, and shrubs. Palustrine marshes are often a prime breeding habitat for waterfowl and other aquatic or semiaquatic vertebrates.

Although wetlands occupy only about 2 percent of the world's area, they are estimated to contain 10 to 14 percent of the carbon (Armentano, 1980). Wetland soils, such as the histosols, may contain up to 20 percent carbon by weight, and of course the peats are even more carboniferous. Draining and conversion to agriculture releases large quantities of CO_2 to the atmosphere, thus contributing to the "CO_2 problem" (Chapter 4, Section 4). The aerobic-anaerobic stratification of wetland sediments (including salt-water marshes) is important far out of proportion to their area for the part they play in global cycling of sulfur, nitrogen, and phosphorus as well as carbon (see Chapter 2, Section 5). Figure A–13 summarizes key aspects of microbial decomposition and recycling in wetland and shallow marine sediments.

During the 1970s, public attitudes toward wetlands changed dramatically as ecological and economic studies revealed previously unrecognized values. No longer are wetlands always viewed as wastelands to be destroyed or modified. Although some progress has been made in preservation, especially of the coastal wetlands, much remains to be done in the legal and political arenas. Much of the new knowledge on wetland values is summarized in the large symposium volume edited by Greeson, Clark, and Clark (1979).

3. Marine Ecosystems (Figure A–14)

The features of the sea that are of major ecological interest may be listed as follows:

1. The sea is big; it covers 70 percent of the earth's surface.
2. The sea is deep, and life extends to all its depths. Although apparently there are no abiotic zones in the ocean, life is much denser around the margins of continents and islands.
3. The sea is continuous, not separated as are land and freshwater habitats. All the oceans are connected. Temperature, salinity, and depth are the chief barriers to free movement of marine organisms.
4. The sea is continuously circulating; differences in air temperature between poles and equator set up strong winds such as the trade winds (blowing steadily in the same direction the year around) that, together with rotation of the earth, create definite currents. Noteworthy are the equatorial currents running east and west and coastal currents running north and south. Well known are the Gulf Stream and North Atlantic Drift, which bring warm water and moderate climate to high latitudes in Europe, and the California Current, which moves cold water southward, creating the fog belt characteristic of the California coast (which is important for the redwoods and other giant conifers). Major currents act as giant pinwheels (or gyres) that run clockwise in the northern hemisphere and counterclockwise in the southern hemisphere.

Figure A-12 Three freshwater ecosystems. *A*. A wetland; freshwater marsh in the Sacramento National Wildlife Refuge in California, where flocks of geese find refuge and shelter in productive aquatic and semiaquatic vegetation. *B*. A lentic ecosystem; natural pond in the grassland region of western Canada. *C*. Lotic ecosystems; convergence of two streams in northern New Jersey. The stream in the foreground flows from a watershed protected by grass and trees; the stream entering from the left is badly polluted with silt as a result of poor agriculture. (*A* and *B*, U.S. Department of Interior, Fish and Wildlife Service; *C*, U.S. Soil Conservation Service.)

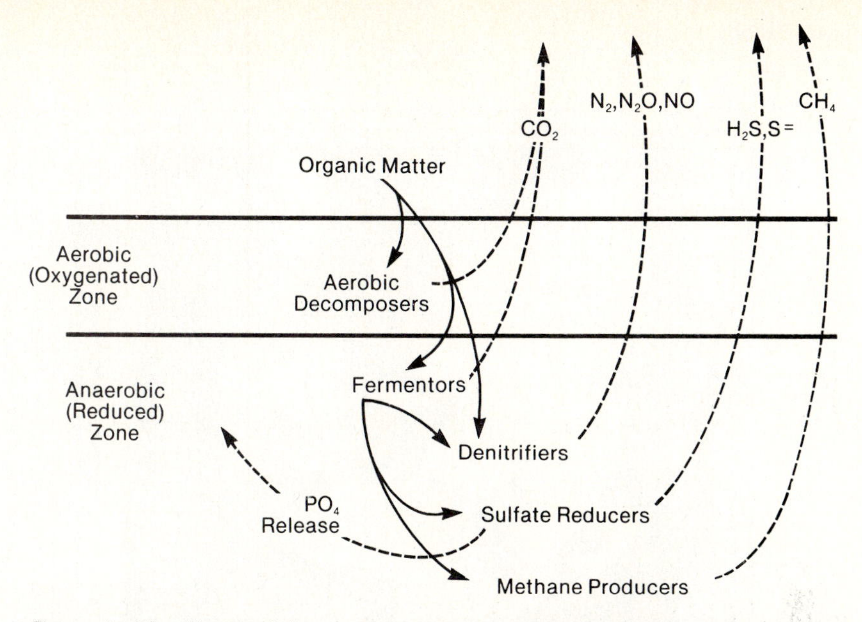

Figure A–13 Microbial decomposition and recycling in wetland sediments. The four major anaerobic decomposers gasify, and thereby recycle into the atmosphere, carbon, nitrogen, and sulfur. Phosphate is also converted from insoluble sulfide forms to soluble forms that are again available to organisms. (After E. P. Odum, 1979.)

Besides the wind-driven currents on the surface, deeper currents result from variations in temperature and salinity, which create differences in density. The interaction of wind stress, Coriolis force, thermohaline currents, and the physical configuration of the basin is very complex and is the focus of extensive research in physical oceanography. So effective is the circulation that oxygen depletion or "stagnation," such as often occurs in freshwater lakes, is comparatively rare in the ocean depths.

5. The sea is dominated by waves of many kinds and by tides produced by the pull of moon and sun. Tides are especially important in the shoreward zones where marine life is often especially varied and dense. Tides are chiefly responsible for the marked periodicities in these communities. Intertidal animals have "lunar-day" rather than "sun-day" biological clocks and entrain their activities with tidal periodicity. Since tides have a periodicity of about 12½ hours, high tides occur in most localities twice daily, being about 50 minutes later on successive days. Every two weeks, when sun and moon are "working together," the amplitude of tides is increased—the so-called **spring** tides, when high tides are very high and low tides very low. Midway in the fortnightly periods, the range between low and high tide is smallest—the so-called **neap tides,** when sun and moon tend to cancel one another. The tidal range varies from less than 1 foot in the open sea to 50 feet in certain narrow or semienclosed bays. Many factors modify tides so that tidal patterns vary in different places over the world. The very first thing a marine ecologist does when working in the coastal zone is to consult the local tide tables.

6. The sea is salty. The average salinity or salt content is 35 parts of salt by weight per 1000 parts of water, or 3.5 percent. This is usually written 35‰ (= parts per

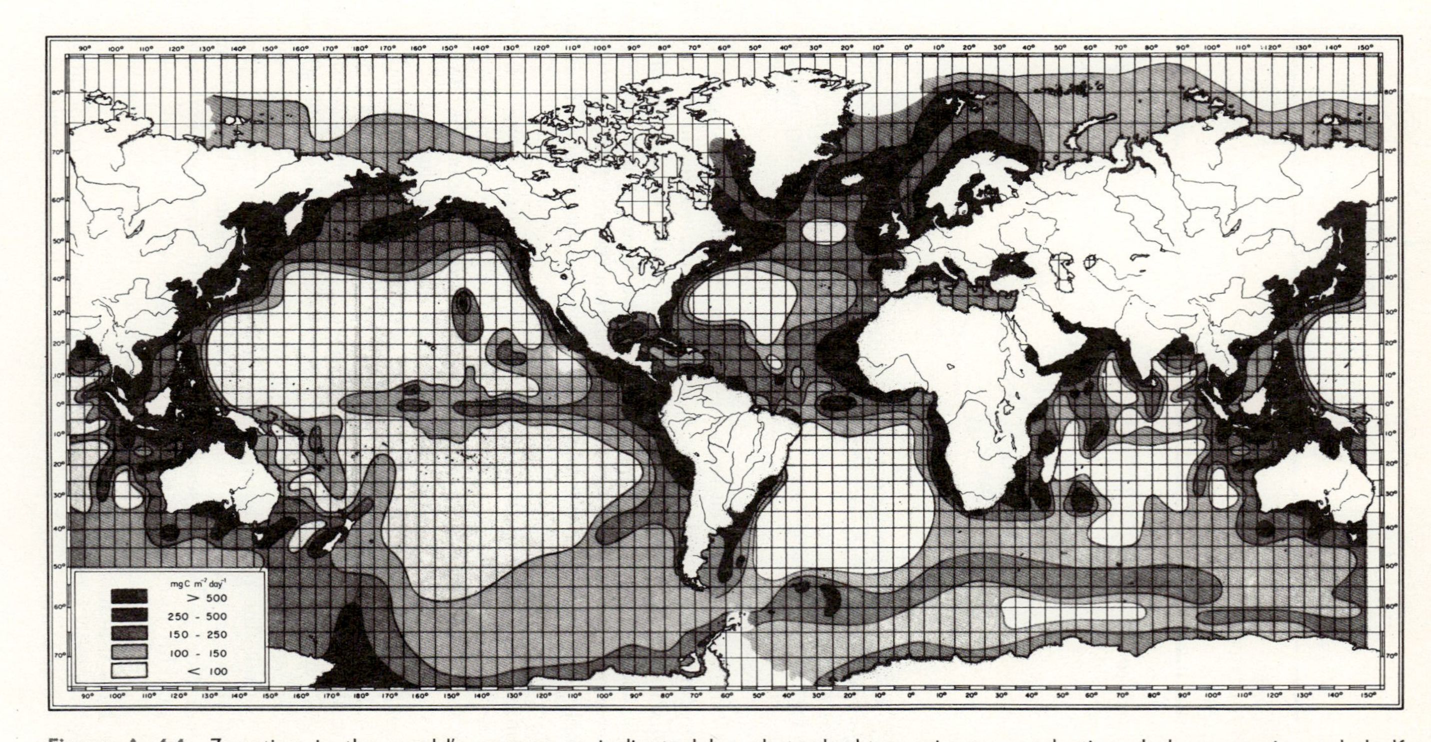

Figure A–14 Zonation in the world's oceans as indicated by phytoplankton primary production. Inshore continental shelf waters and shallow waters around islands have a productivity many times greater than that of the open oceans. The darkest areas on the map support the densest marine life and the most productive fisheries. Some of these coastal zones benefit from upwelling (see Figure A–17). (From Atlas of the Living Resources of the Seas, FAO Fisheries Circular 126, Rev. 1, 1972. Rome, FAO.)

thousand; recall that salinity of fresh water is less than 0.5‰). About 27‰ is sodium chloride; most of the rest consists of magnesium, calcium, and potassium salts. Since the salts dissociate into ions, the best way to picture the chemistry of the sea is as shown in Table A–1. Since the proportion of the radicals remains virtually constant, total salinity may be computed by determining the chloride content (which is easier than determining total salinity). Thus, 19‰ chlorinity approximates 35‰ salinity.

7. The sea is alkaline, strongly buffered but often low in vital nutrients. The electrical dissociation force of the cations exceeds that of the anions (by about 2.4 milliequivalents), which accounts for the alkaline nature of sea water (normally pH = 8.2). Sea water is also strongly buffered (resistant to pH change). Besides the ions listed previously, sea water, of course, contains numerous other elements, theoretically all known ones. Unfortunately, the sea contains increasing amounts of toxic chemicals of industrial origin. Biogenic ions, however, are often in such low concentration that they limit primary production. All these other ions constitute less than 1 percent of the ocean's salinity. As would be expected, the residence time of salts is much longer than for the water itself [estimated by Weyl (1970) to be on the order of 10^7 and 10^4 years, respectively].

Zonation in the sea and something of the complex nature of the ocean floor are shown in Figure A–15. Generally, a continental shelf extends for a distance offshore; beyond that shelf, the bottom drops off steeply as the continental slope then levels off somewhat (the continental rise) before dropping down to a deeper, but more level, plain. The shallow water zone on the continental shelf is the **neritic** ("near shore") zone, and the shore area between high and low tides is the **intertidal** zone (also called the littoral zone). The region of the open ocean beyond the continental shelf is designated the **oceanic** region; the region of the continental slope and rise is the **bathyal** zone, which may be "geologically active" with trenches and canyons subject to underwater erosion and avalanches. The area of the ocean "deeps," or the **abyssal** region, may lie anywhere from 2000 to 5000 meters down. Trenches may drop below 6000 meters (these very deep areas are sometimes known as the **hadal** zone). The abyssal may be the world's largest ecological unit. It is, of course, a heterotrophic ecosystem because the primary energy source lies far above it (except for geothermal rift areas as described on page 34). Midocean ridges are presumed to have been left over after the continents, which were once closer together, drifted apart (see page 477).

Table A–1 The Chemistry of the Sea*

Positive Ions (Cations)		Negative Ions (Anions)	
Sodium	10.7	Chloride	19.3
Magnesium	1.3	Sulfate	2.7
Calcium	0.4	Bicarbonate	0.1
Potassium	0.4	Carbonate	0.007
		Bromide	0.07

* In parts per thousand = grams per kilogram.

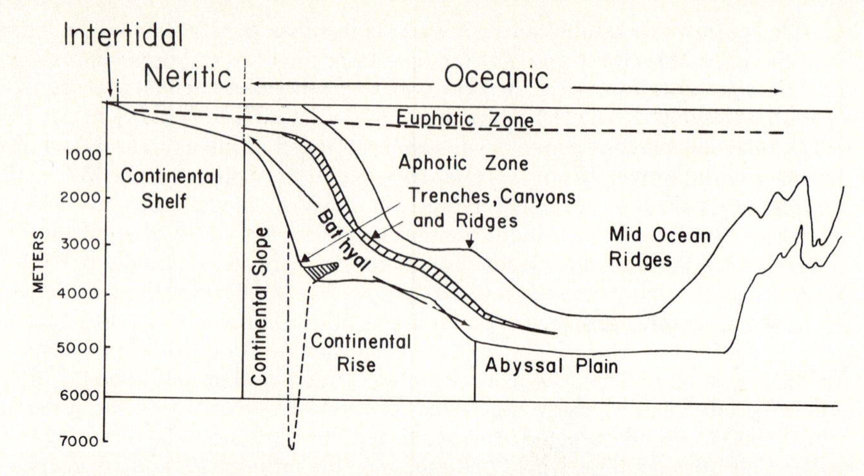

Figure A–15 Horizontal and vertical zonation in the sea. Certain geological features of the ocean floor, such as trenches (which may go down below 6000 meters), canyons, ridges, the abyssal plain, and midocean ridges that rise like great mountain peaks, are also shown in a transect across the western Atlantic. (Diagram based on Heezen, Tarp, and Ewing, 1959.)

The degree of light penetration divides the sea into two horizontal zones: an upper, thin **euphotic** zone of primary production and a lower, vastly thicker **aphotic** zone without sufficient light for photosynthesis. The euphotic zone goes deeper in the clear waters of the oceanic zone, perhaps down 100 to 200 meters, than it does in more turbid (and richer) coastal waters, where effective light penetration rarely exceeds 30 meters.

Oceanic Regions

Compared with inshore waters and estuaries, much of the open ocean is a marine desert because of low concentration of nutrients in the photic zone (see Figure 3–7). Arctic and antarctic seas are more productive than midlatitude seas, as indicated by the larger number of fish and whales to be found in the polar regions.

Until recently, the diatoms, dinoflagellates, and small crustaceans that compose the macroplankton (also called **net plankton** because they are efficiently sampled by a fine mesh silk or nylon plankton net) were assumed to be the primary energy sources in marine food chains. However, recent work points to the **microplankton** or **nannoplankton** (from **nanno,** "dwarf," too small to be collected with nets), both autotrophic and heterotrophic, as the major base for marine food webs. These include tiny green and colorless flagellates and numerous kinds of bacteria (see Pomeroy, 1974). Detritus, both particulate organic matter (POM) and dissolved organic matter (DOM), as in most other ecosystem types, provides the link between autotrophs and heterotrophs. In the sea, however, not only may POM become DOM, but DOM is known to be converted to POM by wave action, formation of fecal pellets, and other mechanisms not fully understood. The aggregates and bacteria that colonize them appear to provide an important source of food for numerous kinds of filter-feeders.

The surprisingly high diversity of the deep sea fauna has been noted. Two theories to explain it were discussed in Chapter 7, Section 6. Deep sea fish are a curious lot; some produce their own light ("lantern fish"); others have a luminous-tipped movable spine that is used as bait to attract prey ("angler fish"); many have enormous mouths and can swallow prey larger than themselves ("viperfish," "gulpers"). Meals are few and far between in the dark depths, but fish are adapted to make the best of their opportunities.

Continental Shelf Region

Marine life is concentrated near shore where nutrient conditions are favorable. No other area has such a variety of life, not even the tropical rain forests. The inshore zooplankton is enriched with many meroplankton (temporary or seasonal) consisting of pelagic larvae of benthic organisms (crabs, marine worms, mollusks, and so on), in sharp contrast to freshwater and the open ocean where most of the floating life is holoplankton (entire life cycle planktonic). Pelagic larvae have been shown to have a remarkable ability to locate the kind of bottom suitable for survival as sedentary adults. When ready to metamorphose, larvae do not settle at random, but settle only in response to particular chemical conditions of the substrate. The benthos has two vertical components: (1) the epifauna, organisms living on the surface either attached or moving freely on the surface, and (2) the infauna, which dig into the substrate or construct a tube and burrows. Benthic aggregations worldwide occur in what Thorson (1955) has called "parallel level bottom communities" dominated by ecologically equivalent species often of the same genus.

The great commercial fisheries of the world are almost entirely located on or near the continental shelf, especially in regions of cold water upwelling (see subsequent section and Figure A–17). A relatively few species make up the bulk of the commercial fishes, including anchovelas, herring, cod, mackerel, pollock, pilchard, flatfishes (flounders, halibut), salmon, and tuna. Most marine fishery biologists believe that world catch has peaked and many areas are now overfished. Fishing, especially involving long-distance trawling or seining, is energy expensive. Increasing food production from the sea may well depend on mariculture (as food or "fish farming" in enclosures in bays and estuaries).

Because of their accessibility and richness of life, seashores are the most studied part of the continental shelf region. No biologist, not to mention the amateur naturalist, considers his or her education complete without study of the seashore. As on mountains, communities in the intertidal zone are arranged in distinct zones. Some aspects of this zonation on the two contrasting seashores, the sandy beach and the rocky shore, are shown in Figure A–16. Ecologists who study community organization on the densely populated intertidal zone are impressed with the role of competition and predation (see Chapter 5, Section 1).

The physical energy level of breaking waves, surf, and tides is a major input factor to which organisms must adapt. A low-energy coast with gentle water flow will be populated by more and different species than will a high-energy coast subject to strong waves.

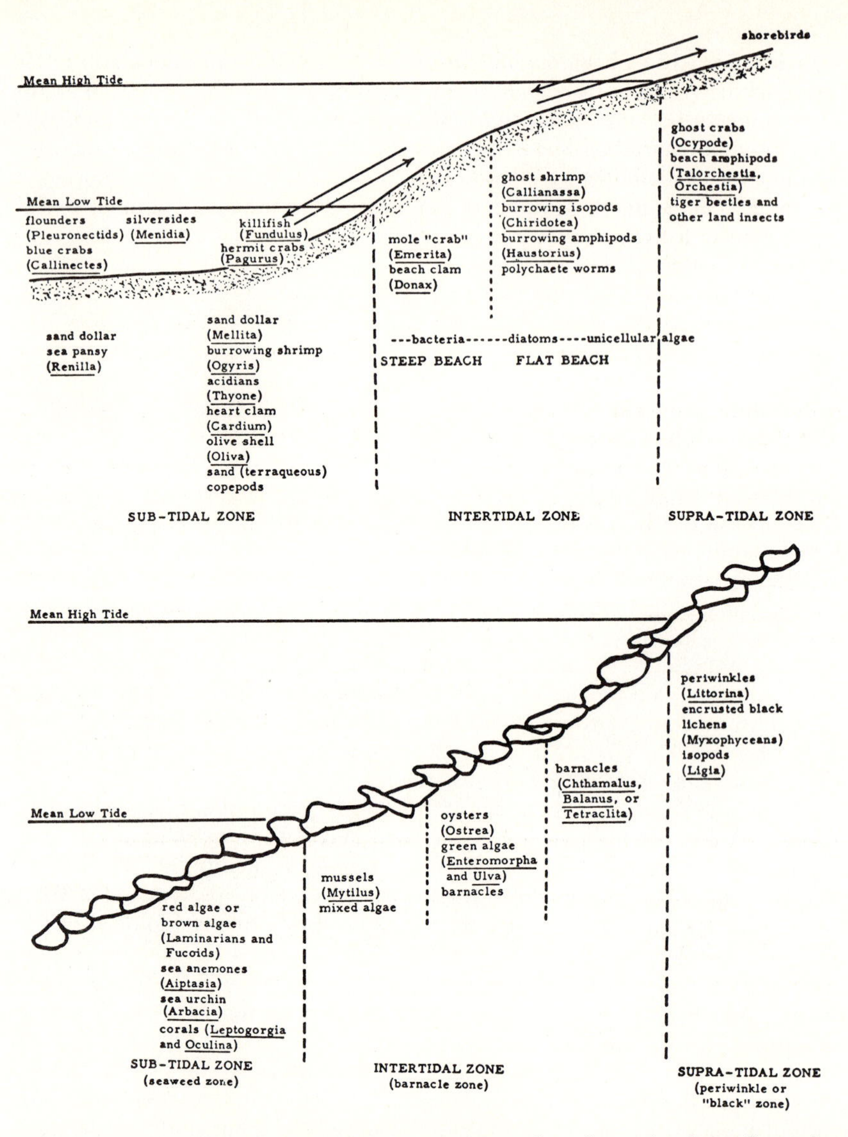

Figure A–16 Transects of a sandy beach (upper) and a rocky shore (lower) at Beaufort, North Carolina, showing zones and characteristic dominants. (Upper diagram based on data of Pearse, Humm, and Wharton, 1942; lower based on data of Stephenson and Stephenson, 1952.)

Upwelling Regions

An important process called **upwelling** occurs where winds consistently move surface water away from precipitous coastal slopes, bringing to the surface cold water rich in nutrients that have been accumulating in the depths. Upwelling creates the most productive of all marine ecosystems that support large fisheries. They are located largely on western coasts, as shown in Figure A–17. Besides fish, upwelling supports large populations of seabirds that deposit countless tons of nitrate- and phosphate-rich guano on coastal shores and islands.

Some characteristic features of the upwelling biomes are as follows:

1. There is a high concentration of nutrients and organisms; pelagic rather than demersal (bottom) fish are dominant.
2. The immense fish (and bird) populations can be attributed not only to high productivity but also to short food chains. Some species of crustacea and fish that are carnivorous in the oceanic region become herbivorous in upwelling regions. Diatoms and clupeid fish dominate the short food chain.
3. Sediment deposited on the sea floor has high organic content and characteristic accretions of phosphate.
4. In contrast to the richness of the sea, adjacent land area is often coastal desert, since, to have upwelling, winds must blow from land to sea (carrying away moisture from the land). Frequent fog, however, may support some vegetation.

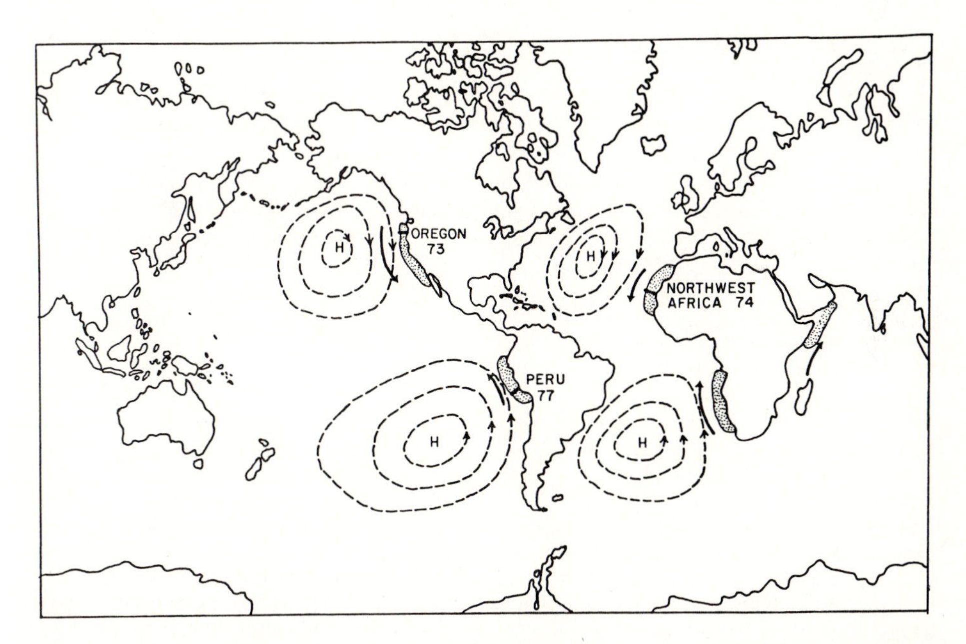

Figure A–17 The five most productive coastal upwelling zones (stippled areas) and the sea-level atmospheric pressure systems (anticyclones) that influence them. Arrows show mean position of major currents. The potential fish yields in millions of tons annually are as follows: Peru-Chile area, 12; Southwest Africa, 5; Oregon-California, 5; Northwest Africa and Arabian Sea, 4 each. Differences are due largely to number of months in the year that effective upwelling occurs, for example, 12 months in the Peru area but only 6 months in the Arabian Sea. (Map reproduced with permission of Richard T. Barber.)

5. The productivity "boom" is subject to periodic "busts" when winds shift, when anoxic conditions develop, or when poisonous dinoflagellate blooms (red tides) occur. Fish die, and fisheries collapse. This catastrophe has happened several times in the Peruvian upwelling zone, generally rated the world's most productive fishery. It is difficult for both humans and nature to maintain very high rates of production continually; cyclic behavior may be a basic property of eutrophicated systems (including industrialized agriculture).

Estuaries (Figure A–18)

An estuary (from *aestus*, "tide") is a semienclosed coastal body of water that has a free connection with the open sea. An estuary is thus strongly affected by tidal action, and within it sea water is mixed (and usually measurably diluted) with fresh water from land drainage. River mouths, coastal bays, tidal marshes, and bodies of water behind barrier beaches are examples. Estuaries could be considered transition zones or ecotones between the freshwater and marine habitats, but many of their most important physical and biological attributes are not transitional, but unique. Furthermore, uses and abuses of this zone by humans are becoming so critical that the unique features of estuaries must become widely understood.

From the geomorphological standpoint, it is convenient to consider five types of estuaries as follows:

1. Drowned river valleys are most extensively developed along coastlines with relatively low and wide coastal plains. Chesapeake Bay, on the mid-Atlantic coast of the United States, is a good example.
2. Fjord-type estuaries are deep, U-shaped coastal indentures gouged out by glaciers and generally having a shallow sill at their mouths formed by terminal glacial deposits. The famous fjords of Norway and similar ones along the coast of British Columbia and Alaska are good examples.
3. Bar-built or barrier island estuaries are shallow basins, often partly exposed at low tide, enclosed by a chain of offshore bars or barriers islands, broken at intervals by

Figure A–18 A barrier island type of estuary in Georgia (Wassaw Island). *A*. Looking in from the sea toward the barrier beach, a series of dunes—the older ones covered with mature forest—and the wide band of salt marsh estuaries that lie between the outer barriers and the mainland. (Photo by Floyd Jillson, Atlanta Journal and Constitution Magazine.) *B*. Close-in view of the *Spartina* salt marsh, showing the tall grass, numerous tidal creeks and sounds, and the algal-rich mud flats (small one shown in center foreground). Note also the first stage in the formation of *Spartina* detritus (right foreground) that eventually will nourish many square miles of water. (Photo courtesy of U.S. Forest Service.)

The upper photo emphasizes the interdependent sequence of sea, beach, sea island, and estuary. The salt marsh estuary is a sediment and nutrient trap that constantly adjusts to currents and sediment loads; without it the beautiful outer white sand beaches would be mudded and eroded. The barriers are constantly changing, eroding here (as on the north end of the island or right side of the picture) and building up there (as in the left foreground of the picture). Without vegetative protection on the dunes the rate of erosion can easily exceed the rate of formation of new beaches.

Figure A–18

inlets (thus ensuring "a free connection with the sea"). Sometimes the sand bars are deposited offshore, but in other instances the barriers may represent former coastal dunes that have become isolated by recent gradual rises in sea levels. The "sounds" behind the "outer banks" of North Carolina (Cape Hatteras National Seashore Park) and the salt marsh estuaries inshore from the South Carolina, Georgia, and Gulf Coast "sea islands" are well-studied examples of the bar-built type of estuary.

4. Estuaries produced by tectonic processes are coastal indentures formed by geological faulting or by local subsidence, often with a large inflow of fresh water. San Francisco Bay is a good example of this type of estuary.

5. River delta estuaries are found at mouths of large rivers such as the Mississippi or the Nile. In this situation, semienclosed bays, channels, and brackish marshes are formed by shifting silt deposits.

Where river flow is strongly dominant over tidal action, a stratified "salt-wedge" circulation pattern develops. Fresh water tends to overflow heavier salt water, which therefore forms a wedge extending along the bottom for a considerable distance upstream. Where freshwater outflow and tidal inflow are more nearly equal, the dominant mixing agent is turbulence, and a partially mixed, moderately stratified low-salinity estuary results. Finally, where tidal action is strongly dominant and vigorous, the water tends to be well mixed from top to bottom, and the salinity approaches that of the open sea.

Important properties of estuaries that are also reasons for their preservation are as follows:

1. Estuaries are generally productive because of water flow subsidies and abundant nutrients. Vertical mixing and turbulence create a sort of nutrient trap so that estuarine plants have access to more phosphorus and other nutrients than are present in adjacent freshwater or marine habitats.

2. Three types of autotrophs provide a variety of food resources for heterotrophs the year around: macrophytes (marsh grasses, seaweeds, sea grasses), benthic algae on and in the sediments (including diatoms that live in tubes and move up and down in response to light and tides), and phytoplankton.

3. Estuaries are nursery grounds for important seafood species. Oysters and edible crabs remain in estuaries throughout their life cycle, and several kinds of shrimp and many commercial and sport fish that are harvested offshore spend their early life in estuaries where abundant food and protection from predators enhance survival and rapid growth at the critical stages in the life history.

The mutualism between fiddler crabs and their plant food source was noted on page 142. The estuary provides a good example of market failure when it comes to economic evaluation, as discussed on pages 172 to 174.

When alterations and construction in estuaries are considered, two precautions should be kept in mind. First, avoid disrupting the normal water-circulation pattern. For example, roads should be built on pilings and bridges, not on earth fills that block water flow. Second, avoid disturbing the shallow-water production zones such as reefs, seagrass beds, and marshes, which are the nursery grounds for valuable seafood. Too often in the past, alterations have been shortsighted and unnecessarily harmful (see W. E. Odum, 1970).

APPENDIX BIBLIOGRAPHY

Major Ecosystem Types of the World

These references are not individually cited in Appendix accounts except as indicated by an asterisk (*).

Biogeography

Daubenmire, Rexford. 1978. *Plant Geography, with Special Reference to North America.* New York, Academic Press.

Good, R. D. 1964. *The Geography of Flowering Plants* (3rd ed.). London, Longmans, 518 pp.

Hunt, Charles B. 1973. *Natural Regions of the United States and Canada.* San Francisco, W. H. Freeman.

Neill, Wilfred T. 1969. *The Geography of Life.* New York, Columbia University Press, 480 pp.

Pielou, E. C. 1979. *Biogeography.* New York, Wiley-Interscience, 351 pp.

Polunin, N. 1960. *Introduction to Plant Geography.* New York, McGraw-Hill, 640 pp.

Udvardy, M. F. D. 1969. *Dynamic Zoogeography, with Special Reference to Land Animals.* New York, Van Nostrand Reinhold, 446 pp.

The Biosphere and International Biological Program (IBP)

* Blair, W. F. 1977. *Big Biology: The US/IBP.* Stroudsburg, Pa., Dowden, Hutchinson and Ross.

Botkin, Daniel B. 1977. Bits, bytes, and IBP. BioScience 27:385–386.

Flessa, Karl W. 1980. Biological effects of plate tectonics and continental drift. BioScience 30:518–523.

Hallam, A. 1972. Continental drift and the fossil record. Sci. Am. 226(4):42–52.

Hutchinson, G. E. (ed.). 1970. The Biosphere. Special issue of Scientific American (Vol. 223, No. 3). (Also published in book form by W. H. Freeman & Co., San Francisco.)

Lieth, H., and R. H. Whittaker (eds.). 1975. *Primary Productivity of the Biosphere.* Berlin and New York, Springer-Verlag.

* Loucks, Orie L. 1983. The U.S. IBP, in perspective after 10 years. In: *Ecosystem Theory and Application,* C.-A. Knox, ed. New York, John Wiley & Sons.

Terrestrial Biomes—General

Carpenter, J. R. 1939. The biome. Am. Midl. Nat. 21:75–91.

Holdridge, L. R. 1947. Determination of world plant formations from simple climatic data. Science 105:267–368.

Jordan, Carl F. 1971. A world pattern of plant energetics. Am. Sci. 59:425–433.

Merriam, C. Hart. 1894. Laws of temperature control of the geographic distribution of terrestrial animals and plants. National Geographic 6:229–238. (See also Life Zones and Crop Zones, U.S.D.A. Bull. No. 10, 1899.)

Odum, Eugene P. 1945. The concept of the biome as applied to the distribution of North American birds. Wilson Bull. 57:191–201.

Pitelka, Frank A. 1941. Distribution of birds in relation to major biotic communities. Am. Midl. Nat. 25:113–137.

Riley, Dennis, and Anthony Young. 1968. *World Vegetation.* Cambridge, England, Cambridge University Press.

Shelford, V. E. 1963. *The Ecology of North America.* Urbana, University of Illinois Press.

Walter, Heinrich, and Elgene Box. 1976. Global classification of natural terrestrial ecosystems. Vegetatio 32:75–81.

Tundra

Bliss, L. C. 1971. Arctic and alpine plant life cycles. Ann. Rev. Ecol. Syst. 2:405–438.

Brown, Jerry, P. C. Miller, L. L. Tiezen, and F. L. Bunnell (eds.). 1981. *An Arctic Ecosystem: The Coastal Tundra at Barrow, Alaska.* New York, Academic Press.

Johnson, P. L. 1969. Arctic plants, ecosystems and strategies. Arctic 22:341–355.

Northern Coniferous Forest

Denison, W. C. 1973. Life in tall trees. Sci. Am. 228(6):74–80.

Edmonds, Robert L. (ed.). 1981. *Analysis of Coniferous Forest Ecosystems in the Western United States.* New York, Academic Press.

Shelford, V. E., and Sigurd Olson. 1935. Sere, climax and influent animals with special reference to the transcontinental coniferous forest of North America. Ecology 16:375–402.

Waring, R. H., and J. F. Franklin. 1979. The evergreen coniferous forests of the Pacific Northwest. Science 204:1380–1386.

Temperate Deciduous Forest

Braun, E. Lucy. 1950. *Deciduous Forests of Eastern North America.* Philadelphia, Blakiston Co.

McCormick, Jack. 1966. *The Life of the Forest* (Our Living World of Nature Series). New York, McGraw-Hill.

Reichle, D. E. (ed.). 1970. *Analysis of Temperate Forest Ecosystems.* Berlin and New York, Springer-Verlag.

Temperate Grassland

Allen, Durward L. 1967. *The Life of Prairies and Plains* (Our Living World of Nature Series). New York, McGraw-Hill.

Carpenter, J. R. 1940. The grassland biome. Ecol. Monogr. 10:617–684.

Clements, F. E., and V. E. Shelford. 1939. The North American grassland. In: *Bio-Ecology,* F. E. Clements and V. E. Shelford. New York, John Wiley & Sons.

Hanson, H. D. 1950. Ecology of the grassland. II. Bot. Rev. 16:283–360.

Love, R. M. 1970. The rangelands of the western United States. Sci. Am. 272(2):88–96.

* Morello, Jorge. 1970. Modelo de relaciones entra pastizales y lenosas colonzodoras en el Chaco-Argentino. (A model of relationship between grassland and wood colonizer plank in the Argentine Chaca.) Idia 276:31–51, Dec. 1970.

Weaver, J. E., and F. W. Albertson. 1956. *Grasslands of the Great Plains: Their Nature and Use.* Lincoln, Neb., Johnsen Publ. Co.

Tropical Grassland and Savanna

Bell, Richard H. V. 1971. A grazing ecosystem in the Serengeti. Sci. Am. 225(1):86–93.

Bourliere, Francois, and Malcolm Hadley. 1970. The ecology of tropical savannas. Ann. Rev. Ecol. Syst. 1:125–152.

Sinclair, A. R. E., and M. Norton-Griffiths (eds). 1979. *Serengeti: Dynamics of an Ecosystem.* Chicago, University of Chicago Press.

Chaparral

Castri, F. di, and H. A. Mooney (eds). 1973. *Mediterranean Type Ecosystems.* Ecol. Studies, Vol. 7. Berlin and New York, Springer-Verlag.

Cogswell, Howard L. 1947. Chaparral country. Audubon 49:75–81.

Cooper, William S. 1922. The broad-sclerophyll vegetation of California: An ecological study of the chaparral and its related communities. Washington, D.C., Carnegie Inst., 319:1–124.

Desert

Gore, Rick. 1979. The desert; an age-old challenge grows. Nat. Georg. 156:594–639.

McGinnis, W. G., B. J. Goldman, and P. Paylore (eds.). 1969. *Deserts of the World.* Tucson, University of Arizona Press.

Niering, W. A., R. H. Whittaker and C. H. Lowe. 1963. The saguaro: A population in relation to environment. Science 142:15–23.

Noy-Meir, Imanuel. 1973. Desert ecosystems: Environment and producers. Ann. Rev. Ecol. Syst. 4:25–51.

______. 1974. Desert ecosystems: Higher trophic levels. Ann. Rev. Ecol. Syst. 5:195–214.

Shantz, H. L. 1942. The desert vegetation of North America. Bot. Rev. 8:195–246.

Sutton, Anne, and Myron Sutton. 1966. *The Life of the Desert* (Our Living World of Nature Series). New York, McGraw-Hill.

* Walter, H. 1954. Le facteur eau dans les regions arides et sa signification pour l'organisation de la vegetation dans les contrees sub-tropicales. In: *Les Divisions Ecologiques du Monde.* Paris, Centre Nationale de la Recherche Scientifique, pp. 27–39.

Tropical Forests

Golley, F. B., J. T. McGinnis, R. G. Clements, G. I. Child, and M. J. Duever. 1969. The structure of tropical forests in Panama and Columbia. BioScience 19:693–696; 697–700.

Golley, F. B., and E. Medina (eds.). 1975. *Tropical Ecological Systems.* Ecol. Studies No. 11. Berlin and New York, Springer-Verlag.

Gomez-Pompa, A., C. Vasquez-Yanes, and S. Guaria. 1972. The tropical rain forest, a non-renewable resource. Science 177:762–765.

* Jordan, Carl F. 1982. Amazon rain forests. Am. Sci. 79:394–401.

National Academy of Science. 1980. Conversion of Tropical Forests. Committee Report, Norman Meyers, Chairman. Washington, D.C., National Academy Press.

Richards, P. W. 1952. *The Tropical Rain Forest.* New York, Cambridge University Press.

______. 1973. The tropical rain forest. Sci. Am. 229(6):58–67.

Aquatic Ecosystems—General

Barnes, R. S. K., and K. H. Mann (eds.). 1980. *Fundamentals of Aquatic Ecosystems.* Oxford, Blackwell, 229 pp.

Mann, K. H. 1969. The dynamics of aquatic ecosystems. In: *Advances in Ecological Resarch,* J. B. Cragg, ed. Vol. 6. New York, Academic Press, pp. 1–81.

Russell-Hunter, W. D. 1970. *Aquatic Productivity: An Introduction to Some Basic Aspects of Biological Oceanography and Limnology.* New York and Toronto, Macmillan.

Freshwater Ecosystems—General

Coker, Robert E. 1954. *Streams, Lakes, Ponds.* Chapel Hill, University of North Carolina Press.

Gerking, S. C. (ed.). 1967. *The Biological Basis of Freshwater Fish Production.* Oxford, Blackwell, 495 pp.

Gibbs, R. J. 1970. Mechanisms controlling water chemistry. Science 170:1088–1090.

Haynes, H. B. N. 1969. *The Biology of Polluted Waters.* Liverpool, England, Liverpool University Press.

* Livingstone, D. A. 1963. Chemical composition of rivers and lakes. U.S. Geological Survey Prof. Paper 440-G.

Porter, Karen. 1977. The plant-animal interface in freshwater ecosystems. Am. Sci. 65:159–170.

Lentic Ecosystems

Beeton, A. M. 1969. Changes in the environment and biota of the Great Lakes. In: *Eutrophication: Causes, Consequences and Corrections.* Washington, D.C., National Academy of Science, pp. 150–187.

Bennett, G. W. 1962. *Management of Artificial Lakes and Ponds.* New York, Reinhold.

Brinkhurst, R. O. 1974. *The Benthos of Lakes.* London and New York, Macmillan.

Brylinsky, M., and K. H. Mann. 1973. An analysis of factors governing productivity in lakes and reservoirs. Limnol. Oceanogr. 18:1–14.

Deevy, Edward S., Jr. 1951. Life in the depths of a pond. Sci. Am. 185:68–72.

Lowe-McConnell, R. H. (ed.). 1966. *Man-Made Lakes.* New York, Academic Press.
Ragotzkie, Robert A. 1974. The Great Lakes rediscovered. Am. Sci. 62:454–464.
Richey, J. E., et al. 1978. Carbon flow in four lake ecosystems: A structural approach. Science 202:1183–1186.

Lotic Systems

* Allen, K. R. 1951. The Horokiwi stream: A study of a trout population. New Zealand Mar. Dept. Fish Bull. No. 10.
Cummins, Kenneth W. 1974. Structure and function of stream ecosystems. BioScience 24:631–641.
Fisher, Stuart G., and Gene E. Likens. 1972. Stream ecosystem: Organic energy budget. BioScience 22:33–35.
* Horton, P. A. 1961. The bionomics of brown trout in a Dartmoor stream. J. Anim. Ecol. 30:331–338.
Hynes, H. B. N. 1970. *The Ecology of Running Waters.* Toronto, University of Toronto Press.
* Mann, K. H. 1964. The pattern of energy flow in the fish and invertebrate fauna of the River Thames. Vert. Int. Ver. Limnol. 15:485–495. (See also J. Am. Ecol. 34:253–275, 1965.)
Patrick, Ruth. 1970. Benthic stream communities. Am. Sci. 58:546–549.
Vannote, R. L., G. W. Menshall, K. W. Cummins, J. R. Sedell, and C. E. Cushing. 1980. The river continuum concept. Can. J. Fish. Aquatic Sci. 37:130–137.
Whitten, B. A. (ed.). 1975. *River Ecology.* Two vols. Berkeley, University of California Press.

Wetlands

* Armentano, Thomas V. 1980. Drainage of organic soils as a factor in the world carbon cycle. BioScience 30:825–830.
Brinson, Mark M., Ariel E. Lugo, and Sandra Brown. 1981. Primary productivity, decomposition and consumer activity in freshwater wetlands. Ann. Rev. Ecol. Syst. 12:123–162.
Cowardin, L. M., Virginia Carter, and F. C. Golet, 1979. Classification of wetlands and deepwater habitats of the United States. FWS/OBS-79/31. Washington, D.C., Fish and Wildlife Service, U.S. Dept. of Interior.
Good, Ralph E., D. F. Whigham, and R. L. Simpson, 1978. *Freshwater Wetlands: Ecological Processes and Management Potential.* New York, Academic Press.
* Greeson, Phillip E., John R. Clark, and Judith E. Clark (eds.). 1979. *Wetland Functions and Values: The State of Our Understanding.* Minneapolis, American Water Resources Assoc.
* Odum, E. P. 1979. The value of wetlands; a hierarchical approach. In: *Wetland Functions and Values,* Greeson, Clark, and Clark, eds. Minneapolis, American Water Resources Assoc.
Niering, William A. 1966. *The Life of the Marsh* (Our Living World of Nature Series). New York, McGraw-Hill.
Teal, John, and Mildred Teal. 1969. *Life and Death of the Salt Marsh.* Boston, Little, Brown.
Wharton, C. H., W. M. Kitchens, E. C. Pendleton, and T. W. Sipe. 1982. The ecology of bottomland hardwood swamps of the Southeast: A community profile. FWS/OBS-81/37. Washington, D.C., U.S. Fish and Wildlife Service, Biol. Serv. Program.

The Oceans—General

Berrill, N. J. 1966. *The Ocean* (Our Living World of Nature Series). New York, McGraw-Hill.
Coker, Robert E. 1947. *This Great and Wide Sea.* Chapel Hill, University of North Carolina Press.
Cushing, D. H., and J. J. Walsh (eds.). 1976. *The Ecology of the Seas.* Philadelphia, W. B. Saunders.
Falkowski, P. G. (ed.). 1980. *Primary Productivity in the Sea.* New York, Plenum Press. (See review in Science 212:794, 1981.)
Goldreich, Peter. 1972. Tides and the earth-moon system. Sci. Am. 226(4):42–52.
Heezen, Bruce C., M. Tarp, and M. Ewing. 1959. The floors of the ocean. I. North Atlantic. Geol. Soc. Amer. Spec. Paper 65, 122 pp.
MacIntyre, Ferren. 1970. Why the sea is salt. Sci. Am. 223(5):104–115.
Ravelle, Roger (ed.). 1969. The Oceans. Special issue of Scientific American (Vol. 221, No. 3).
Steele, John H. 1974. *The Structure of Marine Ecosystems.* Cambridge, Mass., Harvard University Press, 128 pp.
Sverdrup, H. U., Martin W. Johnson, and Richard Fleming. 1942. *The Oceans: Their Physics,*

Chemistry, and General Biology. Englewood Cliffs, N.J., Prentice-Hall.
Walsh, J. J. 1972. Implications of a systems approach to oceanography. Science 176:969–975.

Oceanic Region—The Deep Sea

Hardy, A. C. 1957. *The Open Sea: The World of Plankton.* New York, Houghton-Mifflin.
Menzies, R. J., R. Y. George, and G. T. Rowe. 1973. *Abyssal Environment and Ecology of the World Oceans.* New York, John Wiley & Sons, 488 pp.
* Pomeroy, L. R. 1974. The ocean's food web, a changing paradigm. BioScience 24:299–504.
Rex, Michael A. 1981. Community structure in the deep-sea benthos. Ann. Rev. Ecol. Syst. 12:331–354.

Continental Shelf, Coastal Zone, and Seashores

Amos, William H. 1966. *The Life of the Seashore* (Our Living World of Nature Series). New York, McGraw-Hill.
Bahr, L. M., and W. P. Lanier. 1981. The ecology of intertidal oyster reefs of the South Atlantic coast: A community profile. FWS/OBS-81/15. Washington, D.C., U.S. Fish and Wildlife Service, Office of Biological Services, 105 pp.
Barnes, R. S. K. (ed.). 1977. *The Coastline.* New York, Wiley-Interscience, 356 pp.
Carson, Rachel. 1955. *The Edge of the Sea.* New York, Houghton-Mifflin.
Dolan, Robert, Bruce Hayden, and Harry Lins. 1980. Barrier islands. Am. Sci. 65:16–25.
Jackson, T. C., and Diana Reische. 1981. Coast Alert: Scientists Speak Out. San Francisco, Friends of the Earth.
Mann, K. H. 1982. *Ecology of Coastal Waters: A Systems Approach.* Berkeley, University of California Press.
———. 1973. Seaweeds: Their productivity and strategy for growth. Science 182:975–981.
Odum, H. T., B. J. Copeland, and E. A. McMahan. 1974. *Coastal Ecological Systems of the United States.* Washington, D.C., The Conservation Foundation.
Odum, William E., C. C. McIvor, and T. J. Smith III. 1982. The ecology of mangroves of South Florida: A community profile. Washington, D.C., U.S. Fish and Wildlife Services, Office of Biological Services.
Paulson, Alan C. 1980. The coral atoll: An oasis in the desert. Sea Frontiers, Jan.-Feb., pp. 37–43.
Pearse, A. S., H. J. Humm, and G. W. Wharton. 1942. Ecology of sand beaches at Beaufort, N.C. Ecol. Monogr. 12:136–190.
Peterson, C. H., and N. M. Peterson. 1979. The ecology of intertidal flats of North Carolina: A community profile. FWS/OBS-79/39. Washington, D.C., U.S. Fish and Wildlife Service, Office of Biological Services.
Stephenson, T. A., and Anne Stephenson. 1949. The universal features of zonation between tidemarks on rocky coasts. J. Ecol. 37:289–305.
———. 1973. *Life Between Tidemarks on Rocky Shores.* San Francisco, W. H. Freeman.
* Thorson, Gunnar. 1956. Marine level-bottom communities of recent seas, their temperature adaptation, and their "balance" between predators and food animals. Trans. N.Y. Acad. Sci. 18:693–700.

Upwelling Regions

Barber, R. T., and R. L. Smith. 1980. Coastal upwelling ecosystems. In: *Analysis of Marine Ecosystems,* A. R. Longhurst, ed. New York, Academic Press.
Boje, R., and M. Tomczak (ed.). 1978. *Upwelling Ecosystem.* New York, Springer-Verlag.

Estuaries

Cronin, L. Eugene (ed.). 1975. *Estuarine Research.* Two volumes. New York, Academic Press.
Kennedy, Victor S. (ed.). 1980. *Estuarine Perspectives.* New York, Academic Press.
Lauff, George A. (ed.). 1967. Estuaries. Publ. No. 83. Washington, D.C., American Association for the Advancement of Science, 757 pp.
McHugh, J. L. 1966. Management of estuarine fisheries. In: *A Symposium on Estuarine Fisheries.* Washington, D.C., Am. Fish. Soc.
McLusky, Donald. 1981. *The Estuarine Ecosystem.* New York, Halsted Press, John Wiley & Sons.

Odum, Eugene P. 1961. The role of tidal marshes in estuarine production. The Conservationists, June–July, pp. 12–15. Albany, N.Y. Dept. of Conservation.

Odum, E. P. 1980. The status of three ecosystem-level hypotheses regarding salt marsh estuaries. In: *Estuarine Perspectives,* V. S. Kennedy, ed. New York, Academic Press.

* Odum, William E. 1970. Insidious alteration of the estuarine environment. Trans. Am. Fish. Soc. 99:836–847.

Ragotzkie, R. A. 1960. Marine marsh. In: *McGraw-Hill Encyclopedia of Science and Technology.* New York, McGraw-Hill, pp. 217–218.

Turner, R. E. 1977. Intertidal vegetation and commercial yields of penaeid shrimp. Trans. Am. Fish. Soc. 106:411–416.

Warner, William W. 1976. *Beautiful Swimmers; Watermen, Crabs and Chesapeake Bay.* New York, Penguin Books.

BIBLIOGRAPHY

Note: In addition to the references which are actually cited in the text, a number of other books, review papers, and symposium volumes that contain useful bibliographies are included in this list. Selected references pertaining to major ecosystem types are listed separately in the Appendix.

Aaronson, S. 1970. *Experimental Microbial Ecology.* New York, Academic Press, 598 pp.

Abele, L. G., and K. Walters. 1979. The stability-time hypothesis: Reevaluation of the data. Am. Nat. 114:559–568.

Abrahamson, W. G., and M. Gadgil. 1973. Growth form and reproductive effort in goldenrods (Solidago, Compositae). Am. Nat. 107:651–661.

Adams, Charles C. 1935. The relation of general ecology to human ecology. Ecology 16:316–335.

Adkisson, Perry L., G. A. Niles, J. K. Walker, L. S. Bird, and H. B. Scott. 1982. Controlling cotton's insect pests: A new system. Science 216:19–22.

Agreñ, G. I., and B. Axelsson. 1980. Population respiration: A theoretical approach. Ecol. Modelling 11:39–54.

Ahlgren, I. F., and C. E. Ahlgren. 1960. Ecological effects of forest fires. Bot. Rev. 26:483–533.

Albertson, F. W., G. W. Tomanek, and Andrew Riegel. 1957. Ecology of drought cycles and grazing intensity on grasslands of Central Great Plains. Ecol. Monogr. 27:27–44.

Alexander, Martin. 1964. Biochemical ecology of soil micro-organisms. Am. Rev. Microbiol. 18:217–252.

———. 1971. *Microbial Ecology.* New York, John Wiley & Sons, 511 pp.

Alexander, Richard D., and Gerald Borgia. 1978. Group selection, altruism, and the levels of organization of life. Ann. Rev. Ecol. Syst. 9:449–474.

Allee, W. C. 1931. *Animal Aggregations: A Study in General Sociology.* Chicago, University of Chicago Press.

———. 1951. *Cooperation Among Animals with Human Implications.* New York, Schuman. (Revised edition of *Social Life of Animals*, New York, W. W. Norton, 1938.)

Allee, W. C., A. E. Emerson, Orlando Park, Thomas Park, and Karl P. Schmidt. 1949. *Principles of Animal Ecology.* Philadelphia, W. B. Saunders Co., 837 pp.

Allen, T. F. H. and Thomas B. Starr. 1982. *Hierarchy: Perspectives for Ecological Complexity.* Chicago, University of Chicago Press.

American Public Health Association. 1975. *Standard Methods for Examination of Water and Wastewater Including Bottom Sediments and Sludges* (14th ed.). New York, American Public Health Association.

Anagnostakis, Sandra L. 1982. Biological control of chestnut blight. Science 215:466–471.

Anderson, F. R., A. V. Kneese, P. D. Reed, S. Taylor, and R. B. Stevenson. 1977. *Environmental Improvement Through Economic Incentives.* Baltimore, Johns Hopkins Press.

Anderson, G. R. 1955. Nitrogen fixation by pseudomonas-like soil bacteria. J. Bacteriol. 70:129–133.

Anderson, R. M., B. D. Turner, and L. R. Taylor (eds.). 1979. *Population Dynamics.* Oxford, Blackwell.

Andrewartha, H. G. 1961. *Introduction to the Study of Animal Populations.* Chicago, University of Chicago Press, 281 pp.

Andrewartha, H. G., and L. C. Birch. 1954. *The Distribution and Abundance of Animals.* Chicago, University of Chicago Press.

Ardrey, Robert. 1967. *The Territorial Imperative.* New York, Atheneum, 390 pp.

Aruga, Y., and M. Monsi. 1963. Chlorophyll amount as an indicator of matter production in biocommunities. Plant Cell Physiol. 4:29–39.

Ashby, W. R. 1963. *An Introduction to Cybernetics.* New York, John Wiley & Sons, 295 pp.

Auclair, Allan N. 1976. Ecological factors in the development of intensive management ecosystems in the midwestern United States. Ecology 57:431–444.

Auclair, Allan N., and F. G. Goff. 1971. Diversity relations of upland forests in the western Great Lakes area. Am. Nat. 105:499–528.

Auerbach, S. I., D. A. Crossley, and M. D. Engelman. 1957. Effects of gamma radiation on collembola population growth. Science 126:614.

Auerbach, S. I., J. S. Olson, and H. D. Waller. 1964. Landscape investigations using cesium 137. Nature 201:761–764.

Austin, G. T., and C. S. Tomoff. 1978. Relative abundance in bird populations. Am. Nat. 112:695–699.

Axelrod, Robert, and W. D. Hamilton. 1981. The evolution of cooperation. Science 211:1390–1396.

Ayala, F. J. 1968. Genotype, environment and population numbers. Science 162:1453–1459.

———. 1969. Experimental invalidation of the principle of competitive exclusion. Nature 224:1076–1079.

———. 1972. Competition between species. Am. Sci. 60:348–357.

Ayres, Robert V. 1978. *Resources, Environment and Economics. Applications of the Materials/Energy Balance Principle.* New York, Wiley-Interscience, 207 pp.

Ayres, Robert V., and A. V. Kneese. 1971. Economic and ecological effects of a stationary economy. Ann. Rev. Ecol. Syst. 2:1–22.

Azevedo, J., and D. L. Morgan. 1974. Fog precipitation in coastal California forests. Ecology 55:1135–1141.

Baes, C. F., Jr., H. E. Goeller, J. S. Olson, and R. M. Rotty. 1977. Carbon dioxide and climate: The uncontrolled experiment. Am. Sci. 65:310–320.

Bakelaar, R. Gary, and E. P. Odum. 1978. Community and population level responses to fertilization in an old-field ecosystem. Ecology 59:660–665.

Baker, H. G. 1961. The adaptation of flowering plants to nocturnal and crepuscular pollinators. Quart. Rev. Biol. 36:64–73.

———. 1963. Evolutionary mechanisms in pollination biology. Science 139:877–883.

Bakuzis, E. V. 1969. Forestry viewed in an ecosystem perspective. In: *The Ecosystem Concept in Natural Resource Management,* G. Van Dyne, ed. New York, Academic Press, pp. 189–257.

Ballard, R. D., and J. F. Grassle. 1979. Return to oases of the deep; strange world without sun. Nat. Geogr. 156:680–703.

Baltensweiler, W. 1964. *Zeiraphera griseana* Hubner (Lepidoptera: Tortricidae) in the European Alps. A contribution to the problem of cycles. Can. Entomol. 96:792–800.

Bardach, John E. 1968. *Harvest of the Sea.* New York, Harper & Row.

———. 1969. Aquaculture. Science 161:1098–1106.

Barkley, Paul W., and D. W. Seckler. 1972. *Economic Growth and Environmental Decay.* New York, Harcourt Brace Jovanovich.

Barnett, H. W. and C. Morse. 1963. *Scarcity and Growth—The Economics of Natural Resource Availability.* Baltimore, Johns Hopkins University Press.

Barney, G. O. 1980. The Global 2000 Report. See under *Council on Environmental Quality.*

Barr, Terry N. 1981. The world food situation and global grain prospects. Science 214:1087–1095.

Barrett, Gary W. 1969. The effects of an acute insecticide stress on a semienclosed grassland ecosystem. Ecology 49:1019–1035.

Barrett, G. W., and R. Rosenberg (eds.). 1981. *Stress Effects on Natural Ecosystems.* New York, John Wiley & Sons.

Bates, Marston. 1964. *Man in Nature* (2nd ed.). Englewood Cliffs, N.J., Prentice-Hall.

Batie, S. S., and R. G. Healy (eds.). 1980. *The Future of American Agriculture as a Strategic Resource.* Washington, D.C., The Conservation Foundation, 291 pp.

Bator, Francis M. 1958. The anatomy of market failure. Quart. J. Econ. 72:351–379.

Batra, Suzanne W. T. 1982. Biological control in agroecosystems. Science 215:134–139.

Baylor, E. R., and Sutcliffe, W. H., Jr. 1963. Dissolved organic matter in seawater as a source of particulate food. Limnol. Oceanogr. 8:369–371.

Bazzaz, F. A. 1975. Plant species diversity in old-field successional ecosystems in southern Illinois. Ecology 56:485–488.

Beals, E. W. 1969. Vegetational change along altitudinal gradients. Science 165:981–985.

Beck, Stanley D. 1960. Insects and the length of the day. Sci. Am. 202(2):108–118.

Beecher, William J. 1942. Nesting birds and the vegetation substrate. Chicago, Chicago Ornithological Society.

Beery, J. A. 1975. Adaptation of photosynthetic processes to stress. Science 188:644–650.

Beeton, A. M. 1961. Environmental changes in Lake Erie. Trans. Am. Fish. Soc. 90:153–159.

Begon, Michael, and Martin Mortimer. 1981. *Population Ecology: A Unified Study of Animals and Plants.* Sunderland, Mass., Sinauer Associates.

Belt, C. B., Jr. 1975. The 1973 flood and man's constriction of the Mississippi River. Science 189:681–684.

Bennett, George W. 1962. *Management of Artificial Lakes and Ponds.* New York, Reinhold.

Bennett, Ivan L. (Panel Chairman). 1967. The World Food Problem. Report of the Panel on World Food

Supply to President's Science Advisory Committee. Three vols. Washington, D.C., Supt. of Documents.

Benton, G. S., R. T. Blackburn, and V. O. Snead. 1950. The role of the atmosphere in the hydrological cycle. Trans. Am. Geophys. Union 31:61–73.

Bergmann, G. 1944. Holism, historicism and emergence. Philos. Sci. 11:209–221.

Berkner, L. V., and L. C. Marshall. 1964. The history of growth of oxygen in the earth's atmosphere. In: *The Origin and Evolution of Atmospheres and Oceans*, D. J. Brancazio and A. G. W. Cameron, eds. New York, John Wiley & Sons, pp. 102–126.

———. 1965. History of major atmospheric components. Proc. Natl. Acad. Sci. USA 53:1215–1226.

———. 1966. The role of oxygen. Saturday Review, May 7, 1966, pp. 30–33.

Berlinski, D. 1976. *On Systems Analysis*. Cambridge, Mass., MIT Press.

Bernardi, Riccardo (ed.). 1979. Biological and Mathematical Aspects in Population Dynamics. Memorie dell "Instituto di Idrobiologia." Vol. 37 Suppl. Pallanza, Italy.

Bernstein, Brock B. 1981. Ecology and economics: Complex systems in changing environments. Ann. Rev. Ecol. Syst. 12:309–330.

Bertalanffy, Ludwig von. 1950. An outline of general systems theory. Brit. J. Philos. Sci. 1:139–164.

———. 1957. Quantitative laws in metabolism and growth. Quart. Rev. Biol. 32:217–231.

———. 1968. *General Systems Theory: Foundations, Development, Application*. New York, George Braziller, 295 pp. (Revised edition, 1975.)

Best, R. 1962. Production factors in the tropics. In: "Fundamentals of Dry-Matter Production and Distribution." Special issue of Neth. J. Agr. Sci. 10(5):347–353.

Beverton, R. J. H., and S. J. Holt. 1957. On the dynamics of exploited fish populations. Great Brit. Min. Agr. Fish, Food, Fish. Invest. Ser. 2–19:1–533.

Beyers, Robert J. 1962. Relationship between temperature and the metabolism of experimental ecosystems. Science 136:980–982.

———. 1963. The metabolism of twelve aquatic laboratory microecosystems. Ecol. Monogr. 33:281–306.

———. 1964. The microcosm approach to ecosystem biology. Am. Biol. Teacher 26:491–498.

Billings, W. D. 1952. The environment complex in relation to plant growth and distribution. Quart. Rev. Biol. 27:251–265.

———. 1957. Physiological ecology. Ann. Rev. Plant Physiol. 8:375–392.

———. 1968. *Plants, Man and the Ecosystem*. Belmont, Calif., Wadsworth, 154 pp.

Birch, L. C. 1948. The intrinsic rate of natural increase of an insect population. J. Anim. Ecol. 17:15–26.

Birge, E. A. 1915. The heat budgets of American and European lakes. Trans. Wisconsin Acad. Sci. Arts Letters 18:166–213.

Bjorkman, J. 1966. The effect of oxygen concentration on photosynthesis in higher plants. Physiol. Plantarium 19:618–633.

Bjorkman, O. 1973. Comparative studies on photosynthesis in higher plants. In: *Photophysiology*, A. C. Giese, ed. Vol. 8. New York, Academic Press.

Bjorkman, O., and J. Beery. 1973. High efficiency photosynthesis. Sci. Am. 229(4):80–93.

Black, C. A. 1968. *Soil-Plant Relationships* (2nd ed.). New York, John Wiley & Sons, 792 pp.

Black, C. C. 1971. Ecological implications of dividing plants into groups with distinct photosynthetic capacities. Adv. Ecol. Res. 7:87–114.

Black, J. N. 1971. Energy relations in crop production: A preliminary survey. Ann. Appl. Biol. 67:272–278.

Bliss, L. C. 1966. Plant productivity in Alpine microenvironments. Ecol. Monogr. 36:125–155.

Blum, Murray S. 1969. Alarm pheromones. Ann. Rev. Entomol. 14:57–80.

Bodenheimer, F. S. 1937. Population problems of social insects. Biol. Rev. 12:393–430.

Bolin, Bert. 1977. Changes of land biota and their importance for the carbon cycle. Science 196:613–615.

Borgstrom, Georg. 1965. *The Hungry Planet*. New York, Macmillan, 487 pp.

———. 1969. *Too Many: An Ecological Overview of Earth's Limitations*. New York, Collier Books, 400 pp.

———. 1979. Ecological constraints on global food production. In: *Growth without Ecodisasters*, Nicholus Polunin, ed. London, Macmillan Press, pp. 293–323.

Bormann, F. H., and G. E. Likens. 1967. Nutrient cycling. Science 155:424–429.

———. 1979. *Pattern and Process in a Forested Ecosystem*. New York, Springer-Verlag, 253 pp.

Bormann, F. H., G. E. Likens, and J. M. Melillo. 1977. Nitrogen budget for an aggrading northern hardwood ecosystem. Science 196:981–983.

Botkin, D. B., G. M. Woodwell, and Neal Tempel. 1970. Forest productivity estimated from carbon dioxide uptake. Ecology 51:1057–1060.

Botkin, J. W., M. Elmandjra, and M. Malitza. 1980. *No Limits to Learning. Bridging the Human Gap*. A Report to the Club of Rome. New York, Pergamon Press, 159 pp.

Boulding, Kenneth E. 1962. *A Reconstruction of Economics*. New York, Science Editions.

———. 1964. *The Meaning of the 20th Century: The Great Transition.* New York, Harper & Row.

———. 1966. The economics of the coming spaceship earth. In: *Environmental Quality in a Growing Economy.* A Resources for the Future Book. Baltimore, Johns Hopkins Press, 314 pp.

———. 1966a. Economics and ecology. In: *Future Environments of North America,* F. F. Darling and J. P. Milton, eds. Garden City, N.Y., The Natural History Press, pp. 225–234.

———. 1978. *Ecodynamics: A New Theory of Societal Evolution.* Beverly Hills, Calif., Sage.

———. 1982. Knowledge, resources and the future. A review of Simon's *The Ultimate Resource* and Brown's *Building a Sustainable Society.* BioScience 32:343–344.

Bouvonder, B. 1979. Impact of green revolution in India. Tech. Forecasting and Social Change 13:297–313.

Bouwer, Herman. 1968. Returning wastes to the land, a new role for agriculture. J. Soil Water Cons. 23:164–168.

Bowen, G. D. 1973. Mineral nutrition of ectomycorrhizae. In: *Ectomycorrhizae: Their Ecology and Physiology,* G. C. Marks and T. T. Kozlowski, eds. New York, Academic Press.

Bowman, K. O., K. Hutcheson, E. P. Odum, and L. R. Shenton. 1970. Comments on the distribution of indices of diversity. In: *International Symposium on Statistical Ecology,* Vol. 3. University Park, The Pennsylvania State University Press.

Box, Elgene, 1978. Geographical dimensions of terrestrial net and gross productivity. Radiat. Environ. Biophys. 15:305–322.

Braun-Blanquet, J. 1932. *Plant Sociology: The Study of Plant Communities,* G. D. Fuller and H. C. Conard, trans. and eds. New York, McGraw-Hill.

———. 1951. *Pflanzensoziologie.* Vienna, Springer-Verlag.

Bray, J. R. 1956. Gap phase replacement in a maple-basswood forest. Ecology 37:598–600.

———. 1958. Notes towards an ecologic theory. Ecology 30:770–776.

Bray, J. R., and E. Gorham. 1964. Litter production in forests of the world. Adv. Ecol. Res. 2:101–157.

Breit, William. 1982. *The Academic Scribblers* (2nd ed.). Hinsdale, Ill., Dryden Press.

Brian, M. W. 1956. Exploitation and interference in interspecies competition. J. Anim. Ecol. 25:335–347.

Brill, Winston J. 1977. Biological nitrogen fixation. Sci. Am. 236(3):68–81.

———. 1979. Nitrogen fixation: Basic to applied. Am. Sci. 67:458–466.

Brillouin, L. 1949. Life, thermodynamics and cybernetics. Am. Sci. 37:354–368.

Brink, R. A., J. W. Densmore, and G. A. Hill. 1977. Soil deterioration and the growing world demand for food. Science 197:625–630.

Brock, T. D. 1967. Relationship between primary productivity and standing crop along a hot spring thermal gradient. Ecology 48:566–571.

Brock, T. D., and M. L. Brock. 1966. Temperature options for algal development in Yellowstone and Iceland hot springs. Nature 209:733–734.

Brody, Samuel. 1945. *Bioenergetics and Growth.* New York, Reinhold.

Broecker, W. S., T. Takahashi, H. J. Simpson, and T. H. Peng. 1979. Fate of fossil fuel carbon dioxide and the global carbon budget. Science 206:409–418.

Brooks, John L., and S. I. Dodson. 1965. Predation, body size and composition of plankton. Science 150:28–35.

Brower, L. P., W. N. Ryerson, L. L. Coppinger, and S. C. Glazier. 1968. Ecological chemistry and the palatability spectrum. Science 161:1349–1350. (See also Zoologica 49:137, 1964.)

Brown, Frank A., J. W. Hastings, and J. D. Palmer. 1970. *The Biological Clock—Two Views.* New York, Academic Press, 94 pp.

Brown, Harrison S. 1977. World Food and Nutrition Study. See under National Academy of Science.

———. 1978. *The Human Future Revisited.* New York, Norton.

Brown, Jerram L. 1969. Territorial behavior and population regulation in birds, a review and reevaluation. Wilson Bull. 81:293–329.

Brown, Jerram L., and G. H. Arians. 1970. Spacing patterns in mobile animals. Ann. Rev. Ecol. Syst. 1:239–262.

Brown, L. L., and E. O. Wilson. 1956. Character displacement. Syst. Zool. 5:49–64.

Brown, Lester R. 1975. The world food prospect. Science 190:1053–1059.

———. 1978. The global economic prospect: New sources of economic stress. Worldwatch Paper No. 20. Washington, D.C., Worldwatch Institute, 56 pp.

———. 1978a. The worldwide loss of cropland. Worldwatch Paper No. 24. Washington, D.C., Worldwatch Institute, 48 pp.

———. 1980. Food or fuel: New competition for the world croplands. Worldwatch Paper No. 35. Washington, D.C., Worldwatch Institute.

———. 1981. *Building a Sustainable Society.* New York, Norton, 433 pp.

———. 1981a. World population growth, soil erosion and food security. Science 214:995–1002.

Brown, Sandra. 1981. A comparison of the structure, primary productivity and transpiration of cypress ecosystems in Florida. Ecol. Monogr. 51(4):403–427.

Brown, William L. 1961. Mass insect control programs: Four case histories. Psyche 68:75–109.

Brugam, R. B. 1978. Human disturbance and the historical development of Linsley Pond. Ecology 59:19–36.

Bryson, Reid A. 1974. A perspective on climatic change. Science 184:753–760.

Buckman, H. O., and N. C. Brody. 1974. *The Nature and Properties of Soil* (8th ed.). New York, Macmillan.

Budyko, M. I. 1955. *Atlas of the Heat Balance.* Leningrad. See: The heat balance of the earth's surface. Translated by N. A. Stepannova.

Buell, M. F. 1956. Spruce-fir and maple-basswood competition in Itasca Park, Minn. Ecology 37:606.

Bullock, T. H. 1955. Compensation for temperature in the metabolism and activity of poikotherms. Biol. Rev. 30:311–342.

Burkholder, Paul R. 1952. Cooperation and conflict among primitive organisms. Am. Sci. 40:601–631.

Burris, R. H. (ed.). 1978. Future of biological N_2 fixation. BioScience 28:563–592.

Butzer, Karl W. 1964. *Environment and Archaeology, An Introduction to Pleistocene Geography.* Chicago, Aldine Press, 524 pp.

———. 1980. Civilizations: Organisms or systems? Am. Sci. 68:517–523.

Cahn, Robert. 1978. *Footprints on the Planet: A Search for an Environmental Ethic.* New York, Universe Books.

Cain, Stanley A. 1944. *Foundations of Plant Geography.* New York, Harper & Bros., 556 pp.

———. 1950. Lifeforms and phytoclimate. Bot. Rev. 16:1–32.

Cairns, John (ed.). 1970. The structure and function of freshwater microbial communities. Research Division Monograph 3. Blacksburg, Va., Va. Polytechnic Institute and State University, 300 pp.

———. 1980. *The Recovery Process in Damaged Ecosystems.* Ann Arbor, Mich., Ann Arbor Science, 167 pp.

———. 1982. *Artificial Substrates.* Ann Arbor, Mich., Ann Arbor Science, 300 pp.

Cairns, John, Jr., and Kenneth L. Dickson. 1971. A simple method for the biological assessment of the effects of water discharges on aquatic bottom dwelling organisms. J. Water Pollution Control Fed. 43:755–772.

Cairns, John, K. L. Dickson, and E. E. Herrick (eds.). 1977. *Recovery and Restoration of Damaged Ecosystems.* Charlottesville, University of Virginia Press, 531 pp.

Calhoun, J. B. 1962. Population density and social pathology. Sci. Am. 206(2):1399–1408.

Callahan, Daniel. 1972. Ethics and population limitations. Science 175:487–494.

Callicott, J. B. 1979. Elements of an environmental ethic: Moral considerability and the biotic community. Environ. Ethics 1:71–81.

Calvin, Melvin. 1969. *Chemical Evolution; Molecular Evolution Towards the Origin of Living Systems on the Earth and Elsewhere.* New York, Oxford University Press, 278 pp.

Cameron, Austin W. 1964. Competitive exclusion between the rodent genera *Microtus* and *Clethrionomys.* Evolution 18:630–634.

Cannon, Walter B. 1932. *The Wisdom of the Body.* New York, Norton. (2nd Ed., 1939.)

Carpenter, E. J. 1969. A simple, inexpensive algal chemostat. Limnol. Oceanogr. 14:720–721.

Carpenter, F. L., and H. F. Recher. 1979. Pollination, reproduction and fire. Am. Nat. 113:871.

Carpenter, J. R. 1940. Insect outbreaks in Europe. J. Anim. Ecol. 9:108–147.

Carson, H. L. 1958. Response to selection under different conditions of recombination of *Drosophila.* Cold Spring Harbor Symp. Quant. Biol. 23:291–306.

Carson, Rachel. 1962. *Silent Spring.* Boston, Houghton Mifflin.

Castle, Emory M. 1964. The market mechanism, externalities and land economics. J. Farm Economics 47(3):546.

Caswell, Hal, F. Reed, S. N. Stephenson, and P. A. Werner. 1973. Photosynthetic pathways and selected herbivory: A hypothesis. Am. Nat. 107:465–480.

Caughley, Graeme. 1970. Eruption of ungulate populations with emphasis on Himalayan thor in New Zealand. Ecology 51:53–72.

Chambers, K. L. (ed.). 1970. *Biochemical Coevolution.* Twenty-ninth Biology Colloquium. Corvallis, Oregon State University Press.

Chambers, R. S., R. A. Herendeen, J. J. Joyce, and P. S. Penner. 1979. Gasohol: Does it or doesn't it produce positive net energy? Science 206:789–795.

Chapman, R. N. 1928. The quantitative analysis of environmental factors. Ecology 9:111–122.

———. 1931. *Animal Ecology, with Special Reference to Insects.* New York, McGraw-Hill.

Charnov, E. L., and W. M. Schaffer. 1973. Life-history consequences of natural selection: Cole's result revisited. Am. Nat. 107:791.

Cheng, T. C. (ed.). 1971. *Aspects of the Biology of Symbiosis.* Baltimore, University Park Press.

Chew, R. M. 1974. Consumers as regulators of ecosystems. Ohio J. Sci. 74:359–370.

Chitty, Dennis. 1960. Population processes in the vole and their relevance to general theory. Can. J. Zool. 38:99–113.

_______ . 1967. The natural selection of self-regulatory behavior in animal populations. Proc. Ecol. Soc. Austr. 2:51–78.

Christensen, A. M., and McDermott. 1958. Life history and biology of the oyster crab, *Pinnotheres ostreum* Say. Biol. Bull. 144:146–179.

Christian, John J. 1950. The adreno-pituitary system and population cycles in mammals. J. Mammal. 31:247–259.

_______ . 1963. Endocrine adaptive mechanisms and the physiologic regulation of population growth. In: *Physiological Mammalogy*, W. V. Mayer and R. G. van Gelder, eds. New York, Academic Press, pp. 189–353.

_______ . 1970. Social subordination, population density, and mammalian evolution. Science 168:84–90.

Christian, J. J., and D. E. Davis. 1964. Endocrines, behavior and populations. Science 146:1550–1560.

Clark, Colin W. 1981. Bioeconomics. In: *Theoretical Ecology* (2nd ed.), R. M. May, ed. Sunderland, Mass., Sinauer Associates, pp. 387–418.

Clark, John R. 1969. Thermal pollution and aquatic life. Sci. Am. 220(3):3–11.

Clark, L. R. 1964. The population dynamics of *Cardiaspina albitextura* (Psyllidae). Austr. J. Zool. 12:362–380.

Clark, L. R., P. W. Geier, R. D. Hughes, and R. F. Morris. 1967. *The Ecology of Insect Populations in Theory and Practice*. London, Methuen, 232 pp.

Clark, W. C., D. D. Jones, and C. S. Holling. 1979. Lessons for ecological policy design: A case study of ecosystem management. Ecol. Model. 7:153.

Clarke, F. W. 1924. The data of geochemistry. U.S. Geol. Surv. Bull. No. 228.

Clarke, George L. 1954. *Elements of Ecology*. New York, John Wiley & Sons. (Revised printing, 1965.)

Clatworthy, J. N., and J. L. Harper. 1962. The comparative biology of closely related species living in the same area. J. Exp. Bot. 13:307–324.

Clements, F. E. 1905. *Research Methods in Ecology*. Lincoln, Neb., Univ. Publ. Co., 199 pp.

_______ . 1916. Plant succession: Analysis of the development of vegetation. Washington, D.C., Publ. Carnegie Inst. 242:1512. (See also reprinted edition, 1928, entitled *Plant Succession and Indicators*. New York, Wilson.)

Clements, Frederic E., and V. E. Shelford. 1939. *Bioecology*. New York, John Wiley & Sons.

Cleveland, L. R. 1924. The physiological and symbiotic relationships between the intestinal protozoa of termites and their host, with special reference to *Reticulitermes fluipes* Kollar. Biol. Bull. 46:177–225.

_______ . 1926. Symbiosis among animals with special reference to termites and their intestinal flagellates. Quart. Rev. Biol. 1:51–60.

Cloud, Preston E., Jr. (ed.). 1969. *Resources and Man*. San Francisco, W. H. Freeman, 259 pp.

_______ . 1978. *Cosmos, Earth, and Man: A Short History of the Universe*. New Haven, Yale University Press.

Coase, R. H. 1960. The problem of social cost. J. Law Econ. 3:1–44.

Cobb, James C., and Thomas Dyer. 1979. Environmental protection or economic prosperity: Georgia, Tennessee and the Tennessee Copper Companies, 1903–1975. Presented at Annual Meeting, Organization of American Historians, New Orleans. 26 pp (to be published by University of Tennessee Press).

Cody, Martin L. 1966. A general theory of clutch size. Evolution 20:174–184.

_______ . 1974. Optimization in ecology. Science 183:1156–1164.

Cody, Martin L., and M. Diamond (eds.). 1975. *Ecology and Evolution of Communities*. Cambridge, Mass., Belknap Press, Harvard University.

Coffin, C. C., F. R. Hayes, L. H. Jodrey, and S. G. Whiteway. 1949. Exchanges of materials in a lake as studied by the addition of radioactive phosphorus. Can. J. Res., Section C, 27:207–222.

Cohen, Mark Nathan, Roy S. Malpass, and Harold G. Klein. 1980. *Biosocial Mechanisms of Population Regulation*. New Haven, Yale University Press.

Cole, LaMont C. 1946. A theory for analyzing contagiously distributed populations. Ecology 27:329–341.

_______ . 1946a. A study of the cryptozoa of an Illinois woodland. Ecol. Monogr. 16:49–86.

_______ . 1951. Population cycles and random oscillations. J. Wildl. Mgmt. 15:233–251.

_______ . 1954. Some features of random cycles. J. Wildl. Mgmt. 18:107–109.

_______ . 1954a. The population consequences of life history phenomena. Quart. Rev. Biol. 29:103–137.

_______ . 1957. Sketches of general and comparative demography. Cold Spring Harbor Symp. Quant. Biol. 22:1–15.

_______ . 1958. The ecosphere. Sci. Am. 198(4):83–92.

_______ . 1966. Protect the friendly microbes. In: "The Fragile Breath of Life," 10th Anniversary Issue, Science and Humanity Supplement. Saturday Review, May 7, 1966, pp. 46–47.

Coleman, David C. 1976. Energy flow and partitioning in selected man-managed and natural ecosystems. Agro-Ecosystems 3:45–54.

Coleman, David C., C. V. Cole, H. W. Hunt, and D. A. Klein. 1978. Trophic interactions in soil as they affect energy and nutrient dynamics. Microb. Ecol. 4:345–349; 373–380.

Coleman, David C., C. P. P. Reid, and C. V. Cole. 1983. Biological strategies of nutrient cycling in soil systems. Adv. Ecol. Res. 13 (in press).

Colwell, R. K. 1973. Competition and coexistence in a simple tropical community. Am. Nat. 107:737–760.

Colwell, R. N., et al. 1963. Basic matter and energy relationships involved in remote reconnaissance. Photogr. Engr. 29:761–799.

Comar, C. L. 1955. *Radioisotopes in Biology and Agriculture: Principles and Practice.* New York, McGraw-Hill.

________. 1965. The movement of fallout radionuclides through the biosphere and man. Ann. Rev. Nucl. Sci. 15:175–206.

Commoner, Barry. 1971. *The Closing Circle: Nature, Man and Technology.* New York, Alfred A. Knopf.

Connell, Joseph H. 1961. The influence of interspecific competition and other factors on the distribution of the barnacle, *Chthamalus stellatus.* Ecology 42:133–146.

________. 1972. Community interactions on marine rocky intertidal shores. Ann. Rev. Ecol. Syst. 3:169–192.

________. 1978. Diversity in tropical rain forests and coral reefs. Science 199:1302–1310.

Connell, Joseph H., and E. Orias. 1964. The ecological regulation of species diversity. Am. Nat. 98:399–414.

Connell, J. H., and R. O. Slayter. 1977. Mechanism of succession in natural communities and their role in community stability and organization. Am. Nat. 111:1119–1144.

Conner, W. H., and J. W. Day. 1976. Productivity and composition of a bald cypress water tupelo site and a bottomland hardwood site in a Louisiana swamp. Am. J. Bot. 63:1354–1364.

Connor, E. F., and E. D. McCoy. 1979. The statistics and biology of the species-area relationship. Am. Nat. 113:791–833.

Cooke, G. Dennis. 1967. The pattern of autotrophic succession in laboratory microecosystems. BioScience 17:717–721.

Cooke, G. Dennis, R. J. Beyers, and E. P. Odum. 1968. The case for the multi-species ecological system, with special reference to succession and stability. In: *Bioregenerative Systems.* NASA Special Publication 165:129–139.

Cooper, Charles R. 1961. The ecology of fire. Sci. Am. 204(4):150–160.

Copeland, B. J., and T. C. Dorris. 1962. Photosynthetic productivity in oil refinery effluent holding ponds. J. Water Poll. Control Fed. 34:1104–1111.

Costanza, Robert. 1980. Embodied energy and economic valuation. Science 210:1219–1224.

Council on Environmental Quality (CEQ). 1972. Integrated Pest Management. Washington, D.C., U.S. Government Printing Office.

________. Environmental Quality. 1978, Ninth Annual Report; 1979, Tenth Annual Report; 1980, Eleventh Annual Report; 1981, Twelfth Annual Report. Washington, D.C., U.S. Government Printing Office.

________. 1980. The Global 2000 Report to the President. Entering the Twenty-first Century. Vols. 1, 2, and 3. Washington, D.C., U.S. Government Printing Office. (Textbook edition published by Pergamon Press, Elmsford, N.Y.)

________. 1981. Global Future—Time to Act, G. Speth, ed. Washington, D.C., U.S. Government Printing Office.

________. 1981. Global Energy Futures and the Carbon Dioxide Problem. Washington, D.C., U.S. Government Printing Office, 92 pp.

________. 1981. Contamination of Ground Water by Toxic Organic Chemicals. Washington, D.C., U.S. Government Printing Office.

Cowgill, U. M., and G. E. Hutchinson. 1964. Cultural eutrophication in Lago Monterosi during Roman antiquity. Proc. Int. Assoc. Theoret. Appl. Limnol. 15(2):644–645.

Cowles, H. C. 1899. The ecological relations of the vegetation of the sand dunes of Lake Michigan. Bot. Gaz. 27:95–117; 167–202; 281–308; 361–391.

Cox, G. W., and M. D. Atkins. 1979. *Agricultural Ecology.* San Francisco, W. H. Freeman, 721 pp.

Craighead, J. J., and F. C. Craighead, Jr. 1956. *Hawks, Owls and Wildlife.* Harrisburg, Pa., The Stackpole Co.

Crisp, D. (ed.). 1964. Grazing in terrestrial and marine environments. Oxford, Blackwell, 322 pp.

Crocker, Robert L. 1952. Soil genesis and the pedogenic factors. Quart. Rev. Biol. 27:139–168.

Crombie, A. C. 1947. Interspecific competition. J. Anim. Ecol. 16:44–73.

Crossley, D. A. 1963. Consumption of vegetation by insects. In: *Radioecology,* V. Schultz and A. W. Klement, eds. New York, Reinhold, pp. 427–430.

________. 1964. Biological elimination of radionuclides. Nuclear Safety 5:265–268.

Crossley, D. A., and K. K. Bohnsack. 1960. The oribated mite fauna in pine litter. Ecology 41:785–790.

Crossley, D. A., R. N. Coulson, and C. S. Gist. 1973. Trophic level effects on species diversity in arthropod communities of forest canopies. Environ. Ent. 2:1097–1100.

Crossley, D. A., and Mary P. Hoglund. 1962. A litterbag method for the study of microarthropods inhabiting leaf litter. Ecology 43:571–573.

Crossley, D. A., and Martin Witkamp. 1964. Forest soil mites and mineral cycling. Acarologia, special issue, Proceedings 1st Int. Conf. Acrology, pp. 137–145.

Crowell, K. L. 1962. Reduced interspecific competition among the birds of Bermuda. Ecology 43:75–88.

Currie, R. I. 1958. Some observations on organic production in the northeast Atlantic. Rapp. Proc. Verb. Cons. Int. Explor. Mer. 144:96–102.

Curtis, J. T. 1955. A prairie continuum in Wisconsin. Ecology 36:558–566.

Curtis, J. T., and R. P. McIntosh. 1951. An upland forest continuum in the prairie forest border region of Wisconsin. Ecology 32:476–496.

Dale, M. B. 1970. Systems analysis and ecology. Ecology 51:2–16.

Dales, R. Phillip. 1957. Commensalism. In: *Treatise on Marine Ecology and Paleoecology*, J. W. Hedgpeth, ed. Vol. 1. Boulder, Col., Geological Society of America, pp. 391–412.

Daly, Herman E. 1973. *Toward a Steady-State Economy.* San Francisco, W. H. Freeman.

Darling, F. Fraser. 1938. *Bird Flocks and the Breeding Cycle.* Cambridge, England, Cambridge University Press.

———. 1951. The ecological approach to the social sciences. Am. Sci. 39:244–254.

Darling, F. Fraser, and J. P. Milton (eds.). 1966. *Future Environments of North America.* Garden City, N.Y., Natural History Press, 265 pp.

Darmstadler, Joel, J. Dunkerley, and J. Alterman. 1977. *How Industrial Societies Use Energy: A Comparative Analysis.* A Resources for the Future Book. Baltimore, Johns Hopkins Press.

Dasmann, R. F. 1968. *A Different Kind of Country.* New York, Macmillan.

Daubenmire, Rexford. 1966. Vegetation: Identification of typal communities. Science 151:291–298.

———. 1968. Ecology of fire in grasslands. In: *Advances in Ecological Research*, J. B. Cragg, ed. Vol. V. New York, Academic Press, pp. 209–266.

———. 1968a. *Plant Communities.* New York, Harper & Row, 300 pp.

———. 1974. *Plants and Environment* (3rd ed.). New York, John Wiley & Sons.

———. 1978. *Plant Geography; with Special Reference to North America.* New York, Academic Press.

Davidson, James. 1938. On the growth of the sheep population in Tasmania. Trans. R. Soc. S. Austr. 62:342–346.

Davidson, James, and H. G. Andrewartha. 1948. Annual trends in a natural population of *Thrips imaginis* (Thysanoptera). J. Anim. Ecol. 17:193–199; 200–222.

Davis, J. J., and R. F. Foster. 1958. Bioaccumulation of radioisotopes through aquatic food chains. Ecology 39:530–535.

Davis, Kingsley (ed.). 1965. Cities. Sci. Am. (Special Issue) 213(3):1–280.

Davis, Margaret B. 1969. Palynology and environmental history during the Quaternary Period. Am. Sci. 57:317–332.

Day, F. P., and C. D. Monk. 1974. Vegetation patterns on a southern Appalachian watershed. Ecology 55:1064–1074.

Dayton, R. K. 1971. Competition, disturbance and community organization: The provision and subsequent utilization of space in a rocky intertidal community. Ecol. Monogr. 41:351–389.

———. 1975. Experimental evaluation of ecological dominance in a rocky intertidal algal community. Ecol. Monogr. 45:137–389.

DeBach, P. 1964. *Biological Control of Insect Pests and Weeds.* New York, Reinhold.

DeBenedictis, P. A. 1974. Interspecific competition between tadpoles of *R. pipiens* and *R. sylvatica:* An experimental field study. Ecol. Monogr. 44:129–151.

Deevey, Edward S., Jr. 1947. Life tables for natural populations of animals. Quart. Rev. Biol. 22:283–314.

———. 1950. The probability of death. Sci. Am. 182:58–60.

———. 1958. The equilibrium population. In: *The Population Ahead*, R. G. Francis, ed. Minneapolis, University of Minnesota Press, pp. 64–86.

de la Cruz, A. A. 1975. Proximate nutritive value changes during decomposition of salt marsh plants. Hydrobiologia 47:475–480.

———. 1979. Production and transport of detritus in wetlands. In: *Wetland Functions and Values: The State of Our Understanding.* Minneapolis, Am. Water Resources Assoc., pp. 162–173.

Delwiche, C. C. 1970. The nitrogen cycle. Sci. Am. 223(5):137–146.

Dendy, J. S. 1945. Predicting depth distribution in three TVA storage type reservoirs. Trans. Am. Fish. Soc. 75:65–71.

Diamond, J. M., and R. M. May. 1981. Island biogeography and the design of natural reserves. In: *Theoretical Ecology* (2nd ed.), R. M. May, ed. Sunderland, Mass., Sinauer Associates.

Dice, Lee R. 1952. Measure of spacing between individuals within a population. Contr. Lab. Vert. Biol., U. of Michigan, 55:1–23.

———. 1952a. *Natural Communities.* Ann Arbor, University of Michigan Press.

Dickinson, Robert E. 1970. *Regional Ecology: The Study of Man's Environment.* New York, John Wiley & Sons.

Dobzhansky, Theodosius. 1968. Adaptedness and fitness. In: *Population Biology and Evolution*, R. C. Lewontin, ed. Syracuse, N.Y., Syracuse University Press, pp. 109–121.

Dokuchaev, V. V. 1889. The zones of nature. (In Russian.) Akad. Nauk Moscow, Vol. 6.

Dommergues, Y. R., and S. V. Krupa. 1978. *Interactions Between Non-pathogenic Soil Microorganisms and Plants.* New York, Elsevier.

Dougherty, E. C. 1959. Introduction to axenic culture

of invertebrate metazoa: A goal. Ann. N.Y. Acad. Sci. 77:27–54.

Drake, Ellen T. (ed.). 1968. *Evolution and Environment.* New Haven, Yale University Press, 478 pp.

Drury, W. H., and I. C. T. Nisbet. 1973. Succession. J. Arnold Arboretum 54:331–368.

Dublin, L. I., and A. J. Lotka. 1925. On the true rate of natural increase as exemplified by the population of the United States, 1920. J. Am. Statist. Assoc. 20:305–339.

Dubos, Rene. 1976. Symbiosis between the earth and humankind. Science 193:459–462.

________. 1980. *The Wooing of Earth: New Perspective on Man's Use of Nature.* New York, Charles Scribner's Sons, 183 pp.

Duckham, A. N., J. G. W. Jones, and E. H. Robert (eds.). 1976. *Food Production and Consumption. The Efficiency of Human Food Chains and Nutrient Cycles.* Amsterdam, North-Holland; New York, Elsevier; 542 pp.

Dugger, W. M., Jr., Jane Koukol, and R. L. Palmer. 1966. Physiological and biochemical effects of atmospheric oxidants on plants. J. Air Poll. Control Assoc. 16:467–471.

Dugsdale, R. C. 1967. Nutrient limitation in the sea: Dynamics, identification, and significance. Limnol. Oceanogr. 12:685–695.

Duncan, Otis Dudley. 1964. Social organizations and the ecosystem. In: *Handbook of Modern Sociology,* R. E. L. Faris, ed. Chicago, Rand McNally.

Duvigneaud, P. 1971. *La Synthese Ecologique. Populations, Communantes, Ecosystemes, Biosphere, Noosphere.* Paris, Boin.

Dyer, M. I., and U. G. Bokhari. 1976. Plant-animal interactions: Studies of effects of grasshopper grazing on blue grama grass. Ecology 57:762–772.

Dyksterhuis, E. J. 1958. Ecological principles in range evaluation. Bot. Rev. 24:253–272.

Echholm, Erik. 1975. The other energy crisis: Firewood. Worldwatch Paper No. 1. Washington, D.C., Worldwatch Institute, 22 pp.

Edmondson, W. T. 1968. Water-quality management and lake eutrophication: The Lake Washington case. In: *Water Resources Management and Public Policy,* T. H. Campbell and R. O. Sylvester, eds. Seattle, University of Washington Press, pp. 139–178.

________. 1970. Phosphorus, nitrogen and algae in Lake Washington after diversion of sewage. Science 169:690–691.

________. 1979. Lake Washington and the predictability of limnological events. Arch. Hydrobiol. Bech. Ergobn. Limnol. 13:234–241.

Edson, M. M., T. C. Foin, and C. M. Knapp. 1981. Emergent properties and ecological research. Am. Nat. 118:593–596.

Edwards, C. A. 1969. Soil pollutants and soil animals. Sci. Am. 220(4):88–92; 97–99.

Edwards, C. A., D. E. Reichle, and D. A. Crossley. 1970. The role of soil invertebrates in turnover of organic matter and nutrients. In: *Analysis of Temperate Forest Ecosystems,* D. E. Reichle, ed. Berlin, Springer-Verlag, pp. 147–172.

Egerton, F. N. 1968. Leeuwenhoek as a founder of animal demography. J. Hist. Biol. 1:1–22.

________. 1969. Richard Bradley's understanding of biological productivity; a study of eighteenth-century ecological ideas. J. Hist. Biol. 2(2):391–410.

Egler, F. E. 1954. Vegetation science concepts. 1. Initial floristic composition—a factor in old-field vegetation development. Vegetatio 4:412–417.

Ehrenfeld, D. 1978. *The Arrogance of Humanism.* New York, Oxford University Press.

Ehrlich, Paul R. 1968. *The Population Bomb.* New York, Ballantine Books.

Ehrlich, Paul R., and Anne Ehrlich. 1970. *Population, Resources, Environment: Issues in Human Ecology.* San Francisco, W. H. Freeman.

Ehrlich, Paul, and Anne Ehrlich. 1981. *Extinction: The Causes and Consequences of Disappearance of Species.* New York, Random House, 305 pp.

Ehrlich, Paul R., and Peter H. Raven. 1965. Butterflies and plants: A study of coevolution. Evolution 18:586–608.

Einarsen, A. S. 1945. Some factors affecting ring-necked pheasant population density. Murrelet 26:39–44.

Eldredge, N., and S. J. Gould. 1972. Punctuated equilibria: An alternative to phyletic gradualism. In: *Models in Paleobiology,* T. J. M. Schopf, ed. San Francisco, Freeman, Cooper, pp. 82–115.

Eliassen, Rolf. 1952. Stream pollution. Sci. Am. 186:17–21.

Ellul, Jacques. 1967. *The Technological Society.* New York, Random House.

Elton, Charles. 1927. *Animal Ecology.* New York, Macmillan. (2nd ed., 1935; 3rd ed., 1947.)

________. 1933. *The Ecology of Animals.* London, Methuen.

________. 1942. *Voles, Mice and Lemmings: Problems in Population Dynamics.* London, Oxford University Press.

________. 1958. *The Ecology of Invasions by Animals and Plants.* London, Methuen, 181 pp.

________. 1966. *The Pattern of Animal Communities.* London, Methuen, 432 pp.

Elwood, J. W., and D. J. Nelson. 1972. Periphyton production and grazing rates in a stream measured with a ^{32}P material balance method. Oikos 23:295–303.

Emanuel, W. R., and R. J. Mulholland. 1976. Linear periodic control with applications to environmental systems. Int. J. Control 24(6):807–820.

Emerson, Alfred E. 1954. Dynamic homeostasis: A unifying principle in organic, social and ethical evolution. Sci. Monthly 78(2):67–85.

Emerson, F. C. (ed.). 1973. The economics of environmental problems. Mich. Business Papers No. 58. Ann Arbor, University of Michigan, 98 pp.

Emery, K. O., and C. O. D. Iselin. 1967. Human food from ocean and land. Science 157:1279–1281.

Emlen, J. Merritt. 1966. The role of time and energy in food preference. Am. Nat. 100:611–617.

———. 1970. Age specificity and ecological theory. Ecology 51:588–601.

Emlen, John T., and M. J. Dejong. 1981. Intrinsic factors in the selection of foraging substrates by pine warblers; a test of a hypothesis. Auk 98:294–298.

Engelberg, J., and L. L. Boyarsky. 1979. The noncybernetic nature of ecosystems. Am. Nat. 114:317–324.

Engelmann, M. D. 1968. The role of soil arthropods in community energetics. Am. Zool. 8:61–69.

Enright, J. T. 1970. Ecological aspects of endogenous rhythmicity. Ann. Rev. Ecol. Syst. 1:221–238.

Epstein, Samuel S. 1974. Environmental determinants of human cancer. Cancer Res. 34:2425–2435.

Errington, Paul L. 1946. Predation and vertebrate populations. Quart. Rev. Biol. 21:144–177; 221–245.

Esch, G. W., and T. C. Hazen. 1978. Thermal ecology and stress: A case history for red-sore disease in large-mouth bass. In: *Energy and Environmental Stress in Aquatic Systems,* Thorp and Gibbons, eds. U.S. Dept. of Energy, Symp. No. 48. Springfield, Va., National Technical Info. Center, pp. 331–363.

Esch, G. W., and McFarlane, R. W. (eds.). 1975. Thermal Ecology II. Energy Research and Development Administration. Springfield, Va., National Technical Info. Center, 404 pp.

Estep, M. F., and H. Dabrowski. 1980. Tracing food webs with stable hydrogen isotopes. Science 109:1537–1538.

Etkins, Robert, and E. S. Epstein. 1982. The rise of global mean sea level as an indication of climate change. Science 215:287–289.

Evans, Francis C. 1956. Ecosystem as the basic unit in ecology. Science 123:1227–1228.

Evans, Francis C., and Stanley A. Cain. 1952. Preliminary studies on the vegetation of an old field community in southeastern Michigan. Contr. Lab. Vert. Biol., U. of Michigan, 51:1–17.

Evans, L. T. (ed.). 1963. *Environmental Control of Plant Growth.* New York, Academic Press, 449 pp.

———. 1980. The natural history of crop yield. Am. Sci. 68:388–397.

Ewel, Katherine, and H. T. Odum. 1978. Cypress swamps for nutrient removal and wastewater recycling. In: *Advances in Water and Wastewater Treatment; Biological Nutrient Removal.* Ann Arbor, Mich., Ann Arbor Science.

Fager, E. W. 1972. Diversity: A sampling study. Am. Nat. 106:293–310.

Falk, J. H. 1976. Energetics of a suburban lawn ecosystem. Ecology 57:141–150.

———. 1977. The frenetic life forms that flourish in suburban lawns. Smithsonian 8:90–96.

Farner, D. S. 1964. The photoperiodic control of reproductive cycles in birds. Am. Sci. 52:137–156.

———. 1964a. Time measurement in vertebrate photoperiodism. Am. Nat. 98:375–386.

Farnworth, Edward G., T. H. Tidrick, C. F. Jordan, and W. M. Smathers. 1981. The value of natural ecosystems. An economic and ecological framework. Environ. Conserv. 8:275–282.

Farvar, M. T., and John P. Milton (eds.). 1972. *The Careless Technology; Ecology and International Development.* Garden City, N.Y., Natural History Press, 1030 pp.

Faulkner, Edward H. 1943. *Plowman's Folly.* Norman, University of Oklahoma Press, 161 pp.

Feener, Donald H. 1981. Competition between ant species: Outcome controlled by parasitic flies. Science 214:815–817.

Fenchel, T. M. 1969. The ecology of marine microbenthos. Part IV. Ophelia 6:1–182.

———. 1978. The ecology of micro- and meiobenthos. Ann. Rev. Ecol. Syst. 9:99–122.

Fenchel, T. M., and S. Kalding. 1979. Habitat selection and distribution patterns of 5 species of the amphropod genus *Gammarus.* Oikos 33:316–322.

Fernandez, Louis. 1980. Managing chemical wastes. In: Nat. Jour. Special Report "Chemical Issues in the News." Washington, D.C.

Fiebleman, J. K. 1954. Theory of integrative levels. Brit. J. Philos. Sci. 5:59–66.

Filzer, Paul. 1956. *Pflanzengemeinschaft und Umwelt, Ergebnisse und Probleme der Botanischen Standortforschung.* Stuttgart, Enke.

Finerty, James Patrick. 1980. *The Population Ecology of Cycles in Small Mammals.* New Haven, Yale University Press.

Finn, John T. 1976. Measures of ecosystem structure and function derived from the analysis of flows. J. Theoret. Biol. 56:363–380.

———. 1978. Cycling index: A general definition for cycling in compartment models. In: *Environmental Chemistry and Cycling Processes,* D. C. Adriano and I. L. Brisbin, eds. U.S. Dept. of Energy Symp. No. 45. Springfield, Va., National Technical Info. Center, pp. 138–164.

Fischer, A. G. 1960. Latitudinal variations in organic diversity. Evolution 14:64–81.

Fisher, A. C., J. V. Krutilla, and C. J. Cicchetti. 1972.

The economies of environmental preservation: A theoretical and empirical analysis. Am. Econ. Rev. 62:605–619.
Flader, Susan L. 1979. *Thinking Like a Mountain: Aldo Leopold and the Evolution of an Ecological Attitude Towards Deer, Wolves and Forest.* Lincoln, University of Nebraska Press.
Fleischer, W. E. 1935. The relation between chlorophyll content and rate of photosynthesis. J. Gen. Physiol. 18:573–597.
Flint, M. L., and Robert van den Bosch. 1981. *Introduction to Integrated Pest Management.* New York, Plenum, 256 pp.
Fogg, G. E. 1965. *Algal Cultures and Phytoplankton Ecology.* Madison, University of Wisconsin Press, 126 pp.
Foin, T. C., and S. K. Jain. 1977. Ecosystem analysis and population biology: Lessons for the development of community ecology. BioScience 27:532–539.
Forbes, S. A. 1887. The lake as a microcosm. Bull. Sc. A. Peoria. Reprinted in Ill. Nat. Hist. Surv. Bull. 15:537–550, 1925.
Forrester, Jay W. 1968. *Principles of Systems.* Cambridge, Mass., Wright-Allen Press.
________ . 1971. *World Dynamics.* Cambridge, Mass., Wright-Allen Press, 142 pp.
Fortescue, J. A. C. 1980. *Environmental Geochemistry: A Holistic Approach.* Ecol. Series 35. New York, Springer-Verlag, 346 pp.
Foster, R. F., and J. J. Davis. 1956. The accumulation of radioactive substances in aquatic forms. Proc. Int. Conf. Peaceful Uses Atomic Energy, Geneva, 13:364–367.
Frank, Peter W. 1957. Coactions in laboratory populations of two species of *Daphnia.* Ecology 38:510–518.
________ . 1965. The biodemography of an intertidal snail population. Ecology 46:831–844.
________ . 1968. Life histories and community stability. Ecology 49:355–357.
Frankenberg, D., and K. L. Smith, Jr. 1967. Coprophagy in marine animals. Limnol. Oceanogr. 12:443–450.
Freeze, R. A., and J. A. Cherry. 1979. *Groundwater.* Englewood Cliffs, N.J., Prentice-Hall.
French, N. R. 1965. Radiation and animal population: Problems, progress and projections. Health Physics 11:1557–1568.
Fretwell, S. D. 1977. The regulation of plant communities by the food chains exploiting them. Perspect. Biol. Med. 20:169–185.
Fried, M., and H. Broeshart. 1967. *The Soil-Plant System.* New York, Academic Press.
Frieden, Earl. 1972. The chemical elements of life. Sci. Am. 227(1):52–60.
Friederichs, K. 1930. *Die Grundfragen und Gesetzmassigkeiten der landund fortswirtschaftlichen Zoologie.* Two vols. Berlin, Paul Parey.
Fritsch, Albert J. 1980. *Environmental Ethics: Choices for Concerned Citizens.* New York, Anchor Books.
Fry, F. E. J. 1958. Temperature compensation. Ann. Rev. Physiol. 20:207–227.
Gabor, D., V. Columbo, A. King, and R. Galli. 1978. *Beyond the Age of Waste.* A Report to the Club of Rome. New York, Pergamon Press.
Gadgil, M., and W. H. Bossert. 1970. Life historical consequences of natural selection. Am. Nat. 104:1–24.
Gadgil, M., and O. T. Solbrig. 1972. The concept of *r* and *K* selection: Evidence from wild flowers and some theoretical considerations. Am. Nat. 106:14–31.
Galbraith, J. K. 1958. *The Affluent Society.* New York, The New American Library.
________ . 1967. *The New Industrial State.* Boston, Houghton Mifflin, 427 pp.
Gardner, M. R., and W. R. Ashby. 1970. Connectance of large dynamic cybernetic systems: Critical values for stability. Nature 288:784.
Garrels, R. M., F. T. Mackenzie, and C. Hunt. 1975. Chemical cycles and the global environments; assessing human influences. Los Altos, Calif., William Kaufmann, 206 pp.
Garren, Kenneth H. 1943. Effects of fire on the vegetation of the southeastern United States. Bot. Rev. 9:617–654.
Gasaway, Charles R. 1970. Changes in the fish population of Lake Francis Case in South Dakota in the first 16 years of impoundment. Tech. Paper No. 56. Washington, D.C., Bureau of Sport Fisheries and Wildlife.
Gates, David M. 1962. *Energy Exchange in the Biosphere.* New York, Harper & Row, 151 pp.
________ . 1963. The energy environment in which we live. Am. Sci. 51:327–348.
________ . 1965. Radiant energy, its receipt and disposal. Metero. Monogr. 6:1–26.
________ . 1965a. Energy, plants and ecology. Ecology 46:1–13.
________ . 1971. The flow of energy in the biosphere. Sci. Am. 224(3):88–99.
________ . 1980. *Biophysical Ecology.* New York, Springer-Verlag, 640 pp.
Gates, J. E., and L. W. Gysel. 1978. Avian nest dispersion and fledgling success in a field-forest ecotone. Ecology 59:871–883.
Gause, G. F. 1932. Ecology of populations. Quart. Rev. Biol. 7:27–46. (See also Quart. Rev. Biol. 11:320–336, 1936.)
________ . 1934. *The Struggle for Existence.* Baltimore, Williams & Wilkins, 163 pp.

Georgescu-Roegen, Nicholas. 1971. *The Entropy Law and the Economic Process*. Cambridge, Mass., Harvard University Press.

_______ . 1977. Bioeconomics. In: *The Political Economy of Food and Energy*, L. Junker, ed. Michigan Business Papers No. 62. Ann Arbor, University of Michigan, pp. 105–134.

_______ . 1977a. The steady-state and ecological salvation: A thermodynamic analysis. BioScience 27:266–271.

_______ . 1977b. *Energy and Economic Myths*. Elmsford, N.Y., Pergamon Press.

Gessner, F. 1949. Der chlorophyllgehalt in see and seine photosynthetische Valenz als geophysikalishes problem. Schweizer Z. Hydrologie 11:378–410.

Giarini, Orio. 1980. *Dialogue on Wealth and Welfare: An Alternate View of World Capital Formation*. A Report to the Club of Rome. New York, Pergamon Press.

Gibbons, J. Whitfield, and Rebecca R. Sharitz. 1974. Thermal alteration of aquatic ecosystems. Am. Sci. 62:660–670.

Gibbons, J. W., and R. R. Sharitz (eds.). 1981. *Thermal Ecology*. U.S. Atomic Energy Commission. Springfield, Va., National Technical Info. Center, 670 pp.

Gibson, David T. 1968. Microbial degradation of aromatic compounds. Science 161:1093–1097.

Giesy, John P. (ed.). 1980. *Microcosms in Ecological Research*. U.S. Dept. of Energy Symp. No. 52. Springfield, Va., National Technical Info. Center, 1110 pp.

Gifford, C. E., and E. P. Odum. 1961. Chlorophyll—a content of intertidal zones on a rocky seashore. Limnol. Oceanogr. 6:83–85.

Gilbert, L. E., and P. H. Raven (eds.). 1975. *Coevolution of Animals and Plants*. Austin, University of Texas Press. (Revised edition, 1980.)

Gilliland, Martha W. 1975. Energy analysis and public policy. Science 180:1051–1056.

_______ . (ed.). 1978. *Energy Analysis: A New Public Policy Tool*. Selected Symposium 9, American Association for the Advancement of Science. Boulder, Col., Westview Press.

Gilpin, M. E. 1975. *Group Selection in Predator-Prey Communities*. Princeton, N.J., Princeton University Press.

Glasser, John W. 1979. The role of predation in shaping and maintaining the structure of communities. Am. Nat. 113:631–641.

_______ . 1982. On the causes of temporal change in communities: Modification of the biotic environment. Am. Nat. 119:375–390.

Glasstone, Samuel (ed.). 1957. *The Effects of Nuclear Weapons*. Washington, D.C., U.S. Atomic Energy Commission.

Gleason, H. A. 1917. The structure and development of the plant association. Bull. Torrey Bot. Club 44:463–481.

_______ . 1926. The individualistic concept of the plant association. Bull. Torrey Bot. Club 53:7–26.

Glemarec, Michel. 1979. Problèmes d'ecologie dynamique et de succession en Baie de Concerneau. Vie Milou, Vol. 28–29, Fasc. 1, Ser. AB, pp. 1–20.

Goldman, C. R. 1960. Molybdenum as a factor limiting primary productivity in Castle Lake, California. Science 132:1016–1017.

Goldman, C. R. (ed.). 1965. Primary productivity in aquatic environments. Berkeley, University of California Press, 464 pp.

Goldschmidt, V. M. 1954. *Geochemistry*. Oxford, Clarendon Press, 730 pp.

Golley, Frank B. 1960. Energy dynamics of a food chain of a old-field community. Ecol. Monogr. 30:187–206.

_______ . 1968. Secondary productivity in terrestrial communities. Am. Zool. 8:53–59.

_______ . 1983. The historical origins of the ecosystem concept in biology. In press.

Golley, Frank B., J. T. McGinnis, R. G. Clements, G. I. Childs, and M. J. Duever. 1975. Mineral cycling in a tropical moist forest ecosystem. Athens, University of Georgia Press.

Goodall, D. W. 1963. The continuum and the individualistic association. Vegetatio 11:297–316.

_______ . 1970. Statistical plant ecology. Ann. Rev. Ecol. Syst. 1:99–124.

Gorden, Robert W. 1969. A proposed energy budget of a soybean field. Bull. Ga. Acad. Sci. 27:4152.

Gorden, R. W., R. J. Beyers, E. P. Odum, and E. G. Eagon. 1969. Studies of a simple laboratory microecosystem: Bacterial activities in a heterotrophic succession. Ecology 50:86–100.

Gordon, H. T. 1961. Nutritional factors in insect resistance to chemicals. Ann. Rev. Entomol. 6:27–54.

Gorham, E., P. M. Vitousek, and W. A. Reiners. 1979. The regulation of chemical budgets over the course of terrestrial succession. Ann. Rev. Ecol. Syst. 10:53–84.

Gornitz, V., S. Lebedeff, and J. Hansen. 1982. Global sea level trend in the past century. Science 125:1611–1614.

Gosselink, J. G., E. P. Odum, and R. M. Pope. 1974. The Value of the Tidal Marsh. LSU-SG-74-03. Center Wetland Resources, Louisiana State University, Baton Rouge, 33 pp.

Gould, Stephen J. 1977. The return of hopeful monsters. Nat. Hist. 86:22–30.

_______ . 1982. Darwinism and the expansion of evolutionary theory. Science 216:380–387.

Gould, S. J., and N. Eldredge. 1977. Punctuated equilibria: The tempo and mode of evolution reconsidered. Paleobiology 3:115–151.

Goulden, Clyde (ed.). 1977. *Changing Scenes in Natural*

Sciences. Special Publication 12. Philadelphia, Academy of Natural Sciences.

Graham, S. A. 1951. Developing forests resistant to insect injury. Sci. Monthly 73:235–244.

Grinnell, Joseph. 1917. Field test of theories concerning distributional control. Am. Nat. 51:115–128.

———. 1928. Presence and absence of animals. Univ. Calif. Chron. 30:429–450.

Gross, A. O. 1947. Cyclic invasion of the snowy owl and the migration of 1945–46. Auk 64:584–601.

Grunchy, Allan G. 1947. *Modern Economic Thought: The American Contribution.* Englewood Cliffs, N.J., Prentice-Hall.

Grzenda, Alfred, G. J. Caver, and H. P. Nicholson. 1964. Water pollution by insecticides in an agricultural river basin. II. The zooplankton, bottom fauna, and fish. Limnol. Oceanogr. 9:318–323.

Gutierrez, L. T., and W. R. Fey. 1975. Feedback dynamics analysis of secondary successional transients in ecosystems. Proc. Natl. Acad. Sci. USA 72:2733–2737.

Gutschick, Vincent P. 1978. Energy and nitrogen fixation. BioScience 28:571–575.

Haag, John. 1981. A yearning for synthesis: Organic thoughts since 1945. Int. J. Social Econ. 8:87–111.

Haagen-Smit, A. J. 1963. The control of air pollution. Sci. Am. 209(1):24–29.

Haagen-Smit, A. J., E. F. Darley, E. F. Zaitlin, M. Hull, and W. Noble. 1952. Investigation of injury to plants from air pollution in the Los Angeles area. Plant Physiol. 27:18–34.

Haines, E. B., and R. B. Hanson. 1979. Experimental degradation of detritus made from salt marsh plants *Spartina alterniflora, Salicornus virginia* and *Juncus roemerianus.* J. Exp. Marine Biol. Ecol. 40:27–40.

Haines, E. B., and C. L. Montague. 1979. Food sources of estuarine invertebrates analyzed using $^{13}C/^{12}C$ ratios. Ecology 60:48–56.

Hairston, N. G. 1959. Species abundance and community organization. Ecology 40:404–416.

Hairston, N. G., F. K. Smith, and L. B. Slobodkin. 1960. Community structure, population control and competition. Am. Nat. 94:421–425.

Halfon, Efraim (ed.). 1979. *Theoretical Systems Ecology: Advances and Case Studies.* New York, Academic Press, 516 pp.

Hall, Charles A. S., and John W. Day. 1977. Systems and models: Terms and basic principles. In: *Ecosystem Modeling in Theory and Practice,* C. Hall and J. Day, eds. New York, John Wiley & Sons, pp. 5–36.

Hamilton, William J. III, and Kenneth E. F. Watt. 1970. Refuging. Ann. Rev. Ecol. Syst. 1:263–286.

Hannon, Bruce. 1973. The structure of ecosystems. J. Theoret. Biol. 41:535–546.

———. 1973a. An energy standard of value. Ann. Am. Acad. Political Social Sci. 410:139–153.

———. 1974. Energy, manpower and the highway trust fund. Science 185:669–675.

———. 1979. Total energy costs in ecosystems. J. Theoret. Biol. 80:271–293.

Hansson, Lennart. 1979. On the importance of landscape heterogeneity in northern regions for breeding population density of homeotherms: A general hypothesis. Oikos 33:182–189.

Hardin, Garrett. 1960. The competitive exclusion principle. Science 131:1292–1297.

———. 1963. The cybernetics of competition. A biologist's view of society. Perspect. Biol. Med. 7:58–84.

———. 1968. The tragedy of the commons. Science 162:1243–1248.

———. 1977. *The Limits of Altruism: An Ecologist's View of Survival.* Bloomington, Indiana University Press.

Hardin, G., and J. Boden. 1977. *Managing the Commons.* San Francisco, W. H. Freeman.

Harlan, J. R. 1974. Our vanishing genetic resources. Science 188:618–621.

Harley, J. L. 1952. Associations between microorganisms and higher plants (mycorrhiza). Ann. Rev. Microbiol. 6:367–386.

———. 1959. *The Biology of Mycorrhiza.* Plant Science Monographs. London, Leonard Hill; New York, Interscience; 233 pp.

Harper, John L. 1961. Approaches to the study of plant competition. In: "Mechanisms in Biological Competition." Symp. Soc. Exp. Biol., Number XV, pp. 1–268.

———. 1969. The role of predation in vegetational diversity. In: *Diversity and Stability in Ecological Systems,* G. M. Woodwell and H. H. Smith, eds. Brookhaven Symp. Biol., Number 22. Upton, N.Y., Brookhaven National Laboratory, pp. 48–62.

———. 1974. Agricultural ecosystems. Agro-Ecosystems 1:1–6.

———. 1977. *Population Biology of Plants.* New York and London, Academic Press, 892 pp. (Paperback edition, 1981.)

———. 1977a. The contributions of terrestrial plant studies to the development of the theory of ecology. In: *Changing Scenes in Natural Science,* C. E. Goulden, ed. Special Publication 12, Academy of Natural Sciences, Philadelphia. Lancaster, Pa., Fulton Press, pp. 139–157.

Harper, J. L., and J. N. Clatworthy. 1963. The comparative biology of closely related species. VI. Analysis of the growth of *Trifolium repens* and *T. fragifesum* in pure and mixed populations. J. Exp. Bot. 14:172–190.

Harper, J. L., and J. White. 1974. The demography of plants. Ann. Rev. Ecol. Syst. 5:419–463.

Harris, E. 1959. The nitrogen cycle in Long Island Sound. Bull. Bingham Oceanogr. Coll. 17:31–65.

Harris, P. 1974. A possible explanation of plant yield increases following insect damage. AgroEcosystems 1:219–225.

Harrison, A. D. 1962. Hydrobiological studies of all saline and acid still waters in Western Cape Province. Trans. R. Soc. S. Afr. 36:213.

Hart, J. S. 1952. Lethal temperatures of fish from different latitudes. Publ. Ontario Fish. Res. Lab. 72:1–79.

Hart, M. L. (ed.). 1962. Fundamentals of dry-matter production and distribution. Neth. J. Agr. Sci. 10:309–444 (special issue).

Hartenstein, Ray. 1981. Sludge decomposition and stabilization. Science 212:743–749.

Harvey, H. W. 1950. On the production of living matter in the sea off Plymouth. J. Marine Biol. Assoc. U.K. n.s. 29:97–137.

Harwell, M. A., W. P. Cropper, and H. L. Ragsdale. 1977. Nutrient cycling and stability: A reevaluation. Ecology 58:660–666.

Haskell, E. F. 1940. Mathematical systemization of "environment," "organism" and "habitat." Ecology 21:1–16.

———. 1949. A clarification of social science. Main Currents in Modern Thought 7:45–51.

Hasler, A. D. 1947. Eutrophication of lakes by domestic drainage. Ecology 28:383–395.

———. 1969. Cultural eutrophication is reversible. BioScience 19:425–431.

——— (ed.). 1975. *Coupling of Land and Water Systems.* New York, Springer-Verlag, 309 pp.

Hawkes, H. A. 1963. *The Ecology of Waste Water Treatment.* New York, Pergamon Press, 203 pp.

Hawkins, Arthur S. 1940. A wildlife history of Faville Grove, Wisconsin. Trans. Wisc. Acad. Sci. Arts Letters 32:39–65.

Hawrylyshyn, Bohdan. 1980. *Road Maps to the Future: Toward More Effective Societies.* A Report to the Club of Rome. New York, Pergamon Press.

Hayes, F. R., and C. C. Coffin. 1951. Radioactive phosphorus and exchange of lake nutrients. Endeavour 10:78–81.

Hazard, T. P., and R. E. Eddy. 1950. Modification of the sexual cycle in the brook trout (*Salvelinus fontinalis*) by control of light. Trans. Am. Fish. Soc. 80:158–162.

Heald, E. J., and W. E. Odum. 1970. The contribution of mangrove swamps to Florida fisheries. Proc. Gulf Carib. Fish. Inst. 22:130–135.

Heath, Robert T. 1979. Holistic study of an aquatic microcosm: Theoretical and practical implications. Int. J. Environ. Studies 13(2):87–93.

Heatwold, Harold, and K. Levine. 1972. Trophic structure, stability and faunal change during recolonization. Ecology 53:531–534.

Hedgpeth, Joel W. (ed.). 1957. *Treatise on Marine Ecology and Paleoecology.* Vol. 1, *Ecology.* New York, The Geological Society of America, 1296 pp.

Hegner, Robert. 1938. *Big Fleas Have Little Fleas,* or *Who's Who Among the Protozoa.* Baltimore, Williams & Wilkins.

Heichel, G. H. 1976. Agricultural production and energy resources. Am. Sci. 64:64–72.

Heinrich, Bernd. 1979. *Bumblebee Economics.* Cambridge, Mass., Harvard University Press, 224 pp.

———. 1980. The role of energetics in bumblebee-flower interrelationships. In: *Coevolution of Animals and Plants,* L. E. Gilbert and P. H. Raven, eds. Austin, University of Texas Press, pp. 141–158.

Heller, Alfred (ed.). 1972. *The California Tomorrow Plan.* Los Altos, Calif., William Kaufmann.

Henderson, G. S. 1975. Letter to editor. BioScience 25:770.

Henderson, Hazel. 1978. *Creating Alternate Futures: The End of Economics.* New York, Berkley, 403 pp.

Henderson, J. V. 1974. Optimum city size: The external diseconomy questions. J. Polit. Econ., Chicago, (82) n.r. 2, p. 373.

Hendrix, P. F., C. L. Langner and E. P. Odum. 1982. Cadmium in aquatic microcosms. Environ. Mgmt. 6:543–553.

Henle, P. 1942. The status of emergence. J. Philos. 39:483–493.

Henry, S. M. (ed.). 1966. *Symbiosis.* Two volumes. New York, Academic Press. Vol. 1, 478 pp.; Vol. 2, 400 pp.

Herrera, R., C. F. Jordan, H. Klinge, and E. Medina. 1978. Amazon ecosystems. Their structure and functioning with particular emphasis on nutrients. Interciencia 3:223–232.

Hibbert, A. R. 1967. Forest treatment effects on water yield. In: *International Symposium on Forest Hydrology,* W. E. Sopper and H. W. Lull, eds. New York, Pergamon Press, pp. 527–543.

Hickey, J. J., and D. W. Anderson. 1968. Chlorinated hydrocarbons and egg shell changes in raptorial and fish-eating birds. Science 162:271–272.

Hickling, C. F. 1948. Fish farming in the Middle and Far East. Nature 161:743–751.

Hill, James, and S. L. Durham. 1978. Input, signals and control in ecosystems. In: Proc. 1978 Conference on Acoustics, Speech and Signal Processing. Tulsa, Okla. New York, Inst. Electrical and Electronics Engineers, pp. 391–397.

Hill, James, and R. G. Wiegert. 1980. Microcosms in ecological modelling. In: *Microcosms in Ecological Research,* J. P. Giesy, ed. U.S. Dept. of Energy Symp. 52. Springfield, Va., National Technical Information Service, pp. 138–163.

Hills, G. A. 1952. The classification and evaluation of site for forestry. Res. Rep. No. 24, Ontario Dept. of Lands and Forests.

Hirsch, Fred. 1978. *Social Limits to Growth.* Cambridge, Mass., Harvard University Press, 208 pp.

Hjort, John. 1926. Fluctuations in the year classes of important food fishes. J. du Conseil Permanent Internationale pour l'Exploration de la Mer 1:1–38.

Ho, M. W., and P. T. Saunders. 1979. Beyond neo-Darwinism, an epigenetic approach to evolution. J. Theoret. Biol. 78:573–591.

Hoch, Irving. 1976. City size: Effects, trends and policies. Science 193:856–863.

Holdgate, M. W., and M. J. Woodman (eds.). 1978. *The Breakdown and Restoration of Ecosystems.* New York, Plenum, 496 pp.

Holdren, John P., and Paul R. Ehrlich. 1974. Human population and the global environment. Am. Sci. 62:282–292.

Holeman, J. N. 1968. The sediment yield of major rivers of the world. Water Res. 4:737–747.

Holling, C. S. 1961. Principles of insect predation. Ann. Rev. Entomol. 6:163–182.

———. 1965. The functional response of predators to prey density and its role in mimicry and population regulation. Mem. Entomol. Soc. Canada, No. 45, 60 pp.

———. 1966. The functional response of invertebrate predators to prey density. Mem. Entomol. Soc. Canada, No. 48, 86 pp.

———. 1973. Resilience and stability of ecological systems. Ann. Rev. Ecol. Syst. 4:1–23.

——— (ed.). 1978. *Adaptive Environmental Assessment and Management.* New York, Wiley-Interscience.

———. 1980. Forest insects, forest fires, and resilience. In: *Fire Regimes and Ecosystem Properties*, H. Mooney, J. M. Bonnicksen, N. L. Christensen, J. E. Lotan, and W. E. Reiners, eds. USDA Forest Service Gen. Tech. Rep., WO-26.

Holt, S. J. 1969. The food resources of the ocean. Sci. Am. 221(3):178–194.

Hopkinson, C. S., and J. W. Day. 1980. Net energy analysis of alcohol production from sugarcane. Science 207:302–304.

Horn, H. S. 1974. The ecology of secondary succession. Ann. Rev. Ecol. Syst. 5:25–37.

———. 1975. Forest succession. Sci. Am. 232:90–98.

———. 1981. Succession. In: *Theoretical Ecology* (2nd ed.), R. M. May, ed. Sunderland, Mass., Sinauer Associates.

Howarth, Robert W., and J. M. Teal. 1979. Sulfate reduction in a New England salt marsh. Limnol. Oceanogr. 24:999–1113.

———. 1980. Energy flow in a salt marsh ecosystem: The role of reduced inorganic sulfur compounds. Am. Nat. 116:862–872.

Hubbell, Stephen P. 1979. Tree dispersion, abundance and diversity in a tropical dry forest. Science 203:1299–1309.

Huber, D. M., H. L. Warren, D. W. Nelson, and C. Y. Tsai. 1977. Nitrification inhibitors—new tools for food production. BioScience 27:523–529.

Huettner, D. A. 1976. Net energy analysis: An economic assessment. Science 192:101–104.

Huffaker, C. B. 1957. Fundamentals of biological control of weeds. Hilgardia 27:101–167.

———. 1959. Biological control of weeds with insects. Ann. Rev. Entomol. 4:251–276.

Huffaker, C. B., and C. E. Kennett. 1956. Experimental studies on predation: Predation and cyclamen-mite populations on strawberries in California. Hilgardia 26:191–222.

Hulbert, M. King. 1971. The energy resources of the earth. Sci. Am. 224(3):60–70.

Humphreys, W. F. 1978. Ecological energetics of *Geolycosa godeffroyi* (Aruneae: Lycosidae) with an appraisal of production efficiency of ectothermic animals. J. Anim. Ecol. 47:627–652.

Hungate, F. P. 1963. Symbiotic associations: The rumen bacteria. In: *Symbiotic Associations.* 13th Symp. Soc. Gen. Microbiol. New York, Cambridge University Press, pp. 266–297.

———. 1966. *The Rumen and Its Microbes.* New York, Academic Press, 533 pp.

———. 1975. The rumen microbial ecosystem. Ann. Rev. Ecol. Syst. 6:39–66.

Hunter, W. S., and W. B. Vernberg. 1955. Studies on oxygen consumption of digenetic trematodes. II. Effects of two extremes in oxygen tension. Exp. Parasit. 4:427–434.

Huston, Michael. 1979. A general hypothesis of species diversity. Am. Nat. 113:81–101.

Hutcheson, Kermit. 1970. A test for comparing diversities based on the Shannon formula. J. Theoret. Biol. 29:151–154.

Hutchinson, G. E. 1944. Nitrogen and biogeochemistry of the atmosphere. Am. Sci. 32:178–195.

———. 1948. On living in the biosphere. Sci. Monthly 67:393–398.

———. 1948a. Circular causal systems in ecology. Ann. N.Y. Acad. Sci. 50:221–246.

———. 1950. Survey of contemporary knowledge of biogeochemistry. III. The biogeochemistry of vertebrate excretion. Bull. Am. Mus. Nat. Hist. 95:554.

———. 1953. The concept of pattern in ecology. Proc. Acad. Nat. Sci. (Phila.) 105:1–12.

———. 1957. *A Treatise on Limnology.* Vol. I, *Geography, Physics and Chemistry.* New York, John Wiley & Sons, 1015 pp.

———. 1957a. Concluding remarks. Cold Spring Harbor Symp. Quant. Biol. 22:415–427.

———. 1959. Homage to Santa Rosalina, or why are there so many kinds of animals? Am. Nat. 93:145–159.

———. 1961. The paradox of the plankton. Am. Nat. 107:406–425.

———. 1964. The lacustrine microcosm reconsidered. Am. Sci. 52:331–341.

———. 1965. The niche: An abstractly inhabited hyper-volume. In: *The Ecological Theatre and the Evolutionary Play*. New Haven, Yale University Press, pp. 26–78.

———. 1967. *A Treatise on Limnology*. Vol. II, *Introduction to Lake Biology and the Limnoplankton*. New York, John Wiley & Sons, 1115 pp.

———. 1967a. Ecological biology in relation to the maintenance and improvement of the human environment. In: *Applied Science and Technical Progress*. Proc. Natl. Acad. Sci. USA, pp. 171–184.

———. 1978. *An Introduction to Population Biology*. New Haven, Yale University Press, 260 pp.

Hutchinson, G. E., and V. T. Bowen. 1948. A direct demonstration of phosphorus cycle in a small lake. Proc. Natl. Acad. Sci. USA 33:148–153.

———. 1950. Limnological studies in Connecticut; quantitative radiochemical study of the phosphorus cycle in Linsley Pond. Ecology. 31:194–203.

Hutner, S. H., and L. Provasoli. 1964. Nutrition of algae. Am. Rev. Plant Physiol. 15:37–56.

Huxley, Julian. 1935. Chemical regulation and the hormone concept. Biol. Rev. 10:427.

Huxley, T. H. 1863. *Evidence as to Man's Place in Nature*. (Reprinted with an introduction by Ashley Montagu, Ann Arbor Paperbacks, University of Michigan Press, 1959, 184 pp.)

Hynes, H. B. N. 1960. *The Biology of Polluted Waters*. Liverpool, Liverpool University Press, 202 pp.

Itô, Yosiaki. 1959. A comparative study on survivorship curves for natural insect populations. Jpn. J. Ecol. 9:107–115.

———. 1971. Some notes on the competitive exclusion principle. Res. Pop. Ecol. XIII:46–54.

Ivlev, V. S. 1945. The biological productivity of waters. In: *Uspekhi Soureminnoi Biologii (Advances in Modern Biology)* 19:98–120. (In Russian; English translation by W. E. Ricker.) J. Fish. Res. Bd. Can. 23:1727–1759, 1966.

Jackson, J. B. C. 1968. Bivalves; spatial and size-frequency distributions of two intertidal species. Science 161:479–480.

Jahoda, Marie. 1982. Wholes and parts; meaning and mechanisms. Nature 296:8–14.

Jannasch, H. W. 1969. Estimations of bacterial growth rates in natural waters. J. Bacteriol. 99:156–160.

Jannasch, H. W., and G. E. Jones. 1959. Bacterial populations in seawater as determined by different methods of enumeration. Limnol. Oceanogr. 4:128–138.

Jansen, D. H. 1966. Coevolution of mutualism between ants and acacias in Central America. Evolution 20:249–275.

———. 1967. Interaction of the bull's horn acacia (*Acacia cornigera* L.) with an ant inhabitant (*Pseudomyrmex ferruginea* F. Smith) in eastern Mexico. Univ. Kansas Sci. Bull. 57:315–558. (See also Ecology 48:26–35, 1967.)

Jantsch, E. 1980. *The Self-Organizing Universe*. Oxford, Pergamon Press, 343 pp.

Jassby, Alan D., and Charles R. Goldman. 1974. A quantitative measure of succession role and its application to the phytoplankton of lakes. Am. Nat. 108:688–693.

Jeffries, H. P. 1979. Biochemical correlates of a seasonal change in marine communities. Am. Nat. 113:643–658.

Jenkins, James H., and T. T. Fendley. 1968. The extent of contamination, detention and health significance of high accumulation of radioactivity in southeastern game populations. Proc. 22nd Annual Conf. Southeast Assoc. Game and Fish Commissioners 22:89–95.

Jennings, T. J., and J. P. Barkham. 1979. Niche separation in woodland slugs. Oikos 33:127–131.

Jenny, H., R. J. Arkley, and A. M. Schultz. 1969. The pygmy forest-podsol ecosystem and its dune associates of the Mendocino coast. Madrono 20:60–75.

Johannes, R. E. 1964. Phosphorus excretion in marine animals; microzooplankton and nutrient regeneration. Science 146:923–924.

———. 1965. The influence of marine protozoa on nutrient regeneration. Limnol. Oceanogr. 10:434–442.

———. 1968. Nutrient regeneration in lakes and oceans. In: *Advances in Microbiology of the Sea*, M. Droop and E. J. Ferguson Wood, eds. Vol. I. New York, Academic Press, pp. 203–213.

Johnson, L. E., L. R. Albee, R. O. Smith, and A. L. Moxon. 1951. Cows, calves and grass. South Dakota Agr. Expt. Sta. Bull. No. 412.

Johnson, M. P., and P. H. Raven. 1970. Natural regulation of plant species diversity. Evol. Biol. 4:127–162.

Johnson, Philip L. (ed.). 1969. *Remote Sensing in Ecology*. Athens, University of Georgia Press, 244 pp.

———. 1970. Dynamics of carbon dioxide and productivity in an arctic biosphere. Ecology 51:73–80.

Johnson, W. Carter, and D. M. Sharpe. 1976. An analysis of forest dynamics in the northern Georgia piedmont. Forest. Sci. 22:307–322.

Johnston, David W., and Eugene P. Odum. 1956.

Breeding bird populations in relation to plant succession on the Piedmont in Georgia. Ecology 37:50–62.

Jones, J. R. Erichsen. 1949. A further ecological study of a calcareous stream in the "Black Mountain" district of south Wales. J. Anim. Ecol. 18:142–159.

Jones, Thomas E. 1977. Current prospects of sustainable economic growth. In: *Goals in a Global Community*. (The original background papers for *Goals of Mankind*, A Report to the Club of Rome.) E. Laszlo and J. Bierman, eds. New York, Oxford, Frankfurt, Pergamon Press, pp. 117–179.

Jordan, Carl F. 1971. A world pattern in plant energetics. Am. Sci. 59(4):425–433.

Jordan, Carl F., and Rafael Herrera. 1981. Tropical rain forests: Are nutrients really critical? Am. Nat. 117:167–180.

Jordan, Carl F., and Jerry R. Kline. 1972. Mineral cycling: Some basic concepts and their application in a tropical rain forest. Ann. Rev. Ecol. Syst. 3:33–50.

Jordan, C. F., J. R. Kline, and D. E. Sasscer. 1972. Relative stability of mineral cycles in forest ecosystems. Am. Nat. 106:237–253.

Jorgenson, B. B., and T. M. Fenchel. 1974. The sulfur cycle of a marine sediment model system. Marine Biol. 24:189–201.

Juday, C. 1940. The annual energy budget of an inland lake. Ecology 21:438–450.

———. 1942. The summer standing crop of plants and animals in four Wisconsin lakes. Trans. Wisc. Acad. Sci. 34:103–135.

Kahl, M. Philip. 1964. The food ecology of the wood stork. Ecol. Monogr. 34:97–117.

Kahn, Alfred E. 1966. The tyranny of small decisions: Market failures, imperfections and the limits of economies. Kyklos 19:23–47.

Kahn, Herman, and A. Wiener. 1967. *The Year 2000*. New York, William Morrow.

Kahn, Herman, William Broun, and Martel Leon. 1976. *The Next 200 Years*. New York, William Morrow.

Kale, Herbert W. 1965. Ecology and bioenergetics of the long-billed marsh wren *Telmatodytes palustris griseus* (Brewster) in Georgia salt marshes. Publ. No. 5. Cambridge, Mass., Nuttall Ornithol. Club, 142 pp.

Kamen, Martin D. 1953. Discoveries in nitrogen fixation. Sci. Am. 188:38–42.

Kamen, Martin D., and H. Gest. 1949. Evidence for a nitrogenase system in the photosynthetic bacterium *Rhodospirillum rubrum*. Science 109:560.

Kardos, L. T. 1967. Waste water renovation by the land: A living filter. In: *Agriculture and Quality of Our Environment*, Brady, ed. Publ. No. 85. Washington, D.C., American Association for the Advancement of Science, pp. 241–250.

Karl, D. M., C. O. Wirsen, and H. W. Jannasch. 1980. Deep-sea primary production at the Galapagos hydrothermal vents. Science 207:1345–1347.

Kaushik, N. K., and H. B. N. Hynes. 1968. Experimental study on the role of autumn-shed leaves in aquatic environments. J. Ecol. 56:229–243.

Keever, Catherine. 1950. Causes of succession on old fields on the Piedmont, North Carolina. Ecol. Monogr. 20:229–250.

———. 1955. *Heterotheca latifolia*, a new and aggressive exotic dominant in Piedmont old-field succession. Ecology 36:732–739.

Keith, Lloyd B. 1963. *Wildlife's Ten-year Cycle*. Madison, University of Wisconsin Press, 201 pp.

Kellogg, Charles E. 1975. *Agricultural Development, Soil, Food, People, Work*. Madison, Wis., Soil Sci. Soc. of America, Inc.

Kellogg, W. W., R. D. Cadle, E. R. Allen, A. L. Lazrus, and E. A. Martell. 1972. The sulfur cycle. Science 175:587–596.

Kendeigh, S. Charles. 1934. The role of environment in the life of birds. Ecol. Monogr. 4:299–417.

———. 1949. Effect of temperature and season on energy resources of the English Sparrow. Auk 66:113–127.

———. 1974. *Ecology*. Englewood Cliffs, N.J., Prentice-Hall, 468 pp.

Kennedy, J. S. 1968. The motivation of integrated control. J. Appl. Ecol. 4:492–499.

Kercher, J. R., and H. H. Shugart. 1975. Trophic structure, effective trophic position, and connectivity in food webs. Am. Nat. 109:191–206.

Kettlewell, H. B. D. 1956. Further selection experiments on industrial melanism in the Lepidoptera. Heredity 10:287–301.

Keyfitz, N., and W. Flieger. 1968. *World Population: An Analysis of Vital Data*. Chicago, University of Chicago Press.

Kimball, James W. 1948. Pheasant population characteristics and trends in the Dakotas. Trans. 13th N. A. Wildl. Conf. 13:291–314.

Kinne, Otto. 1956. Uber den Einfluss des Salzgehaltes und der Temperatur auf Wachstum, Form und Vermehrung bei dem Hydroidpolypen Cordylophoral *caspia* (Pallas), Thecata, Clavidae. Zool. Jahrb. 66:565–638.

Kira, T., and T. Shidei. 1967. Primary production and turnover of organic matter in different forest ecosystems of the western Pacific. Jpn. J. Ecol. 17:70–87.

Kitchell, J. F., R. V. O'Neill, D. Webb, G. W. Gallepp, S. M. Bartell, J. F. Keonce, and B. S. Ausmus. 1979. Consumer regulation of nutrient cycling. BioScience 29:28–34.

Klages, K. H. W. 1942. *Ecological Crop Geography*. New York, Macmillan.

Kleiber, Max. 1961. *The Fire of Life*. New York, John Wiley & Sons.

Klopfer, P. H., and J. P. Hailman. 1967. *An Introduction to Animal Behavior; Ethology's First Century*. Englewood Cliffs, N.J., Prentice-Hall, 297 pp.

Kneese, A. V. 1973. The Faustian bargain. Resources. 44:1–4.

Kneese, A. V., R. V. Ayres and R. D'Argo. 1970. *Economics and Environment*. Baltimore, Johns Hopkins Press.

Koestler, A. 1969. Beyond atomism and holism—the concept of holon. In: *Beyond Reductionism*. The Alpbach Symposium 1968. London, Hutchinson, pp. 192–232.

Komarek, E. V. 1967. Fire and the ecology of man. Proc. 6th Annual Tall Timbers Fire Ecol. Conf., Tallahassee, Fla., pp. 143–170.

________ (ed.). 1969. *Ecological Animal Control by Habitat Management*. Tallahassee, Fla., Tall Timbers Research Station, 244 pp.

Kononova, M. M. 1961. *Soil Organic Matter: Its Nature, Its Role in Soil Formation and in Soil Fertility*. (Translated from Russian by T. Z. Nowankowski and G. A. Greenwood.) New York, Pergamon Press, 450 pp.

Kozlowski, D. G. 1968. A critical evaluation of the trophic level concept. (1) Ecological efficiencies. Ecology 49:48–116.

Kozlowski, T. T. 1964. *Water Metabolism in Plants*. New York, Harper & Row.

________ (ed.). 1968. *Water Deficits and Plant Growth*. Two vols. New York, Academic Press.

________. 1980. Impacts of air pollution on forest ecosystems. BioScience 30:88–93.

Kozlowski, T. T., and C. E. Ahlgren (eds.). 1974. *Fire and Ecosystems*. New York, Academic Press, 538 pp.

Krebs, C. J. 1978. A review of the Chitty hypothesis of population regulation. Can. J. Zool. 56:2463–2480.

Krebs, C. J., and J. H. Meyers. 1974. Population cycles in small mammals. Adv. Ecol. Res. 8:267–349.

Krenkel, R. A., and F. L. Parker (eds.). 1969. *Biological Aspects of Thermal Pollution*. Nashville, Tenn., Vanderbilt University Press.

Krutilla, John V. 1967. Conservation reconsidered. Am. Econ. Rev. 59:777–786.

Kuentzel, L. E. 1969. Bacteria, carbon dioxide and algal blooms. J. Water Poll. Control Fed. 41:1737–1747.

Kuenzler, E. J. 1958. Niche relations of three species of Lycosid spiders. Ecology 39:494–500.

________. 1961. Phosphorus budget of a mussel population. Limnol. Oceanogr. 6:400–415.

________. 1961a. Structure and energy flow of a mussel population. Limnol. Oceanogr. 6:191–204.

Kushlan, James A. 1979. Design and management of continental wildlife reserves: Lesions from the Everglades. Biol. Conserv. 15:281–290.

Kyke, G. H., H. P. Pulliam, and E. L. Charnov. 1977. Optimal foraging: A selective review of theory and test. Quart. Rev. Biol. 52:137–154.

Lack, David L. 1945. Ecology of closely related species with special reference to cormorant (*Phalacrocorax carbo*) and shag (*P. aristotelis*). J. Anim. Ecol. 14:12–16.

________. 1947. *Darwin's Finches*. New York, Cambridge University Press.

________. 1947. The significance of clutch size. Ibis 89:302–352 (continued 90:25–45, 1948).

________. 1954. *The Natural Regulation of Animal Numbers*. New York, Oxford University Press.

________. 1966. *Population Studies of Birds*. Oxford, Clarendon Press, 341 pp.

________. 1969. Tit niches in two worlds; or homage to Evelyn Hutchinson. Am. Nat. 103:43–49.

Landsberg, Hans (ed.). 1979. *Energy: The Next 20 Years*. Report Ford Foundation and Resources for Future. Cambridge, Mass., Ballinger.

Langdale, G. W., A. P. Barnett, L. Leonard and W. G. Fleming. 1979. Reduction of soil erosion by the no-till system in the southern Piedmont. J. Soil, Water Conserv. 34:226–228.

Lange, O. L., W. Kock, and E. D. Schulze. 1969. CO_2-gas exchange and water relationships of plants in the Negev Desert at the end of the dry period. Ber. Dtsch. Bot. Ges. 82:39–61.

Lange, W. 1967. Effect of carbohydrates on the symbiotic growth of planktonic blue-green algae with bacteria. Nature 215:2177.

LaPorte, Leo F. (ed.). 1972. *The Earth and Human Affairs*. Rept. Comm. Geol. Sci. Nat. Acad. Sci., Nat. Res. Council. San Francisco, Canfield Press.

Larkin, P. A. 1963. Interspecific competition and exploitation. J. Fish. Res. Bd. Canada 20:647–678.

Laskey, Amelia R. 1939. A study of nesting eastern bluebirds. Birdbanding 10:23–32.

Laszlo, Ervin. 1972. *The Systems View of the World*. New York, George Braziller, 131 pp.

________ (ed.). 1977. *Goals for Mankind: A Report to the Club of Rome on the New Horizons of Global Community*. New York, Dutton, 374 pp.

________. 1977a. The Club of Rome of the future vs. the future of the Club of Rome. In: *Goals in a Global Community*, E. Laszlo and J. Bierman, eds. New York, Pergamon Press, pp. 281–285.

Laszlo, E., and Margenau, H. 1972. The emergence of integrating concepts in contemporary science. Philos. Sci. 39:252–259.

Lawton, J. H. 1981. Moose, wolves, daphnia and hydra: On the ecological efficiency of endotherms and ectotherms. Am. Nat. 117:782–783.

Leffler, J. W. 1978. Ecosystem responses to stress in aquatic microcosms. In: *Energy and Environmental Stress in Aquatic Systems.* J. H. Thorp and J. W. Gibbons, eds. U.S. Dept. of Energy. Springfield, Va., National Technical Info. Center.

———. 1980. Microcosmology: Theoretical applications of biological models. In: *Microcosms in Ecological Research,* John Giesy, ed. U.S. Dept. of Energy. Springfield, Va., National Technical Info. Center, pp. 14–29.

Lemon, E. R. 1960. Photosynthesis under field conditions. II. An aerodynamic method for determining the turbulent carbon dioxide exchange between the atmosphere and a corn field. Agron. J. 52:697–703.

Lent, C. M. (ed.). 1969. Adaptations of intertidal organisms. Am. Zool. 9:269–426.

Leontief, Wassily. 1966. *Input-Output Economics.* New York, Oxford University Press.

———. 1980. The world economy of the year 2000. Sci. Am. 243(3):206–231.

———. 1982. Academic economics. Letter to the editor. Science 217:104, 107.

Leopold, Aldo. 1933. *Game Management.* New York, Charles Scribner's Sons.

———. 1933a. The conservation ethic. J. Forestry 31:634–643.

———. 1943. Deer irruptions. Wisc. Cons. Bull., August 1943. Reprinted in Wisc. Cons. Dept. Publ. 321:3–11.

———. 1949. The land ethic. In: *A Sand County Almanac.* New York, Oxford University Press.

Leopold, Luna B. 1974. *Water: A Primer.* San Francisco, W. H. Freeman.

Leopold, Luna B., and W. B. Langbein. 1962. The concept of entropy in landscape evolution. U.S. Geological Survey Paper 500 A.

Leopold, Luna B., and Thomas Maddock. 1954. *The Flood Control Controversy: Big Dams, Little Dams, and Land Management.* New York, The Ronald Press.

Lephowski, Will. 1979. The social thermodynamics of Ilya Prigogine. Chem. Engr. News 57(16):30–33.

Leslie, P. H. 1945. On the use of matrices in certain population mathematics. Biometrika 33:183–212.

Leslie, P. H., and Thomas Park. 1949. The intrinsic rate of natural increase of *Tribolium castaneum* Herbst. Ecology 30:469–477.

Leslie, P. H., and R. M. Ranson. 1940. The mortality, fertility, and rate of natural increase of the vole (*Microtus agrestis*) as observed in the laboratory. J. Anim. Ecol. 9:27–52.

Levin, D. A. 1976. Alkaloid-bearing plants: An ecogeographic perspective. Am. Nat. 110:261–284.

Levin, Simon, and David Pimentel. 1981. Selection of intermediate rates of increase in parasite-host systems. Am. Nat. 117:308–315.

Levins, Richard. 1966. The strategy of model building in population biology. Am. Sci. 54:421–431.

———. 1968. *Evolution in Changing Environments.* Princeton, N.J., Princeton University Press.

Lewin, J. C. 1963. Heterotrophy in marine diatoms. In: *Marine Microbiology,* C. H. Oppenheimer, ed. Springfield, Ill., Charles C Thomas, pp. 229–235.

Lewontin, R. C. (ed.). 1968. *Population Biology and Evolution.* Syracuse, N.Y., Syracuse University Press, 205 pp.

Liebig, Justus. 1840. *Chemistry in Its Application to Agriculture and Physiology.* London, Taylor and Walton. (4th ed., 1847.)

Lieth, Helmut (ed.). 1962. *Die Stoffproduktion der Pflanzendecke.* Stuttgart, Gustav Fischer Verlag, 156 pp.

Lieth, Helmut, and R. H. Whittaker (eds.). 1975. *Primary Productivity of the Biosphere.* New York, Springer-Verlag, 340 pp.

Ligon, J. D. 1968. Sexual differences in foraging behavior in two species of *Dendrocopus* woodpeckers. Auk 85:203–215.

Likens, G. E., and F. H. Bormann. 1974. Linkages between terrestrial and aquatic ecosystems. BioScience 24:447–456.

———. 1974a. Acid rain: A serious regional environmental problem. Science 184:1176–1179.

Likens, G. E., F. H. Bormann, R. S. Pierce, J. S. Eaton, and N. M. Johnson. 1977. *Biogeochemistry of a Forested Ecosystem.* New York, Springer-Verlag, 146 pp.

Likens, Gene E., Richard F. Wright, James N. Galloway, and Thomas J. Butler. 1979. Acid rain. Sci. Am. 241(4):43–51.

Lloyd, M., and R. J. Ghelardi. 1964. A table for calculating the equitability component of species diversity. J. Anim. Ecol. 33:421–425.

Lorenz, K. Z. 1966. *Evolution and Modification of Behavior.* London, Methuen.

Lotka, A. J. 1925. *Elements of Physical Biology.* Baltimore, Williams & Wilkins, 460 pp. (Reprinted as *Elements of Mathematical Biology.* New York, Dover, 1956.)

Lovelock, James E. 1979. *Gaia: A New Look at Life on Earth.* New York, Oxford University Press, 157 pp.

Lovelock, J. E., and S. R. Epton. 1975. The quest for Gaia. New Scientist 65:304–306.

Lovelock, J. E., and Lynn Margulis. 1973. Atmospheric homeostasis by and for the biosphere: The Gaia hypothesis. Tellus 26:1–10.

Lovins, Amory B. 1977. *Soft Energy Paths.* Cambridge, Mass., Ballinger.

Lowe, C. H., and W. G. Heath. 1969. Behavioral and physiological responses to temperature in the desert pupfish *Cyprinodon* maculariusi. Physiol. Zool. 42:53–59.

Lowry, A. 1974. A note on emergence. Mind 83:276–277.

Lowry, William P. 1967. The climate of cities. Sci. Am. 217(2):15–23.

Lucas, C. E. 1947. The ecological effects of external metabolites. Biol. Rev. Cambridge Philos. Soc. 22:270–295.

Ludwig, D., D. D. Jones, and C. S. Holling. 1978. Qualitative analysis of insect outbreak systems: The spruce budworm and forest. J. Anim. Ecol. 47:315–332.

Lugo, A. E., and M. H. Brinson. 1978. Calculations of the value of salt water wetlands. In: *Wetland Functions and Value: The State of Our Understanding,* P. E. Greeson, J. R. Clark, and J. E. Clark, eds. Minneapolis, Am. Water Resources Assoc., pp. 120–130.

Lugo, A. E., E. G. Farnworth, D. Pool, P. Jerez, and G. Kaufman. 1973. The impact of the leaf cutter ant *Attica colombia* on the energy flow of a tropical wet forest. Ecology 54:1292–1301.

Lynch, Kevin. 1965. The city as environment. Sci. Am. 213(3):209.

Lyons, N. I. 1981. Comparing diversity indices based on counts weighted by biomass or other importance values. Am. Nat. 118:438–442.

MacArthur, Robert H. 1955. Fluctuations of animal populations and a measure of community stability. Ecology 36:533–536.

________. 1957. On the relative abundance of bird species. Proc. Natl. Acad. Sci. USA 45:293–295.

________. 1958. Population ecology of some warblers of northeastern coniferous forest. Ecology 39:599–619.

________. 1960. On the relative abundance of species. Am. Nat. 94:25–36.

________. 1965. Patterns of species diversity. Biol. Rev. 40:410–533.

________. 1968. The theory of the niche. In: *Population Biology and Evolution,* R. C. Lewontin, ed. Syracuse, N.Y., Syracuse University Press, pp. 159–176.

________. 1972. *Geographical Ecology: Patterns in the Distribution of Species.* New York, Harper & Row, 269 pp.

MacArthur, R. H., and Joseph Connell. 1966. *The Biology of Populations.* New York, John Wiley & Sons, 200 pp.

MacArthur, R. H., and E. R. Pianka. 1966. On optimal use of a patchy environment. Am. Nat. 100:603.

MacArthur, R. H., and E. O. Wilson. 1963. An equilibrium theory of insular zoogeography. Evolution 17:373–387.

________. 1967. *The Theory of Island Biogeography.* Princeton, N.J., Princeton University Press, 208 pp.

MacDonald, D. R. 1975. Biological interactions associated with spruce budworm infestations. In: *Ecological Effects of Nuclear War,* G. Woodwell, ed. Brookhaven National Laboratory Publ. No. 917, pp. 61–68.

MacElroy, R. D., and M. M. Averner. 1978. Space ecosynthesis: An approach to the design of closed ecosystems for use in space. NASA Tech. Memo. 78491. Moffet Field, Calif., National Aeronautics and Space Administration, Ames Research Center, 31 pp.

MacFadyen, A. 1949. The meaning of productivity in biological systems. J. Anim. Ecol. 17:75–80.

________. 1963. *Animal Ecology: Aims and Methods* (2nd ed.). London, Pitman, 344 pp.

Mackereth, F. J. H. 1965. Chemical investigations of lake sediments and their interpretation. Proc. Soc. London, Series B, 161:295–309.

MacLulich, D. A. 1937. Fluctuations in the numbers of the varying hare (*Lepus americanus*). U. of Toronto Studies, Biol. Ser., No. 43.

Major, Jack. 1969. Historical development of the ecosystem concept. In: *The Ecosystem Concept in Natural Resource Management,* G. M. Van Dyne, ed. New York, Academic Press, pp. 9–22.

Malthus, T. R. 1798. *An Essay on the Principle of Population.* London, Johnson. (Reprinted in Everyman's Library, 1914.)

Margalef, Ramon. 1958. Temporal succession and spatial heterogeneity in phytoplankton. In: *Perspectives in Marine Biology,* Buzzati-Traverso, ed. Berkeley, University of California Press, pp. 323–347.

________. 1958a. Information theory in ecology. Gen. Syst. 3:36–71.

________. 1961. Communication of structure in plankton populations. Limnol. Oceanogr. 6(2):124–128.

________. 1963. Successions of populations. Adv. Frontiers Plant Sci. (Institute Adv. Sci. and Culture, New Delhi, India) 2:137–188.

________. 1963a. On certain unifying principles in ecology. Am. Nat. 97:357–374.

________. 1967. Concepts relative to the organization of plankton. Oceanogr. Marine Biol. Ann. Rev. 5:257–289.

________. 1968. *Perspectives in Ecological Theory.* Chicago, University of Chicago Press, 112 pp.

________. 1974. *Ecologia.* Barcelona, Ediciones Omega.

________. 1979. The organization of space. Oikos 33:152–159.

Margulis, Lynn. 1981. *Symbiosis in Cell Evolution; Life and Its Environment on the Early Earth.* San Francisco, W. H. Freeman, 419 pp.

________. 1982. *Early Life.* Boston, Science Books International, 160 pp.

Margulis, Lynn, and J. E. Lovelock. 1974. Biological modulation of the earth's atmosphere. Icarus 21:471–489.

———. 1975. The atmosphere as circulatory system of the biosphere—The Gaia hypothesis. Coevolution Quarterly, Summer 1975.

Margulis, Lynn, and K. Schwartz. 1982. *Five Kingdoms: An Illustrated Guide to the Phyla of Life on Earth.* San Francisco, W. H. Freeman.

Marks, G. C., and T. T. Kozlowski (eds.). 1973. *Ectomycorrhizae, Their Ecology and Physiology.* New York, Academic Press.

Marks, P. L. 1974. The role of the pin cherry (*Prunus pennsylvania*) in the maintenance of stability in northern hardwood ecosystems. Ecol. Monogr. 44:73–88.

Marsh, George Perkins. 1864. *Man and Nature; or Physical Geography as Modified by Human Action.* Reprinted by Harvard University Press, Cambridge, Mass., 1965 (D. Lowenthal, ed.). For an evaluation of Marsh's classic, see Franklin Russell, Horizon 10:17–23, 1968.

Martin, Michael M. 1970. The biochemical basis of the fungus-attine ant symbiosis. Science 1969:16–20.

Maruyama, M. 1963. The second cybernetics. Deviation-amplifying mutual causal processes. Am. Sci. 51:164–179.

Marx, Donald H., and J. L. Ruehle. 1979. Fiber, food, fuel and fungal symbionts. Science 106:419–422.

Mason, W. H., and E. P. Odum. 1969. The effect of coprophagy on retention and bioelimination of radionuclides of detritus-feeding animals. Proc. 2nd Nat. Symposium on Radioecology, D. J. Nelson and F. C. Evans, eds. U.S. Dept. of Commerce. Springfield, Va., Clearinghouse Fed. Sci. Tech. Info., pp. 721–724.

Mattson, W. J. (ed.). 1977. *The Role of Arthropods in Forest Ecosystems.* New York, Springer-Verlag.

Mattson, W. J., and A. Addy. 1975. Phytophagous insects as regulators of forest primary production. Science 190:515–521.

Mauldin, W. P. 1980. Population trends and prospects. In: Science Centennial Issue, P. H. Abelson and R. Kulstad, eds. Science 209:148–157.

May, Robert M. 1973. *Stability and Complexity in Model Ecosystems Monographs in Population Biology.* Princeton, N.J., Princeton University Press, 235 pp.

———. 1979. Production and respiration in animal communities. Nature 282:443–444.

——— (ed.). 1981. *Theoretical Ecology: Principles and Applications* (2nd ed.). Sunderland, Mass., Sinauer Associates, 489 pp.

May, R. M., and G. F. Oster. 1976. Bifurcations and dynamic complexity in simple ecological models. Am. Nat. 110:573–599.

Maynard Smith, J. 1976. Group selection—a commentary. Quart. Rev. Biol. 51:277–283.

Mayr, Ernst. 1982. *The Growth of Biological Thought.* Cambridge, MA. Harvard University Press.

McCarthy, J. J., and J. C. Goldman. 1979. Nitrogenous nutrition of marine phytoplankton in nutrient-depleted waters. Science 203:670–672.

McCormick, F. J. 1969. Effects of ionizing radiation on a pine forest. In: Proceedings of 2nd National Symposium on Radioecology, D. Nelson and F. Evans, eds. U.S. Dept. of Commerce. Springfield, Va., Clearinghouse Fed. Sci. Tech. Info., pp. 78–87.

McCormick, F. J., and F. B. Golley. 1966. Irradiation of natural vegetation—an experimental facility, procedures and dosimetry. Health Physics 12:1467–1474.

McCullough, Dale R. 1979. *The George Reserve Deer Herd. Population Ecology of a K-selected Species.* Ann Arbor, University of Michigan Press, 271 pp.

McHarg, Ian L. 1969. *Design with Nature.* Garden City, N.Y., Natural History Press, 197 pp.

McIntosh, Robert P. 1967. The continuum concept of vegetation. Bot. Rev. 33:130–187.

———. 1975. H. A. Gleason—"individualistic ecologist," 1882–1975. Bull. Torrey Bot. Club 102:253–273.

———. 1980. The relationship between succession and the recovery process in ecosystems. In: *The Recovery Process in Damaged Ecosystems,* John Cairns, ed. Ann Arbor, Mich., Ann Arbor Sciences, pp. 11–62.

McLeese, D. W. 1956. Effects of temperature, salinity and oxygen on the survival of the American lobster. J. Fish. Res. Board Canada 13:247–272.

McMillan, Calvin. 1956. Nature of the plant community. 1. Uniform garden and light period studies of five grass taxa in Nebraska. Ecology 37:330–340.

———. 1969. Ecotypes and ecosystem function. BioScience 19:131–134.

McNab, B. K. 1963. Bioenergetics and the determinations of home range size. Am. Nat. 97:135–140.

McNaughton, S. J. 1966. Ecotype function in the *Typha* community-type. Ecol. Monogr. 36:297–325.

———. 1968. Structure and function in California grasslands. Ecology 49:962–972.

———. 1975. r- and K-selection in *Typha*. Am. Nat. 129:251–261.

———. 1976. Serengeti migratory wildebeest: Facilitation of energy flow by grazing. Science 191:92–94.

———. 1978. Stability and diversity in grassland communities. Nature 279:351–352.

———. 1979. Grazing as an optimization process: Grass-ungulate relationships in the Serengeti. Am. Nat. 113:691–703.

McNeill, S., and J. H. Lawton. 1970. Annual production and respiration in animal populations. Nature 225:472.

Meadows, Dennis L. 1974. *Dynamics of Growth in a Finite World.* Cambridge, Mass., MIT Press.

Meadows, Dennis L., and Donella H. Meadows (eds.). 1973. *Toward Global Equilibrium: Collected Papers.* Cambridge, Mass., Wright-Allen Press, 358 pp.

Meadows, Donella H. 1982. Whole earth models and systems. CoEvolution Quarterly, Summer 1982, pp. 98–108.

Meadows, Donella H., Dennis L. Meadows, J. Randers, and W. W. Behrens. 1972. *The Limits to Growth: A Report for the Club of Rome's Project on the Predicament of Mankind.* New York, Universe Books, 205 pp.

Meadows, Donella H., J. Richardson, and G. Bruckmann (eds.). 1983. *Groping in the Dark; The First Decade of Global Modeling.* New York, John Wiley & Sons.

Meentemeyer, Vernon. 1978. Macroclimate and lignin control of litter decomposition rates. Ecology 59:465–472.

Meentemeyer, Vernon, Elgene O. Box, and Richard Thompson. 1982. World patterns and amounts of terrestrial plant litter production. BioScience 32:125–128.

Menhinick, Edward F. 1963. Estimation of insect population density in herbaceous vegetation with emphasis on removal sweeping. Ecology 44:617–622.

________. 1964. A comparison of some species diversity indices applied to samples of field insects. Ecology 45:859–861.

Menshutkin, V. V. 1962. The realization of elementary models for fish populations on electronic computers (in Russian). Vopr. Ilshtiol. 4:625–631.

Menzel, D. W. 1980. Applying results derived from experimental microcosms to study of natural pelagic ecosystems. In: *Microcosms in Ecological Research,* J. P. Giesy, ed. U.S. Dept. of Energy Symposium No. 52. Springfield, Va., National Technical Info. Center, pp. 742–752.

Menzel, D. W., E. M. Hulbert, and J. H. Ryther. 1963. The effects of enriching Sargasso Sea water on the production and species composition of the phytoplankton. Deep-Sea Res. 10:209–219.

Menzel, D. W., and J. H. Ryther. 1961. Nutrients limiting the production of phytoplankton in the Sargasso Sea, with special reference to iron. Deep-Sea Res. 7:276–281.

Mertz, Walter. 1981. The essential trace elements. Science 213:1332–1338.

Mesarovic, Mihajlo, and Eduard Pestel (eds.). 1974. *Mankind at the Turning Point: The Second Report to the Club of Rome.* New York, Signet, 210 pp.

Mesarovic, M. D., and Y. Takahara. 1975. *General Systems Theory: Mathematical Foundations.* New York, Academic Press.

Metcalf, R. L., and W. Luckman (eds.). 1975. *Introduction to Pest Management.* New York, Wiley-Interscience.

Middleton, J. T. 1961. Photochemical air pollution damage to plants. Ann. Rev. Plant Physiol. 12:431–448.

________. 1964. Trends in air pollution damage. Arch. Environ. Health 8:19–24.

Miles, John. 1979. *Vegetation Dynamics.* New York, John Wiley & Sons; London, Chapman and Hall; 80 pp.

Miller, M. W., and G. G. Berg (eds.). 1969. *Chemical Fallout.* Springfield, Ill., Charles C Thomas.

Miller, R. S. 1967. Pattern and process in competition. Adv. Ecol. Res. 4:1–74.

Milthrope, F. L. 1956. *The Growth of Leaves.* London, Butterworth.

Mishan, E. J. 1967. *The Cost of Economic Growth.* London, Stapes Press; New York, Praeger.

________. 1970. *Technology and Growth: The Price We Pay.* New York, Praeger.

Misra, R., J. S. Singh, and K. P. Singh. 1968. A new hypothesis to account for the opposite trophic-biomass structure on land and in water. Current Sci. (India) 37:382–383.

Mitchell, Roger. 1979. *The Analysis of Indian Agroecosystems.* New Delhi, Interprint.

Mohr, Carl O. 1940. Comparative populations of game, fur and other mammals. Am. Midl. Nat. 24:581–584.

Monk, Carl D. 1966. Ecological importance of root/shoot ratios. Bull. Torrey Bot. Club 93:402–406.

________. 1966a. An ecological significance of evergreenness. Ecology 47:504–505.

________. 1966b. Effects of short-term gamma irradiation on an old field. Rad. Bot. 6:329–335.

Montague, Clay L. 1980. A natural history of temperate western Atlantic fiddler crabs (genus *Uca*) with reference to their impact on the salt marsh. Contr. Mar. Sci. 23:25–55.

deMontbrial, Thierry. 1979. *Energy: The Countdown.* A Report to the Club of Rome. New York, Pergamon Press.

Monteith, J. L. 1962. Measurement and interpretation of carbon dioxide fluxes in the field. Netherlands J. Agr. Sci. 10(special issue):334–346.

Mooney, H. A., O. J. Ehleringer, and J. Berry. 1976. Photosynthetic capacity of *in situ* Death Valley plants. Carnegie Inst. Year Book 75:410–413.

Moore, A. W. 1969. *Azola*: Biology and agronomic significance. Bot. Rev. 35:17–34.

Moore, N. W. (ed.). 1966. Pesticides in the environment and their effect on wildlife. (A supplement to the Journal of Applied Ecology, Vol. 3.) Oxford, Blackwell, 312 pp.

Morehouse, Ward, and J. Sigurdson. 1977. Science, technology and poverty. Bull. Atomic Sci. 33:21–28.

Mudd, J. Brian, and T. T. Kozlowski (eds.). 1975. *Responses of Plants to Air Pollution.* New York, Academic Press.

Muller, C. H. 1966. The role of chemical inhibition (allelopathy) in vegetational composition. Bull. Torrey Bot. Club 93:332–351.

———. 1969. Allelopathy as a factor in ecological process. Vegetatio 18:348–357.

Muller, C. H., R. B. Hanawalt, and J. K. McPherson. 1968. Allelopathic control of herb growth in the fire cycle of California chaparral. Bull. Torrey Bot. Club 95:225–231.

Muller, C. H., W. H. Muller, and B. L. Haines. 1964. Volatile growth inhibitors produced by aromatic shrubs. Science 143:471–473.

Mumford, Lewis. 1967. Quality in the control of quantity. In: *Natural Resources, Quality and Quantity,* Ciriacy-Wantrup and Parsons, eds. Berkeley, University of California Press, pp. 7–18.

Munson, R. D., and J. P. Doll. 1959. The economics of fertilizer use in crop production. In: *Advances in Agronomy XI,* A. G. Norman, ed. New York, Academic Press.

Murie, Adolph. 1944. Dall sheep. In: *Wolves of Mount McKinley.* Washington, D.C., Natl. Parks Service Fauna No. 5.

Murphy, G. I. 1967. Vital statistics of the Pacific sardine (*Sardinops caerulea*) and the population consequences. Ecology 48:731–736.

Murphy, G. I., and J. D. Isaacs. 1964. Species replacement in marine ecosystems with reference to the California Current. Calif. Coop. Oceanic Fish Invest. Marine Res. Comm. Mtg., San Pedro, Doc. VII:1–6.

Murray, Bertram G., Jr. 1979. *Population Dynamics—Alternative Models.* New York, Academic Press, 224 pp.

Muscatine, L. C. 1961. Symbiosis in marine and freshwater coelenterates. In: *The Biology of Hydra and Some Other Coelenterates,* H. M. Lenhoff and W. F. Loomis, eds. Miami, University of Miami Press, pp. 255–268.

Muscatine, L. C., and James Porter. 1977. Reef corals: Mutualistic symbioses adapted to nutrient-poor environments. BioScience 27:454–456.

Myers, Norman. 1979. *The Sinking Ark: A New Look at the Problem of Disappearing Species.* Oxford and New York, Pergamon Press.

Nader, Laura, and Stephen Beckerman. 1978. Energy as it relates to the quality and style of life. Ann. Rev. Energy 3:28. Palo Alto, Calif. Annual Reviews, Inc.

National Academy of Science. 1966. *Waste Management and Control.* Publ. No. 1400. A. Spilhaus, ed., 1966.

———. 1969. *Resources and Man.* P. E. Cloud, Jr., ed. San Francisco, W. H. Freeman.

———. 1969. *Eutrophication: Causes, Consequences and Correctives.* Int. Symp. Eutrophication, Washington, D.C., 661 pp.

———. 1971. *Rapid Population Growth,* R. Revelle, ed. Baltimore, Johns Hopkins Press.

———. 1972. *The Earth and Human Affairs.* L. F. LaPorte, ed. San Francisco, Canfield Press.

———. 1973. Geographical perspective and urban problems. Committee on Geography, Division Earth Sciences, Nat. Res. Council, Washington, D.C.

———. 1977. World Food and Nutrition Study; The Potential Contributions of Research. Harrison S. Brown, Chairman of Study Committee. Washington, D.C., National Research Council, 192 pp. (Also 5 volumes of supporting papers.)

———. 1978. Nitrates: An Environmental Assessment. Environ. Studies Board. Panel on Nitrates, P. L. Brezonik, Chairman. Washington, D.C., National Research Council.

———. 1979. *Carbon Dioxide and Climate, A Scientific Assessment.* Washington, D.C., National Research Council.

———. 1981. Testing for the effects of chemicals on ecosystems. Committee to Review Methods for Ecotoxicology, John Cairns, Chairman. Washington, D.C., Commission on Natural Resources, National Research Council, Nat. Acad. Press.

Neess, John C. 1946. Development and status of pond fertilization in central Europe. Trans. Am. Fish. Soc. 76:335–358.

Nelson, D. J. 1967. Microchemical constituents in contemporary and pre-Columbian clam shells. In: *Quaternary Paleoecology.* Proc. VII Congress Int. Assoc. Quaternary Research, Vol. 7, pp. 185–204.

Nelson, D. J., and D. C. Scott. 1962. Role of detritus in the productivity of a rock-outcrop community in a Piedmont stream. Limnol. Oceanogr. 7:396–413.

Nelson, T. C. 1955. Chestnut replacement in the southern highlands. Ecology 36:353–354.

Newell, Richard. 1965. The role of detritus in the nutrition of two marine deposit feeders, the prosobranch *Hydrobia ulvae* and the bivalve *Macoma balthica.* Proc. Zool. Soc. London 144:25–45.

Newell, S. J., and E. J. Tramer. 1978. Reproductive strategies in herbaceous plant communities in succession. Ecology 59:228–234.

Newland, Kathleen. 1980. City limits; emerging constraints on urban growth. Worldwatch Paper 38. Washington, D.C., Worldwatch Institute, 31 pp.

Nice, Margaret M. 1941. The role of territory in bird life. Am. Midl. Nat. 26:441–487.

Nichols, J. D., W. Conley, B. Batt, and A. R. Tipton. 1976. Temporally dynamic reproductive strategies and the concept of *r*- and *K*-selection. Am. Nat. 110:990–1005.

Nicholson, A. J. 1954. An outline of the dynamics of animal populations. Austr. J. Zool. 2:9–65.

________. 1957. The self-adjustment of populations to change. Cold Spring Harbor Symp. Quant. Biol. 22:153–173.

________. 1958. Dynamics of insect populations. Ann. Rev. Entomol. 3:107–136.

Nicholson, S. A., and C. D. Monk. 1974. Plant species diversity in old-field succession on the Georgia piedmont. Ecology 55:1075–1085.

Nicolis, G., and I. Prigogine. 1977. *Self-organization in Non-equilibrium Systems.* New York, John Wiley & Sons.

Nixon, S. W. 1969. A synthetic microcosm. Limnol. Oceanogr. 14:142–145.

Nordhaus, William. 1979. *The Efficient Use of Energy Resources.* New Haven, Yale University Press.

Nordhaus, William, and J. Tobin. 1972. *Economic Growth: Is Growth Obsolete?* Colloquium V, National Bureau of Economic Research. New York, Columbia University Press, 92 pp.

Norman, A. G. 1957. Soil-plant relationships and plant nutrition. Am. J. Bot. 44:67–73.

Novikoff, A. B. 1945. The concept of integrative levels in biology. Science 101:209–215.

Nutman, P. S. (ed.). 1976. *Symbiotic Nitrogen Fixation.* Cambridge, England, Cambridge University Press, 211 pp.

Nutman, P. S., and B. Masse (eds.). 1963. *Symbiotic Associations.* 13th Symposium for Gen. Microbiology. New York, Cambridge University Press, 356 pp.

Oberlander, G. T. 1956. Summer fog precipitation on the San Francisco Peninsula. Ecology 37:851–852.

Odell, Rice. 1980. *Environmental Awakening.* Washington, D.C., The Conservation Foundation.

Odum, Eugene P. 1957. The ecosystem approach in the teaching of ecology illustrated with sample class data. Ecology 38:531–535.

________. 1961. The role of tidal marshes in estuarine production. The Conservationist (New York State Conservation Dept., Albany) 15(6):12–15.

________. 1963. Primary and secondary energy flow in relation to ecosystem structure. Proc. XVI Int. Congr. Zool., Washington, D.C., pp. 336–338.

________. 1964. The new ecology. BioScience 14:14–16.

________. 1968. Energy flow in ecosystems: A historical review. Am. Zool. 8:11–18.

________. 1968a. A research challenge: Evaluating the productivity of coastal and estuarine water. Proc. 2nd Sea Grant Conf., Grad. School Oceanography, University of Rhode Island, Newport, pp. 63–64.

________. 1969. The strategy of ecosystem development. Science 164:262–270.

________. 1969a. Air–land–water = an ecological whole. J. Soil Water Cons. 24:4–7.

________. 1970. Optimum population and environment: A Georgian microcosm. Curr. Hist. 58:355–359, 365.

________. 1971. Ecological principles and the urban forest. In: Proc. Symp. on Role of Trees in the South's Urban Environment. Athens, Ga., Ga. Center for Continuing Education, pp. 78–81.

________. 1972. Ecosystem theory in relation to man. In: *Ecosystems: Structure and Function,* 31st Biol. Coll., J. Wiens, ed. Corvallis, Oregon State University Press.

________. 1975. *Ecology: The Link Between the Natural and the Social Sciences* (2nd ed.). New York, Holt, Rinehart and Winston, 244 pp.

________. 1975a. Harmony between man and nature: An ecological view. In: *Essays on Alternate Futures.* Yale University School of Forestry, Bull. 88, pp. 43–55.

________. 1976. Diversity as a function of energy flow. In: *Unifying Concepts in Ecology* (Proc. 1st International Congress of Ecology). The Hague, W. Junk Publ.

________. 1976a. Energy, ecosystem development and environmental risk. Risk Insurance 43(1):1–16.

________. 1977. The emergence of ecology as a new integrative discipline. Science 195:1289–1293.

________. 1977a. Ecology—the common sense approach. The Ecologist 7(7):250–253.

________. 1977b. The life support value of forests. In: *Forest for People.* Washington, D.C., Society of American Foresters, pp. 101–105.

________. 1978. Ecological importance of the riparian zone. In: *Strategies for Protection of Floodplain Wetlands and Other Riparian Ecosystems,* B. R. Johnson and J. F. McCormick, eds. Gen. Tech. Rept. WO-12. Washington, D.C., U.S. Forest Service, pp. 2–4.

________. 1979. The value of wetlands: A hierarchical approach. In: *Wetland Functions and Values: The State of Our Understanding,* P. E. Greeson, J. R. Clark, and J. E. Clark, eds. Minneapolis, American Water Resources Association, pp. 16–25.

________. 1979a. Rebuttal of "Economic value of natural coastal wetlands: A critique." Coastal Zone Mgmt. Jr. 5:231–237.

________. 1980. The status of three ecosystem-level hypotheses regarding salt marsh estuaries: Tidal subsidy, outwelling and detritus-based food chains. In: *Estuarine Perspectives,* Kimbel, ed. New York, Academic Press, pp. 485–495.

________. 1980a. Radiation ecology at Oak Ridge. In: *Environmental Sciences Laboratory Dedication.* Oak

Ridge, Tenn., Oak Ridge National Laboratory, pp. 53–57.

———. 1981. The effects of stress on the trajectory of ecological succession. In: *Stress Effects on Natural Ecosystems*, G. W. Barrett and R. Rosenberg, eds. London, John Wiley & Sons, pp. 43–47.

Odum, E. P. and L. Biever. 1983. Resource quality, mutualism and energy partitioning in food chains. (Manuscript submitted for publication.)

Odum, E. P., G. A. Bramlett, A. Ike, J. R. Champlin, J. C. Zieman, and H. H. Shugart. 1976. Totality indices for evaluating environmental impacts of highway alternates. Transportation Record No. 561. Washington, D.C., National Academy of Science, pp. 57–67.

Odum, Eugene P., and Thomas D. Burleigh. 1946. Southward invasion in Georgia. Auk. 63:388–401.

Odum, Eugene P., Clyde E. Connell, and L. B. Davenport. 1962. Population energy flow of three primary consumer components of old-field ecosystems. Ecology 43:88–96.

Odum, E. P., and J. L. Cooley. 1980. Ecosystem profile analysis and performance curves as tools for assessing environmental impacts. In: *Biological Evaluation of Environmental Impacts.* Washington, D.C., Council on Environmental Quality and Fish and Wildlife Service, pp. 94–102.

Odum, Eugene P., and Armando A. de la Cruz. 1963. Detritus as a major component of ecosystems. AIBS Bulletin (now BioScience) 13:39–40.

———. 1967. Particulate organic detritus in a Georgia salt marsh-estuarine ecosystem. In: *Estuarines*, G. Lauff, ed. Am. Assoc. Adv. Sci. Publ. 83:383–388.

Odum, E. P., and M. E. Fanning. 1973. Comparison of the productivity of *Spartina alterniflora* and *S. cynosuroides* in Georgia coastal marshes. Bull. Ga. Acad. Sci. 31:1–12.

Odum, E. P., J. T. Finn, and E. H. Franz. 1979. Perturbation theory and the subsidy-stress gradient. BioScience 29:349–352.

Odum, E. P., and E. H. Franz. 1977. Whither the life-support system. In: *Growth Without Ecodisasters*, N. Polunin, ed. London, MacMillan Press, pp. 264–274.

Odum, Eugene P., and Frank B. Golley. 1963. Radioactive tracers as an aid to the measurement of energy flow at the population level in nature. In: *Radioecology*, V. Schultz and A. W. Klement, eds. New York, Reinhold, pp. 403–410.

Odum, Eugene P., and David W. Johnston. 1951. The house wren breeding in Georgia: An analysis of a range extension. Auk 68:357–366.

Odum, Eugene P., and R. L. Kroodsma. 1977. The power park concept: Ameliorating man's disorder with nature's order. In: *Thermal Ecology II*, G. W. Esch and R. W. McFarlane, eds. Springfield, Va., National Technical Info. Service, pp. 1–9.

Odum, Eugene P., and Edward J. Kuenzler. 1963. Experimental isolation of food chains in an old-field ecosystem with use of phosphorus-32. In: *Radioecology*, V. Schultz and A. W. Klement, eds. New York, Reinhold, pp. 113–120.

Odum, Eugene P., E. J. Kuenzler, and Marion X. Blunt. 1958. Uptake of P^{32} and primary productivity in marine benthic algae. Limnol. Oceanogr. 3:340–345.

Odum, Eugene P., and H. T. Odum. 1972. Natural areas as necessary components of man's total environment. Proc. 37th N.A. Wildl. and Nat. Res. Conf. Washington, D.C., Wildlife Mgmt. Institute, pp. 178–189.

Odum, E. P., S. E. Pomeroy, J. C. Dickinson, and K. Hutcheson. 1973. The effects of late winter burn on the composition, productivity and diversity of a 4-year-old fallow field in Georgia. In: Proc. Annual Tall Timbers Fire Ecol. Conf., E. V. Komarek, ed. Tallahassee, Fla., Tall Timbers Research Station.

Odum, Eugene P., and A. E. Smalley. 1957. Comparison of population energy flow of a herbivorous and a deposit-feeding invertebrate in a salt marsh ecosystem. Proc. Natl. Acad. Sci. 45:617–622.

Odum, Howard T. 1956. Primary production in flowing waters. Limnol. Oceanogr. 1:102–117.

———. 1956a. Efficiencies, size of organisms, and community structure. Ecology 37:592–597.

———. 1957. Trophic structure and productivity of Silver Springs, Florida. Ecol. Monogr. 27:55–112.

———. 1960. Ecological potential and analogue circuits for the ecosystem. Am. Sci. 48:1–8.

———. 1962. Ecological tools and their use—Man and the ecosystem. In: *Proceedings of the Lockwood Conference on the Suburban Forest and Ecology*, Paul E. Waggoner and J. D. Ovington, eds. The Connecticut Agricultural Experiment Station Bull. 652, pp. 57–75.

———. 1963. Limits of remote ecosystems containing man. Am. Biol. Teacher 25:429–443.

———. 1967. Biological circuits and the marine systems of Texas. In: *Pollution and Marine Ecology*, T. A. Olson and F. J. Burgess, eds. New York, Wiley Interscience, pp. 99–157.

———. 1967a. Energetics of world food production. In: *The World Food Problem, A Report of the President's Science Advisory Committee.* Panel on World Food Supply (I. L. Bennett, Chairman). Vol. 3. Washington, D.C., The White House, pp. 55–94.

———. 1968. Work circuits and systems stress. In: *Symposium on Primary Productivity and Mineral Cycling in Natural Ecosystems*, H. E. Young, ed. Orono, University of Maine Press, pp. 81–138.

———. 1971. *Environment, Power and Society.* New York, John Wiley & Sons, 331 pp.

———. 1973. Energy, ecology and economics. Ambio 2(6):220–227.

———. 1975. Energy quality and carrying capacity of the earth. Trop. Ecol. 16(1):1–8.

———. 1978. Energy analysis, energy quality, and environment. In: *Energy Analysis: A New Public Policy Tool*, Martha Gilliland, ed. AAAS Selected Symp. 9. Washington, D.C., American Association for the Advancement of Science, pp. 55–87.

———. 1982. *Systems Ecology*. New York, John Wiley & Sons.

Odum, Howard T., J. E. Cantlon, and L. S. Kornicker. 1960. An organizational hierarchy postulate for the interpretation of species-individuals distribution, species entropy and ecosystem evolution and the meaning of a species-variety index. Ecology 41:395–399.

Odum, Howard T., and Charles M. Hoskin. 1957. Metabolism of a laboratory stream microcosm. University of Texas Publ. Inst. Marine Sci. 4:115–133.

Odum, Howard T., W. M. McConnell, and W. Abbott. 1958. The chlorophyll "A" of communities. University of Texas Publ. Inst. Marine Sci. 5:65–97.

Odum, Howard T., and Elizabeth C. Odum. 1981. *Energy Basis for Man and Nature* (2nd ed.). New York, McGraw-Hill, 337 pp.

Odum, Howard T., and Eugene P. Odum. 1955. Trophic structure and productivity of a windward coral reef community on Eniwetok Atoll. Ecol. Monogr. 25:291–320.

Odum, Howard T., and R. F. Pigeon (eds.). 1970. *A Tropical Rain Forest*. A study of irradiation and ecology at El Verde, Puerto Rico. Springfield, Va., National Technical Info. Service, 1678 pp.

Odum, Howard T., and R. C. Pinkerton. 1955. Times speed regulator, the optimum efficiency for maximum output in physical and biological systems. Am. Sci. 43:331–343.

Odum, Howard W. 1936. *Southern Regions of the United States*. Chapel Hill, University of North Carolina Press, 664 pp.

———. 1951. The promise of regionalism. In: *Regionalism in America*, Merrill Jensen, ed. Madison, University of Wisconsin Press.

Odum, Howard W., and Harry E. Moore. 1938. *American Regionalism*. New York, Henry Holt & Co.

Odum, William E. 1968. Pesticide pollution in estuaries. Sea Frontiers (International Oceanographic Foundation, Miami, Fla.) 14:234–245.

———. 1968a. The ecological significance of fine particle selection by the striped mullet, *Mugil cephalus*, Limnol. Oceanogr. 13:92–98.

———. 1970. Insidious alteration of the estuarine environment. Trans. Am. Fish. Soc. 99:836–847.

———. 1982. Environmental degradation and the tyranny of small decisions. BioScience, 32:728–729.

Odum, William E., and E. J. Heald. 1972. Trophic analysis of an estuarine mangrove community. Bull. Marine Sci. 22(3):671–738.

———. 1975. The detritus-based food web of an estuarine mangrove community. In: *Estuarine Research*, G. E. Cronin, ed. Vol. 1. New York, Academic Press, pp. 265–286.

Odum, William E., G. M. Woodwell, and C. F. Wurster. 1969. DDT residues absorbed from organic detritus by fiddler crabs. Science 164:576–577.

Office of Technology Assessment (OTA), U.S. Congress. 1982. Global Models, World Futures and Public Policy. A Critique. Washington, D.C., Office of Technology Assessment.

Ogburn, W. F. 1922. Social change with respect to culture and original nature. New York, Heubsch.

Oliver, Chadwick D. 1981. Forest development in North America following major disturbances. Forest Ecol. Mgmt. 3:153–168.

Oliver, Chadwick D., and E. P. Stephens. 1977. Reconstruction of a mixed-species forest in central New England. Ecology 58:562–572.

Olson, J. S. 1958. Rates of succession and soil changes on southern Lake Michigan sand dunes. Bot. Gaz. 119:125–170.

———. 1963. Energy storage and the balance of producers and decomposers in ecological systems. Ecology 44:322–332.

———. 1964. Gross and net production of terrestrial vegetation. J. Ecol. 62:99–118.

Omernik, J. M. 1977. Non-point source stream nutrient level relationship: A nationwide study. EPA Ecol. Res. Ser. 600/3-77-105, Corvallis, Oregon.

O'Neill, Gerald K. 1977. *The High Frontier; Human Colonies in Space*. New York, William Morrow, 288 pp.

O'Neill, R. V. 1967. Niche segregation in seven species of diplopods. Ecology 48:983.

———. 1976. Ecosystem persistence and heterotrophic regulation. Ecology 57:1244–1253.

O'Neill, R. V., and D. E. Reichle. 1980. Dimensions of ecosystem theory. In: *Forests*. R. W. Waring (ed.). Corvallis, Oregon State Press.

Oosting, Henry J. 1942. An ecological analysis of the plant communities of Piedmont, North Carolina. Am. Midl. Nat. 28:126.

Oparin, A. I. 1938. *The Origin of Life*. New York, Macmillan.

Ophel, Ivan L. 1963. The fate of radiostrontium in a freshwater community. In: *Radioecology*, V. Schultz and W. Klement, eds. New York, Reinhold, pp. 213–216.

Osborn, Fairfield. 1948. *Our Plundered Planet*. Boston, Little, Brown.

Overgaard-Neilsen, C. 1949. Freeliving nematodes and

soil microbiology. Proc. 4th Int. Congr. Microbiology, Copenhagen, 1947, pp. 283–284.

Ovington, J. D. 1961. Some aspects of energy flow in plantations of *Pinus sylvestris*. Ann. Bot. n.s. 25:12–20.

________. 1962. Quantitative ecology and the woodland ecosystem concept. In: *Advances in Ecological Research*, J. B. Cragg, ed. Vol. 1. New York, Academic Press, pp. 103–192.

Owen, D. F., and R. G. Wiegert. 1976. Do consumers maximize plant fitness? Oikos 27:489–492.

Page, Talbot. 1977. *Conservation and Economic Efficiency*. A Resources for the Future Book. Baltimore, Johns Hopkins Press.

Paine, R. T. 1966. Food web diversity and species diversity. Am. Nat. 100:65–75.

Palmgren, P. 1949. Some remarks on the short-term fluctuations in the numbers of northern birds and mammals. Oikos 1:114–121.

Paris, O. H., and F. A. Pitelka. 1962. Population characteristics of the terrestrial isopod *Armedillidium vulgare* in California grassland. Ecology 43:229–248.

Park, Thomas. 1934. Studies in population physiology: Effect of conditioned flour upon the productivity and population decline of *Tribolium confusum*. J. Exp. Zool. 68:167–182.

________. 1954. Experimental studies of interspecific competition. II. Temperature, humidity and competition in two species of *Tribolium*. Physiol. Zool. 27:177–238.

Park, T. B., B. G. Ginsburg and S. Horwitz. 1945. Ebony, a gene affecting the body color and fecundity of *Tribolium confusum*. Physiol. Zool. 18:35–52.

Parker, J. R. 1930. Some effects of temperature and moisture upon *Melanoplus mexicanus* and *Camnula pellucida* Scudder (Orthoptera). Bull. Univ. Montana Agr. Exp. Sta. 223:1–132.

Patrick, Ruth. 1954. Diatoms as an indication of river change. Proc. 9th Industrial Waste Conference, Purdue University Engineering Extension Series 87:325–330.

________. 1967. Diatom communities in estuaries. In: *Estuaries*, G. H. Lauff, ed. Publication No. 83. Washington, D.C., American Association for the Advancement of Science, pp. 311–315.

Patrick, W. H., S. Gotoh, and B. G. Williams. 1973. Strengite dissolution in flooded soils and sediments. Science. 179:564–565.

Patrick, W. H., and R. A. Khalid. 1974. Phosphate release and sorption by soils and sediments: effect of aerobic and anaerobic conditions. Science. 186:53–55.

Pattee, Howard H. 1973. *Hierarchy Theory: The Challenge of Complex Systems*. New York, George Braziller.

Patten, Bernard C. 1966. Systems ecology; a course sequence in mathematical ecology. BioScience 16:593–598.

________. (ed.). 1971. *Systems Analysis and Simulation in Ecology*. Vol. I. New York, Academic Press. (See also Vol. II, 1972; Vol. III, 1974; Vol. IV, 1976.)

________. 1978. Systems approach to the concept of environment. Ohio J. Sci. 78:206–222.

________. 1981. Environs: The superniches of ecosystems. Am. Zool. 21:845–852. (See also Am. Nat. 119:179–219, 1982.)

Patten, B. C., and G. T. Auble. 1981. The systems theory of the ecological niche. Am. Nat. 118:345–369.

Patten, B. C., and E. P. Odum. 1981. The cybernetic nature of ecosystems. Am. Nat. 118:886–895.

Patten, B. C., and Witkamp, M. 1967. Systems analysis of ^{134}cesium kinetics in terrestrial microcosms. Ecology 48:813–824.

Patton, D. R. 1975. A diversity index for quantifying habitat "edge." Wildl. Soc. Bull. 3:171–173.

Paul, E. A., and R. M. N. Kucey. 1981. Carbon flow in plant microbial associations. Science 213:473–474.

Payne, W. J. 1970. Energy yields and growth of heterotrophs. Ann. Rev. Microbiol. 24:17–51.

________. 1981. *Denitrification*. New York, John Wiley & Sons.

Peakall, David B. 1967. Pesticide-induced enzyme breakdown of steroids in birds. Nature 216:505–506.

Peakall, D. B., and P. N. Witt. 1976. The energy budget of an orb web-building spider. Comp. Biochem. Physiol. 54A:187–190.

Pearl, Raymond. 1927. The growth of populations. Quart. Rev. Biol. 2:532–548.

Pearl, Raymond, and S. L. Parker. 1921. Experimental studies on the duration of life: Introductory discussion of the duration of life in *Drosophila*. Am. Nat. 55:481–509.

Pearl, Raymond, and L. J. Reed. 1930. On the rate of growth of the population of the United States since 1790 and its mathematical representation. Proc. Natl. Acad. Sci. (Wash.) 6:275–288.

Pearse, A. S. 1939. *Animal Ecology* (2nd ed.). New York, McGraw-Hill.

Pearse, P. H. 1968. A new approach to the evaluation of non-priced recreational resources. Land Economics 44(1):87–99.

Peet, Robert K. 1974. The measurement of species diversity. Ann. Rev. Ecol. Syst. 5:285–308.

________. 1978. Ecosystem convergence. Am. Nat. 112:441–444.

Penfound, William T. 1956. Primary production of vascular aquatic plants. Limnol. Oceanogr. 1:92–101.

Penman, H. L. 1956. Weather and water in the growth of grass. In: *The Growth of Leaves*, F. L. Milthrope, ed. London, Butterworth.

Peterman, R. M. 1978. The ecological role of mountain pine beetle in lodgepole pine forests, and the insect as a management tool. In: *Mountain Pine Beetle Management in Lodgepole Pine Forests*, Berryman, Stark, and Amman, eds. Moscow, University of Idaho Press.

Peters, Gerald A. 1978. Blue-green algae and algal associations. BioScience 28:580–585.

Peterson, B. J., et al. 1980. Salt marsh detritus: An alternative interpretation of stable carbon isotope ratios and the fate of *Spartina alterniflora*. Oikos 34:173–177.

Petrides, George A. 1950. The determination of sex and age ratios in fur animals. Am. Midl. Nat. 43:355–382.

________. 1956. Big game densities and range carrying capacity in East Africa. N.A. Wildl. Conf. Trans. 21:525–537.

Petrusewicz, K. (ed.). 1967. Secondary productivity of terrestrial ecosystems. Inst. Ecol. Polish Acad. Sci. Int. Biol. Program (Warsaw). Vol. I, 367 pp.; Vol. II, 879 pp.

Pfennig, Norbert. 1967. Photosynthetic bacteria. Ann. Rev. Microbiol. 21:285–324.

Phillips, E. A. 1959. *Methods of Vegetation Study*. New York, Holt, Rinehart and Winston.

Phillips, Ronald E., R. L. Blevins, G. W. Thomas, W. W. Frye, and S. H. Phillips. 1980. No-tillage agriculture. Science 208:1108–1113.

Pianka, E. R. 1967. Lizard species diversity. Ecology 48:333–351.

________. 1970. On r- and K-selection. Am. Nat. 104:592–597.

________. 1978. *Evolutionary Ecology* (2nd ed.). New York, Harper & Row.

Pickett, S. T. A. 1976. Succession: An evolutionary interpretation. Am. Nat. 110:107–119.

Picozzi, N. 1968. Grouse bags in relation to the management and geology of heather moors. J. Appl. Ecol. 5:483–488.

Pielou, E. C. 1966. The measurement of diversity in different types of biological collections. J. Theoret. Biol. 13:131–144.

________. 1966a. Species-diversity and pattern-diversity in the study of ecological succession. J. Theoret. Biol. 10:370–383.

________. 1975. *Ecological Diversity*. New York, Wiley-Interscience, 165 pp.

Pigou, A. C. 1920. *The Economics of Welfare* (4th ed., 1932). London, Macmillan, 976 pp.

Pilson, M. E. Q., and S. W. Nixon. 1980. Marine microcosms in ecological research. In: *Microcosms in Ecological Research*, J. P. Giesy, ed. U.S. Dept. of Energy Symp. No. 52. Springfield, Va., National Technical Info. Center, pp. 724–741.

Pimentel, David. 1961. Animal population regulation by the genetic feedback mechanism. Am. Nat. 95:65–79.

________. 1961a. Species diversity and insect population outbreaks. Ann. Entomol. Soc. Am. 54:76–86.

________. 1968. Population regulation and genetic feedback. Science 159:1432–1437.

Pimentel, D., L. E. Hurd, A. C. Bellotti, et al. 1973. Food production and the energy crisis. Science 182:443–449.

Pimentel, David, M. A. Moran, S. Fast, et al. 1981. Biomass energy from crop and forest residues. Science 212:1110–1115.

Pimentel, David, W. Pritschilo, J. Krummel, and J. Kutzman. 1975. Energy and land constraints in food protein production. Science 190:754–760.

Pimentel, David, and F. A. Stone. 1968. Evolution and population ecology of parasite-host systems. Can. Entomol. 100:655–662.

Pimentel, David, E. C. Terhune, R. Dyson-Hudson, et al. 1976. Land degradation: Effects on food and energy resources. Science 194:149–155.

Pimm, S. L., and J. H. Lawton. 1977. Number of trophic levels in ecological communities. Nature 268:329–331.

Pippenger, Nicholas. 1978. Complexity theory. Sci. Am. 238(6):114–124.

Pirt, S. J. 1957. The oxygen requirement of growing cultures of an *Aerobacter* species determined by means of the continuous culture technique. J. Gen. Microbiol. 16:59–75.

Pitelka, Frank A. 1964. The nutrient recovery hypothesis for arctic microtine cycles. In: *Grazing in Terrestrial and Marine Environments*, A. J. Crisp, ed. Oxford, Blackwell.

________. 1973. Cyclic patterns in lemming populations near Barrow, Alaska. In: *Alaska Arctic Tundra*, M. E. Britton, ed. Tech. Paper 25. Washington, D.C., Arctic Inst. N. Am., pp. 119–215.

Platt, Robert B. 1965. Ionizing radiation and homeostasis of ecosystems. In: *Ecological Effects of Nuclear War*, G. M. Woodwell, ed. Upton, N.Y., Brookhaven National Laboratory Publ. No. 917, pp. 39–60.

Pollard, H. P., and S. Gorenstein. 1980. Agrarian potential, population, and the Tarascan state. Science 209:274–277.

Polunin, N. (ed.). 1980. *Growth Without Ecodisasters*. Proc. 2nd Int. Conf. on Environ. Future, Iceland, 1977. London, Macmillan Press, 675 pp.

Polynov, B. B. 1937. *The Cycle of Weathering*. Trans. from Russian by Muir A. Murby. London, 230 pp.

Pomeroy, L. R. 1959. Algal productivity in Georgia salt marshes. Limnol. Oceanogr. 4:386–397.

________. 1960. Residence time of dissolved phosphate in natural waters. Science 131:1731–1732.

________. 1970. The strategy of mineral cycling. Ann. Rev. Ecol. Syst. 1:171–190.

________ (ed.). 1974. *Cycles of Essential Elements*. Benchmark Papers in Ecology. Stroudsburg, Pa.,

Dowden, Hutchinson and Ross; New York, Academic Press, 373 pp.

_______ . 1974a. The ocean's food web, a changing paradigm. BioScience 24:299–304.

_______ . 1979. Secondary production mechanisms of continental shelf communities. In: *Ecological Processes in Coastal and Marine Systems*, R. J. Livingston, ed. New York, Plenum, pp. 163–186.

_______ . 1980. Detritus and its role as a food source. In: *Fundamentals of Aquatic Ecosystems*, Barnes and Mann, eds. Oxford, Blackwell, pp. 84–192.

Pomeroy, L. R., R. E. Johannes, E. P. Odum, and B. Roffman. 1969. The phosphorus and zinc cycles and productivity of a salt marsh. In: Proc. 2nd Symp. Radioecology, D. J. Nelson and F. C. Evans, eds. Springfield, Va., Clearinghouse Fed. Sci. Tech. Info., pp. 412–419.

Pomeroy, L. R., H. M. Mathews, and H. S. Min. 1963. Excretion of phosphate and soluble organic phosphorus compounds by zooplankton. Limnol. Oceanogr. 8:50–55.

Pomeroy, L. R., E. P. Odum, R. E. Johannes, and B. Roffman. 1966. Flux of ^{32}P and ^{65}Zn through a salt marsh ecosystem. Proceedings of the Symposium on the Dispersal of Radioactive Wastes into Seas, Oceans and Surface Waters, Vienna, pp. 177–188.

President's Science Advisory Committee. 1967. *The World Food Problem.* Three vols. Washington, D.C., U.S. Government Printing Office.

Preston, F. W. 1948. The commonness and rarity of species. Ecology 29:254–283.

_______ . 1960. Time and space and the variation of species. Ecology 41:611–627.

Prigogine, Ilya. 1962. *Introduction to Nonequilibrium Thermodynamics.* New York, John Wiley & Sons.

_______ . 1978. Time structure and fluctuations. Science 201:777–785.

Prigogine, Ilya, G. Nicolis, and A. Babloyantz. 1972. Thermodynamics and evolution. Physics Today 25(11):23–38; 25(12):138–144.

Prigogine, N. G., and I. Prigogine. 1977. *Self Organization in Nonequilibrium Systems.* New York, John Wiley & Sons.

Pritchard, D. W. 1967. What is an estuary: Physical viewpoint. In: *Estuaries*, G. H. Lauff, ed. Publ. No. 83. Washington, D.C., American Association for the Advancement of Science, pp. 3–5.

Prosser, C. L. (ed.). 1967. *Molecular Mechanisms of Temperature Adaptation.* Washington, D.C., American Association for the Advancement of Science.

Provasoli, Luigi. 1958. Nutrition and ecology of protozoa and algae. Ann. Rev. Microbiol. 12:279–308.

_______ . 1963. Organic regulation of phytoplankton fertility. In: *The Sea*, M. N. Hill, ed. Vol. 2. New York, Wiley Interscience, pp. 165–219.

Pyke, G. H., H. R. Pulliam, and E. L. Charnov. 1977. Optimal foraging: A selective review of theory and tests. Quart. Rev. Biol. 52:137–154.

Randall, Paul L. 1972. Market solutions to externality problems. Amer. J. Agr. Econ. 54:175–183.

Randolph, P. A., J. C. Randolph, and C. A. Barlow. 1975. Age-specific energetics of the pea aphid. Ecology 56:359–369.

Randolph, P. A., J. C. Randolph, K. Mattingly, and M. M. Foster. 1977. Energy costs of reproduction in the cotton rat, *Sigmodon hispidus.* Ecology 58:31–45.

Rapoport, A., and A. M. Chammah. 1970. *Prisoner's Dilemma.* Ann Arbor. University of Michigan Press.

Rappaport, R. A. 1967. *Pigs for the Ancestors.* New Haven, Yale University Press.

_______ . 1971. The flow of energy in an agricultural society. Sci. Am. 224(3):116–132.

Rasmussen, D. I. 1941. Biotic communities of Kaibab Plateau, Arizona. Ecol. Monogr. 11:229–275.

Raunkaier, C. 1934. *The Life Form of Plants and Statistical Plant Geography.* Oxford, Clarendon Press.

Ravera, O. 1969. Seasonal variation of the biomass and biocoenotic structure of plankton of the Bay of Ispra (Lago Maggiore). Verh. Int. Ver. Limnol. 17:237–254.

Rawson, D. S. 1952. Mean depth and the fish production of large lakes. Ecology 33:513–521.

Reardon, P. O., C. L. Leinweber, and L. B. Merrill. 1972. The effect of bovine saliva on grasses. J. Anim. Sci. 34:897–898.

_______ . 1974. Response of sideoats grama to animal saliva and thiamine. J. Range Mgmt. 27:400–401.

Redfield, Alfred C. 1958. The biological control of chemical factors in the environment. Am. Sci. 46:205–221.

Reichle, David (ed.). 1970. *Analyses of Temperate Forest Ecosystems.* Heidelberg, Berlin, Springer-Verlag, 304 pp.

_______ . 1975. Advances in ecosystem analysis. BioScience 25:257–264.

Reichle, D. E., J. F. Franklin, and D. W. Goodall (eds.). 1975. Productivity of World Ecosystems. Washington, D.C., National Academy of Science.

Reif, Arnold E. 1981. The causes of cancer. Am. Sci. 69:437–447.

Reifsnyder, W. E., and H. W. Lull. 1965. Radiant energy in relation to forests. Tech. Bull. No. 1344. Washington, D.C., U.S. Dept. of Agriculture, Forest Service, 111 pp.

Rensberger, Boyce. 1981. Evolution of evolution. Mosaic (NSF Journal) 12(5):14–22.

_______ . 1982. Evolution since Darwin. Science-82 3(3):41–45.

Revelle, Roger (ed.). 1966. Population and food supplies: The edge of the knife. In: *Prospects of the World Food Supply: A Symposium.* Washington, D.C., National Academy of Science, pp. 24–47.

———. 1971. *Rapid Population Growth.* National Academy of Science Study. Baltimore, Johns Hopkins Press.

Rice, Elroy L. 1974. *Allelopathy.* New York, Academic Press, 368 pp.

Rice, E. L., and S. K. Pancholy. 1972. Inhibition of nitrification by climax vegetation. Am. J. Bot. 59:1033–1040; 60:691–702 (1973); 61:1094–1103 (1974).

Rich, Peter H. 1978. Reviewing bioenergetics. BioScience 28:80.

Rich, P. H., and R. G. Wetzel. 1978. Detritus in the lake ecosystem. Am. Nat. 112:57–71.

Richards, B. N. 1974. *Introduction to the Soil Ecosystem.* New York, Longman, 266 pp.

Richmond, R. C., M. E. Gilpin, S. P. Salas, and F. J. Hyala. 1975. A search for emergent competitive phenomena: The dynamics of multispecies *Drosophila* systems. Ecology 56:709–714.

Ricker, W. E. 1946. Production and utilization of fish populations. Ecol. Monogr. 16:373–391.

———. 1958. Handbook of computations for biological statistics of fish populations. Fish. Res. Bd. Canada Bull. 119:1–300.

Ricker, W. E., and R. E. Forester. 1948. Computation of fish production. Bull. Bingham Oceanogr. Coll. 11:173–211.

Ricklefs, R. E. 1970. Clutch size in birds; outcome of opposing predator and prey adaptations. Science 168:599–600.

———. 1979. *Ecology* (2nd ed.). New York, Chiron Press, 966 pp.

Ridker, Ronald G. 1973. To grow or not to grow: That's not the relevant question. Science 182:1315–1318.

Riechert, Susan E. 1981. The consequences of being territorial: Spiders, a case study. Am. Nat. 117:871–892.

Rifkin, Jeremy. 1980. *Entropy, Entropy, Entropy; A New World Review.* New York, The Viking Press, 305 pp.

Rigler, F. H. 1961. The uptake and release of inorganic phosphorus by *Daphnia magna* Straus. Limnol. Oceanogr. 6:165–174.

Riley, Gordon A. 1944. The carbon metabolism and photosynthetic efficiency of the earth. Am. Sci. 32:132–134.

———. 1952. Phytoplankton of Block Island Sound, 1949. Bull. Bingham Oceanogr. Coll. 13:40–64.

———. 1956. Oceanography of Long Island Sound, 1952–54. IX. Production and utilization of organic matter. Bull. Bingham Oceanogr. Coll. 15:324–344.

———. 1963. Organic aggregates in sea water and the dynamics of their formation and utilization. Limnol. Oceanogr. 8:372–381.

Ripper, W. E. 1956. Effect of pesticides on the balance of arthropod populations. Ann. Rev. Entomol. 1:403–438.

Robertson, G. P., and P. M. Vitousek. 1981. Nitrification potential in primary and secondary succession. Ecology 62:376–386.

Rodhe, W. 1955. Can plankton production proceed during winter darkness in subarctic lakes? Proc. Int. Assoc. Theoret. Appl. Limnol. 12:117–122.

Root, Richard B. 1967. The niche exploitation pattern of the blue-gray gnatcatcher. Ecol. Monogr. 37:317–350.

———. 1969. The behavior and reproductive success of the blue-gray gnatcatcher. Condor 71:16–31.

Rosenzweig, M. L. 1968. Net primary production of terrestrial communities; prediction from climatological data. Am. Nat. 102:67–74.

Roughgarden, J. 1971. Density-dependent natural selection. Ecology 52:453–468.

Rounsefell, G. A. 1946. Fish production in lakes as a guide for estimating production in proposed reservoirs. Copeia 1946:29–40.

Rovira, A. D. 1965. Interaction between plant roots and soil microorganisms. Ann. Rev. Microbiol. 19:241–266.

Rudd, R. I. 1964. *Pesticides and the Living Landscape.* Madison, University of Wisconsin Press, 320 pp.

Ruehle, J. L., and D. H. Marx. 1979. Fiber, food, fuel and fungal symbionts. Science 206:419–422.

Russell, E. J., and E. W. Russell. 1950. *Soil Conditions and Plant Growth.* London, Longmans.

Russell, Franklin, 1968. The Vermont Prophet: George Perkins Marsh. Horizon 10:17–23.

Ryther, John H. 1954. The ecology of phytoplankton blooms in Moriches Bay and Great South Bay, Long Island, New York. Biol. Bull. 106:198–209.

———. 1956. Photosynthesis in the ocean as a function of light intensity. Limnol. Oceanogr. 1:61–70.

———. 1969. Photosynthesis and fish production in the sea. Science 166:72–76.

Ryther, J. H., and C. S. Yentsch. 1957. The estimation of phytoplankton production in the ocean from chlorophyll and light data. Limnol. Oceanogr. 2:281–286.

Saarinen, Eliel. 1943. *The City: Its Growth, Its Decay, Its Future.* Cambridge, Mass., MIT Press, 380 pp.

Sale, Kirkpatrick. 1978. The polis perplexity: An inquiry into the size of cities. *Working Papers,* Jan.–Feb. 1978.

Salt, George W. 1979. A comment on the use of the term *emergent properties.* Am. Nat. 113:145–148.

Samuelson, Paul A. 1950. Evaluation of real national income. The Oxford Economic Papers N.S., 2:1–29.

Sanders, F. E., B. Mosse, and P. B. Tinker (eds.). 1975. *Endomycorrhizas.* New York and London, Academic Press.

Sanders, H. L. 1968. Marine benthic diversity: A comparative study. Am. Nat. 102:243–282.

San Pietro, A., F. A. Greer, and T. J. Army (eds.). 1967. *Harvesting the Sun.* New York, Academic Press, 342 pp.

Santos, P. F., J. Phillips, and W. G. Whitford. 1981. The role of mites and nematodes in early stages of buried litter decomposition in a desert. Ecology 62:664–669.

Santos, P. F., and W. G. Whitford. 1981. The effects of microarthropods on litter decomposition in a Chihuahuan desert ecosystem. Ecology 62:654–663.

Sauer, C. O. 1950. Grassland, climax, fire and man. J. Range Mgmt. 3:16–21.

________. 1963. Fire and early man. In: *Land and Life: A Selection from the Writings of Carl Ortwin Sauer,* J. Leighly, ed. Berkeley, University of California Press, pp. 228–299.

Saunders, D. S. 1976. The biological clock of insects. Sci. Am. 234(2):114–121.

Saunders, George W. 1957. Interrelationships of dissolved organic matter and phytoplankton. Bot. Rev. 23:389–409.

Saunders, J. F. (ed.). 1968. *Bioregenerative Systems.* NASA. SP165. Washington, D.C., Superintendent of Documents, U.S. Govt. Printing Office, 153 pp.

Saunders, P. T. 1978. Population dynamics and length of food chains. Nature 272:189–190.

Saunders, P. T., and M. W. Ho. 1976. On the increase in complexity in evolution. J. Theoret. Biol. 63:375–384.

Schaffer, W. M. 1974. Selection for optimal life histories: The effects of age structure. Ecology 55:291–303.

Schelske, C. L. 1977. Trophic status and nutrient leaching for Lake Michigan. In: North American Project—A Study of U.S. Water Bodies. EPA Report 600/3-77-086, Corvallis, Oregon.

Schelske, Claire L., and E. P. Odum. 1961. Mechanisms maintaining high productivity in Georgia estuaries. Proc. Gulf Carib. Fish. Inst. 14:75–80.

Schindler, D. W. 1977. Evolution of phosphorus limitation in lakes. Science 195:260–262.

Schmidt-Nielsen, Bodil, and Knut Schmidt-Nielsen. 1952. Water metabolism of desert mammals. Physiol. Rev. 32:135–166.

Schmidt-Nielsen, Knut. 1964. *Desert Animals: Physiological Problems of Heat and Water.* London, Oxford University Press, 270 pp.

Schoener, T. W. 1969. Optimal size and specialization in constant and fluctuating environments: An energy-time approach. In: *Diversity and Stability in Ecological Systems,* G. M. Woodwell and H. H. Smith, eds. Brookhaven Symp. Biol. No. 22. Upton, N.Y., Brookhaven National Laboratory, pp. 103–114.

________. 1971. Theory of feeding strategies. Ann. Rev. Ecol. Syst. 2:369–404.

________. 1974. Resource partitioning in ecological communities. Science 185:27–39.

________. 1974a. Competition and the form of habitat shift. Theor. Pop. Biol. 6:265–307.

Schroder, G. D., and M. L. Rosenzweig. 1975. Perturbation analysis of competitions and overlap in habitat utilization between *Dipodomys ordi* and *Dipodomys mernami.* Oecologia 19:9–28.

Schrödinger, E. 1945. *What Is Life? The Physical Aspects of the Living Cell.* Cambridge, England, Cambridge University Press, 91 pp.

Schultz, A. M. 1964. The nutrient recovery hypothesis for arctic microtine cycles. II. Ecosystem variables in relation to arctic microtine cycles. In: *Grazing in Terrestrial and Marine Environments,* D. J. Crisp, ed. Oxford, Blackwell, pp. 57–68.

________. 1969. A study of an ecosystem: The arctic tundra. In: *The Ecosystem Concept in Natural Resource Management,* G. Van Dyne, ed. New York, Academic Press, pp. 77–93.

Schumacher, G. F. 1973. *Small Is Beautiful.* New York, Harper & Row.

Schurr, S. H. (ed.). 1972. *Energy, Economic Growth and the Environment.* A Resources for the Future Book. Baltimore, Johns Hopkins Press.

________ (Project Director). 1979. *Energy in America's Future: The Choices Before Us.* A Resources for the Future Book. Baltimore, Johns Hopkins Press.

Scott, J. O. (ed.). 1956. Allee memorial number. Series of papers on social organization in animals. Ecology 37:211–273.

Sears, Paul B. 1935. *Deserts on the March.* Normal, University of Oklahoma Press.

________. 1957. *The Ecology of Man.* Condon Lectures. Corvallis, University of Oregon Press. 61 pp.

Selye, Hans. 1955. Stress and disease. Science 122:625–631.

Shabman, L., and S. S. Batie. 1978. Economic value of natural coastal wetlands: A critique. Coastal Zone Mgmt. J. 4:231–247.

Shannon, Claude. 1950. Memory requirements in a telephone exchange. Bell Tech. J. 29:343–347.

Shannon, C. E., and W. Weaver. 1949. *The Mathematical Theory of Communication.* Urbana, University of Illinois Press, 117 pp.

Shantz, H. L. 1917. Plant succession on abandoned roads in eastern Colorado. J. Ecol. 5:19–42.

Shelford, V. E. 1911. Physiological animal geography. J. Morphol. 22:551–618.

________. 1911a. Ecological succession: Stream fishes and the method of physiographic analysis. Biol. Bull. 21:9–34.

———. 1911b. Ecological succession: Pond fishes. Biol. Bull. 21:127–151.

———. 1913. *Animal Communities in Temperate America*. Chicago, University of Chicago Press.

———. 1943. The abundance of the collared lemming in the Churchill area, 1929–1940. Ecology 24:472–484.

Sheridan, D. 1981. The underwatered west: Overdrawn at the well. Environment 23:6–12; 30–33.

Shugart, H. H., and R. V. O'Neill (eds.). 1979. *Systems Ecology*. Benchmark Papers in Ecology No. 9. Stroudsburg, Pa., Dowden, Hutchinson and Ross.

Silliman, R. P. 1969. Population models and test populations as research tools. BioScience 19:524–528.

Silverstein, Robert M. 1981. Pheromones: Background and potential for use in insect pest control. Science 213:1326–1332.

Simberloff, D. S. 1974. Equilibrium theory of island biogeography and ecology. Ann. Rev. Ecol. Syst. 5:161–182.

———. 1976. Trophic structure determination and equilibrium in an arthropod community. Ecology 57:395–398.

Simberloff, D. S., and L. G. Abele. 1976. Island biogeography theory and conservation practice. Science 191:285–286; 193:1032.

Simberloff, D. S., and E. O. Wilson. 1969. Experimental zoogeography of islands: The colonization of empty islands. Ecology 50:278–296.

Simon, Herbert A. 1962. The architecture of complexity. Proc. Am. Phil. Soc. 106:467–482. Reprinted in *Sciences of the Artificial* (2nd ed.), 1981, Cambridge, Mass., MIT Press.

———. 1973. The organization of complex systems. In: *Hierarchy Theory: The Challenge of Complex Systems*, H. H. Pattee, ed. New York, George Braziller, pp. 3–27.

Simon, Julian L. 1981. *Ultimate Resource*. Princeton, N.J., Princeton University Press, 415 pp.

Simpson, E. H. 1949. Measurement of diversity. Nature 163:688.

Simpson, George Gaylord. 1969. The first three billion years of community evolution. In: *Diversity and Stability in Ecological Systems*, G. M. Woodwell and H. H. Smith, eds. Brookhaven Symposium in Biology, No. 22. Upton, N.Y., Brookhaven National Laboratory, pp. 162–177.

Sinclair, A. R. E., and M. Norton-Griffiths (eds.). 1979. *Serengeti: Dynamics of an Ecosystem*. Chicago, University of Chicago Press, 389 pp.

Sinden, John A., and A. C. Worrell. 1979. *Unpriced Values: Decisions Without Market Prices*. New York, Wiley-Interscience, 511 pp.

Skellum, J. G. 1952. Studies in statistical ecology. I-spatial pattern. Biometrica 39:346–362.

Slesser, M. 1978. *Energy in the Economy*. New York, Macmillan.

Slobodkin, L. B. 1954. Population dynamics in *Daphnia obtusa* Kurz. Ecol. Monogr. 24:69–88.

———. 1960. Ecological energy relationships at the population level. Am. Nat. 95:213–236.

———. 1962. *Growth and Regulation of Animal Populations*. New York, Holt, Rinehart and Winston, 184 pp.

———. 1964. Experimental populations of hydrida. In: Brit. Ecol. Soc. Jubilee Symp. Suppl. to J. Ecol. 52 and J. Anim. Ecol. 33:1–244. Oxford, Blackwell, pp. 131–148.

———. 1968. How to be a predator. Am. Zool. 8:43–51.

Small, Maxwell. 1976. Marsh/pond sewage treatment plants. In: *Proceedings of Symposium on Freshwater Wetlands and Sewage Effluent Disposal*, D. L. Tilton, R. H. Kodlec, and C. J. Richardson, eds. Ann Arbor, University of Michigan Press, pp. 197–213.

Smith, A. D. 1940. A discussion of applications of climatological diagrams, the hythergraph, to distribution of natural vegetation types. Ecology 21:184–191.

Smith, Christopher. 1976. When and how much to reproduce: The trade-off between power and efficiency. Am. Zool. 16:763–774.

Smith, Frederick E. 1954. Quantitative aspects of population growth. In: *Dynamics of Growth Processes*, E. J. Boell, ed. Princeton, N.J., Princeton University Press.

———. 1969. Today the environment, tomorrow the world. BioScience 19:317–320.

———. 1969a. Effects of enrichment in mathematical models. In: *Eutrophication*. Washington, D.C., National Academy of Science, pp. 631–645.

———. 1970. Ecological demand and environmental response. J. Forestry 68:752–755.

Smith, N. G. 1968. The advantage of being parasitized. Nature 219:690–694.

Smith, Ray F., and R. van den Bosch. 1967. Integrated control. In: *Pest Control: Biological, Physical and Selected Chemical Methods*, Kilgore and Doutt, eds. New York, Academic Press, pp. 295–340.

Smith, Roland F. (ed.). 1966. A symposium on estuarine fisheries. Trans. Am. Fish. Soc. (Suppl.) 95(4):1–154.

Smith, S. H. 1966. Species succession and fishery exploitations in the Great Lakes. Symposium on Overexploited Animal Populations. Washington, D.C., American Association for the Advancement of Science, 28 pp. (mimeographed).

Smith, V. K., and J. V. Krutilla. 1979. Resource and environmental constraints to growth. Am. J. Agr. Econ. 61:395–408.

Smith, W. H. 1981. *Air Pollution and Forests: Interaction Between Air Contaminants and Forest Ecosystems.* New York, Springer-Verlag.

Snow, C. P. 1959. *The Two Cultures and the Scientific Revolution.* New York, Cambridge University Press.

Solbrig, Otto T. 1971. The population biology of dandelions. Am. Sci. 59:686–694.

Somero, G. N. 1969. Enzymic mechanisms of temperature compensation. Am. Nat. 103:517–530.

Sopper, William E. 1968. Waste water renovations for reuse: Key to optimum use of water resources. In: *Water Research.* Vol. 2. New York, Pergamon Press, pp. 471–480.

Sorokin, J. T. 1964. On the trophic role of chemosynthesis in water bodies. Int. Rev. Ges. Hydrobiol. 49:307–324.

Soule, Michael E., and Bruce A. Wilcox (eds.). 1980. *Conservation Biology: An Evolutionary-Ecological Perspective.* Sunderland, Mass., Sinauer Associates, 396 pp.

Sparrow, A. H. 1962. The role of the cell nucleus in determining radiosensitivity. Brookhaven Lecture Series No. 17. Brookhaven National Laboratory Publication No. 766.

Sparrow, A. H., L. A. Schairer, and R. C. Sparrow. 1963. Relationship between nuclear volumes, chromosome numbers, and relative radiosensitivities. Science 141:163–166.

Sparrow, A. H., and G. M. Woodwell. 1962. Prediction of the sensitivity of plants to chronic gamma irradiation. Rad. Bot. 2:9–26.

Spedding, C. R. W. 1975. *Biology of Agricultural Systems.* New York, Academic Press.

Spilhaus, Athelstan (ed.). 1966. *Waste Management and Control.* Publication No. 1400. Washington, D.C., National Academy of Science, 257 pp.

Sprugel, Douglas G., and F. H. Bormann. 1981. Natural disturbance and the steady state in high altitude balsam fir forests. Science 211:390–393.

Spurlock, J. M., and M. Modell. 1978. Technology requirements and planning criteria for closed life support systems for manned space missions. Washington, D.C., Office of Life Sciences, NASA.

Stanford Research Institute. 1975. City Size and the Quality of Life. Report of Subcommittee on Rural Development of Committee on Agriculture and Forestry, U.S. Senate. (Published as a Congressional Hearing in 1978.)

Stanley, S. M. 1979. *Macroevolution: Pattern and Process.* San Francisco, W. H. Freeman, 332 pp.

________. 1981. *The New Evolutionary Timetable.* New York, Basic Books, 222 pp.

Starr, Chauncey (ed.). 1971. Energy and Power. Sci. Am. Special Issue, Vol. 224, No. 3.

Stearns, Stephen C. 1976. Life-history tactics; a review of ideas. Quart. Rev. Biol. 51:3–47.

________. 1977. The evolution of life history traits. Ann. Rev. Ecol. Syst. 8:145–172.

Steele, John H. (ed.). 1970. *Marine Food Chains.* London, Oliver and Boyd, 552 pp.

Steemann-Nielsen, E. 1952. The use of radioactive carbon (C^{14}) for measuring organic production in the sea. J. Cons. Int. Explor. Mer. 18:117–140.

________. 1954. On organic production in the ocean. J. Cons. Int. Explor. Mer. 49:309–328.

Steinhart, John S., and C. E. Steinhart. 1974. Energy use in the U.S. food system. Science 184:307–316.

Steinwascher, Kurt. 1978. Interference and exploitation competition among tadpoles of *Rana utricularia.* Ecology 59:1039–1046.

Stern, A. C. (ed.). 1968. *Air Pollution* (2nd ed.). Three vols. New York, Academic Press.

Stewart, Paul A. 1952. Dispersal, breeding, behavior, and longevity of banded barn owls in North America. Auk 69:277–285.

Stickel, Lucille F. 1950. Populations and home range relationships of the box turtle, *Terrapene c. carolina* (Linnaeus). Ecol. Monogr. 20:351–378.

Stobaugh, Robert, and D. Yergin (eds.). 1979. *Energy Future: Report of the Energy Project at the Harvard Business School.* New York, Random House.

Stoddard, D. R. 1965. Geography and the ecological approach. The ecosystem as a geographical principle and method. Geography 50:242–251.

Stoddard, Herbert L. 1936. Relation of burning to timber and wildlife. Proc. 1st N.A. Wildl. Conf. 1:1–4.

Stokes, Allen W. (ed.). 1974. *Territory.* Benchmark Papers in Animal Behavior. Stroudsburg, Pa., Dowden, Hutchinson and Ross, 416 pp.

Stonehouse, Bernard, and Christopher Perrins (eds.). 1977. *Evolutionary Ecology.* London and New York, Macmillan Press.

Straffa, P. 1973. *The Production of Commodities by Means of Commodities. Prelude to a Critique of Economic Theory.* Cambridge, England, Cambridge University Press.

Stugren, Von Dr. Bogdan. 1965. *Ecologie generala.* Bucharest, Editura Dedochea Sipedogogua.

________. 1972. *Grundlagen der allgemernen Okologie.* Jena, Veb Gustav Fischer Verlag.

Sukachev, V. N. 1944. On principles of genetic classification in biocenology. (In Russian.) Zur Obshchei Biol. 5:213–227. Translated by F. Raney and R. Daubenmire, Ecology 39:364–376.

________. 1959. The correlation between the concepts "forest ecosystem" and "forest biogeocoenose" and their importance for the classification of forests. Proc. IX Int. Bot. Congr., Vol. II, page 387 (abstract). (See also Silva Fennica 105:94, 1960.)

Sukachev, V. N., and N. V. Dylis (eds.). 1964. *Fundamentals of Forest Biogeocoenology.* Moscow, Bot. Inst. Lab. For. Sci., 474 pp. (In Russian; see review in Science 148:868, 1965.)

Sutherland, J. P. 1974. Multiple stable points in natural communities. Am. Nat. 108:859–873.

Svardson, Gunnar. 1949. Competition and habitat selection in birds. Oikos 1:157–174.

Svenson, B. H., and R. Söderlund (eds.). 1976. Nitrogen, phosphorus and sulfur—global cycles. Ecol. Bull./NFR 22. Stockholm, Royal Swedish Academy of Science.

Swank, W. T., and N. H. Miner. 1968. Conversion of hardwood-covered watershed to white pine reduces water yield. Water Resources Res. 4:947–954.

Sweeney, J. R. 1956. Responses of vegetation to fire. Univ. Calif. Publ. Bot. 28:143–250.

Swift, M. J., O. W. Heal, and J. M. Anderson. 1979. *Decomposition in Terrestrial Ecosystems.* Studies in Ecology, Vol. 5. Berkeley, University of California Press, 372 pp.

Swingle, H. S., and E. V. Smith. 1947. Management of farm fish ponds (rev. ed.). Alabama Polytechnic Inst. Agr. Exp. Sta. Bull. No. 254.

Taber, R. D., and Raymond Dasmann. 1957. The dynamics of three natural populations of the deer, *Odocoileus hemionus columbianus.* Ecology 38:233–246.

Takahashi, M., and S. Ichimura. 1968. Vertical distribution and organic matter production of photosynthetic sulfur bacteria in Japanese lakes. Limnol. Oceanogr. 13:644–655.

Talbot, L. M., and M. H. Talbot. 1963. The high biomass of wild ungulates on East African savanna. Trans. N.A. Wildl. Conf. 28:465–476.

Tansley, A. G. 1935. The use and abuse of vegetational concepts and terms. Ecology 16:284–307.

Tappen, Helen. 1968. Primary production, isotopes and the atmosphere. In: *Palaeogeography, Palaeoclimatology, Palaeoecology.* Vol. 4. Amsterdam, Elsevier, pp. 187–210.

Taub, Freida B. 1969. A biological model of a freshwater community in a gnotobiotic ecosystem. Limnol. Oceanogr. 14:136–142.

———. 1974. Closed ecological systems. Ann. Rev. Ecol. Syst. 5:139–160.

Taylor, M. W. 1977. A comparison of three edge indexes. Wildl. Soc. Bull. 5:192–193.

Taylor, O. C., W. M. Dugger, Jr., E. A. Cardiff, and E. F. Darley. 1961. Interaction of light and atmospheric photochemical (smog) within plants. Nature 192:814–816.

Taylor, W. R. 1964. Light and photosynthesis in intertidal benthic diatoms. Helgd. Wiss. Meeresunters 10:29–37.

Teal, John M. 1957. Community metabolism in a temperate cold spring. Ecol. Monogr. 27:283–302.

———. 1958. Distribution of fiddler crabs in Georgia salt marshes. Ecology 39:185–193.

———. 1962. Energy flow in the salt marsh ecosystem of Georgia. Ecology 43:614–624.

Teitelbaum, Michael S. 1975. Relevance of demographic transition theory for developing countries. Science 188:420–425.

Theobald, Robert (ed.). 1979. *Challenge in Renewable Natural Resources. A Guide to Alternate Futures.* Washington, D.C., U.S. Department of Agriculture, 123 pp.

Thienemann, August. 1929. Der Nahrungskreislauf im Wasser. Verh. deutsch. Zool. Ges. 31:29–79.

———. 1939. Grundzuge einer allgemeinen Oekologie. Arch. Hydrobiol. 35:267–285.

Thomas, Jack W., C. Maser, and E. Rodiek. 1979. Edges. In: *Wildlife Habitat in Managed Forests,* J. W. Thomas, ed. Agricultural Handbook No. 53. USDA Forest Service. Washington, D.C., Supt. of Documents, U.S. Govt. Printing Office, pp. 48–59.

Thomas, Moyer D., and George R. Hill. 1949. Photosynthesis under field conditions. In: *Photosynthesis in Plants,* James Franck and Walter E. Loomis, eds. Ames, Iowa State College Press, pp. 19–52.

Thomas, William A. 1969. Accumulation and cycling of calcium by dogwood trees. Ecol. Monogr. 39:101–120.

Thomas, William L. (ed.). 1956. *Man's Role in Changing the Face of the Earth.* Chicago, University of Chicago Press.

Thompson, David H., and G. W. Bennett. 1939. Fish management in small artificial lakes. Trans 4th N.A. Wildl. Conf. 4:311–317.

Thornthwaite, C. W. 1931. The climates of North America according to a new classification. Geogr. Rev. 21:633–655.

———. 1948. An approach to a rational classification of climate. Geogr. Rev. 38:55–94.

Thornthwaite, C. W., and J. R. Mather. 1957. Instructions and tables for computing potential evapotranspiration and water balance. Drexel Inst. Technol. Lab. Climatol. Publ. Climatol. 17:231–615.

Thorson, Gunnar. 1955. Modern aspects of marine level-bottom animal communities. J. Marine Res. 14:387–397.

Tilly, L. J. 1968. The structure and dynamics of Cone Spring. Ecol. Monogr. 38:169–197.

Tilton, D. L., R. H. Kedlec, and C. J. Richardson (eds.). 1976. *Proceedings of Symposium on Freshwater Wetlands and Sewage Effluent Disposal.* Ann Arbor, University of Michigan, 343 pp.

Tinbergen, Jan (ed.). 1977. *The Rio: Reshaping the*

International Order. A Report to the Club of Rome Coordinator. New York, Signet, 432 pp.

Tinbergen, N. 1968. On war and peace in animals and man. Science 160:1411–1418.

Tjephema, J. D., and L. J. Winship. 1980. Energy requirement for nitrogen fixation in actinorhizal and legume root nodules. Science 209:279–281.

Torrey, J. G. 1978. Nitrogen fixation by actinomycete-nodulated angiosperms. BioScience 28:586–592.

Tounsend, C. R., and P. Calow (eds.). 1981. *Physiological Ecology: An Evolutionary Approach to Resource Use.* Sunderland, Mass., Sinauer Associates, 401 pp.

Toynbee, Arnold. 1961. *A Study of History.* New York, Oxford University Press.

Tramer, E. J. 1969. Bird species diversity; components of Shannon's formula. Ecology 50:927–929.

Transeau, E. N. 1926. The accumulation of energy by plants. Ohio J. Sci. 26:1–10.

Trautman, Milton B. 1942. Fish distribution and abundance correlated with stream gradients as a consideration in stocking programs. Trans. 7th N.A. Wildl. Conf. 7:221–223.

Tribe, H. T. 1957. Ecology of microorganisms in soil as observed during their development upon buried cellulose film. In: *Microbial Ecology,* Williams and Spicer, eds. Cambridge, Cambridge University Press.

———. 1963. The microbial component of humus. In: *Soil Organisms,* J. Packser and J. van der Drift, eds. Amsterdam, North-Holland Publishing Co.

Triplett, Glover B., Jr., and David M. VanDoren, Jr. 1977. Agriculture without tillage. Sci. Am. 636(1):28–33.

Turner, F. B. (ed.). 1968. Energy flow in ecosystems (Refresher Course, Am. Soc. Zool.). Am. Zool. 8:10–69.

Udall, Stewart L. 1965. *The Quiet Crisis.* New York, Holt, Rinehart and Winston, 209 pp.

Udvardy, M. F. D. 1959. Notes on the ecological concepts of habitat biotope and niche. Ecology 40:725–728.

Ulanowicz, Robert E. 1980. A hypothesis on the development of natural communities. J. Theoret. Biol. 85:223–245.

United States Department of Agriculture. 1979. *Challenge in Renewable Natural Resources. A Guide to Alternate Futures,* R. Theobald, ed. Washington, D.C., U.S. Dept. of Agriculture, 123 pp.

United States Water Resources Council. 1978. *The Nation's Water Resources—1975–2000.* Two vols. Washington, D.C., U.S. Govt. Printing Office.

Uvarov, B. P. 1957. The aridity factor in the ecology of locust and grasshoppers of the old world. In: *Human and Animal Ecology.* Arid Zone Res. VIII. Paris, UNESCO.

Valentine, James W. 1968. Climatic regulation of species diversification and extinction. Bull. Geol. Soc. Am. 79:273–276.

Valerio, Carlos E. 1975. A unique case of mutualism. Am. Nat. 109:235–238.

Valiela, Ivan. 1971. Food specificity and community succession: Preliminary ornithological evidence for a general framework. Gen. Syst. 16:77–84.

Vallentyne, J. R. 1960. Geochemistry of the biosphere. In: *McGraw-Hill Encyclopedia of Science and Technology,* Vol. 2. New York, McGraw-Hill, pp. 239–245.

———. 1962. Solubility and the decomposition of organic matter in nature. Arch. Hydrobiol. 58:423–434.

Van den Bosch, Robert. 1971. Biological control of insects. Ann. Rev. Ecol. Syst. 2:45–66.

Vandermeer, J. H. 1972. Niche theory. Ann. Rev. Ecol. Syst. 3:107–132.

———. 1981. *Elementary Mathematical Ecology.* New York, Wiley-Interscience.

Van Doblen, W. H., and R. H. Lowe-McConnell (eds.). 1975. *Unifying Concepts in Ecology.* Report of First International Congress of Ecology. The Hague, W. Junk B. V. Publishers.

Van Dyne, G. M. (ed.). 1969. *The Ecosystem Concept in Natural Resource Management.* New York, Academic Press, 383 pp.

Vannote, R. L., G. W. Minshall, K. W. Cummins, J. R. Sedell, and C. E. Cushing. 1980. The river continuum concept. Can. J. Fish. Aquat. Sci. 37:130–137.

Van Valen, Leigh. 1965. Morphological variation and width of ecological niche. Am. Nat. 99:377–390.

Van Voris, Peter, R. V. O'Neill, W. R. Emanuel, and H. H. Shugart. 1980. Functional complexity and ecosystem stability. Ecology 61:1352–1360.

Varley, G. C. 1947. The natural control of population balance in the knapweed gall-fly (*Urophora jaceana*). J. Anim. Ecol. 16:139–187.

———. 1949. Population changes in German forest pests. J. Anim. Ecol. 18:117–122.

Varley, G. C., and R. I. Edwards. 1957. The bearing of parasite behaviour on the dynamics of insect host and parasite populations. J. Anim. Ecol. 26:471–477.

Vayda, A. P., and Roy Rappoport. 1968. Ecology, cultural and non-cultural. In: *Introduction to Cultural Anthropology.* J. Clifton (ed.). Boston, Houghton-Mifflin.

Verhulst, P. F. 1838. Notice sur la loi que la population suit dans son accroissement. Corresp. Math. Phys. 10:113–121.

Vernadsky, V. I. 1945. The biosphere and the noosphere. Am. Sci. 33:1–12.

Vernberg, F. John. 1978. Multiple-factor and synergistic stresses in aquatic systems. In: *Energy and Environmental Stress in Aquatic Systems*, T. H. Thorp and J. W. Gibbons, eds. U.S. Dept. of Energy, Symp. No. 48. Springfield, Va., Technical Info. Center, pp. 726–747.

Vernberg, F. J., and W. B. Vernberg. 1970. *The Animal and the Environment*. New York, Holt, Rinehart and Winston, 416 pp.

Vernberg, W. B., P. J. De Coursey, and J. O. Hara. 1974. Multiple environment factor effects on physiology and behavior of the fiddler crab, *Uca pugilator*. In: *Pollution and Physiology of Marine Organisms*. New York, Academic Press, pp. 381–425.

Verner, T. 1977. On the adaptive significance of territoriality. Am. Nat. 111:769–775.

Vickery, P. J. 1972. Grazing and net primary production of a temperate grassland. J. Appl. Ecol. 9:307–314.

Vida, Gabor. 1978. Genetic diversity and environmental future. Environ. Cons. 5:127–132.

Viosca, Percy, Jr. 1936. Statistics on the productivity of inland waters, the master key to better fish culture. Trans. Am. Fish. Soc. 65:350–358.

Vitousek, P. M., and W. A. Reiners. 1975. Ecosystem succession and nutrient retention: A hypothesis. BioScience 25:376–381.

Vogl, R. J. 1980. The ecological factors that produce perturbation dependent ecosystems. In: *The Recovery Process in Damaged Ecosystems*, J. Cairns, Jr., ed. Ann Arbor, Mich., Ann Arbor Science, pp. 63–69.

Vogt, William. 1948. *Road to Survival*. New York, Sloane.

Volterra, Vito. 1926. Variations and fluctuations of the number of individuals in animal species living together. In: *Animal Ecology*, R. N. Chapman, ed. New York, McGraw-Hill, pp. 409–448.

________. 1928. Variations and fluctuations of the number of individuals in animal species living together. J. Cons. Inst. Explor. Mer. 3:3–51.

Wager, J. Alan. 1970. Growth versus the quality of life. Science 168:1170–1184.

Wagner, Frederic H. 1969. Ecosystem concepts in fish and game management. In: *The Ecosystem Concept in Natural Resource Management*, G. M. Van Dyne, ed. New York, Academic Press, pp. 259–307.

Waksman, Selman A. 1952. *Soil Microbiology*. New York, John Wiley & Sons.

Walker, James C. G. 1977. *Evolution of the Atmosphere*. London, Collier and MacMillan, 318 pp.

Wallace, Bruce, and A. M. Srb. 1961. *Adaptation*. Englewood Cliffs, N.J., Prentice-Hall, 113 pp.

Waloff, Z. 1966. The upsurges and recessions of the desert locust; an historical survey. Antilocust Mem. No. 8, London, 111 pp.

Wangersky, P. J., and W. J. Cunningham. 1956. On time lags in equations of growth. Proc. Nat. Acad. Sci. (Wash.) 42:699–702.

Ward, Barbara. 1966. *Spaceship Earth*. New York, Columbia University Press.

________. 1976. *The Home of Man*. New York, W. W. Norton.

Warrington, Robert. 1851. Notice of observation on the adjustment between animal and vegetable kingdoms. Quart. J. Chem. Soc. London 3:5254.

________. 1857. On the aquarium. Notices Proc. Royal Inst. 2:403–408.

Watson, Vicki, and O. L. Loucks. 1979. An analysis of turnover times in a lake ecosystem and some implications for system properties. In: *Theoretical Systems Ecology: Advances and Case Studies*, E. Halfon, ed. New York, Academic Press, pp. 355–383.

Watt, A. S. 1947. Pattern and process in the plant community. J. Ecol. 35:1–22.

Watt, Kenneth E. F. 1963. How closely does the model mimic reality? Can. Entomol. Mem. 31:109–111.

________. 1963a. Dynamic programming, "look ahead programming," and the strategy of insect pest control. Can. Entomol. 95:525–536.

________. 1966. *Systems Analysis in Ecology*. New York, Academic Press, 276 pp.

________. 1968. *Ecology and Resource Management: A Quantitative Approach*. New York, McGraw-Hill, 450 pp.

Watt, Kenneth E. F., L. F. Molloy, C. K. Varshney, D. Weeks, and S. Wirosardjono. 1977. *The Unsteady State: Environmental Problems, Growth, and Culture*. Honolulu, East-West Center, The University Press of Hawaii, 287 pp.

Weaver, J. E., and F. E. Clements. 1929. *Plant Ecology*. (2nd ed., 1938.) New York, McGraw-Hill.

Webster, J. R., and B. C. Patten. 1979. Effect of watershed perturbation on stream potassium and calcium dynamics. Ecol. Monogr. 49:51–72.

Weinberg, Alvin M., and R. P. Hammond. 1970. Limits to the use of energy. Am. Sci. 58:412–418.

Weisz, Paul B., and J. F. Marshall. 1979. High-grade fuels from biomass farming: Potentials and constraints. Science 206:24–29.

Welch, Harold. 1967. Energy flow through the major macroscopic components of an aquatic ecosystem. Ph.D. Dissertation, University of Georgia, Athens.

________. 1968. Relationships between assimilation efficiencies and growth efficiencies for aquatic consumers. Ecology 49:755–759.

Wellington, W. G. 1957. Individual differences as a factor in population dynamics: The development of a problem. Can. J. Zool. 35:293–323.

________. 1960. Qualitative changes in natural populations during changes in abundance. Can. J. Zool. 38:289–314.

Wells, B. W. 1928. Plant communities of the coastal plain of North Carolina and their successional relations. Ecology 9:230–242.

Went, Fritz W. 1957. *The Experimental Control of Plant Growth.* Waltham, Mass., Chronica Botanica Co.

Went, F. W., and N. Stark. 1968. Mycorrhiza. BioScience 18:1035–1039.

Werner, E. E., and D. J. Hall. 1974. Optimal foraging and size selection of prey by bluegill sunfish. Ecology 55:1042–1052.

Wesley, J. P. 1974. *Ecophysics: the Application of Physics to Ecology.* Springfield, Ill., Charles C Thomas, 340 pp.

West, D. C., H. H. Shugart, and D. Botkin (eds.). 1981. *Forest Succession: Concepts and Applications.* Ecol. Studies Vol. 41. New York, Springer-Verlag.

Westman, W. E. 1975. Letter to editor. BioScience 25:770.

———. 1977. How much are nature's services worth? Science 197:960–964.

———. 1978. Measuring the inertia and resilience of ecosystems. BioScience 28:705–710.

White, Gilbert F. 1980. Environment. In: Science Centennial Issue, P. H. Abelson and R. Kulstad, eds. Science 209:183–190.

White, Lynn. 1967. The historical roots of our ecological crisis. Science 155:1203–1207.

———. 1980. The ecology of our science. Science-80 1:72–76.

Whitehead, F. H. 1957. Productivity in alpine vegetation (abstract). J. Anim. Ecol. 26:241.

Whittaker, R. H. 1951. A criticism of the plant association and climatic climax concepts. Northwest Sci. 25:17–31.

———. 1954. Plant populations and the basis of plant indication. In: Angewandte Pflanzensoziologie, Veroffentlichungen des Karntner Landesinstituts fur angewandte Pflanzensoziologie in Klagenfurt, Festschrift Aichinger, Vol. 1.

———. 1957. Recent evolution of ecological concepts in relation to eastern forests of North America. Am. J. Bot. 44:197–206.

———. 1960. Vegetation of the Siskiyou Mountains, Oregon and California. Ecol. Monogr. 30:279–338.

———. 1965. Dominance and diversity in land plant communities. Science 147:250–260.

———. 1967. Gradient analysis of vegetation. Biol. Rev. 42:207–264.

———. 1969. New concepts of kingdoms of organisms. Science 163:150–160.

———. 1970. The biochemical ecology of higher plants. In: *Chemical Ecology*, E. Sondheimer and J. B. Simeone, eds. New York, Academic Press, pp. 43–70.

———. 1972. Evolution and measurement of species diversity. Tarpon 21:213–251.

———. 1975. *Communities and Ecosystems* (2nd ed.). New York, Macmillan, 385 pp.

Whittaker, R. H., and P. P. Feeny. 1971. Allelechemics; chemical interaction between species. Science 171:757–770.

Whittaker, R. H., S. A. Levin, and R. B. Root. 1973. Niche, habitat and ecotope. Am. Nat. 107:321–338.

Whittaker, R. H., and G. E. Likens (eds.). 1973. The primary production of the biosphere. Human Ecol. 1:301–369.

Whittaker, R. H., and G. M. Woodwell. 1969. Structure, production and diversity of the oak-pine forest at Brookhaven, New York. J. Ecol. 57:155–174.

———. 1972. Evolution of natural communities. In: *Ecosystem Structure and Function*, J. A. Wiens, ed. Corvallis, Oregon State University Press, pp. 137–156.

Wiebes, J. T. 1979. Coevolution of figs and their insect pollinators. Ann. Rev. Ecol. Syst. 10:1–12.

Wiegert, R. G. 1965. Energy dynamics of the grasshopper populations in old-field and alfalfa field ecosystems. Oikos 16:161–176.

———. 1968. Thermodynamic considerations in animal nutrition. Am. Zool. 8:71–81.

———. 1974. Competition: A theory based on realistic, general equations of population growth. Science 185:539–542.

———. 1975. Simulation models of ecosystems. Ann. Rev. Ecol. Syst. 6:311–338.

Wiegert, R. G., D. C. Coleman, and E. P. Odum. 1970. Energetics of the litter-soil subsystem. In: *Methods of Study in Soil Ecology.* Proc. Paris Symp. Int. Biol. Prog. Paris, UNESCO, pp. 93–98.

Wiegert, R. G., and F. C. Evans. 1967. Investigations of secondary productivity in grasslands. In: *Secondary Productivity of Terrestrial Ecosystems*, K. Petrusewicz, ed. Vol. II. Warsaw, Polish Acad. Sci., pp. 499–518.

Wiegert, R. G., and E. P. Odum. 1969. Radionuclide tracer measurement of food web diversity in nature. In: *Proceedings of 2nd National Symposium on Radioecology*, D. J. Nelson and F. C. Evans, eds. Springfield, Va., Clearinghouse Fed. Sci. Tech. Info., pp. 709–710.

Wiegert, R. G., E. P. Odum, and J. H. Schnell. 1967. Forb-arthropod food chains in a one-year experimental field. Ecology 48:75–83.

Wiegert, R. G., and D. F. Owen. 1971. Trophic structure, available resources and population density in terrestrial vs. aquatic ecosystems. J. Theoret. Biol. 30:69–81.

Wiener, N. 1948. *Cybernetics.* (2nd ed., 1961.) Cambridge, Mass., MIT Press.

Wiens, John A. 1976. Population responses to patchy environments. Ann. Rev. Ecol. Syst. 7:81–120.

———. 1977. On competition and variable environments. Am. Sci. 65:590–591.

Wilbur, H. M., D. W. Tinkle, and J. P. Collins. 1974. Environmental and resource availability in life history evolution. Am. Nat. 108:805–817.

Wilde, S. A. 1968. Mycorrhizae and tree nutrition. BioScience 18:482–484.

Wilhm, J. L. 1967. Comparison of some diversity indices applied to populations of benthic macroinvertebrates in a stream receiving organic wastes. J. Water Poll. Control Fed. 39:1673–1683.

Wilhm, J. L., and T. C. Dorris. 1968. Biological parameters for water quality criteria. BioScience 18:477–481.

Williams, C. B. 1964. *Patterns in the Balance of Nature; and Related Problems in Quantitative Ecology.* New York, Academic Press, 324 pp.

Williams, Carroll M. 1967. Third-generation pesticides. Sci. Am. 217(1):13–17.

Williams, G. C. 1966. *Adaptation and Natural Selection.* Princeton, N.J., Princeton University Press.

Wilson, Carroll L. 1979. Nuclear energy: What went wrong? Bull. Atomic Sci. 35(6):13–17.

Wilson, David Sloan. 1975. Evolution on the level of communities. Science 192:1358–1360.

———. 1977. Structured denes and the evolution of group-advantageous traits. Am. Nat. 111:157–185.

———. 1980. *The Natural Selection of Populations and Communities.* Menlo Park, Calif., and Reading, Mass., Benjamin/Cummings.

Wilson, E. O. 1965. Chemical communications in the social insects. Science 149:1064–1071.

———. 1969. The species equilibrium. In: *Diversity and Stability in Ecological Systems*, G. M. Woodwell and H. H. Smith, eds. Brookhaven Symposia in Biology, No. 22. Upton, N.Y., Brookhaven National Laboratory, pp. 38–47.

———. 1973. Group selection and its significance for ecology. BioScience 23:631–638.

———. 1975. *Sociobiology: The New Synthesis.* Cambridge, Mass., The Belknap Press of Harvard University Press. (Abridged edition, 1980.)

Wilson, E. O., and W. H. Bossert. 1971. *A Primer of Population Biology.* Stamford, Conn., Sinauer Associates.

Wilson, E.O., and E. O. Willis. 1975. Applied biogeography. In: *Ecology and Evolution of Communities*, M. L. Cody and J. M. Diamond, eds., Cambridge, Mass., Harvard University Press, pp. 522–534.

Wilson, J. Tuzo (ed.). 1972. *Continents Adrift. Readings from Scientific American.* San Francisco, W. H. Freeman.

Winogradsky, S. 1949. *Microbiologie du Sol: Problemes et Methodes.* Paris, Masson et Cie., 861 pp.

Witherspoon, J. P. 1965. Radiation damage to forest surrounding an unshielded fast reactor. Health Physics 11:1637–1642.

———. 1969. Radiosensitivity of forest tree species to acute fast neutron radiation. In: *Proceedings of 2nd National Symposium on Radioecology*, D. J. Nelson and F. C. Evans, eds. Springfield, Va., Clearinghouse Fed. Sci. Tech. Info., pp. 120–126.

Witkamp, M. 1963. Microbial populations of leaf litter in relation to environmental conditions and decomposition. Ecology 44:370–377.

———. 1966. Decomposition of leaf litter in relation to environmental conditions, microflora and microbial respiration. Ecology 47:194–201.

———. 1971. Soils as components of ecosystems. Ann. Rev. Ecol. Syst. 2:85–110.

Witkamp, M., and J. S. Olson. 1963. Breakdown of confined and nonconfined oak litter. Oikos 14:138–147.

Witt, J. M., and S. W. Gillett (eds.). 1978. *Terrestrial Microcosms and Environmental Chemistry.* Corvallis, Oregon State University Press.

Wohlschlag, Donald E. 1960. Metabolism of an antarctic fish and the phenomenon of cold adaptation. Ecology 41:287–292.

Wolcott, G. N. 1937. An animal census of two pastures and a meadow in northern New York. Ecol. Monogr. 7:1–90.

Wolfanger, Louis A. 1930. *The Major Soil Divisions of the United States.* New York, John Wiley & Sons.

Wolfenbarger, D. O. 1946. Dispersion of small organisms. Am. Midl. Nat. 35:1–152.

Wollast, R., G. Billen, and F. T. Mackenzie. 1975. Behavior of mercury in natural systems and its global cycle. In: *Ecological Toxicity Research*, A. D. McIntyre and C. F. Mills, eds. New York, Plenum Press.

Wolman, A. 1965. The metabolism of cities. Sci. Am. 213:179–190.

Wood, E. J. Ferguson. 1967. *Microbiology of Oceans and Estuaries.* Amsterdam, Elsevier, 319 pp.

Wood, J. M. 1974. Biological cycles for toxic elements in the environment. Science 183:1049–1058.

Woodmansee, R. G. 1978. Additions and losses of nitrogen in grassland ecosystems. BioScience 28:448–453.

Woodruff, L. L. 1912. Observations on the origin and sequence of the protozoan fauna of hay infusions. J. Exp. Zool. 12:205–264.

Woodwell, G. M. 1962. Effects of ionizing radiation on terrestrial ecosystems. Science 138:572–577.

———. 1963. The ecological effects of radiation. Sci. Am. 208(6):1–11.

——— (ed.). 1965. *Ecological Effects of Nuclear War.* Upton, N.Y., Brookhaven National Laboratory Publ. No. 917, 72 pp.

———. 1967. Toxic substances and ecological cycles. Sci. Am. 216(3):24–31.

———. 1977. Recycling sewage through plant communities. Am. Sci. 65:556–562.

Woodwell, G. M., and W. D. Dykeman. 1966. Respiration of a forest measured by CO_2 accumulation during temperature inversions. Science 154:1031–1034.

Woodwell, G. M., and E. V. Pecan (eds.). 1973. *Carbon and the Biosphere.* U.S. Atomic Energy Commission. Springfield, Va., National Technical Info. Center.

Woodwell, G. M., and H. H. Smith (eds.). 1969. *Diversity and Stability in Ecological Systems.* Upton, N.Y., Brookhaven National Laboratory Publ. No. 22, 264 pp.

Woodwell, G. M., and R. H. Whittaker. 1968. Primary production in terrestrial communities. Am. Zool. 8:19–30.

Woodwell, G. M., R. H. Whittaker, W. A. Reiners, G. E. Likens, C. C. Delwiche, and D. B. Botkin. 1978. The biota and the world carbon budget. Science 199:141–146.

Woodwell, G. M., C. F. Wurster, and P. A. Isaacson. 1967. DDT residues in an east coast estuary: A case of biological concentration of a persistent insecticide. Science 156:821–824.

Wright, Sewall. 1945. Tempo and mode in evolution; a critical review. Ecology 26:415–419.

Wurster, Charles F. 1969. Chlorinated hydrocarbon insecticides and the world ecosystem. Biol. Cons. 1:123–129.

Wynne-Edwards, V. C. 1959. The control of population through social behavior: A hypothesis. Ibis 101: 436–441.

———. 1962. *Animal Dispersion in Relation to Social Behavior.* New York, Hafner.

———. 1964. Population control in animals. Sci. Am. 211(2):68–72.

———. 1965. Self-regulating systems in populations of animals. Science 147:1543–1548.

———. 1965a. Social organization as a population regulator. Symp. Zool. Soc. London 14:173–178.

Zelitch, I. 1975. Improving the efficiency of photosynthesis. Science 188:626–633.

Zeuthen, E. 1953. Oxygen uptake and body size in organisms. Quart. Rev. Biol. 28:1–12.

Zippin, Calvin. 1958. The removal method of population estimation. J. Wildl. Mgmt. 22:82–90.

Zunuska, J. A. 1971. Cooperation or chaos: The challenge to natural resource biologists. BioScience 21:726–728.

INDEX

Numbers in **boldface type** indicate pages on which terms and concepts are most fully defined and explained. Key groups of organisms are listed, but species names are not indexed except in connection with the illustration of ecological principles. Names of persons are not indexed; see Bibliography for names of authors.